THE SOLID LETTER

CONTRIBUTORS

Bernhard Böschenstein
Stanley Corngold
Jean-François Courtine
Peter Fenves
Aris Fioretos
Hans-Jost Frey
Christopher Fynsk
Cyrus Hamlin
Philippe Lacoue-Labarthe
Arne Melberg
Rainer Nägele
Jean-Luc Nancy
Edgar Pankow
Thomas Schestag
Arnaud Villani
Hent de Vries
Andrzej Warminski

Cultural Memory
in
the
Present

Mieke Bal and Hent de Vries, Editors

THE SOLID LETTER

Readings of Friedrich Hölderlin

Edited by Aris Fioretos

STANFORD UNIVERSITY PRESS
STANFORD, CALIFORNIA

Stanford University Press
Stanford, California

Printed and bound by CPI Group (UK) Ltd, Croydon, CR0 4YY
CIP data appear at the end of the book.

Acknowledgments

The essays in this volume were finalized in 1993–94. Only the Bibliography and references in a few endnotes have since been slightly updated. Some of the contributions have previously appeared elsewhere, in part or in whole. Peter Fenves, "Measure for Measure: Hölderlin and the Place of Philosophy," first appeared in *Philosophy Today* 37 (1993), 369–82 (© *Philosophy Today*); Arnaud Villani, "Figures of Duality: Hölderlin and Greek Tragedy," appeared as "Figures de la dualité: Hölderlin et la tragédie grecque" in a special Hölderlin issue of *Les cahiers de l'Herne*, ed. Jean-François Courtine, 52 (1989): 277–96 (© Editions de l'Herne); Andrzej Warminski, "Monstrous History: Heidegger Reading Hölderlin," appeared in *Yale French Studies* 77 (1990): 193–209 (© Yale University Press); Aris Fioretos, "Color Read: Hölderlin and Translation," appeared in an earlier version as "Sorgearbete (Hölderlin)," in *Det kritiska ögonblicket: Hölderlin, Benjamin, Celan* (Stockholm: Norstedts, 1991), 28–70, and in an English version in *Compar(a)isons* 1 (1995): 109–30; Bernhard Böschenstein, "'Brod und Wein': From the 'Classical' Final Version to the Later Revision," appeared as "'Brod und Wein': Von der 'klassischen' Reinschrift zur späten Überarbeitung," in *Turm-Vorträge 1989/90/91. Hölderlin: Christentum und Antike*, ed. Valérie Lawitschka (Tübingen: Hölderlinturm, 1991), 173–200; and Thomas Schestag, "The Highest," appeared as "Das Höchste" in *parerga—Friedrich Hölderlin, Carl Schmitt, Franz Kafka, Platon, Schleiermacher, Walter Benjamin, Jacques Derrida—zur literarischen Hermeneutik* (Munich: Boer, 1991), 15–50.

A.F.
Berlin-Charlottenburg
November 1997

Contents

Sources and Abbreviations

References to Hölderlin's works are given to one of six abbreviations, followed by Arabic numerals designating volume and page. Where further references occur after a semicolon, they refer to an English translation identified earlier in the endnotes. In a few cases, additional abbreviations have been introduced within single chapters.

FA *Sämtliche Werke.* Große Frankfurter Ausgabe. Ed. D. E. Sattler et al. Frankfurt am Main: Stroemfeld/Roter Stern, 1975–.

FF *Friedensfeier.* Bibliotheca Bodmeriana. Ed. Friedrich Beißner. Stuttgart: Kolhammer, 1954.

HF *Sämtliche Werke.* Homburger Folioheft. Facsimile edition. Ed. D. E. Sattler and Emery E. George. Frankfurt am Main: Stroemfeld/Roter Stern, 1986.

KT *Sämtliche Werke.* Kritische Textausgabe. Ed. D. E. Sattler et al. Darmstadt and Neuwied: Luchterhand, 1979–88, 11 vols.

SA *Sämtliche Werke.* Große Stuttgarter Ausgabe. Ed. Friedrich Beißner et al. Stuttgart: Kolhammer, 1943–85, 8 vols.

SF *Sämtliche Werke.* Stuttgarter Foliobuch. Facsimile edition. Ed. D. E. Sattler and Hans Gerhard Steiner. Frankfurt am Main: Stroemfeld/Roter Stern, 1989.

In general, the German original of Hölderlin's texts is from *Sämtliche Werke*, Große Stuttgarter Ausgabe, ed. Friedrich Beißner et al. (Stuttgart: Kolhammer, 1943–85), 8 vols. (© W. Kolhammer Verlag); or from *Sämtliche Werke*, Große Frankfurter Ausgabe, ed. D. E. Sattler et al. (Frankfurt am Main: Stroemfeld/Roter Stern, 1975–) (© Stroemfeld/Roter Stern). In a few cases, the incomplete (and now discontinued) "critical text edition" has been used, since it includes reproductions of materials based on corrections against the *Frankfurter Ausgabe*: *Sämtliche Werke*, Kritische

Textausgabe, ed. D. E. Sattler et al. (Darmstadt and Neuwied: Luchterhand, 1979–88), 11 vols. (© Hermann Luchterhand Verlag). In some other cases, references are given to the facsimile editions of the folios, both contained in the *Sämtliche Werke* published by Sattler: the *Homburger Folioheft*, ed. D. E. Sattler and Emery E. George (Frankfurt am Main: Stroemfeld/Roter Stern, 1986) and the *Stuttgarter Foliobuch*, ed. D. E. Sattler and Hans Gerhard Steiner (Frankfurt am Main: Stroemfeld/Roter Stern, 1989). Finally, references to the late hymn "Feast of Peace" are to Friedrich Hölderlin, *Friedensfeier*, Bibliotheca Bodmeriana, ed. Friedrich Beißner (Stuttgart: Kolhammer, 1954). Unless otherwise noted, translations are the authors' own.

Facsimiles are taken from Friedrich Hölderlin. *Sämtliche Werke. Frankfurter Ausgabe.* (FHA) Historisch-Kritische Gesamtausgabe, ed. D. E. Sattler. Copyright © Verlag Stroemfeld/Roter Stern, Basel/Frankfurt am Main 1975 ff. Reprinted by permission. Information on this Hölderlin edition with all Hölderlin manuscripts in facsimile: Stroemfeld Verlag, Holzhausenstr.4, D-60322 Frankfurt am Main / Germany, e-mail: stroemfeld@t-online.de.

Permission to reprint is in all cases gratefully acknowledged.

Contributors

BERNHARD BÖSCHENSTEIN is Professor of German and Comparative Literature at the University of Geneva and editor of the *Hölderlin-Jahrbuch*. Among his books are *Hölderlins Rheinhymne*, *Leuchttürme: Von Hölderlin zu Celan*, and most recently, *"Frucht des Gewitters": Zu Hölderlins Dionysos als Gott der Revolution*. He is currently completing an edition of Paul Celan's "Der Meridian."

STANLEY CORNGOLD is Professor of German and Comparative Literature at Princeton University. Among his books are *The Commentators' Despair*, *The Fate of the Self: German Writers and French Theory*, *Franz Kafka: The Necessity of Form*, *Borrowed Lives*, and *Complex Pleasure: Forms of Feeling in German Literature*. He is currently completing *The Melancholy Object of Consumption*.

JEAN-FRANÇOIS COURTINE is Professor of Philosophy at the Ecole normale supérieure in Paris. The author of, among other works, *Heidegger et la phénoménologie*, *Extase de la raison: Essais sur Schelling*, and *Suarez et le système de la métaphysique*, his most recent publications include two edited volumes, *Le dernier Schelling: Raison et positivité* and *On the Sublime: Presence in Question*, as well as a special 1989 issue of the *Cahiers de l'Herne* devoted to Hölderlin.

PETER FENVES is Professor of German and Comparative Literature at Northwestern University. Author of *A Peculiar Fate: Metaphysics and World History in Kant* and *"Chatter": Language and History in Kierkegaard*, he is also translator and editor of *Raising the Tone of Philosophy: Late Essays by Kant, Transformative Critique by Jacques Derrida*.

ARIS FIORETOS is Senior Researcher in the Department of Comparative Literature at the University of Stockholm. The author of, among other works, *Det kritiska ögonblicket: Hölderlin, Benjamin, Celan*, he is the translator into Swedish of Hölderlin's "patriotic songs," *Hymner*, as well as editor of *Word Traces: Readings of Paul Celan*.

HANS-JOST FREY is Professor of Comparative Literature at the University of Zürich. He is the author of several books of literary criticism in German, the latest being *Der unendliche Text*, and of *Studies in Poetic Discourse: Mallarmé, Baudelaire, Rimbaud, Hölderlin*.

CHRISTOPHER FYNSK is Professor of Comparative Literature and Philosophy at Binghamton University. He is the author of *Heidegger: Thought and Historicity*, *Language and Relation: . . . that there is language*, and *Infant Figures*.

CYRUS HAMLIN is Professor of German and Comparative Literature at Yale University. The author of numerous essays and articles on German eighteenth- and nineteenth-century literature, philosophy, and aesthetics, he is the general editor of the complete edition of Goethe's work in English and of the Norton Critical Edition of Goethe's *Faust*. He is also the author of a collection of essays, *The Hermeneutics of Form: Romantic Poetics in Theory and Practice*.

PHILIPPE LACOUE-LABARTHE is Professor of Philosophy at the University of Strasbourg and Visiting Professor of French at the University of California, Berkeley. The author of numerous books and essays, his most recent publications in English include *Heidegger, Art, and Politics*, *Musica Ficta: Figures of Wagner*, and *Typography: Mimesis, Philosophy, Politics*. He has also published a French translation of Hölderlin's German rendition of *Antigone*.

ARNE MELBERG is Professor of Comparative Literature at the University of Oslo. The author of numerous books and articles, his most recent publications are *Theories of Mimesis* and *Några vändningar hos Rilke*.

RAINER NÄGELE is Professor of German at the Johns Hopkins University. His books include *Literatur und Utopie: Versuche zu Hölderlin*, *Text, Geschichte und Subjektivität in Hölderlins Dichtung: "Uneßbarer Schrift gleich"*, *Reading after Freud: Essays on Goethe, Hölderlin, Habermas, Nietzsche, Brecht, Celan, and Freud*, *Theater, Theory, Speculation: Walter Benjamin and the Scenes of Modernity*, as well as, most recently, *Echoes of Translation: Reading between Texts*.

JEAN-LUC NANCY is Professor of Philosophy at the University of Strasbourg and Visiting Professor of French at the University of California, Berkeley. The author of numerous books and essays, his most recent publications in English include *The Birth to Presence, The Experience of Freedom*, and *The Inoperative Community*, as well as *The Literary Absolute: The Theory of Literature in German Romanticism* and *The Title of the Letter: A Reading of Lacan*, both co-authored with Philippe Lacoue-Labarthe.

EDGAR PANKOW is Assistant Professor of Comparative Literature at the University of Chemnitz. He has published essays and articles in, among others, *Penser après Heidegger, Nietzsche und die Künste, Zeitkritik nach Heidegger, Poetica*, and the *Schopenhauer-Jahrbuch*, and is also the editor, with Günter Peters, of *Prometheus: Mythos der Kultur*.

THOMAS SCHESTAG teaches German and comparative literature at the Johann Wolfgang Goethe University, Frankfurt am Main. The author of several articles and books, among them *para-* and *Parerga*, he has recently edited *"geteilte Aufmerksamkeit": zur Frage des Lesens*.

ARNAUD VILLANI teaches philosophy in Nice (Première Supérieure) The author of *"Le chateau" de Kafka: Kafka et l'ouverture de l'existant*, as well as the co-translator into French of A. N. Whitehead's *Process and Reality*, he has published numerous essays in journals and multiauthor volumes, among them *Nouvelles lectures de Nietzsche, Être et temps de Heidegger*, and *Tausend Plateaus*.

HENT DE VRIES is Professor of Metaphysics and Its History at the University of Amsterdam. The author of *Theologie im pianissimo* and *Philosophy and the Turn to Religion* (forthcoming), he is also the co-editor of *Violence, Identity, and Self-Determination, Die Aktualität der "Dialektik der Aufklärung": Zwischen Moderne und Postmoderne*, and *Enlightenments: Encounters between Critical Theory and Contemporary French Thought*.

ANDRZEJ WARMINSKI is Professor of Comparative Literature at the University of California, Irvine. The author of *Readings in Interpretation: Hegel, Hölderlin, Heidegger* and *Material Inscriptions*, he has also edited Paul de Man, *Aesthetic Ideology*.

. . . daß gepflegt werde
Der veste Buchstab . . .

Introduction

Aris Fioretos

I drink wine from two glasses
and comb through
the king's caesura
like that one
with Pindar,

God turns over the tuning-fork
alone of the small
just ones,

from the fate-engine falls
our measure.[1]

Included in the posthumous 1976 collection *Zeitgehöft*, Paul Celan's "I Drink Wine" is not the first, nor by any means the last, literary document to take Friedrich Hölderlin, the poet, translator, and theoretician (1770–1843), as its subject matter. Yet with its crystalline allusiveness and prismatic brevity, the text might be one of the more acute treatments of the highly particular relationship to writing that characterizes much of Hölderlin's oeuvre. On repeated accounts, Celan dealt with his precursor in both poetry and prose, the best known instance being the frequently quoted and much interpreted poem "Tübingen, Jänner," from the early 1960's, which contains a clear if complex *Auseinandersetzung* with the figure and writings of Hölderlin.[2] In "I Drink Wine," the roles are less certain, the identification with the older poet neither as prompt nor as peculiar, and the radical compaction at work in the text in greater evidence. Even the relation between the poem's two pronominal instances—"I" and "that one" (*Ich* and *Jener*)—is relativized, not only by the conjunctive "like" (*wie*), but also, of course, through its bearing on the earlier poet's relation to one of *his* precursors, Pindar.

In certain respects, this triangulation—coordinating antiquity, the dawn of modernity, and its postwar aftermath—is characteristic for the realm of thought within which the writings of Hölderlin have come to be perceived toward the end of the present century.[3] His preoccupation with the concepts and convictions of the ancients, as well as his always keen, invariably subtle meditations on the differences between antiquity and modernity—notably present in the work that deals with (the translation of) tragedy—cannot be read today without causing reflections also upon the particular, and particularly vicious, caesura of the years in Germany between 1933 and 1945. With the conceivable exception of Kleist, no other German writer of comparable stature has received treatment similar to Hölderlin's during this disastrous period, which Celan, in a poetological statement, once referred to as being marked by both "terrifying silence" and "the thousand darknesses of murderous speech."[4] Although, during the 1930's and 1940's, the internal dissonances, persuasive parataxis, and many modifiers of mood, creed, and orientation that characterize Hölderlin's writings were straightforwardly ignored in favor of extolling him as the poet of the Fatherland (his poems were disseminated in so-called *Feldausgaben* intended to edify foot soldiers at the front), it is nonetheless obvious that few if any of his peers have faced, with such critical acumen, the fraught issue of what is contained in the name of Germany.

Later, less wayward readings of Hölderlin—of which, encouragingly, there are some—have tended to sidestep this "question of Germany," in the words of one of the contributors to the present volume. Rather than addressing the politically precarious and ideologically overdetermined resonances of Hölderlin's preoccupation with what he variously termed *Vaterland*, *Abendland*, and *Hesperien*, postwar scholarship has tended to focus instead on issues of textual (thus purportedly more neutral) character.[5] And, admittedly, distinguished by an evermore rigorous philological scrutiny and historical-critical accuracy, recent work has removed scholarship far from the stage when Hölderlin's writings were read as tracts of truth or professions of faith prefiguring the alleged probity of the Third Reich.

After a brief period of notoriety—during which Hölderlin, recognized as a man with unsettling poetic and intellectual resources, was read and supported by the likes of Hegel, Schelling, Goethe, and Schiller—for almost a century his work was, by and large, depreciated and neglected. Notable exceptions can be found, of course, such as Mörike and Nietzsche, who read him with care and considerable empathy. Not until the circle of friends, poets, and philologists formed around the commanding figure of Stefan George championed him as the poet of poets, however, was Hölderlin rescued from unconcern and oblivion.[6] Until then, if his poetry was known at all, it was considered as compositions of near-transcendental quality, but also, and more damaging, it was deemed too brittle in its claim to equipoise, too close to incoherence for hermeneutical comfort, communicated by a mad writer of rare promise but of even more destructive talent. The commendable edition initiated in the 1910's by Norbert von Hellingrath, a scholar belonging to the George circle, and completed after his death by Friedrich Seebaß and Ludwig von Pigenot, launched a new, philologically sounder period of interest—one which, however, soon became marked by the ideologically biased adulteration and tantamount disregard for the letter characteristic of wartime scholarship. From this time dates the image of Hölderlin as the *Dichter der Deutschen*, a poet torn between ire and sobriety, finally embracing the twilight of insanity as the only remedy for over exposure to heavenly light.

During a long period which came to an end only in the first postwar decades, Hölderlin's texts were indeed studied, even scrupulously, but with rare, scattered exceptions (one thinks, most immediately, of Benjamin's early essay),[7] the particular understanding that emerged from these works was mediated by one of several dominant fictions concerning his distraught life and pitiable fate. Hagiographic accounts and nationalistically phrased interpretations did not begin to lose their hold on the poet's received image until Friedrich Beißner, a classicist who had started as a young wartime celebrator of Hölderlin, set out to replace the Hellingrath edition with the immense philological labor of love known today as the *Große Stuttgarter Ausgabe.* Commenced already during the war (in 1943), and completed some forty years later (in 1985), the edition put an effective end to the less hermeneutically controlled efforts at interpretation which thus far had been the fate of Hölderlin's writings. Beißner's work also laid the ground for what was to present, from the 1950's and on, an unprecedented amount of exegetical effort—comparable, perhaps, only to the one bestowed on Kafka during the same period. Gradually discrediting the notion of the poet as the front figure of a certain *Deutschheit,* Hölderlin scholarship soon bloomed into an industry of its own, marked in equal measures by reverence and solemnity. The need to rescue Hölderlin's writings from the grips of *Blut und Boden* philosophy; the desire to read it as an achieved body of work displaying every appropriate sign of coherence and monumentality; the will to reinstate him as one of the few truly regal writers of German literature, on a pair with Goethe and Schiller; these and similar scholarly urges are both meritorious and understandable.

Yet the texts presented by Beißner—organized as trajectories leading from partial notations, glosses, and scattered ruminations, over first, second, and third drafts, to finished products marked by the signal characteristics of completed masterpieces—presuppose a critical doctrine guided by assumptions of a systematized whole which might strike today's critical awareness as, in matching parts, naive and conceited. Not even the claims made by Pierre Bertaux and others, in the late 1960's and early 1970's, that Hölderlin had been a radical revolutionary as well as a stout Jacobin, perhaps even involved in serious political plots, a notion that forced scholarship to recast the framework within which his writings had been understood, managed to change the fundamental assumptions about Hölderlin's working methods or attitudes toward the poetical process. Even less did these claims tackle the internal rifts and discontinuities at play on all levels in his texts, fractures which, far from questioning the solidity of the poems, might in fact account for their exceptional energy and truly compelling character.

Begun by way of an "introduction" in 1975, Dieter Eberhard Sattler's historical-critical *Frankfurter Ausgabe* (the third major edition of Hölderlin's work in less than three-quarters of a century), set out with the explicit purpose of laying bare the faulty premises of the Stuttgart edition, by then the long-established standard. Here an editorial policy was presented which, in vital regards, ran counter to the scholarly premises taken for granted by Beißner. While thoroughly in sympathy with the cardinal virtues of philology—rigor, truthfulness, accuracy—the aim of Sattler's edition was, and remains, to avoid establishing an internal textual hierarchy and instead to respect the processual character of the poems. Textual layers are not ranked in accordance with their distance to a presupposed finality; rather, their palimpsestic qualities and graphematically adventurous features are acknowledged, making possible a perusal of documents which remains focused and yet multifaceted, singular but still sprawling, an activity, thoroughly historical in nature, for which collaboration in the texts' manner of making sense is not merely encouraged but inevitable. Framed in this way, the reader's participation in the production of a poem is constantly emphasized and the vicissitudes of understanding brought to the fore with exemplary efficacy. In Sattler's edition, texts are offered which are neither definitive nor thing-like, but, rather, occurrence, movement, constellation—philological reflections, it might be argued, of the perhaps apocryphal account that Hölderlin (who, judged mentally incompetent in 1807, was confined for the second half of his life to a room in a tower next to the river Neckar in Tübingen) covered his walls with maps from all corners of the world.

Yet the reading practice required by such textual rarities is, of course, by no means obvious. The complexities involved in interpreting Hölderlin's work remain considerable and the challenge his imbricated poems pose to perceived notions of cognizance cannot be ignored. While Hölderlin, the poet, translator, and theoretician, has, by now, been so eruditely treated by scholarship as to spawn tertiary literature and even a website,[8] it is less clear to what extent some of the principal cruxes thus signaled by his writerly practice (to take only two examples: those of the problematic relationship between history and materiality, reference and aporia) have received proper, or even minimal, critical scrutiny. Although a scholarly cliché, hence devoid of the epistemic acuity needed to spur exegetical readiness into action, it is difficult to avoid the impression that we have only really begun to read Hölderlin in the manner his texts demand.

Since 1913, when von Hellingrath undertook to publish the first critical edition, Hölderlin's writings have provided one of the main bodies of work by which the disciplines of German literary criticism, theory, and historiography have defined themselves. Today available in two massive publications of rather different persuasion—the one directed by Beißner completed, the one headed by Sattler stalled momentarily—his oeuvre and its importance for any contemporary understanding of the place, value, and significance of poetry, is by no means limited to the academic disciplines traditionally considered closest to literature. Over the last eighty years, philosophy and aesthetics both have been shaped, particularly in their Continental form, as extended interpretations of central texts in Hölderlin's work. His role, for example, in the authorship of the so-called "Oldest System-Program" (*SA* 4; 1: 297–99), considered the founding document of German Idealism, has been treated and debated extensively; the intricate issues raised by *Hyperion* (*FA* 10 and 11) concerning notions of epistolarity and aesthetic edification, the vexed relationship between *Bildungsroman* and *Geistesgeschichte*, as well as more or less questionable doctrines of nationality and fatality, have been scrutinized repeatedly; and the importance of Hölderlin's views on the interrelationship between antiquity and modernity have been significant not only for his contemporaries, but also for later generations of thinkers, such as Nietzsche, Heidegger, Benjamin, and Adorno. One may even argue, as one of the contributors to the present volume does, that Hölderlin has become, in the course of the present century, the poet of philosophers.

Yet, and despite early and consistent translations into English of his main poetical works, including *Hyperion*, Hölderlin's oeuvre has been treated almost exclusively within the institutional confines of German studies.[9] By and large considered the author of an epistolary novel and some dozen or so late hymns, as well as the translator of mainly Greek verse and drama into German, Hölderlin the thinker on history, poetics, and philosophy has received scant attention from critics writing in English or working within disciplines with tangential relation to *Germanistik*. (One American exception of note is the multiauthor volume edited by Emery E. George in 1972.)[10] Only in the 1980's—particularly due to the early efforts of Paul de Man—did his philosophically oriented writings, as well as the theoretical presuppositions of much of his poetry, receive close critical consideration. Interpretative efforts by Timothy Bahti, Adrian Del-Caro, Véronique Fóti, Eckart Förster, Rainer Nägele, Eric Santner, and Andrzej Warminski, among others, as well as translations of essays by German and French scholars (among them Maurice Blanchot, Dieter Henrich, Philippe Lacoue-Labarthe, Jean-Luc Nancy, and Peter Szondi) have been instrumental in establishing the crucial pertinence of these writings not only for an understanding of the history of German and/or Romantic literature, but also for a less local, more concise consideration of the relationship between poetry, philosophy, and history. Although not uniformly persuasive in style or conceptual consistency, recent translations by Jeremy Adler and Thomas Pfau have further broadened the textual foundation from which Hölderlin's daunting oeuvre may now be addressed.[11]

Although different in focus and dissimilar in nature, thus revealing the shifting allegiances among a group of critics at the cutting edge of theory, the essays gathered in *The Solid Letter* are written in the context of this rejuvenated—theoretically informed and critically sophisticated—interest in Hölderlin's work. The first consolidated attempt, in English, to affirm as well as to discuss fundamental philosophical aspects of his writings, the volume aims to be wide in scope but succinct in nature, thus hoping to appeal to specialists not only in German studies, but also in modern poetry, comparative literature, literary theory, and philosophy. Striving to be philologically responsible, it endeavors to bring together into a constellation related, though diverse, contributions, with the explicit purpose of both acknowledging and demonstrating the relevance of Hölderlin's writings for thinking about history, culture, language, and difference today.

While previous scholarship has tended to define Hölderlin either as a poet, *or* an epistolary novelist, *or* a philosopher, *or* a translator, this volume focuses on him as a writer who cares for "the solid letter" to which the late hymn "Patmos" urges attention (*SA* 2; 1: 172). Detailed in their concerns yet broad-minded in scope, wide-ranging in choices of readings though consistently focused on textual singularities, the essays collected here attempt not only to offer a prismatic understanding of Hölderlin's finished or published work, but also to treat the processual character of his writing. By discussing, for example, the interrelationship between various unpublished variants, between theoretical and poetic texts, as well as between different conceptions of the distinction between theory and practice itself, the contributions provide a propitious opportunity to reassess the categories by which humanistic study presently is defined. Furthermore, by addressing the theological-political implications of Hölderlin's notion of history, as well as the philosophical stakes involved in his concepts of "measure" and the "alternation of tones," the significance of marginalia, notations, and seemingly auxiliary or secondary materials, as well as kinds of text not commonly considered intrinsic to an author's oeuvre (such as translations and letters), the essays provide analyses of the formidable multifacetedness of Hölderlin's writerly practice. Thriving on a variety of approaches and benefiting from the scholarly authority of the contributions, *The Solid Letter* wishes to go a certain way, then, toward reinstating for serious debate questions concerning the political and historiographical implications of reading.

Although the prismatic effects created by the proximity of varying approaches are an explicit objective, the volume also aims to provide a stable framework within which the reader may orient him- or herself according to more traditional criteria of interest. Thus the essays are grouped into three principal parts, each focusing on a particular set of texts, tropes, and concerns. Whereas the first section deals mainly with the philosophical, poetological, and theological aspects of Hölderlin's writings, the secondcenters on his often radical theory and practice of translation. In addition to containing a treatment of the textual intricacies involved in Hölderlin's so-called Pindar comments, the third and final part of the volume is largely devoted to his poetry, ranging from the early poems and midcareer elegies to the late, often uncompleted and always heterogeneous hymns.

In his contribution to the first section of the volume, Peter Fenves offers a reading of the particular place accredited to philosophy in Hölderlin's writings. Entitled "Measure for Measure," the essay argues that the philosophical character of his poetry consists not of its discursivity, as previous scholarship generally has claimed, but of its digressiveness—a certain strategic "besideness" which is captured only partly in the famous phrase concerning "an eccentric path" (*SA* 3; 1: 83). Juxtaposing the famed interpretations of Heidegger and Adorno, and proposing to consider them as theoretical inversions of each other, Fenves pinpoints the realm of meter—where measure is all—as the peculiar domain in which Hölderlin's thinking about poetry and philosophy, antiquity and modernity, is articulated both most alluringly and with greatest consequence. As a "differentiation in language of language," measure is shown to be not that which gives mass, volume, or even form to anything, but rather that which distinguishes language from itself, and in so doing, gives rise to different languages. Hence, Fenves argues, it may also account for "the immense difference" between the poetries of Hellas and Hesperia, so important to Hölderlinian historiography, as well as the distinct difference between poetry and philosophy often treated—not to mention mistreated—in Hölderlin scholarship.

The second essay, Jean-Luc Nancy's "The Calculation of the Poet," continues Fenves's line of inquiry by investigating two hypotheses, neither particularly favored by earlier scholarship: first, that the poet is more important to Hölderlin than poetry; second, that the task of the poet resides in a calculus. Rather than proposing or even animating a theory of literature or an aesthetics, Hölderlin's thinking about poetry thus assumes an *ars poetica* in the stricter sense of the term, that is, as a technics of composition. The poet's task is neither equivalent to nor translatable into that of the philosopher: the objective of his endeavors is a synopsis, as Nancy demonstrates, not a synthesis. Whereas the philosopher treats time as an occasion on which to elaborate thoughts and in which to unfold the path of thinking, the poet must measure and calculate time in view of precision and exactness. Only then is the merging of self and other, familiar and foreign, man and gods, possible, even if not attained. By establishing exactness—and thus, by implication, determination, definition, and even destination—as the objective of the poet's calculation, Nancy is able to lay bare the kairology implicit in Hölderlin's writings, as well as to reframe and to reinterpret his difficult, often misconstrued remarks concerning prosody, rhythm, tone, and tact.

In "Poetry's Courage," the third essay of the opening section, Philippe Lacoue-Labarthe treats one of the implications of measure and calculation with particular attentiveness: destination. Beginning with an evaluation of Heidegger's Hölderlin interpretations from the 1930's and 1940's, in which determination and destination are considered along "theological-political" lines and the courage of the poet is identified as that of someone who assumes a (national) vocation, Lacoue-Labarthe turns to Benjamin's celebrated analysis of "Dichtermut." In this poem, later entitled "Blödigkeit" (*FA* 5: 683–700), the motif of the poet's vocation is dealt with in association with a theory of *das Gedichtete*, or "that which is poetically written," a terminological innovation of considerable conceptual complexity. In contrast to Heidegger, Benjamin theorizes the topoi of mission and vocation as a courage to turn to *das Gedichtete* rather than to what might be covered by terms such as *Volk* or *Nation*—thereby conceiving of Hölderlin as the poet of poetry, rather than as the poet of Germany. An entire politics is determined by this difference, Lacoue-Labarthe argues, who concludes by treating its implications for the contemporary discussion of the political stakes of theory.

In the fourth essay, "*Winke*: Hölderlin, Heidegger, Nancy," Hent de Vries shifts the focus, but only slightly, by attempting to trace the concern with the topos of the divine prevalent in both Hölderlin and Heidegger. Framing the inquiry as an examination of Jean-Luc Nancy's propositions regarding "divine places," and focusing on the fragmentary texts "Was ist Gott?" (*SA* 2; 1: 210) and the perhaps spurious "In lieblicher Bläue" (*SA* 2; 1: 372–74), as well as on the interpretations of them offered by Heidegger and, subsequently, by Derrida, de Man, and Nancy, de Vries is able to circumscribe the divine "nod" or "sign" as the elusive gesture by which poets function as heralds in Hölderlin's poetry. Yet rather than supposing a soteriological scheme in which the poet acts as mediator, de Vries proposes to conceive of such *Wink* as an indication which can be neither inscribed into a negative theology, nor conceptualized as part of the gestural vocabulary of linguistic mysticism. Rather, he suggests, the singular rigor and alluring integrity of Hölderlin's texts emancipate the question "What Is God?" from the dilemmas of linguistic ineffability, thereby helping, by implication, to elucidate questions of naming and placing the divine that preoccupy much of contemporary philosophy.

Continuing the line of inquiry opened by Lacoue-Labarthe and de Vries, Jean-François Courtine further treats the vexed issue of "What Is God?" in an essay devoted to "Hölderlin's Christ." Emphasizing that the figure of Christ, in Hölderlin's writings, changes the course of time by opening it onto another scansion, and underscoring that the turn from Greece to Hesperia signaled in much of his work is articulated as a prolonged meditation on the nature of the tragic, Courtine traces elements of Christology in three late poems in particular: "Der Einzige," "Friedensfeier," and "Patmos," including earlier versions and variants (*SA* 2; 1: 153–64, *FF*, and *SA* 2; 1: 165–87, respectively). In this reading, the figure of Christ emerges not as the conventional, if unique, mediator between the divine and man, but rather as the one who accomplishes mediation itself. The last hero in a succession of demigods, he does not simply represent an "image of God," but in him, God is seen as having arrived on Earth. The transition from ancient Greece to modern Germany is thus accomplished as a nonreversible trajectory that allows for little nostalgia for the past and much concern about the present. Accordingly, Courtine suggests, any attempt to conceptualize Hölderlin's view of antiquity must attend to the particular temporality characteristic of this trajectory—an unorthodox path, marked by disjunctiveness, for which the care of the *veste Buchstab* remains crucial and "that which endures" must be "well interpreted" (*SA* 2; 1: 172).

The sixth and last essay of the opening section, Edgar Pankow's "Epistolary Writing, Fate, Language: Hölderlin's *Hyperion*," concludes the treatment of fate and destination in Hölderlin's oeuvre by turning to a text for which these problems are both immanent and generic: the epistolary novel *Hyperion*, published in 1797 and 1799. Focusing on the much neglected importance of letter writing in Hölderlin's life and work, Pankow is able to discern the peculiar form of addressing that informs his notion of correspondence. Hölderlin departs from conventional concepts of the epistolary, based on structures of exchange between distinct and stable subjects. Rather, the "psyche among friends," mentioned by Hölderlin in the second letter to his friend and colleague Casimir Ulrich von Böhlendorff (*SA* 6; 1: 433), seems to evoke a mutual exploration of otherness, which requires a refined notion of writing and temporality. Pinpointing the implications of such *Psyche*, a Greek word introduced into a German phrase, Pankow thus delineates the presence of "a kind of foreign language" in Hölderlin, which cannot be reduced simply to a national language or a question of terminology. More appropriately, it is the language in which the "psyche"—and, in other contexts, *die Seele*—chooses to speak: that is, the language of the other. Establishing the origin of thinking in conversation and correspondence, Hölderlin appears to consider the originality of thought as something possible to attain only in speaking with another. Pursuing such conceptualization of the language of the soul, Pankow traces the consequences of this notion not only for an understanding of the status of *Hyperion* within German literary history, but also for the tropes which traditionally have secured the stability of the relation of modernity to antiquity—among them return, repetition, and naming. Through close readings of chosen passages, he is able to make clear a particular uncertainty concerning destination (and thus also relationality) that, far from undermining Hölderlin's enterprise in *Hyperion*, conveys to it its exemplariness.

In the first contribution to the second section of the volume, "Figures of Duality: Hölderlin and Greek Tragedy," Arnaud Villani discerns the manners in which figures of duality and discord in Greek drama are conceived by Hölderlin in his remarks to the translations of *Oedipus* and *Antigone* (*FA* 16: 247–58 and 409–21, respectively). Arguing that the texts of Greek tragedy already contain the complex logic of doubling which previous scholarship had confined, at best, to the transition from ancient original to modern version, Villani demonstrates not only the importance of these texts for Hölderlin's conception of the relationship between antiquity and modernity, but also the manner in which their rendition in German doubles and displaces the original duplicity. Through this doubling of doubling, the translations offered by Hölderlin and theorized in his adjacent notes to the plays become texts in which the stakes of modernity remain open—their significance suspended in a temporality most akin to what the remarks on *Ödipus* term the "caesura" (*FA* 16: 250).

In "Monstrous History: Heidegger Reading Hölderlin," Andrzej Warminski continues this line of inquiry by discussing Heidegger's interpretation of Hölderlin's translation of the so-called *polla ta deina* choral ode of *Antigone* (*FA* 16: 298–303). Analyzing the ways in which translation in Hölderlin's understanding is not merely confined to the transition between languages, but also occurs within one language, as its difference from itself, Warminski focuses on Heidegger's translation of Sophocles' *deinon* as *unheimlich*, in contrast to Hölderlin's rendition of the word as *ungeheuer*. By treating also Heidegger's explication of the poem "Der Ister" in close detail (*SA* 2; 1: 190–92)—an analysis in which the turn to the native, and thus familiar, is enacted as a turn to that which is *heimisch*[12]—Warminski is able to demonstrate how Heidegger's interpretation of Hölderlin's *ungeheuer* as *nicht geheuer* leaves something behind: the word's "self-translating difference from itself" loses the very monstrosity, he argues, which, for Hölderlin, alone accounts for the radical disjunction between Hesperians and the Orient which *is* the Greeks.

In "Disowning Contingencies in Hölderlin's *Empedocles*," Stanley Corngold shifts the focus by turning to Hölderlin's sole, if thrice repeated attempt to write a modern tragedy. Discussing Sattler's recent reconstructions of Hölderlin's *Empedocles* manuscripts and Andrzej Warminski's readings of portions of the third version of the play,[13] Corngold proposes a new interpretation of the drama. Paying particular attention to the "field of relations" suggested by the logic of *Verläugnen* ("denying," "disowning"), that he argues is part of Hölderlin's conception of tragedy, Corngold attempts to understand the form a tragedy might take when it tries rigorously to deny contingency. In this drama, Hölderlin constructs an aporia the thrust of which is to represent the experience of an empirical personality in images and signs so severe as to border on the invisible. His tragedy is caught up, Corngold underscores, in a logic in which the rigorous elimination of accidentals will finally affect the conditions of its structure of presentation (*Darstellbarkeit*). Hence, struggling to bring his drama into conformity with a conception of tragedy as the visible agon of the heroic individual and the mob, the embodied will of his adversary, Hölderlin is forced to turn Empedocles into the author or *Darsteller* of his own fate. In so doing, however, the need for poetic autonomy must give rise to doubts concerning its representability. Disowning contingent imagery, the play impedes representation at the same time as it transforms consciousness into a stage, that is, into a scene of representation. Envisioning death as a scene in which "the constraining environment of the sensory *Bild* is cast off," in Corngold's reading, Hölderlin's *Empedocles* becomes a tragedy of incomplete self-deification—or, in other words, a tragedy of hubris. The punishment for the failure of his tragic hero is at once not to die and not to find another way. Thus, the aporia at the core of the play must result in the abandonment of the *Empedocles* project, the price paid for an insight into the insistence of the material image.

In "Reading the *Poetics* after the 'Remarks,'" the fourth contribution to the second section, Christopher Fynsk returns to Hölderlin's concern with the object of tragic representation, in particular as it is spelled out in the remarks on his Sophocles translations. Focusing on the crucial reference to Aristotle's *Poetics*, a subject scarcely treated in previous scholarship, he proposes to reread not Hölderlin in view of Aristotle, but the Greek text in light of the German. In this context, Fynsk offers a reconsideration of a central poetological tenet in view of its far-ranging reconceptualization in the work of one of the most serious readers of the texts of antiquity. By analyzing Hölderlin's description of the "tragic transport" as a rhythmic structure (*FA* 16: 250), Fynsk argues, a solution may be found to what has traditionally been perceived as the fundamental crux of the *Poetics*: the question of catharsis. Maintaining that catharsis must proceed from a certain movement—"an engagement in the tragic action . . . that involves something more than contemplation"—he establishes rhythm as the element in which the visualization and gestures of speech, fundamental to the Aristotelean doctrine of tragic representation, is articulated, and in which the effect of truth is produced. Fynsk concludes that for Hölderlin, by contrast, the presentation of the separation of man and nature's powers, the object of tragedy, occurs in a structure that may follow the calculus of rhythm which Aristotle termed the soul, but it can do so only by way of that *vaterländische Umkehr* in which beginning and end no longer rhyme.

In his "Ancient Sports and Modern Transports: Hölderlin's Tragic Bodies," Rainer Nägele augments this reading by proposing to understand the difference between ancient and modern means of conveying the object of tragic representation in relation to wordplay, or as a tension between trope and meaning. Long avoided by Hölderlin scholars—or, if acknowledged at all, quickly dismissed as a case of literary contingency—the pun has traditionally been deemed a linguistic casualty in Hölderlin's search for a deeper meaning that critics may term "truth." Discussing not only such reduction of the subject of literary criticism to a "mere" vehicle, with "real presences" as its true and legitimate ground, but also the treatment of the interrelationship between the letter and its interpretation in Hölderlin, Nägele considers tragedy as an exemplary instance of textuality for signaling the limits of language. Doing so with and within language, however, "in the . . . display of the wounded body and being that speaks," tragedy is also the genre in which the historical difference between antiquity and modernity may be observed most acutely. A variance residing primarily in different understandings of the relationship of body and language, this historical difference is traced in a careful analysis of the crucial importance of "transport" for Hölderlin. Focusing on the Sophocles translations, as well as on remarks concerning tragedy contained in letters and adjacent materials, Nägele is able to elucidate, in Hölderlin's renditions, the very transport which is that of the *Darstellung* itself. Presupposing a sacrifice (Hölderlin takes literally the German *Opfer*, derivative of the Latin *offere*, "to offer," as well as "to present"), these German translations of Sophocles become not only the vehicle, but the object as well, of a certain—tragically ironic—undoing of *Darstellung*.

The sixth and concluding essay of the second part of *The Solid Letter*, Aris Fioretos's "Color Read: Hölderlin and Translation," traces some of the implications of the earlier essays contained in this section by proposing to treat Hölderlin's *Antigone* rendition not only as a translation of tragedy, but also as a text documenting the tragedy of verbal transfer. In particular, Fioretos offers a close reading of the moment in the drama which Hölderlin defined as its constitutive core or "caesura." Identified formally as the speech of Tiresias, in the Greek text, it is articulated as a failed rendition of material events into celestial meaning, whereas in Hölderlin's German version, the translation of sign into significance appears successful. On a second level of interpretation, however, Hölderlin does not translate the Greek. Pursuing the historical-political consequences of this disjunction for the rupture that, for Hölderlin, marked the relationship between antiquity and modernity, Fioretos suggests that such "transport of non-transport"—providing a repetition of what occurred in the original without ever taking place—signals a "traur'ge Arbeit," or a "mournful work" (*FA* 16: 265), of exemplary linguistic transmutation.

In "The Philosophy of Poetic Form: Hölderlin's Theory of Poetry and the Classical German Elegy," the opening contribution to the third and final section of the volume, Cyrus Hamlin attempts to shift the attention from the late hymns, long a dominant preoccupation among Hölderlin scholars, to the earlier elegies written before 1800. Tracing Hölderlin's development as a poet within the larger context of Idealist philosophy during the latter part of the 1790's, Hamlin appeals for a reconsideration of the early poetry in light of the poet's subsequent theory of tonal variation. He first discusses Hölderlin's implicit reconception of Schiller's definition of a "sentimental" elegy in the fragmentary text "Über die Verfahrungsweise des poetischen Geistes" (*FA* 14: 303–22), then treats his subsequent critical reaction to Fichte in the brief fragment labeled "Urtheil und Seyn" (*SA* 4; 1: 216–17) (or also "Seyn Urtheil Möglichkeit" [*FA* 17: 156]), before analyzing the poet's radicalization of both Schillerian and Hegelian correlations of the work of art with its emotional affect in the elegy "Menons Klagen um Diotima" (earlier entitled simply "Elegie") (*KT* 6: 74–78). Here, Hölderlin's theory of poetic structure is shown to be primarily philosophical in nature, but fundamentally musical in its affects. In no way dependent on—or derivative of—a philosophically informed theory of poetic composition, Hamlin concludes, Hölderlin's poetry must receive priority as an example of poetic language, rather than as an illustration of philosophical insights concerning its formal arrangement or distribution of cognitive material.

In the second essay, "'Brod und Wein': From the 'Classical' Final Version to the Later Revision," Bernhard Böschenstein chronicles both the genetic progress of the elegy "Bread and Wine" (*FA* 6: 222–33) and its intrinsic narrative development. Analyzing the passage from final version to later revision, and focusing in particular on the fact that Hölderlin hardly revised the first triad of the text, although he rewrote both second and third strophes to significant extents, Böschenstein argues against previous editorial attempts to integrate different layers of the text organically. Rather, he suggests, the text must be considered a heterogeneous work, in no manner concluded, whose expressive thrust resides in the very fact that it resists any "equalizing systematization" of different phases of development, as well as different levels of thematic organization and narrative ordering.

In the subsequent contribution to the volume, "Turns and Echoes: Two Examples of Hölderlin's Poetics," Arne Melberg is able to augment Böschenstein's argument by showing how Hölderlin's famous remarks in *Hyperion* concerning beauty—containing his reinscription of the Heraclitean ἓν διαφέρον ἑαυτῷ, "the one in itself differing" (*SA* 3: 81), which turns a passive verb construction into an active one—gives way to the later, post-1800 poetics, for which division and difference seem constitutive. Proposing to consider Hölderlin's concept of beauty in relation to his attempts to define "primordial division" in the seminal fragment on "Seyn Urtheil Möglichkeit," Melberg provides two examples of the peculiar poetic practice that this interrelationship effects. Focusing in particular on the tropes of turning and echo, he offers close readings of a crucial passage from the revised version in "Brod und Wein," as well as the late hymnic fragment variously entitled "Mnemosyne" and "Die Nymphe Mnemosyne" (e.g., *SA* 2; 1: 193–98 and *FA* Einleitung, 55–70). By tracing their implications, Melberg is able to demonstrate, in conclusion, that for Hölderlin, poetry and poetics seem to supersede philosophical meaning. Hence, by extension, he argues, rhythmical patterns appear to be primary to any paraphrasable content. Indeed, rhythm, which Hölderlin is reported to have claimed "is all,"[14] appears to be the most inclusive term to designate his late, radicalized poetics—and thus his notion of language and history, too, as well as their interrelationship in that which, Melberg suggests, can only be termed humanity.

In his essay on "Hölderlin's Marginalization of Language," Hans-Jost Frey localizes some of the issues at stake in Melberg's reading by investigating the complex manners in which Hölderlin's texts are engendered. Focusing on the relationship between the supposed body of a text and its assumed margins, he addresses two fragments in particular, both written in the margins of manuscripts, in which language is crucially at issue: "Aber die Sprache . . ." (*FA* 4: 131) and "Im Walde" (*SF* 34). Discerning the textual consequences of these notations, as well as the theoretical implications ensuing from such indicative marginalization of language about language, Frey proposes to read the very place of inscription—a border, a division—as constitutive of meaning. Although deeming too hasty any allegorical interpretation of the textual frontier with reference to Hölderlin's explicit intention, he is nonetheless able to demonstrate the impossibility of ignoring the potential significance of a text's margins for the particular ways in which it makes sense. The clearer marginalia circumscribe the assumed body of a text, it is suggested, the more open their status as borders must become. Hölderlin's manuscripts reveal a writerly practice, Frey thus affirms, scarcely addressed and rarely theorized, the far-reaching implications of which may account for its pivotal importance for literary criticism today.

The fifth and last essay of the third part of *The Solid Letter*, Thomas

Schestag's "The Highest," continues the pursuits of previous contributions, in particular those of Melberg and Frey, by examining the fifth of Hölderlin's so-called "Pindar Fragments" (*FA* 15: 355). These texts—they consist, in toto, of nine—demonstrate an intricate and often paradoxical interplay between original, translation, and adjacent commentaries. Undermining the customary faith in translation as a substitution of the words and turns of phrase of one language with those of another, Hölderlin's *Pindar-Kommentare* offer an altercation, according to Schestag's philologically scrupulous reading, in which the laws of two languages, cultures, and beliefs, as well as the laws of difference itself, are enacted instructively in all their theological-linguistic radicality.

By focusing on a writerly practice of such rigorously differentiating, albeit hermeneutically demanding, nature, Schestag's essay also completes *The Solid Letter* on a note shared by all the essays included: that of an interpretative respect for the consistency of letters whose meaning, while not given, continue to offer food for thought. Participating in an enterprise which never seems *in situ* nor *in asitu*, but remains perhaps, more properly, *in parasitu* (a place, or rather predicament, where languages intersect but also diverge, and where understanding is invariably at stake), such sharing is always also a parasitical reading (or partaking) of that which Hölderlin, in a late fragment of "Patmos," termed an "inedible writing" (*SA* 2; 1: 185) —a trace and remainder, ever material, of what, in the final analysis, cannot be assimilated into meaning. As the collected contributions to this volume suggest, to care for the solid letter might mean this, too: to continue to engage in aporetic structures that teach us the fraught art of endurance.

The volume concludes with a select bibliography of Hölderlin in English, listing all book-length translations of his literary writings, the more significant translations of his theoretical texts and letters, as well as most critical studies, available in English, devoted in part or in whole to Hölderlin.

When the subject of Celan's "I Drink Wine" claims to toil with the "king's caesura / like that one / with Pindar,"[15] it is also, among other things, hinting at the *summum*, or "the highest," about which Hölderlin's Pindar commentaries speak and in relation to which they claim that "strict mediacy . . . is the law" (*FA* 15: 355). For Celan, writing a century and a half as well as two world wars later, such mediacy is not merely binding, but has become cutting and critical:

Ich trink Wein aus zwei Gläsern
und zackere an
der Königszäsur
wie Jener
am Pindar,

Gott gibt die Stimmgabel ab
als einer der kleinen
Gerechten,

aus der Lostrommel fällt
unser Deut.[16]

Organized according to a formal pattern initiated by *Ich* and ending on *Deut,* putting the mention of *Gott* at the exact center of the poem, with fifteen words coming before and after, Celan's poem might be read as a careful attempt to bring out the difference between two forms of calculating poetry. While divided into 15 + 1 + 15 words, stressing a concentric disposition of the lexical material, the text is also arranged into three stanzas of five, three, and two lines, respectively—a pattern, moving in ever more concentrated fashion from the opening "I" to the concluding "measure," that corresponds to the classical, indeed royal, formula of the golden section (5:3:2).[17] For a poem so preoccupied with fate and allotment, subjectivity and divinity, justice and measure, it is hardly surprising, although of course dazzling, to discover that "Ich trink Wein" thus includes two contradictory means of poetical organization: one emphasizing a concentric order, the other delineating a more eccentric sequence. If God, at least alphabetically, rests at the exact middle of Celan's text, affirming the harmony of formal order, the subject of his poem seems to move according to a course—repeated by the reader—that is decidedly less tranquil.

This distinction between the path as well as position of finite and infinite beings might be brought to bear on the differentiation that Hölderlin introduced in his fifth Pindar commentary—or, rather, already in his translation of the fragment—between mortals and immortals:

Das Gesez,
Von allen der König, Sterblichen und
Unsterblichen; das führt eben
Darum gewaltig
Das gerechteste Recht mit allerhöchster Hand.
(*FA* 15: 355)

The law,
King of all, mortals and
Immortals; just for this reason it wields
Powerfully
The most rightful law with the very highest hand.

In the original, Pindar had allowed mortals and immortals (*θανατῶν . . . καὶ ἀθανάτων*) to cohabit the same line, thus also to share the same rhythmic measure; by contrast, in his translation, Hölderlin observes what the notes to his translations of Sophocles term a "counterrhythmic rupture" (*FA* 16: 250),[18] introducing a break that dissociates the two entities metrically. The statement of the law is interrupted by a caesura—all the more abrupt since it introduces a split in the cardinal form of syntactic union: the conjunction *and*. In the "Remarks," Hölderlin had defined such a breach as the moment when Tiresias interrupts Creon:

He enters the course of fate as the custodian of the natural power which, in a tragic manner, removes man from his own life-sphere, the center of his inner life into another world and into the eccentric sphere of the dead. (*FA* 16: 251; *Essays*, 102)

For Hölderlin, the caesura is the moment when the chosen representative of the law is confronted with the law of "natural power." Appearing in the guise of the blind prophet, this other law is forcefully "counteractive" (*FA* 16: 251; *Essays*, 102). Already at an early stage of *Ödipus der Tyrann*, Oedipus had addressed Tiresias with the emphatic "O König" (*FA* 16: 111), and the prophet's word does indeed carry the entire weight of royalty. It should come as no surprise, then, that, for the subject of Celan's poem, who toils "like that one / with Pindar," such counteractive interruption has become the property of a "king."

In his poetological speech "The Meridian," Celan had proposed to understand the exclamation "Long live the king!," Lucile Desmoulins's last words in Georg Büchner's *Dantons Tod*, as a tangible *Gegenwort*: ". . . a word against the grain, the word which cuts the 'string,' which does not bow to the 'bystanders and old warhorses of history.' It is an act of freedom. It is a step."[19] Such emancipatory effect is, indeed, the attribute of the caesura in Hölderlin. In it, "the change of representation does not appear, but the representation itself" (*FA* 16: 250; *Essays*, 102 [trans. modified]). Itself expressing nothing, neither an intention nor its representation, the critical power of the caesura effects no more and no less than the *form* of representing. It is a "pure word" (*FA* 16: 250; *Essays*, 102), according to Hölderlin, in which "nothing exists but the condition of time or space" (*FA* 16: 258; *Essays*, 108 [trans. modified]).[20] A "counter-word", to use Celan's own description, it is an opening-up in the form of a keeping-apart, a rupture which both differentiates and distances. Or: a disjunctive connective, it imparts. As such, the caesura is the condition of possibility of mediacy, and offers, as Peter Fenves notes in his contribution to this volume, "the dialectical opposite of fate, namely freedom." It announces "the problem of measure, or of meter," or in any case, "the problem of finding the right measure when measure is no longer, or not at all, given."

While God remains at the exact middle of the text, centered and unmovable, the subject of Celan's poem moves—according to an eccentric path, therefore always already in mediated manner—toward a *Deut.* Signifying "measure," but also "token" or "allotment," the word refers, more narrowly, to a coin of little or negligible value (viz., a "tuppence"). Designating the smallest difference conceivable, it nonetheless decides between head and tail, gain and loss, victory and ruin—and ultimately, of course, between life and death. In addition, in a poem forcibly preoccupied with poetry and history, *Deut* obviously also indicates an activity which, in Hölderlin, and crucially in "Patmos," is described as one of well elucidating "what endures." The verb *deuten* carries several meanings, among them "to interpret," "to indicate," "to announce," and "to bring to light," but the word is also, more problematically, related etymologically to *zu deutsch*, "to make understandable, Germanizing, for the people, the Germans."[21]

Freedom and fate, *Zäsur* and *Deut*: the movement traced between these two principles describes the peculiar path of *zackern.* "Toiling" with the king's caesura—but also "plowing," "furrowing," and "combing" it "through"—and, figuratively, even "repeatedly working it over"—the subject of Celan's poem delineates the back-and-forth rhythm that is the gestural blueprint of any engagement with Hölderlin's texts. That this undertaking should project us "into the eccentric sphere of the dead," as the remarks to Sophocles has it, is only appropriate. When drastically initialized, contracted according to the prismatic logic of crystallization that "Ich trink Wein" seems to intimate, Celan's two-noun *Königszäsur* becomes *KZ*, the abbreviation for "concentration camp"—that industry of deadly transport for which no words, however measured, will ever account properly. Celan's *Deut*, only a syllable, barely a word, indicates a legibility-yet-to-come and makes us aware of the equivocal perspective from which, today, measure for measure. Hölderlin's writings must be approached. In the words of "Patmos":

> Vieles wäre
> Zu sagen davon.
> (*SA* 2; 1: 167)

> Much could be
> Said thereof.

PART I

Measure for Measure

HÖLDERLIN AND THE PLACE OF PHILOSOPHY

Peter Fenves

Hölderlin's works have been the occasion for so many philosophical investigations into the nature of poetry that, in contrast to the often cited description of him as the "poet of poets," he has become over the last eighty years the poet of philosophers. This has caused no small consternation among those for whom philosophy is seen to exercise no rights in the realm of poetry and, conversely, among those for whom the pride of philosophy is its independence from poetic pretensions. The sign under which the independence of philosophy from poetry has been proclaimed is the one Plato hung at the entrance to the Academy: "No one who does not know geometry may enter." The sign with which this geometrical injunction has been replaced by more than one modern philosopher and, above all, by those who bring the modernity of philosophy into question is: "Entry refused to those unfamiliar with Hölderlin."

The consternation among those who fear the confusion of philosophy with poetry as well as among those who grant philosophy no say in poetic matters can be understood as a reaction to this barely hidden injunction. The work of a poet replaces the practice of geometry, and this replacement brings to the surface the question of the place of philosophy, the place philosophy can claim to measure. Familiarity with Hölderlin's work does not so much grant a would-be philosopher entrance into the sanctuary of dialectics as give philosophers who have experienced the saturation of this sanctuary the chance, perhaps the one chance, to leave it behind. However complex the specific controversies concerning the relation of Hölderlin to philosophy may appear—and no scholarly controversy is as complex as the one concerning Hölderlin's role in the authorship of the document said to be the "founding" of German Idealism[1]—one controversy concerning Hölderlin is utterly simple: whether his work constitutes a sign at the threshold of another mode of philosophical thought and, in turn, whether his poetry serves as the "geometry" through which the place of this thought is delimited.

Hölderlin as a sign at the threshold—perhaps no phrase captures in a more succinct manner how the relation of his poetry to philosophical thought has often been represented. Within a certain spirit of Kantianism it is natural to inquire into the conditions for the possibility of philosophy, and if this spirit is combined with a strong attraction to Greek art—"Kant and the Greeks are almost my only reading," Hölderlin writes to Hegel[2]—one conclusion is inescapable: the condition for the possibility of philosophy is poetry. Justifications for this representation of the relation of poetry to philosophy can then seek support in the writings of Schelling and Hegel, each of whom conceives of poetry as an anticipation of systematic philosophy that enables such systematization to occur in the first place. And certain texts of Hölderlin appear to be luminous expositions of this thesis. Hyperion, for instance, goes beyond himself and thus begins to live up to his name (*hyper-ion*) when, during a trip to Athens, he exalts the founding function of poetry: "Poetry, I said, certain of my matter at hand, is the beginning and end of this [that is, philosophical] science" (*SA* 3; 1: 81). Hyperion defends poetry from its academic assault by insisting on the all-inclusive and self-differentiating character of its beauty. Poetry appears as the condition for the possibility of a systematic philosophy and so, at least for Hölderlin, it—and not geometry—constitutes the prolegomenon for any future philosophy that wishes to come forward as science.

But this contention would only be good for Hölderlin or, more precisely, for Hyperion on his way to Athens. So little conviction does this contention secure that, for all Hyperion's "certainty," Diotima, the one whom he claims as his inspiration, finds the remark somewhat off the point: "But you're rambling," she says (*SA* 3; 1: 81).[3] And there is little reason to trust Hyperion at this point, for he is not yet, and perhaps never will be, a genuine poet, one who no longer speaks of the relation between poetry and philosophy in the same manner as the British scholars whom he meets discuss the objects they encounter in Athens, namely, as mere objects, as ruins. Hyperion may speak of the foundational role of poetry "certain of [his] matter at hand," but this certainty, which is only of the matter *he* has at hand and is therefore marked as subjective from the start, stands in stark contrast to the certainty geometricians can claim. Far from demonstrating the foundational role of poetry in the development of philosophical knowledge, Hyperion's discourse gives reason to suspect that the relation of one to the other cannot be presented according to the category—itself borrowed from an already established philosophy—in which certainty is generally sought: the category of ground and consequence. Indeed, Hyperion's own presentation of philosophical knowledge as ἕν διαφέρον ἑαυτῷ (the self-differentiating self-relation) leaves no place for this category and, as a result, defeats his own argument. The failure of Hyperion may then deflate enthusiasm for statements like "poetry is the foundation of philosophy" or "the origin of philosophy is poetic," but this failure to secure a ground for philosophy in poetry does not in addition imply that their relation succumbs to mere indeterminacy. The fault does not necessarily reside in the Heraclitean phrase but could perhaps lie in the self-defeating manner in which it is prematurely deployed: as an intimation of what Schelling called "intellectual intuition," it expresses a self-relation that defies the category of ground and consequence. A "mature" deployment of this phrase would no longer be a deployment at all. Unless this phrase, or its equivalent, were itself deployed to stand for "first philosophy," however, the relation of poetry to philosophy could only be expressed in poetry or as philosophy, and therefore in neither case with conviction. No wonder Diotima becomes irritated by Hyperion's speech: every discourse on this relation, including one in which a self-differentiating self-relation is named, turns out to be beside the point.

Yet the relation of poetry to philosophy, for Hölderlin, may lie in precisely this: being, or speaking, beside the point. The philosophical character of his poetry would then consist of its "besideness"—its digressiveness, not its discursivity. Such "besideness" is not altogether captured by the famous phrase with which Hölderlin, in a discarded preface to *Hyperion*, describes the process of education (*Bildung*): "an eccentric path" (*SA* 3; 1: 83). Emphasis on the eccentricity of the path tends to conceal a philosophically more significant digressivity: the poetic statement can take place only on a path and therefore is from the outset mediated, placed in a medium and subordinated to certain calculable means. The philosophical exposition of Hölderlin's poetry cannot simply elucidate the various eccentricities through which the process of education travels—eccentricities that can always be recentered and are not infrequently the very points of thematic concentration—but must, instead, describe how a poem broaches a path in the first place. The means of this breach lie in the technical character of a poetic work. Speaking beside the point, describing the path of a poem, not its "content," ascribing the putative content of a poem to the workings of its technique—all this does more to elucidate the relation of poetry to philosophy in Hölderlin than a vast survey of the philosophical theses that his poetry, in the tradition of a Haller or a Schiller, seems to advance.

No one has expressed this point with greater force than Theodor W. Adorno, whose "Parataxis" attacks those, like Heidegger, who wish to elicit philosophical "statements" from Hölderlin without sufficient consideration for the specificity of their poetic medium. Although Adorno occasionally draws on Hölderlin's poetological reflections, he does not use them to arrive at something like Hölderlin's "philosophy." What Adorno calls the "truth content" (*Wahrheitsgehalt*) of the poetry will manifest itself only in the poems, nowhere else. Like Kantian critique, which seeks to avoid the twin threats of skepticism and dogmatism, the critical interpretation Adorno pursues in "Parataxis" sets out to escape the threat of philosophical—and in particular, Heideggerian—dogma, on the one hand, and the skepticism of mere philological exegesis, on the other. The analysis of "Friedensfeier" in the opening pages of "Parataxis" stands as a model of this mode of critical interpretation. "Friedensfeier," according to Adorno, releases its otherwise inaccessible "truth content" and thus falls into the hands of critique at the precise moment when it breaks the "law of fate" (*Schicksalgesetz*) that it sets down as a philosophical "statement." The caesura of this law, its "counter-rhythmic interruption" (*SA* 5; 1: 196), takes place when the caesura in the poetic line interrupts the statement of the law, and this interruption constitutes the dialectical opposite of fate, namely freedom or, to use the word Adorno discovers at this point, "thanks":

> But two strophes earlier [than the statement of the "law of fate"] fate is handled in this way:
>
> Denn schonend rührt des Maßes allzeit kundig
> Nur einen Augenblick die Wohnungen der Menschen,

> Ein Gott an, unversehn, und keiner weiß es, wenn?
> Auch darf alsdann das Freche drüber gehn,
> Und kommen muß zum heilgn Ort das Wilde
> Von Enden fern, übt rauhbetastend den Wahn,
> Und trifft daran ein Schicksal, aber Dank,
> Nie folgt der gleich hernach dem gottgegebnen Geschenke.

By having the key work *Dank* (thanks) follow *Schicksal* (fate) at the end of these lines, mediated by an *Aber* (but), a caesura is posited, and the linguistic configuration determines the "thanks" as the antithesis to "fate" or, in Hegelian language, as the qualitative leap, which, answering fate, leads out of it.[4]

Or, in other words, "thanks" constitutes a leap that does not lead to a savage freedom but to a thankful one, a freedom to say "thanks" to a present (*Geschenk*) that is no longer *something given* and thus no longer a matter of fate (*Schicksal*). Philosophical interpretation shows Hölderlin to have abrogated fate and to have broached in "thanks" what will never be given.

But Adorno does not rest satisfied with the discovery of "thanks" in the caesura of "the law of fate." The rest of "Parataxis" seeks to show why "thanks," as a word for unforced reconciliation, interrupts what always denies reconciliation: "fate" as something given, as the always already dictated, eternally returning restoration of identity. And "thanks" erupts out of fate in a stanza of "Friedensfeier" because this poem, like all of Hölderlin's mature ones, is an aesthetic model of reconciliation. Nature and spirit, matter and rationality are reconciled, but since this reconciliation takes place in a poetic medium, it never sheds its illusory, *scheinhaft* character. The poems claim to be nothing more than *models* of true reconciliation. As models, they can show the degree to which "celebrations of peace"—or conceptual claims to reconciliation—are themselves illusory. The ability of a poem to maintain itself as an illusion lets it escape the thoroughgoing domination of the concept and, to this extent, lets it surpass every philosophical work, including those that seek to expose its "truth content." As models of reconciliation, Hölderlin's poetry thus constitutes the preeminent geometry of modernity: not a purely spiritual geometry, not a geometry subordinated to mere rationality and immune to history, but an historical geometry whose "paratactic" procedures, accomplished in the suspension of fate, lead its students into the domain of philosophical thought: a domain—a space, not an academy, perhaps a utopia—no longer subject to domination. Not without reason does Adorno, citing Walter Benjamin, liken Hölderlin's language to "trigonometric signs."[5]

Critical interpretation of a poem surprises anyone who is only familiar with the poems and their philological exegesis, for what the critique exposes has never been set down in the work or contained in its allusions. A "law of fate" is asserted in "Friedensfeier," and the word "thanks" is integrated into its workings. Being surprised by "thanks" is, in a sense, the criterion of a critical exposition. Adorno's discovery of an interrupted "law of fate" is not therefore an arbitrary example of critique: it, strangely enough, obeys a rule, perhaps even its own fateful law. And nothing is less surprising than the course of Adorno's interpretation: "Friedensfeier" does not simply anticipate his own *Dialectic of Enlightenment*; it "concretizes" this dialectic by virtue of its paratactical—that is, "negatively dialectical"—procedure. Thus, at the end of "Parataxis" Adorno affirms that "this dialectic is sedimented in 'Friedensfeier.'"[6] Concentration on the technical character of Hölderlin's poetry cannot save Adorno from the fate into which, according to his analysis, Heideggerian dogma falls: discovering in the poetry what has already been given out as philosophy; making the poet into the exponent of the very same philosophical theses—whether they belong to *Seinsphilosophie* or to "negative dialectics"—that the philosopher elsewhere lays down; turning the poet into a witness for philosophy. Adorno becomes the specular double of the Heidegger whom he attacks.

This turn of events does not necessarily occur because Adorno brings Hölderlin's poetic technique to the forefront of philosophical interpretation but is perhaps the result of his having abandoned this attempt too early. In the lines of "Friedensfeier" where Adorno finds "thanks" as the counterpart to "fate," the problem of poetic technique emerges: it appears as the problem of measure, or of meter, in any case as the problem of finding the right measure when measure is no longer, or not at all, given. If, as Hölderlin writes in "Hyperions Schicksalslied" ("Hyperion's Song of Fate"), the gods are *schicksallos*, this does not imply they are free but means in terms of Hölderlin's later lyrics that they are "at all times expert in measure": *des Maßes allzeit kundig*. The measure in which the god is expert cannot be dissociated from the poetic meter of "Friedensfeier," for in both cases a temporary lack of expertise in measure means that all handiwork, including that of poetry, falls prey to "crude treatment" (*rauhbetastend*). Hölderlin does not simply describe in this stanza of "Friedensfeier" certain distant historical occurrences but also, and in a more decisive manner, determines the place toward which the "savage" must come "from ends remote." And where else can this savage, responding to a conclusive question (*wenn?*), end up but at the end points of the line? The problem of a merely temporary expertise in poetic measure announces itself at the end of every line: when to stop? This problem is then concretized in the very place toward which Adorno's interpretation is drawn: in the counter-rhythmic interruption of the poetic measure, "das, was man im Sylbenmaase Cäsur heißt, das reine Wort" ("that which one calls 'caesura' in the meter, the pure word") (*SA* 5; 1: 196).

The word "meter" (*Sylbenmaas*) does not belong to Hölderlin's technical vocabulary. But this does not mean that meter has nothing to do with the technical character of a poem. On the contrary, the problematic status of meter in "Hesperian" poetry—that is, in a poetry unable to establish a stable prosody—gives rise to the need for poetological reflection in the first place. "Measure," moreover, cannot designate a specific technical function as long as it sets out the terms in which poetic technique in general are cast. One of the most often deployed terms of Hölderlin's major poetological reflections, *harmonischentgegensetzte* ("opposed in harmony"), as a revision of the Heraclitean ἓν διαφέρον ἑαυτῷ,[7] serves to explicate what the late poems call "measure," *das Maas*. Whereas the gods are "always expert" in measure, the poet is not: only "when the poet is powerful" does such expertise present itself in and as a poem (*SA* 4; 1: 241). Metrical terms for the poem of an expert, in turn, serve as metaphors not only for the general character of a poetic text but also, at least in the case of "caesura," for the condition by virtue of which the very movement of metaphorization ("der tragische *Transport*") can be presented. If poetic measure, *das Sylbenmaas*, did not name the problem of poetic technique in general—a problem whose solution does not lie in some putative freedom but, on the contrary, in a mechanical calculation, a "lawful calculus"[8]—its terms could never play such decisive roles in the determination of poetic expertise.

The appeal to a poetic *mechanē* issued in the opening paragraphs of the "Remarks on *Oedipus*" addresses—but does not solve—the problem of a merely temporary expertise in measure. Perhaps for this reason Hölderlin replaces his appeal for a modern mechanics of poetry in the "Remarks on *Oedipus*" with a renewed discussion of the relation between philosophical and poetic "faculties" (*Vermögen*) in the "Remarks on *Antigone*." Expertise in measure in any case is a problem with which both "faculties" are always concerned, since the power of each "faculty" lies in its ability to present itself as a whole: and that means to give itself out as measure. The poetic counterpart to logic, according to Hölderlin, is "rhythm in the higher sense, or the calculable law" (*SA* 5; 1: 265). Rhythm can claim a "higher" status precisely because its measures can be calculated in a lawlike manner; it constitutes a metric by means of which an interconnection of poetic parts presents itself as a whole and therefore as a poem. The metric of a poem is thus its "logic." Little wonder, then, that the word "measure" in Hölderlin's poetry attracts the question of its relation to philosophical discourse. And nowhere does it do so with greater clarity than in the penultimate stanza of his great geographical poem, "Der Rhein." The memory (*Gedächtnis*) of the poet contrasts with the "faculty" (*Vermögen*) of the philosopher, and the term around which this contrast organizes itself is precisely "measure":

> Die ewigen Götter sind
> Voll Lebens allzeit; bis in den Tod
> Kann aber ein Mensch auch
> Im Gedächtniß doch das Beste behalten,
> Und dann erlebt er das Höchste.
> Nur hat ein jeder sein Maas.
> Denn schwer ist zu tragen
> Das Unglük, aber schwerer das Glük.
> Ein Weiser aber vermocht es
> Vom Mittag bis in die Mitternacht,
> Und bis der Morgen erglänzte,
> Beim Gastmahl helle zu bleiben.
> (*SA* 2; 1: 148)

> The eternal gods are
> Full of life always; until death
> Can however a man also
> Keep the best in memory
> And then he experiences the highest.
> Only each one has its measure.
> For it is hard to bear
> Unhappiness, but even harder happiness.
> Yet a wise one could do it
> From midday until midnight,
> And until the morning glowed,
> Remaining bright at the banquet.

Hölderlin's poetry has often been accused of obscurity—and this accusation has then served as an excuse for self-declared "philosophical" explications—but these lines are so undeniably clear, so perfectly "sober," indeed so clearly prosaic that they discourage every interpretative effort, especially those allied with philosophy. For in these lines the philosopher outlasts the poets "during the banquet," and his "faculty" (*Vermögen*) of outlasting lies in interpretation itself: Socrates, as a "wise one" who denies knowing anything, suggests at daybreak that a tragedian could be a comedian as well; and since both tragedian and comedian are drunk, they remain oblivious to the wisdom of his interpretation. In contrast to this philosophical "faculty," the ability of the poet does not consist in the power to show the unity of a multiplicity; rather, according to the earlier lines in this stanza, it can justly claim only an ability to "keep the best in memory." The philosopher may outlast the poets, and the poets may even be unaware of his insights into their own work; indeed, the poets may remain in the dark, and anything but sober, even as "the morning glowed." But this last line—"until the morning glowed"—exceeds, or simply escapes, the philosophical "faculty" that the stanza otherwise extols. For the aorist of this line does nothing more than keep "the best in memory." The philosopher doubtless has greater "facility" than the poets, but this facility, which consists in critical interpretation, presents itself only in poetry. The presentation of the philosophical faculty does not simply issue into claims of negative knowledge—this is the prerogative of the "wise one" and constitutes his wisdom—but, instead, takes place in the suspension of the disjunction between knowledge and ignorance, that is, in *Gedächtnis* ("memory"). The slumbering poets are oblivious but not without memory; the sober philosopher, by contrast, may stay awake all night long, but his awareness, or the awareness of his ignorance, gives him no access to *Gedächtnis*. Poetry does not, as Hyperion contended, make philosophical science, or even philosophy as negative knowledge, possible; rather, it keeps the very possibility of philosophy, its *Vermögen*, in "memory," and this keeping in "memory" escapes the categories of knowledge. The language of this escapade in "Der Rhein" bears such a remarkable resemblance to prose that only one trait can be said to distinguish it from a philosophical discourse: its meter. And the formula for this differentiation in language of language—and therefore of poetry from philosophy—asserts itself in the very line that separates the ability of the poet from the faculty of the "wise one": "Nur hat ein jeder sein Maas."

Each one has its measure. Philosophy therefore has its measure, and it is not that of poetry. Incommensurable measures meet—but it's only for a time, the time of a line—in the simplicity of a poetic verse that appears as though it were a philosophical statement: "Nur hat ein jeder sein Maas." "The wise one" and the poets do not themselves share a measure, but their language shares in meter because language, too, has its own measure. The measure of language, meter, belongs neither to poetry nor to philosophy but, distinguishing itself from both, constitutes the technical, "calculable" character of poetry. The philosophical explication of Hölderlin's poetry, as a result, turns on its meter, its *Sylbenmaas*, and in particular, "*that which one calls 'caesura' in the meter*, the pure word." As the measure of language and measure in language, meter does not give measure to anything; rather, it distinguishes language from itself and, in so doing, gives rise to different languages. From this self-distinguishing one—which is a possible translation of the Heraclitean *ἓν διαφέρον ἑαυτῷ* or of the *Harmonischentgegensetzten*—comes the immense difference between ancient and "Hesperian" poetry as well as between poetry and philosophy.

"Only each one has its measure." This is the difficulty, the *Schwerpunkt*, toward which Hölderlin's poetry gravitates. The difficulty is irresistible. "Each one" has a singular gravity, not a specific one, and this gravity is inescapable: "Denn *schwer* ist zu tragen / Das Unglük, aber schwerer das Glük." If *only* each one has its measure—"*Nur* hat ein jeder sein Maas"—then there is no measure against which anything in particular can be measured. No general measure, no universal meter, no metric system can be firmly established. The criteria for something being what it is are not thereby lost, as if the absence of a universal meter implied an undifferentiated mass; rather, "only each one" has its own criteria, which means that the criteria for something being what it is cannot be found in anything but the thing itself, and perhaps not even there. Poetry does not collapse into prose, as if "prose" were itself the name for language as undifferentiated mass, but the criteria by virtue of which poetry distinguishes itself from prose cannot be anchored in anything other than the poem itself. The language of a work without such anchors deserves the title "poetic prose" (*SA* 4; 1: 265).[9] If, in short, "only each one has its measure," then the being of anything lies precisely in its being "only each": being something, being some one thing, is being "only," or being alone.

The difficulty of Hölderlin's poetry—its weight and its plummeting upward into the abyss of a lovely blue sky—is bound up with the experience that "only each one has its measure." Whereas, for instance, the third stanza of "Brod und Wein" asserts that measure is *Allen gemein* ("common to all"),[10] the experience of the poem soon denies the certainty and then even the plausibility of this all-too-hopeful assertion. Instead of a definite assertion, the sixth stanza, which reiterates the third, asks in desperation "Aber wo sind wir?" ("But where are we?"), and the ninth stanza, which tries—and fails—to bring the poem to a close, can make only a paradoxical bid for "belief": "Glaube, wer es geprüft" ("Believe, whoever proves it") (*SA* 2; 1: 95). The mockery elicited by the assertion of a measure "common to all" in the third stanza is not therefore refuted; on the contrary, the point of this mockery—that there is no measure "common to all"—is confirmed so thoroughly that something like mockery, which always depends on general criteria and customary standards, proves self-defeating. If the experience of "Brod und Wein" is the undoing of the apodictic assertion that measure is "common to all," and if, as "Der Rhein" asserts, "only each one has its measure," only one statement is warranted: measure cannot be given. The very existence of measure is in turn made questionable: "Gibt es ein Maß?" ("Is there, or does it give, a measure?"). The fact that each one has a measure not only does not preclude such a question but gives any question of measure its urgency. Nowhere is this urgency more strongly felt than in the matter of poetic meter. Each poem, each stanza, each line of poetry doubtless has its measure, but—at least for the "Hesperian"—no measure is given. The date of modernity could in fact be determined by the absence of this, a metrical datum. Even if this absence were to be constitutive of poetry as such—and nothing in Hölderlin's poetological writings goes against this statement—"Hesperian" poets are the ones who are unable to escape the question of this absence: this absence is the unavoidable *technical* difficulty of each poem, each stanza, each line. In each case Hölderlin's poetry, or more precisely, its technical character, responds to the difficulty that "only each one" has its measure, and for precisely this reason, no measure can be given.

Measure is not simply absent: "only each one has its measure," and so the only place to seek the measure of each one is in "oneliness" itself: in aloneness, in singularity.[11] If the space of coming together and falling apart is called "the Earth," then the questionableness of measure as such turns into a geo-metric question, or a geometric question-and-answer, that runs counter to, and perhaps even negates, the axiomatic geometry through which students once gained entrance to the Academy: "Gibt es auf Erden ein Maaß? Es gibt keines" ("Is there a measure on Earth? There is none.") (*SA* 2; 1: 372). These lines from the disputed text "In lieblicher Bläue . . ." ("In Lovely Blueness . . .") are well-known less because they give insight into Hölderlin's poetry than because they have been the site of innumerable philosophical appropriations.[12] But so questionable is the givenness of a measure on Earth that the very question this text raises is itself questionable: no one knows if Hölderlin wrote them, who Hölderlin was when they were written, whether they are even lines of poetry and, if so, what meter they have. Is there a measure for "Is there measure on Earth"? There is none, or at least none is given. So in place of an exposition of the meter that these lines have, which would respond to a purely technical problem of poetics, there arises a temptation—it might be called the temptation of philosophy—to make them into lines of thought, indeed to line them up with the "inceptive" thought of a Heraclitus or the "conclusive" thought of a Hegel. And in reading the peculiar text from which these lines—along with the phrase ". . . dichterisch wohnet der Mensch . . ."—are drawn, this is precisely what Heidegger appears to do, as if he were uncannily cognizant of Adorno's indictment of his interpretations of Hölderlin and wished to confirm its every article. Heidegger brings "the poet" so closely in line with "the thinker" that the technical character of the poetry goes unquestioned even as the very question of poetic technique—"Gibt es ein Maß?"—is posed. When faced with such a question, the temptation of philosophy is to take measure for measure, or better yet, to take measure for meter: measuring the Earth for measuring a verse; taking a human measure for poetic meter; taking measure in general for the metric of a particular poem. Heidegger's response to this temptation says as much about the attractiveness Hölderlin exercises over philosophy as about the manner in which Heidegger distills the poetic work.

Although Heidegger's essay ". . . dichterisch wohnet der Mensch . . ." concerns itself with "construction" (*Bauen*) no less than with "dwelling" (*Wohnen*), it has nothing to say of the poetic construction from which its very title is drawn and to which its ellipses point: the construction, or reconstruction, of a verse version of "In lieblicher Bläue . . ."[13] Heidegger does not simply delimit a poetic text on the basis of his interpretation, as he does with "Wie wenn am Feiertage." In place of interpretation he gives—or gives into the reconstruction of—a meter for a prose fragment whose central concern is precisely the non-giveness of measure.[14] The neglect of the metrical dimension of a particular text goes hand in hand with an exposition of poetry as measuring and indeed as the measure-taking in which a "dimension" is first opened up: "The measurement of the human essence with regard to its commensurate dimension brings dwelling into its ground breaking structure. . . . Doing poetry is measuring [Dichten ist Messen]."[15] While ignoring the particular technical features of a poem—and a disputed "poem" at that—Heidegger presents "doing poetry" in terms of its singular technique of appropriation: not only does it take measure, it is the only appropriate measure-taking, or the measure-taking that alone resists the appropriation of modern modes of technological measurement and standardization.[16] In this way Heidegger's estrangement of philosophy from its customary occupations ironically follows a standard philosophical procedure: specificity is elided in favor of generality, the specific problem of poetic technique is elided in favor of a generality about measuring as such, and the result is the emphatic statement, "Dichten ist Messen." But Heidegger does not follow this procedure as a whim; it corresponds to a very specific historical-philosophical problem: the problem of metaphysics or, more specifically, Kant and the problem of metaphysics. What traverses the limit of German Idealism and therefore shows the line at which the modernity of philosophy founders? Answer: Hölderlin's absolutely finite "doing poetry," not his "works," but the "measure-taking" that constitutes the essential act of the poet.

If the limit of modern philosophy lies in overcoming self-imposed limitations—and the vicissitudes of its systemization attest to such a limit—Hölderlin's measure-taking, according to Heidegger's account, amounts to a suspension, or caesura, of this immense effort. Heidegger's exposition of poetry as measure-taking then takes issue with what Kant called "receptivity," for the problem of finitude, and therefore of being at the limit, always resolves itself into a question of taking in: of *per-cipere*, *per-ception*, *Ver-nunft*. Berkeley's *esse est percipi* means in this regard: being is taking in. According to Heidegger's reading of Kant, a "transcendental" taking in gives out the space—or the dimension, the playroom, the *Spielraum*—in which something can be given in the first place.[17] Kant renewed the Platonic conception of geometry as the propaedeutic for philosophy once he saw in geometric figures those "pure intuitions" through which the self takes in, or receives, what has not been empirically given. Something similar happens with the measure-taking of the poet, for, like pure intuition, it takes place in the absence of a given, but unlike pure intuition, it implies no given whatsoever: neither the givenness of things-in-themselves as the basis of empirical objects, nor the "factuality" of geometry as a scientific discipline. Poetic measure-taking, moreover, cannot be confused with "intellectual intuition" or the self-conceiving "concept" by means of which the limitations of Kantian thought were supposed to be transcended. The absence of a given measure does not imply that poetry *gives itself* measure and, as a result, produces itself as infinite spirit. Poetry, rather, insists only on the absence of a given. And this insistence, more than anything else, makes Hölderlin's poetry attractive to any philosophy that sets out to deny itself givens, grounds, and foundations.

Poetry as measure-taking takes place without a given, including a self-given: it is therefore radically finite, and the recognition of such finitude sends Hölderlin into the abyss from which Kant, to say nothing of his followers, shrank. Whereas, according to Heidegger's famous "violent" interpretation of his closest philosophical predecessor, Kant sought shelter from the abyssal "ground" he first discovered in the transcendental imagination—and, anticipating his followers, sought shelter in reason (*Vernunft*) as the grounding act of the self[18]—Hölderlin seeks no such shelter; instead, he allows a transcendental imagination, or a non-empirical image (*Bild*), to appear as groundless "foundation,"[19] and thus finds himself in an unparalleled position to say what it means to be without shelter: "dichterisch wohnet der Mensch auf dieser Erde." To speak in terms more familiar to Kant, although in precisely the terms from which, according to Heidegger, he shrank: "Doing poetry is the fundamental faculty [Grundvermögen] of human dwelling."[20] Dwelling, or existing, is possible only insofar as measure is taken in the absence of a given measure, and the one who takes measure in this and no other way is the poet. Geometricians, therefore, can measure the Earth only after a poet takes measure in the absence of a given. And this goes for philosophers—whether Platonic or Kantian—who anywhere subordinate their thought to geometric standards. Poetry does not found philosophy, as Hyperion contended; rather, poetry removes the idea of givenness on which philosophy grounds itself, undoes the foundation of ancient as well as modern philosophy, and—as long as one measure is taken for another—first lets thought come into its own.

If poetry is measure-taking without a given, and if givenness defines fate (*Schicksal*), then poetry abrogates fate: its "reception" of measure does not amount to an acceptance of *Schicksal*.[21] Since the absence of givenness reaches the realm of the self-given, this abrogation of fate cannot call itself "freedom," or the freedom of this abrogation does not square with the conception of freedom on whose basis philosophy has claimed its modernity, that is, the conception of freedom as self-legislation and self-determination. Heidegger then goes in search of a term for the abrogation of fate that is nevertheless not freedom, and like Adorno, he finds something like "thanks" as the solution to Hölderlin's fate:

> When and for how long is doing poetry proper [das eigentliche Dichten]? Hölderlin says it in the already read verses (26–29). Their discussion was held back on purpose until now. The verses read:
>
> . . . So lange die Freundlichkeit noch
> Am Herzen, die Reine, dauert, misset
> Nicht unglüklich der Mensch sich
> mit der Gottheit . . .
>
> "Die Freundlichkeit"—what is this? A harmless word, but one named by Hölderlin with a capitalized adjective "die Reine." "Die Freundlichkeit"—this word [Wort] is, if we take it literally [wörtlich nehmen], Hölderlin's magnificent translation for the Greek word *charis*. . . . As long as this arrival of favor [Huld: Heidegger's retranslation of *charis*] lasts, it happens, and this is happiness [es glückt], that man measures himself with the divinity."[22]

Just as Adorno found "thanks" in "Friedensfeier," Heidegger finds favor in "In lieblicher Bläue," and the diverging ways these two discoveries come about are not as different from one another it has sometimes been suggested.[23] Both arrive at a certain favor, or at an abrogated fate, by seeking out one measure in favor of another. Favor accrues to what "In lieblicher Bläue" calls "the measure of man" (*des Menschen Maas*), not that of the poem. And this favor owes its origin to a certain taking, not precisely a measure-taking of poetry but its philosophical counterpart: *wörtlich nehmen*, taking in a "wordly" manner, taking words given elsewhere, taking literally, if "literality" is here defined by words, not by syllables or letters. Heidegger thus "takes" the word *Freundlichkeit*, "wordly," *verbatim*.[24] Taking the word in this way takes it out of one language and transfers it into the language of "thought," which is not, of course, an independent language but the language claimed by the "essence" of language. Heidegger does not obscure anything when he engages in an act of double translation; rather, he makes this taking and its transference explicit. Taking one word in favor of another founds a language of "thought" that may very well have given up foundations, especially those grounded on a given, but this language nevertheless finds it difficult to give up a taken, and the taking that cannot be given up at any cost is taking a word in a "wordly" manner, taking "favor," for example, in favor of "friendliness"—which is a perfectly fine word. Taking a word in this way does not simply honor a favorite philosophical principle; it gives back something like the given, namely, a word that is solid enough for it to be taken in favor of another. Taking a solid word—which is not the same as Hölderlin's "solid letter" (*SA* 2; 1: 172)—in favor of another cannot help but arrive at a certain "favor" as the solution to the absence of a given: favor is found in taking one measure for another, the "measure of man" for the nonhuman—and, to this extent, nontechnological—technical measure of poetry.

Just as Hölderlin never contrasts "thanks" with the "law of fate," he does not call poetry "measure-taking." The latter is the imposition of a certain "thinking," or taking, on the part of Heidegger. The degree to which this thinking imposes itself on Hölderlin's poetry can be measured when it is contrasted with the line from "Der Rhein" Heidegger prefers not to cite: "Nur hat ein jeder sein Maas." *Having* a measure is not the same as *taking one*.[25] It is doubtless difficult to think of a having without a prior taking, and it is perhaps impossible to *think* of a having without making it into a taking, a prior appropriation, the making proper of what is foreign. But this impossibility is precisely what is at stake in the very phrase of "In lieblicher Bläue" that denies the insight it seems to deliver: "des Menschen Maas ist's" ("It's the measure of man"). In response to the promise of insight that this line holds out to thought Heidegger mimics the text he seeks to distill:

> The *revelation* of God, not He himself, is full of secrets. For this reason the poet immediately asks the next question: "Ist er offenbar wie der Himmel?" [Is he revealed like the sky?]. Hölderlin answers: "Dieses / glaub' ich eher" [This I rather believe].
>
> Why, now we ask, does the supposition of the poet incline this way? The immediately connected words answer. They read succinctly: "Des Menschen Maaß ist's" [It's the measure of man]. What is the measure for human measuring? God? No! Heaven? No! The revelation of the heavens? No! The measure consists in the way that the god who remains unknown is revealed as this through the heavens.[26]

This bizarre colloquy would hardly be necessary, and Heidegger's answer would doubtless appear less arbitrary, if someone, perhaps Hölderlin, had made it clear what *it* refers to in the ever promising line, "It's the measure of man." The referential unclarity of the contraction gives rise to an expanded community of questioners: "Now we ask." So unclear is *it* that Heidegger, as the representative of this unknown but nevertheless manifest "we," initiates a colloquy and arrives at an answer that finally makes *it* thinkable. Even if one takes Heidegger to task for the specific answer he gives to the question "What is 'it'?"—and it is the great merit of Philippe Lacoue-Labarthe to have done so[27]—the task of interpretation remains the same: it is still the discovery of what *it* is, as if the philosophical question concerning the "measure of man" were of the same order as the technical question of the place that the contraction *it's* occupies in a text. But these two questions are commensurate only after one has taken *it's* and made it into *it is*, that is, taken *it's* literally, taken it for the words for which it stands: *it is*. What, then, is *it*? It's not even *it*: it's only the *s* of *it's*.[28]

There is doubtless no difference in thought between it is and *it's* but there is—or, to borrow something from Trakl, it is—a difference: a poetological, not an ontological difference; a difference, however, that cannot be reduced to the rhetorical one between literal and figurative as long as it concerns a metrically enforced contraction; a meaningless difference, in sum, that a linguistic—and therefore nonhuman—measure makes. The difference between *ist es* and *ists* is doubtless a sheer technicality, but as a technicality, it exposes the peculiar mode in which "poetic" language exists: there is none, and yet *it's*. The technical character of poetry, its technicality, lies in denying the measure it otherwise—in philosophy, for thought—promises.

"It's the measure of man" defies the question through which philosophy has from its Platonic inception established its claim to knowledge: τί ἐστίν, "What is it?" But the defiance of this question does not imply that "the measure of man" recedes into the unknown, nor that the manifestation of this measure proceeds out of unknowingness. "It's the measure of man" *has* its measure—and that is why *it's* is there in the first and last place—but its measure makes *it's* impossible to think; or, if thought is given to the *s* of *ists*, it takes *ists* in a "wordly" manner, separates it into the two words *ist es* and arrives at an unthinkable *es*. Call this *es* the concealment that takes place as the ever receding condition of disclosure, or call it an unconscious that gives rise to consciousness; in either case—and therefore in the two most powerful challenges to Platonic thought of this century—it stays the same, for it remains an *it* of thought, an *it* thought in relation to "the measure of man." Although *it's*—and not it, "the measure of man"—escapes the grasp of knowledge, it does not seek shelter in the unknown: rather, *it's* poses a disjunction altogether different from, and far more unyielding than, the disjunction between knowledge and nonknowledge, concealment and unconcealment, consciousness and unconsciousness. It is the disjunction between "poetic" measure and human measure: either the line "It's the measure of man" measures up to its poetic meter, or this line, taken in a "wordly" manner and therefore taken as the elsewhere given words *it is*, gives into the question of another measure, "the measure of man," a measure that poetry—this would define its task—could take. "It's the measure of man" has its measure, therefore, but it is not a measure of man. As long as the human is measured against the divine, the measure of this line cannot be divine either. Someone—let the name be "Hölderlin," but proper names are never more questionable than here—runs into this disjunction and cannot escape the decision it poses: either a human measure or a poetic one. Little insight into the poetry of "Scardanelli" is required to see which measure won out.

Hölderlin has exercised a strong power of attraction over those philosophers who set out to explicate the fate of modernity for at least one reason: his poetry gives up all reliance on a given. The attraction to such a poetry then expresses itself in a taking: one measure—"the measure of man"—is taken in favor of another measure. Although the other measure has been called a "poetic" one, this perhaps says too much: the other measure is poetic only insofar as it is linguistic, and it is linguistic only if language is no longer the medium in which the human measures itself against the divine; the other measure, the nonhuman one, is a measure of language's contraction from the human-divine interplay. The advantage of taking a human measure in favor of the other measure cannot be gainsaid: it lets a philosophical interpreter resolve the disjunction so powerfully contained in, or contracted into, the line, "It's the measure of man." Such an interpreter may subject every metaphysical thesis and every philosophical theme to a thoroughgoing critique or an epochal "destruction," but something of these theses and themes remains in place: words are still taken in a "wordly" manner; taking words "literally" still means neglecting letters and syllables; words, as a result, take the place of givens.[29] Only a given word—and who gives his word with more pathos than the "mad" poet who promises "the measure of man"?—can be taken in a "wordly" manner, and so philosophers who may have left the saturated space of dialectics behind cannot help but grant the word the status of a given. And no word functions better as a given than one that is not given in any conventional sense but must give itself up—to critique, to thought. In place of the given as ground arises the word, and in the place of a given geo-metry arises the word of poetry as measure-taking. If Hölderlin still can be thought of as a poet of irrevocable non-givenness—but perhaps this cannot be thought—there can be no room for such taking. And so the place in which one measure is taken for another contracts to nil. This, too, takes place in the astounding and, for that reason, almost forgettable "word" *ist's*.

The Calculation of the Poet

Jean-Luc Nancy

1. Two—partial and limited—hypotheses will serve here as axioms for a brief foray into Hölderlin's poetics.

First: for Hölderlin, the poet matters more than poetry.

Second: the task of the poet is above all a calculation.

These two hypotheses are connected: calculation refers to the activity and decision of the poet before referring to the arrangement (*dispositio*) of the poem, in which this activity and this decision inscribes only its result. This makes Hölderlinian poetics something other than a "poetic art," something other than a literary and aesthetic theory. Yet at the same time it gives pride of place in Hölderlin to the *ars poetica*, to the technique of composition, and to the one who is its technician, the one who "knows." Poetry is something other than poetry, and also something other than thought. Hence poetry is, quite precisely, this double limit of itself: the poet and his calculation.[1]

2. The material distribution of an oeuvre in which the essays, together with *Hyperion* and the translations, occupy an important place, even if they are outnumbered by the poems, already attests to this double hypothesis. So would, as I read it, the ill-defined but persistent impression of Hölderlin's relative disinterest in the poem itself. Doubtless the word is too strong and the exact nuance easily misunderstood; nonetheless, everything takes place almost as if the poet tended toward a certain abandonment of his poems—left them, in short, to themselves and to a

certain monotony of their flow, without too much care for completion (for the completion of the piece being worked on)—while in the "theoretical" undertakings (as in the constant variations, often left hanging), he worries, strains, pulls up short, starts again, and often dries up, all chasing a precision which conceals itself as much as it is imperiously demanded.

In this regard, it might be tempting to think that, on the level of thought—and to think also that this thought concerns the *ars poetica* or else properly philosophical motifs—Hölderlin often runs out of breath in his search for a dialectical and speculative construction for which, unlike his two closest philosophical friends, Schelling and Hegel, with whom he shared the common post-Kantian ideal of the "system," he finds, properly speaking, neither the exact figure nor the exact process. We might see here a sort of awkwardness and impasse along the philosophical path. For Hölderlin has no real idea how to go about an *ars philosophica*.

This is certainly a matter of temperament, but it is also, and above all, because Hölderlin does not simply or only pose a philosophical question. Rather, he poses precisely the question of another way of approaching "questions" in general, the question of another *ars* or *technē*. Which is why it is not enough to say that he is a thinker: it must immediately be added that he does not think like a thinker, that he does not think according to the *ars* of thought. Hölderlin's first thought is perhaps this: that philosophers have to elaborate a system, in the strongest, most organic sense of the word, but the poet has to touch on something other than a synthetic unity, albeit a living and articulatory one. The poet has to touch upon an absolute point of exactitude, which is more the stake of a calculation than that of a construction, production, or begetting. He must touch on this point, or settle on it, or have it in sight: this is something different from assembling the elements of an articulated totality. The point escapes assemblage. The object of the philosopher is synthesis—the project of the poet is synopsis. The former means above all operation, the latter apprehension. The former demands the time of elaboration and of deliberation, the latter demands the temporal space of a sighting and an aiming, the instant—and with it, perhaps, syncopation.

3. What characterizes the poet for Hölderlin is: "the constant precision of consciousness with which the poet looks at a *whole*" ("die durchgängiger Bestimmtheit des Bewußtseyns, womit der Dichter auf ein Ganzes blickt"; *SA* 4; 1: 156).[2] *Bestimmtheit* can be rendered as "determina-

tion" or as "precision": the poet has an absolute and constantly determined consciousness, one complete and unified, without remainder, which holds nothing back and gives itself entirely over to his look. Clearly this is not exactly what the philosopher understands by "consciousness." It comprises less the moment of self-relation, of re-presentation as such, than the single moment of the look—not even the intention, but the opening, the look turned or thrown on (*Blick auf*), before, out of the self.

This look falls on a *whole*: the totality of this whole is thus touched beyond—or outside—all composition or synthesis, at the center, at the heart, at the joint which does not totalize, but which is the whole. It is the being-whole of the whole which is directly and perfectly sighted and seen. And since it is the *whole*—Hölderlin does not emphasize the word for nothing—this center is no more interior than it is exterior. It is identically and immediately the outline and the periphery. It is not the concept but the figure and existence of the whole: the whole revealing itself as whole. The whole showing itself for a look, by a look fixed exactly on it—for and by a look which, in the final analysis, is only the showing-itself of the whole itself.

The poet: the clearly present whole. The incontestable and incontestably punctual presence of the whole. The act which calculates exactly the moment—the instant, the thrust, the passing—of the presence of the whole. The act which thus leaves nothing out: neither the background of an "intention" nor that of a "thing in itself." But the thing itself in the presence of the look itself, in clarity itself—and the distancing of this "itselfness," its exact calculation.

And nothing else—although here it would be more accurate to say "no one else." The poet is not the "subject" of the representation of the whole, but the place of the envisaging of the whole *in persona*. Hölderlin also calls this "the pure" (*SA* 4; 1: 235; *Essays*, 47), or "pure (poetic) individuality" (*SA* 4; 1: 254; *Essays*, 73).

4. Hölderlinian purity comprises neither admixture nor alterity. It is, on the contrary, the pure coincidence of the same and the other, of the ownmost and the foreign, of the "human" and the "divine": this coincidence in which the look looks at the thing—and is concerned with it—only to the extent to which it is, and becomes, the appearing of the thing. The unit of measure here is thus exactitude itself, *pure* co-incidence.

"Conciseness," "sobriety," these other key words of Hölderlin, designate nothing other than this. With them, it is less a matter of an economy of means than of an extreme precision of end—of this extremity in which precision itself disappears in favor of rigorous exactitude: the whole, each time *this* whole (of the world, of man, of community, of an epoch, of a form, of a country, of a river, of a god . . .). Whence the economy of means: namely the brevity of what one would elsewhere call an "image," and which becomes a "look" when Hölderlin writes, for example: "Since we have been a dialogue" ("Seit ein Gespräch wir sind"; *SA* 3: 536).

This is not an image, but a strict predicative proposition that says we are, in fact, a dialogue (a conversation, a speaking-together or with-one-another, a *Gespräch*, a together which speaks). It says, not that we are in dialogue, but that we are our dialogue. "We": all of us, those who speak, defined in their being by this speaking-between-themselves. Thus, the whole of the indeterminate "we" is absolutely determined: we are this between-us that is language, and, likewise, language is the between-us.

Calculation consists in exactly fixing or sighting on this being and nothing else, without approximation, without comparison or metaphor. While the philosopher is free to gloss this "between" or this *ge-* of the *Gespräch*, what matters for the poet is that it be stated straight out, straight to the point, and that is to say straight to the source: the poet who says "we" and who thus already speaks *for* us, that is, to us and in our place, but thus, also, in the place of the "between," exactly in the "between" itself. The linguist would say: the line performs what it states. It already speaks between *us*, it has taken us, calculated us, placed us at the source of the singular plural emanation of the conversation.

It is the same with "*Since* we have been." "Since" is the exact calculation of a indeterminacy. The poet does not tell us since when, because this (like the "we") cannot be measured. The moment "since" which we have been a conversation is as old as the conversation itself, which, in turn, is as old as any possible antiquity—called "morning" in the previous line—*and* as recent as the moment of its utterance. It is absolutely immemorial *and* contemporary. "Since" calculates the entire interval between the immemorial and the actual present, an interval both null and infinite. "Since" thus measures a cadence of eternity.

But the measure has to be heard as it is calculated in German. "Seit ein Gespräch wir sind": the ordinary grammatical order is not respected. It would have to be translated: "Since a dialogue we have been," and it could even be heard: "Since a dialogue we are." Instead of being woven together in the syntax (*seit wir ein Gespräch sind*), "a dialogue" and "we are" are positioned as two blocks, exactly juxtaposed. Moved forward from its ordinary place, "we" carries the stress: what a *Gespräch* is is precisely what *we* are. Reading the line in its entirety, this being becomes even clearer: *Seit ein Gespräch wir sind und hören voneinander* means: "Since we have been a dialogue and have heard from one another," but just as much: "Since a dialogue we are and hear one another"; we are *that* we hear one another; we are that we hear from one another what we are; what we are we learn from one another as our conversation itself.

5. Each time, then, what is calculated is properly an incalculable. Now that which "cannot be calculated" is "the living sense" (*SA* 5: 195; *Essays*, 101).[3] Calculation consists in relating this incalculable sense to the "type" or the "law" of "equilibrium" (*SA* 5: 196; *Essays*, 101–2), which is that of the work.

The work is not a given measure to which the poet would submit the incommensurable, as if to a yoke or a limitation. Rather, the measure that the work *is* allows the relation of the incalculable (of sense) to a fleeting break between words to be calculated. It is always, in some way, an infinitesimal calculation, the calculation of an exact point of flight or tangency: the coincidence of incalculable sense and a brief word.

It is not the unsayable which would come to be said in some transient or analogical, and hence oblique way. For calculation, there is neither the unsayable nor the sayable, only the exact saying. Which is also why, if calculation is "lawful calculation" (*SA* 4; 1: 195; *Essays*, 101), this does not mean that it is subordinate to the work as product (in Romantic fashion: as the finite production of the infinite; hence, no theme of the fragment in Hölderlin). Rather, the work is, let us say, the site of an exact saying: the point of passage, intonation, inflection.

Exactitude is the proper of calculation. It is at once infinite extremity and the reversal of approximation. Calculation does not happen as vague approximation, either by excess or by default. It admits of no margin (because it has no practical end). Either it works out exactly or not at all. No approximation, no fuzziness of sense, just the coincidence, albeit only at one isolated point, of the saying and the said, without spillage or remainder.

This exactitude is first of all that of the poets themselves:

Everything depends on those who are exceptional and those who are beautiful not excluding from themselves what is inferior and barbaric, but also on their not mingling too much with it, *on their knowing certainly and dispassionately the distance between them and the others, and on their working and suffering on the basis of this knowledge.* (*SA* 4; 1: 236; *Essays*, 48)

Which means, precisely: the poet is *bestimmt,* in a manner at once determined, clearly defined, cut off, depending on whether a voice (*Stimme*) fixes and attunes him. What one calls "precise tone," nothing more, nothing less. Precisely the tone of the one who finds the tone, and also the manner in which this tone attunes him with the others in separating him from them.

(Here, of course, commentary shudders. One can—this is undeniable—hear Hölderlin's lines in the terrifying tones of the Third Reich. One can also hear them in the tone of a simple and precise measure, a measure that allows a tone to be distinguished over the indistinction of a murmur. Here it is a question of Germany, not of Hölderlin alone, and I am not going to speak of this. Except to remark that the measure of the measure, the measure of "sobriety" and "courage," the measure of exact "calculation," the appropriate unity, is not given, and it must not be thought that it is or could ever be so. For Hölderlin, nothing is given, not even that which he calls "German." It is always as a "stranger" ("Fremdlingin") that the "voice that shapes man" ("menschenbildende Stimme") comes (*SA* 2; 1: 126). When Hölderlin identifies "Germany," "Stuttgart," "Heidelberg," or the "Neckar," it is a matter of the exactitude and clarity of the look, of the precise places upon which it is fixed, rather than of the sanctuaries of a fate. Moreover, he also names "the Charente," "Provence," and the "Gascon regions," as well as "Genes" or "Lisbon." The divine is always *wandelnd,* always "changing" or "walking," and for the "homeless soul" ("heimatlose Seele") the only "possession" ("Eigentum") and "friendly sanctuary" ("freundlich(en) Asyl") is to be found in "song" ("Gesang"; *SA* 1; 1: 307). If the home was given, the entire poetics of calculation would be unnecessary.)

6. Exactitude necessitates the exclusion of whatever persists in an indeterminate way. The unity of the whole has to be grasped in passing, as passing, and not followed along its movement. Saying only what is said, and nothing more, suspends discourse. Thus the precision of linkage, of

"rhythmic succession" stems from its interruption, from the "caesura," in which "the pure word, counter-rhythmic suspension" (*SA* 5: 196; *Essays*, 102) also consists.[4] Rhythm comes from counter-rhythm, as the figure of its severance, and that is why "the succession of calculation and of rhythm is divided" (*SA* 5: 196; *Essays*, 102).

The course of sense has to be interrupted in order for sense take place, in order for it to be grasped in passing—in order that the unity of a whole which is more than and something other than the sum of its moments (being, rather, their common scansion and their syncopation) be grasped. Here, then, is the entire calculation of the poet, and his strained, unappeasable gesture, the gesture of "wrenching" (*SA* 5: 196; *Essays*, 102).

Wrenching—away from everything which would mean carrying away, growth, the outgrowth of meanings, away from everything which would mean the summation and totalization of the whole, instead of leaving it to its single caesura—wrenching has to be done "at the right time and in the right place" (*SA* 4; 1: 234; *Essays*, 47). This necessitates "quick comprehension" ("schneller Begriff"; *SA* 4; 1: 235; *Essays*, 47), since it is the whole, the purity of the whole as such, which has to be grasped and scanned. Yet there is no predetermined criteria for this, no guarantee. The unity of the whole, the sense of sense, is not a part that one could deduce from the sum of other parts. Calculation is not a deduction; it is a sighting. In sighting on the unity of the whole, there is no measure outside of the sighting itself. The right time and place, the *kairos* of the poet, thus has to be the whole itself, its sheer coincidence with itself. The whole, every whole, and even the whole of all wholes, is defined not by a fulfilment, but by a coincidence.

So as to get straight to the point, the opportune moment is not an outside circumstance. It makes up part of the whole. It is the unity which presents itself (and which is doubtless nothing other in itself than this coming to presence, this exposition, this offering and this scansion). Which is why calculation does assess the means, the ways of access. It does not calculate in this sense. It is, rather, the essence of calculation: the accounting of unity, of the *one* posited, given, thrown as one. Whence the necessity of not "lingering fearfully" ("scheu . . . verweilst"; *SA* 4; 1: 235; *Essays*, 47). One who lingers (would this be the philosopher?) is afraid, afraid of not measuring up to unity, or else afraid of unity itself. He thus thinks unity too small or too large, and so, as always, ungraspable—whereas it is, in fact, grasping itself, but a grasping that releases itself in unity.

Equally, "it is out of joy that you must understand the pure as such." Understanding is "always oblique" (*SA* 4; 1: 235; *Essays*, 47). Joy, or love, is not a greater or more profound understanding. It is an understanding in which quickness understands, because quickness grasps the ungraspable and so releases it. It alone circumscribes the whole, its unity, its passing sense. Quickness does not go "faster"; it is speed itself, a going straight to the point. Likewise, love is not a more passionate feeling or a stronger relation: neither feeling nor relation, it is the exact calculation of the "one" of the other, its precise caesura.

This circumscription, this caesura, outlines the difference of unity. Without difference, unity could not be unity. Which is why one cannot "differentiate with spirit alone," except by "turning back into pure being" (*SA* 4; 1: 237; *Essays*, 49). Pure being is in-difference, and is not yet "the pure." "The pure" has to be the different, the singular, the whole as different from every other whole, and simultaneously the whole differing from itself and offering itself as such, as *one*. "One" is the difference of the whole from the whole. It is this alone which counts itself one, one every time, without linkage or addition, one without summation, the suspension of continuity, the truth of sense, its coming, its event.

7. Thus there must be "harmonious alternation and tension" ("harmonischen Wechsels und Fortstrebens"; *SA* 4; 1: 241; *Essays*, 62). If harmony is accord, accord supposes difference, change, and distance. Exactitude is the precise fixation of distance—of the distance between tones, between genres, between metric qualities and resonant stresses, as well as between the saying and the said—whereas approximation, even when precise, is a flight of distance into distance. Exactitude proceeds, by contrast, from a precise measure of the divergence necessary in order for there to be concurrence, coincidence, one-to-one correspondence between the *one* of the said and the *one* of the saying. The calculation of the poet is first of all an uncompromising and immediate look at the clear distinction of the one: the clear consciousness of what isolates it, of what suspends it, of what holds it back in the face of the look.

There has to be distinction at the heart of the "community and unified simultaneity," at the heart of the "kinship of all parts" which properly comprises "spiritual content." Community and kinship (or "affinity": *Verwandtschaft*) can apply neither on the level of union nor on that of communion. For "spiritual content" would be reduced to a "petty shadow-play" if the "spiritual" did not in itself differ from itself and thus "come out of itself" (*SA* 4; 1: 241; *Essays*, 62).

While "affinity" concerns "content"—and takes place according to "immobility"—"form" necessitates "alternation" and, with it, movement, "progression," getting out in front (*Fortstreben*). The *one* as such has then to leave itself in order to be presented/scanned in the alternation or the inflection of tones, the cadence of rhythm, and the caesura of sense. The movement of leaving, the ex-position of the "spiritual" in exteriority, necessitates this caesura and its scansion: poetry is the scanner of sense, the poet exactly adjusting its lines of sight.

The poetics of calculation or of exactitude is thus the thought of originary difference. And, as has to be the case, it is so first of all as the originary difference of thought in itself. Which means, needless to say: it *is* this thought *only* in the originary difference of thought in itself. Which further means: it has to be understood as the difference in itself of originary thought—as "poetry."

If thought in general, which is what philosophy represents for us, is necessarily understood as the ultimate appropriation of the thing—the absorption of its "object" in the "subject" that it is, the "subject" itself being the principal structure of the self-appropriation of the object—or if it is understood as the elevation of the thing, of this so-called thing "itself," to the status and visibility of the Idea (in this sense, no thought is not "idealist," and this as much in Hölderlin-the-thinker as in any other Hölderlin), then poetics, that is to say, the thought of poetry or, more precisely, the thought of *poetic* thought, is understood here as the other of thought, as non-Idealism itself. Poetry, or the un-doing of Idealism.

It is not that poetry has to respond, as in Hegel, to the demand for an external and sensible presentation of the Idea. Despite many indisputable similarities, here it is not a matter of a logic of manifestation or of presentation or representation (in this sense, one could say that Hölderlin situates himself outside *mimesis*, at least up to a certain point, or, rather, on the basis of a certain point). Coming to presence is not (re)presentation. The movement of the spiritual outside itself is not its manifestation; it is its opposition, its opposing ex-position.

Hölderlin is, no doubt, equally engaged in a dialectical process of resolving opposition. Nowhere, however, is this resolution proposed as a resolution in the full and proper sense of the term. This is what leads to the extreme, even infinite complication of the analysis which he tries to elaborate (in "On the Procedures of Poetic Spirit"), without ever managing to formalize it in a satisfactory manner. Where there is resolution, the sensible manifestation of the spiritual is correlative to a spiritualization of the sensible, and so, for Hegel, poetry leads de jure to the dissolution of art by the sublimating dissolution of the sensible element and of exteriority. Hegelian poetry is, de jure, self-resolvent or self-dissolving just as it is, in principle, self-poetic, self-modeling, self-mimetic.[5] In Hölderlin, however, it is not a matter of this mutual alignment of the "spiritual" and "sensible" orders, not, that is, a matter of this mimetic self-manifestation of "content" or of "spirit." The order of "alternation" and "tension" is not the exterior *of* the "spiritual" interior; it is the outside, the out as such, the *partes extra partes*. "Content" is not, finally, the true form of the form. Form is the uncontainable of content, the irreconcilable of reconciliation: the one of the whole, as caesura and not as totality. The meaning of the poem "unites the extremes by setting them in opposition through their coming into contact with one another" (*SA* 4; 1: 246; *Essays*, 66). Contact is not confusion, but its contrary. In logical parlance one might say that it is a matter here not of "contradiction" but of "contrariness": the two sides present do not exclude one another; they can be false at the same time. One is not the negation of the other, and the other is not the manifestation of one. But in opposing, they touch, and the point of contact is as much the point of the *one* as it is the point of its envisaging, the scansion of its "outside," the instant of its coming to presence, of a presence infinitely and exactly opposed to itself.

The procession of the Idea in the visible is replaced by this contact in which the two run up against one another, cutting into and capturing one another. Such is "the meaning . . . everywhere set against itself" (*SA* 4; 1: 245–46; *Essays*, 66), the caesura and passage of sense—rhythm heard literally. Spirit, Hölderlin also says, equalizes all opposites, whereas poetic "meaning" disjoins and puts in contact. In other words, poetic "meaning" cannot be content with "meaning" something; it has still to distinguish itself from itself, has still to be posed and opposed outside, in a sensible individuality. Between meaning and this impenetrable point outside, there is contact, coincidence, and caesura. In this way, then, through poetic cadence there can be the instantaneous comprehension which is also outside understanding: not "non-understanding," but "love" of this *outside* which has to be touched on, in the *kairos*, by the grasping-releasing of the one.

The point of contact, the touching-on-the-opposite, is the opportune exactitude of the calculation which delivers sense, or which delivers to sense, in suspending the uninterrupted course of sense.

8. In this way the poem can be a "separate world with respect to form, a world within the world, and thus the voice of the eternal to the eternal" (*SA* 4; 1: 250; *Essays*, 70). This world within a world is not a microcosm. It is not a scaled-down image of the other world, and it does not respond to the logic of a mimeticism. Not the reproduction of an other, but the same in the face of the same, in its opposition to the same: the point in the face of totality, the whole leaving itself as one. The point of the totality, or the point of totality, outside of it in the face of it—the *punctum proximum* of the poet's look. Not the totality assembled on a single point, in the manner of a synthesis or subsumption (henceforth dialectical and, suppressing exteriority, leading it back to the whole as whole), but the totality touched in its point of totality, in the point of its unity: *touched*, which is to say neither penetrated nor embraced nor sublimated, but reached *in its sustained*, and thus properly *felt exteriority*. To touch is not to sublate the difference between what touches and what is touched; it is, on the contrary, to set out that difference as such, in its alternation and its tension.

This is also why the poem, the work as such, is not privileged, and why Hölderlin rarely practices a poetics of rules, genres, and examples. When he examines "the difference between kinds of poetry" (*SA* 4; 1: 266; *Essays*, 83), he does so in order to propose a series of variations upon the basic principle of the "alternation of tones" (*SA* 4; 1: 238), according to the necessity of their distinctions and their mutual references. *Tone* marks the separation and the distinction of the tension inherent to touch.

What matters is the "procedure of poetic spirit," its "process" or its "manner of proceeding" (*Verfahrungsweise*), which is ultimately to say its technique, and consequently its art: this act by which spirit touches on "the infinite unity" (*SA* 4; 1: 251; *Essays*, 71) and, in touching on it—on this unity which forms as much the "point of the separation of the unified" as it does the "point of unification"—touches itself. The themes of contact and of the one (*das Einige*, the one as reunited unity) go hand in hand, because touch—the touch of the look—isolates the point of contact. And it does so in two ways: on the one hand, it touches on a single point; on the other, it leaves the touch and the touched facing one another. Nothing disappears in the over-indifference of an indistinction, nothing drowns in a sea of feeling. Everything is distinct, the whole distinct as such.

What is touched is a body. Not a body as organic assumption and internal finality, but this opposite of the organic that Hölderlin calls the *aorgic* (*SA* 4; 1: 269; *Essays*, 85), a materiality, a divisibility, a body as extension and distinction: an "objective coherence . . . , but also a felt and feeling coherence and an identity which obtains in the alternation of oppositions" (*SA* 4; 1: 251; *Essays*, 71), the unity of divisibility itself, of the "actual separation" that responds to "*the arbitrariness of Zeus*" (*SA* 4; 1: 269; *Essays*, 86), that is to say, to the law of sensible and individual distinction.

It is not out of the question to claim that Hölderlin's poetry is physical and erotic, if not in its themes, at least not ostensibly, then in its type and in its gesture, in its address to the thing itself, to its presence, to its sensible existence in a distinct *hic et nunc*.

Thus—like a desire to paint and for the contact of painting (the importance of painting in Hölderlin's work cannot be exaggerated; it will become more pronounced over the years):

Mit gelben Birnen hänget
Und voll mit wilden Rosen
Das Land in den See,
Ihr holden Schwäne,
Und trunken von Küssen.
(*SA* 2; 1: 117)

With yellow pears suspended
And full of wild roses
The countryside in the lake,
You sweet swans,
And drunk with kisses.

grad aus falbem Stein
Die Wasser silbern rieseln
Und heilig Grün sich zeigt
Auf feuchter Wiese der Charente.
(II: 237–38; PF 585–56)

straight from the dusky stone
The waters trickle silver

And show themselves holy green
On the damp meadow of Charente.

But even more, the touch of language itself, the touch of the timbres and rhythms of words, which can only be quoted in Hölderlin's language:

Kommt eine Fremdlingin sie
Zu uns, die Erwekerin,
Die menschenbildende Stimme.
(*SA* 2; 1: 126)

One translation:

A strange woman comes to us,
She who disturbs sleep, the
Voice shaping men.

—in which the assonance and the essential rhythm are lost. One could try:

From elsewhere comes one
To us, the awakener,
The voice making men.

I cannot speak with great technical precision about either Hölderlin's language or his prosody, which is ultimately to say, his *ars poetica*. But one cannot emphasize too strongly the fact that it is there, in language, prosody, rhythmics, through words and song, that the tone and the tact of his poetics—that is to say of his thought, the outside of his thought, his thought outside of thought—are determined. One must also bear in mind what he himself writes at the head of "Celebration of Peace":

Sollten aber dennoch einige eine solche Sprache zu wenig konventionell finden, so muß ich ihnen gestehen: ich kann nicht anders. An einem schönen Tage läßt sich ja fast jede Sangart hören, und die Natur, woven es her ist, nimmts auch wieder. (*SA* 3: 532).

But if, however, some should find this language too unconventional, I must admit: I cannot do otherwise. On a beautiful day, almost every mode of song makes itself heard, and Nature, from which that singing issues, also continues it.

The language is singular, it has to be: not in the sense of a reserved and affected idiom—there is no mannerism here—but in the sense of a language of the extremity which touches on the impossibility of the real or actual, on its distance, and on the correlative distancing of all languages between them, of all song. Just as every poetic language has to be one, so every point has to be touched. For Hölderlin, "song" ("Gesang"), that ordinary metaphor/metonymy of the poem, more precisely names that which "wraps" ("zu fassen)" the "Father's ray" ("des Vaters Strahl"), the flash or bolt of "God's thunderstorms" ("Gottes Gewittern"), which the poet, "bareheaded" ("mit entblößtem Haupte"), grasps "with his own hands" ("mit eigner Hand"; *SA* 2; 1: 118–19). Song does not conceal the light; it brings one into contact with it.

9. Because it necessitates measure, touch belongs to calculation. The measureless distinguishes neither the places nor the surfaces that touch necessitates.

> Denn schonend rührt des Maases allzeit kundig
> Nur einen Augenblik die Wohnungen der Menschen
> Ein Gott an, unversehn, und keiner weiß es, wenn?
> (*SA* 3; 1: 534)

> For carefully touches, always expert in measure,
> Just a moment, a god unexpectedly
> The dwellings of men, and who knows—when?

Measure is the word, the word and the motif, upon which Hölderlin's thought and poetry very exactly intersect—sharing themselves (*sich mit-teilen*) in every sense of the term. Because it is essentially a thought of measure, his thought has to be a poetics, in order to think it has to touch on its poetic outside.

This measure does not limit, any more than it balances. It does not find a precise middle. Hölderlinian measure is precise in that it measures the divergence by which the whole and unity are possible as whole and as unity. Calculation makes the following links: Whole = Whole, Whole = One, One = One, but "=" equals the divergence necessary in order to posit and count the One facing itself, faced with itself, in the clarity of its distinction. Thus, poetics considers less the equality of the Whole with itself, or of the One with itself, than equality in itself, the equality of equality as it were, "=" = "=" or its measure, that is to say, the distancing, the spacing necessary simply in order to posit equality, its meter, its interval and its scansion.

The idealism of the uni-totality immediately goes beyond all measure, knowing only the self-correspondence in itself of the absolute. It thinks in terms not of equality, but of identity (and, in truth, it thinks identity in terms of the recovery of the self by the self, and not in terms of the exposition of the self to the self). But the poetics of contact necessitates the distance that constitutes the essence of touch. What touches is distinct from what is touched, touch is *discreet*, or it is nothing. Feeling is possible only through the distance of a *concurrence*—an accord and a modesty, the measure of one by the other.

Thus:

> Hier bewegt' er ihr nicht mit dem Sonnenblike den Busen,
>   Und in Reegen und Thau sprach er nicht freundlich zu ihr;
> Und mich wunderte deß und thörig sprach ich: o Mutter
>   Erde, verlierst du denn immer, als Witwe, die Zeit?
> . . .
> Aber vielleicht erwarmst du dereinst am Strale des Himmels
>   Aus dem dürftigen Schlaf schmeichelt sein Othem dich auf;
> Daß, wie ein Saamkorn, du die eherene Schaale zersprengest,
>   Los sich reißt und das Licht grüßt die entbundene Welt.
> (*SA* 2; 1: 80–81)

> Here he does not move her breast with the sun's ray,
>   And in rain and dew he did not speak kindly to her;
> And I wondered at that and foolishly I spoke: O Mother
>   Earth, will you, as widow, be always bereft by time?
> . . .
> But perhaps one day you will warm yourself in the ray of heaven,
>   Out of your poor sleep his breath will coax you,
> So that, like a sown seed, you break your old husk,
>   And an unbound world tears itself loose to greet the light.

—here again, hear how the point of this accord, the spring of the earth, is, in the last line cited, given in the calculation of a tight cadence: "*Los sich reißt und das Licht grüßt die entbundene Welt.*" "*Los*" triggers and discharges the burst, the release—at once in sense, sound, and rhythm. One could say that this line, along with those which follow it, is the onomatopoeia of spring, or of love:

> All' die gesammelte Kraft aufflammt in üppigem Frühling,
>   Rosen glühen und Wein sprudelt im kärglichen Nord.
> (*SA* 2; 1: 81)

> All the assembled force flames out in luxuriant spring,
>   Roses glow and wine bubbles forth in the meager North.

—which is not an imitation, but the touch of language, which distances it from itself and from sense, and which, in between, makes sense, as the beat of the interval. There is no onomatopoeia of the "living sense," but there is the blow of a resonant touch, the blow of an accord of this divergence of sense: poetry is not sense, it neither makes sense nor expresses it, but accords with its distance, or tunes its divergence. Whence its need to be "rhymed with joy" ("Sei zur Freude gereimt"; *SA* 2; 1: 66). "Rhyming" is here a metaphor for "attuning" only in also being the metonymy of poetry, and "joy" is "thought"—or is "said"—as the accord that it is (or of which it is, in turn, the "metaphor"), only through the resonant interval of poetic accord. (Or else, one would have to say that this rhythmic movement of stresses and assonance—*glühen/sprudelt/kärglich*, etc.—is still an imitation, but an imitation of what, exactly? Is it that the rose imitates the glowing embers, or vice versa? That the word *glühen* imitates incandescence, or vice versa? Is it that *glühen*, *sprudelt*, and *kärglich* imitate or respond to one another? Everything can be turned in all directions, around the same point: how sense touches on sound and sound touches on sense—which is what "song" means.)

The point of contact, therefore, is itself a distance or a divergence. This point is not a geometrical point, devoid of dimensions; it is itself the dimension, the distension of the poetic relation to unity and to totality. It is, in short, the absolute distance of the absolute. This absolute distance is also a proximity; it does not measure according to the quantity of distantiation, but according to the simple nature and intensity of the interval as such.

> Fest bleibt Eins; es sei um Mittag oder es gehe
>   Bis in die Mitternacht, immer bestehet ein Maas,
> Allen gemein, doch jeglichem auch ist eignes beschieden.
> (*SA* 2; 1: 91)

> One thing remains; whether at noon or going
>   Deep into the night, there is always a measure,
> Common to all, though to each is also his own allotted.

Distantiation, the spacing of times and places, space-time as such, renders presence and its touch possible. Measure measures—that is to say, establishes and states the concurrence of—"a *determined* [bestimmte], actual infinity" (*SA* 4; 1: 260; *Essays*, 78). Determination is not here the limitation of infinity. It is not finitization, in the sense of a loss or a fall within a finite horizon; it is, on the contrary, finishing, exact completion according to "the firm point from which the mode of relation for the sequence of the drawing and the character and degree of the local color and illumination are determined" (*SA* 4; 1: 261; *Essays*, 79).

Thus measure gives the *tone*, heard absolutely, the tension and the composure of the contact: through the unity of the whole, and before it, against it, in the very instant of its cadence and its blow. Measure is "beautiful precision" ("schöne Bestimmtheit"; *SA* 4; 1: 264; *Essays*, 82), not the precision—determination or exactitude—that would be beautiful anyway, but the precision which is beautiful in itself because it touches *precisely* on the other face of beauty, on "the infinite beautiful reflection which, in the constant delimitation, is at once constantly relating and unifying" (*SA* 4; 1: 265; *Essays*, 82).

Beautiful precision gives the tone, the contour, the color of what localizes the infinite, of what shows it and what touches it—this is all one and the same gesture—at the instantaneous point of its presence, that is to say, of its passing, at the point of what cuts it off and isolates it as a site of enjoyment.

"Measure, . . . the beautiful precision and unity and firmness of its infinite accord, . . . its infinite identity and individuality, and composure, . . . its poetic prose of an all-delimiting moment" (*SA* 4; 1: 264–65; *Essays*, 82). The *allbegrenzender Moment* is as much the moment (read: the suspension) that delimits the whole as the moment that delimits everything. Measure at once encompasses everything, the totality of the whole, and rigorously circumscribes the narrow space-time of the point and the touch.

But this simultaneity and this conjuncture in no way signify an overstepping of one by the other, nor of one in the other. No effusion of the one in the whole, no infusion of the whole in the one. Measured language—meter—does not leave its measure (all *technē* consists in standing firm), while that which measures it and against which it measures itself, brute presence seized firmly in its passing, remains outside of it, though measured language bespeaks this outside, this before, this facing it. There is only the conjunction—the rhyme—of this face-to-face: the one and the look held out toward it, the double inscription of the poem: that is to say, its double exscription. This is the unspeakable itself—but for the poet, and in him, "the unspeakable speaks" (*SA* 4; 1: 155; *Essays*, 55). Such is the effect of the "constant precision of consciousness with which the poet considers a *whole*" (*SA* 4; 1: 156; *Essays*, 56).

Schwer ist worden indeß von Früchten dunkel mein Kirschbaum,
  Und der pflükenden Hand reichen die Zweige sich selbst.
. . .
Aber ich ahn' es schon, in heilige Fremde dahin sind
  Nun auch sie mir.
(*SA* 2; 1: 82)

My cherry tree, meanwhile, has become heavy with fruit, somber,
  And the branches reach, on their own, to plucking hands.
. . .
Already I foresee it, into holy foreign lands have they
  Now also gone, away from me.

10. Meter is properly the divine: that is to say, the incommensurable in its strictly determined precision, the exactitude of the impossible. Thus:

auch hier sind Götter und walten,
Groß ist ihr Maas, doch es mißt gern mit der Spanne der
Mensch.
(*SA* 2; 1: 80)

—There are gods here too, and they reign,
Great is their measure, yet man too readily measures it with
  his span.

The poet does not keep to this poor human measure, the measure of his palm, the span, the measure of which he is himself the unity. Rather, he measures divinely, that is to say, according to a measure that nothing can measure, according to absolute greatness. This consists of the conjunction of the incommensurable and proximity, of the two extremes of the infinite, what is farthest away—"I stood and looked out into the barren African plains" ("stand ich und sah in die Afrikanischen dürren Ebnen hinaus"; SA 2; 1: 80)—and what is closest—"there lives some silent thanks" ("es lebt stille noch einiger Dank"; *SA* 2; 1: 94).

Absolute greatness is the exactitude of the reunion of a concurrence and a divergence: the concurrence of the divergence through which a presence, being-itself, presents itself, holds itself there, manifest, and slips away in the same place, in the same instant, in the one and only passing of sense.

Meter measures the same thing, this "vestige of the gods who have fled" ("Spur der entflohenen Götter"; *SA* 2; 1: 94), which is neither pure loss nor a sign referring to what is distant, but the trace of a passing as true presence, definitively actual in this present which withdraws into the distance upon being touched. Thus, the speech of the poets "is true" or "speaks rightly," "*sagen mit Recht*" (*SA* 2; 1: 94), not through the sense of what it states or of what it evokes, but through the measure which is—materially—the sensible and exact remnant of the passing of the gods, that is to say, of this passing that the gods *are*.

Which does not mean that the poet's truth consists in anything accompanied by a metrics. It means that the poet has to say what is said—truth, precision, sense—in its condition of unity and passing, under the touch of its passing.

The divine is passing and only that: which is why it is essentially a remnant and not an image, at once material remainder and fading trace of a presence which is not subsistence, not being-present-there, stable and manifest, but passing, coming and going, happening, taking place and departing.

> Göttliches Feuer auch treibet, bei Tag und bei Nacht,
> Aufzubrechen.
> (*SA* 2; 1: 91)

> Night and day, divine fire too urges
> Bursting forth.

The divine lies in immanence and slipping away, in approach, passing, and withdrawal; its withdrawal is not its loss, but the condition of its transience, which is itself the condition of its sense. For the divine lies in passing insofar as it points toward nothing other than its own passing. The place that combines "Olympus" and "Parnassus," this place "from which and toward which signals the coming god" ("dorther kommt und züruk deutet der kommende Gott"; *SA* 2; 1: 91), is the coming and passing, it is this touch which—like Diotima: "You alone, your light . . . holds in light" ("Dich nur, dich erhält dein Licht . . . im Lichte"; *SA* 2; 1: 78)—itself illuminates and illuminates only itself, thus illuminating the one. The "coming god" is not "to come," and in this sense he is not "God," but the coming, the whole coming and nothing but that: he is nothing but the blow, the beat, and the metric, syntactic, and semantic stance of this line that turns the "coming" upon itself: "Dorther kommt und züruk deutet der kommende Gott." "From which and toward which signals the coming god": that he comes (*kommt*) gives the assonance or the rhyme of the line—*dort/kommt/komm/Gott*—and this coming turns around this sign or this sense (*deutet*) turned back (*zurück*) toward the *dort*, leading the *Gott* back to the *dort*. *Zurück* turns the coming back upon itself, in the middle of the line and the muted assonance (*und/deutet*) of sense, itself set into the strong assonance of the extremities *Dort/Gott*. The god is only the place, the place is the place of departure and return, of the coming that withdraws and thus makes sense. And does so according to the line, according to the scansion of the line. Poetry: the material calculation of this atheist passing. (To read Hölderlin today is also to tear him from the Romantic imagery which is forcibly his own and with which "the gods" are a piece.)

(One could even understand the line in this way: the coming god comes and utters "Return!" as a sign, ordering the reversal of the line, that is to say, the reading of its scansion.)

The measure to be taken is thus always that of a passing as such. To take measure is the act of the poet, an act which consists in measuring himself against the divine. To measure against the divine is not to confront it in an unlikely rivalry; it is to measure against the incommensurability of the passing, against the incommensurable distancing of the *place* of passing, to grasp it as incommensurable, but to grasp it exactly, with the absolute precision of the incommensurable, at the very point of its incommensurability. That is to say, at its very manifestness, its open patency, at the manifest point of the presence of the passing:

So lange die Freundlichkeit noch am Herzen, die Reine, dauert, misset nicht unglücklich der Mensch sich mit der Gottheit. Ist unbekannt Gott? Ist er offenbar wie der Himmel? dieses glaub' ich eher. Des Menschen Maaß ist's. (*SA* 2; 1: 372)

As long as friendliness, the pure, lasts in his heart, man does not unhappily measure himself against the divine. Is God unknown? Is he manifest like the heavens? This rather I believe. It's the measure of man.

The measure of man—what measures him—is the incommensurable manifestness of the divine, that is to say, in reality, the divinely a-theist "divine" of incommensurable manifestness. This manifestness is manifest "like the heavens" because it is neither the sensible manifestness of an object, nor the intelligible manifestness of an Idea, any more than it is the self-manifestation of a subject. It is different from these three forms of manifestation, or else it is like all three together. It is manifest or is manifestly of the "open" as such. The heavens: the place of the manifest as the manifestness of place. The divine place because the "divine" is precisely this manifestness of place, of all places. (Place is locality, the separation and discretion of passing, the place where the coming to presence each time necessarily takes place—whether this be "the bottom of the valley" or the "banks of Lake Bienne," whether it be the "room" with "tables" or "the desert.") There is no generality of presence, only exactly delimited passing places, right down to the place of the "gravedigger":

> Da soll er alles
> Hinausführen
> Außer den Langen
> An eine reine Stätte,
> Da man die Asche
> Hinschüttet, und solls
> Verbrennen auf dem Holz mit Feuer.
> (*SA* 2; 1: 335)

> There must he
> Lead it all
> Beyond the lengths
> To a pure site,
> Where one pours
> Ash and must
> Burn it on wood with fire.

Likewise, there is no generality of the incommensurable "living sense." There are only the places of its passing, each time a word, a phrase, each time a meter, a scansion. Sense is the distancing of language, its truth each time local.

If, under these conditions, there is no "measure on earth" ("auf Erden ein Maaß"; *SA* 2; 1: 372), it is because all measure is "heavenly." This, in turn, has to be understood in two ways: first, all taking measure takes the measure of the entire, incommensurably open "heavens," and, second, every act of measure comes from or returns to the heavens, *zurück*, with the very thing which renders the act and the encounter possible: measure itself, meter.

The withdrawal of the gods, the passing of sense, is thus at once that which is to be measured, that against which we must measure ourselves, and that which gives the measure, the unity of measure, as well as its instrument, the instrument of measure which is no longer an instrument, namely meter, which is an end in itself.

The precision of meter, which is the concern of the poet, thus responds, very exactly, to the "failing of sacred names":

Schweigen müssen wir oft; es fehlen heilige Nahmen,
  Herzen schlagen und doch bleibet die Rede zurück?
Aber ein Saitenspiel leiht jeder Stunde die Töne,
  Und erfreuet vielleicht Himmlische, welche sich nahn.
Das bereitet und so ist auch beinahe die Sorge
  Schon befriediget, die unter das Freudige kam.
Sorgen, wie diese, muß, gern oder nicht, in der Seele
  Tragen ein Sänger und oft, aber die anderen nicht.

(*SA* 2; 1: 99)

We must often stay silent; the holy names are lacking,
  Hearts beat yet speech lags behind?
To each hour, though, a lyre lends the tone,
  And perhaps divine ones, those who come near, rejoice.
That is preparing, and, also near, the care
  that came amidst our joy is near replete.
Cares like these, in his soul, whether he will or no,
  A singer must carry so often, even if others do not.

11. Meter responds to the failing of the divine. Its exactitude is the putting into contact, the point of tangency of two approaches, of two approximations: that of the heavenly ones (*Himmlische, welche sich nahn*), and that of song (*beinahe . . . befriediget*). *Beinahe* ought to be translated as "by the near." "*Sich nahn . . . beinahe*": nearing the nearest. This is proximity *itself*, that is to say, divergence, but the divergence necessary in order to touch on the thing itself—and this thing itself is a nearing, an imminence. Neither an immanence nor a transcendence, but an imminence, an infinite proximity which passes so close it touches the beating heart. This is the touch of "the living sense" on "nul sense," the point of unity of the whole, and the space-time of this point.

Meter measures the openness of the manifest, measures it in its proper place, at the point of its coming, in "the beautiful instant" (*SA* 4; 1: 52), that is to say, in its passing. Meter responds to the manifest: to the clarity of the vacuum, no doubt, but equally to the opening of clarity. Meter measures the unmeasured, the plenitude of the open. The eye trained on the visible at the point of its totality. Which means:

Allein zu reden
Zu Gott.
(*SA* 2; 1: 128)

To speak alone
To God.

Meter speaks only to divine meter. Calculation has this exactitude in sight: to respond to the springing up and instantaneous passing of the incommensurable. This can only be spoken by "words arisen like flowers" (*SA* 2; 1: 93)—but here again, the language and the restrained measure have to be heard in German:

Nun, nun müssen dafür Worte, wie Blumen, entstehn.

Now, now must words for it arise like flowers.

The measure of meter responds to the without-measure: it bursts forth as its caesura. The uninterrupted continuity of "the living sense" can become sensible only in its interruption. Not because the interruption would simply open an anguished openness of sense, but because it sets back sense still further, seizes it in passing and sends it back to sense. The caesura again calls for sense—and sense satisfies itself in the caesura.

Take these two lines:

Drinn in den Alpen ists noch helle Nacht und die Wolke,
Freudiges dichtend, sie dekt drinnen das gähnende Thal.
(*SA* 2; 1: 96)

The translation (here by Michel Deguy) reads:

Au coeur des Alpes, nuit claire encore, et la nuée,
Source du poème de joie, elle couvre là-bas la vallée béante.[6]

And another (this time by Michael Hamburger):

There in the Alps a gleaming night still delays and, composing
Portents of gladness, the cloud covers a valley agape.

My intention is not to propose a better translation, but instead to bring out what remains untranslatable, even into the German language itself—at least if "translation" is limited to "restoring a meaning." In fact, the lines would have to be heard thus:

There in the Alps, it is still clear night and the clouds,
Pouring joy into verse, cover the gaping valley there.

Drin—drinnen calls and recalls from one line to the other, driving home and losing its assonance in this insistence, whereas *die Wolke* / . . . *sie* (a double caesura: that of the line and that of the prolepsis) holds the cloud above what it covers, and over its *dichten* (poetic composition, fiction, formation of figures of the mist). A sending back and caesura, a setting back and return over and above the caesura, but thus through the caesura itself. *Drinnen* will be repeated twice more in the same strophe, and the whole poem (which ends with the lines already cited on the "cares" of "the singer") can be read as the poet's penetration into the depths or as his raising from them. The sense is thus organized on the basis of the opening *Drin*, an incisive word (itself a contraction of *darin*), being both re-caesuraed at the beginning of the line: *Drin in* . . . as well as being rhymed at the beginning of the third line: *Dahin, dorthin* . . . . The entire strophe is scanned by this "*in*" and resonates from it—the penultimate line speaks of the "echo" that "sounds all round" ("tönet umher")—which is to say that it resonates as much from the direction, the momentum, or the initial address as from within, from the covered and yet open depths: sense in every sense. Sense is here the attacking blow and the rhyme, the measure as well as the envisaged end—

Denn bacchantischer zieht drinnen der Morgen herauf.
(*SA* 2; 1: 96)

For, more Bacchic, morning brews from within.

—but this is, quite precisely, only the infinite end, which itself consists in opening other ends, in opening itself as an end and as its only resonant end, its only prolonged echo, *Freudiges dichtend.*

At every step, Hölderlin's poetry—perhaps, in this regard, one of the most monotonous there is (but it would be necessary to rework the entire habitual evaluation of "monotony")—repeats this same thing; that is to say, it repeats itself in the following way: a phrase which moves forward and which is suspended on its own sense, which surprises its sense in advance or after the fact, and whose meter measures this gap that is each time infinite.

Thus:

so stürzt
Der Strom hinab, er suchet die Ruh.
(*SA* 2; 1: 49)

so hurries
The river down, looking for rest.

—or rather, without looking to translate:

so stürzt
Der Strom hinab, er suchet die Ruh, es reißt,
Es ziehet wider Willen ihn von
Klippe zu Klippe den Steuerlosen
Das wunderbare Sehnen dem Abgrund zu.

thus plunges
The river down, seeking rests, it travels,
Drawn against its will, from
Boulder to boulder, the guideless,
By that wonderful yearning for the abyss.

The poet seeks the rest of revealed sense—"to interpret the holy legends" ("die heiligen auszulegen"; *SA* 2; 1: 53) of which the end of the poem speaks—and, as he does, falls into the abyss of sense. But this is neither a precipitation into senseless insanity, nor a magic evocative of arcane meanings. This poetry has no secrets. If the poet ceaselessly leads it anew toward an incompletion of sense, if this is how he calculates it, there is no mystery and effusion in that. It takes place, rather, through a holding back of sense, in an acute consciousness of its excess and the transience of its passing, and through the exact measure that allows the clarity of this consciousness.

To follow clearly the movement of sense, its truth, not as a path replete with a goal and a progressiveness, but as a cadence which holds its own without ever being a progression—

wie die Quelle dem Strome folgt,
Wohin er denkt, so muß ich fort und
Folge dem Sicheren auf der Irrbahn.
(*SA* 2; 1: 55)

as the source follows the stream,
wherever its thought goes, so must I go and
Pursue the one who is surer though wandering paths.

—or even:

> [. . .] sieh! und das edle Licht
> Gehet, kundig des Wandels,
> Gleichgesinnet hinab den Pfad.
>
> (*SA* 2; 1: 64)

> [. . .] look! and the noble light
> Goes, expert in change,
> Like-minded, down the path.

The "living sense" (living up until the point at which it dies: this is no less its measure)—

> Das Fest verhallt, und jedes gehet morgen
> Auf schmaler Erde seinen Gang.
>
> (*SA* 2; 1: 114)

> The celebration dies away, and tomorrow each will go
> On the narrow earth, its own way.

—the incalculable, is always what goes beyond sense, to the point that one can speak neither of a loss nor a gain of sense, but of only its truth, a truth which is at once both "joy" and "mourning." The conjunction of joy and mourning—surely the Hölderlinian motif par excellence—responds to the conjunction of sense and its suspension:

> wenn einen dann die heilige Wolk umschwebt,
> Da staunen wir und wissens nicht zu deuten.
> Ihr aber würzt mit Nektar uns den Othem.
>
> (*SA* 2; 1: 128)

> when the holy cloud envelops one of us
> We are amazed and do not know the meaning.
> But with nectar you spice our breath.

Othem ("breath," a religious—Lutheran—form of *Athem*) rhymes with *deuten* ("to mean, to interpret"). The rhythm of breath responds to the interruption of sense. Meter is the proper measure of this: neither as the expression (or mimesis) nor the knowledge of this interruption of sense, but as its type and tone.

12. Hence, "Ein Räthsel ist Reinentsprungenes" (*SA* 2; 1: 143), which is to say: "A riddle is what purely springs forth," or "A riddle is pure spring-

ing forth." Sense is hatched, not in linkage, but in springing forth and the bound.

> Ein Räthsel ist Reinentsprungenes. Auch
> Der Gesang kaum darf es enthüllen.

> A riddle is pure springing forth. Even
> Song can scarcely unveil it.

Here, the caesura separates the song from its riddle, and at the same time touches on the springing forth itself. The act of the poet, the act of measure, is to be accorded this "surprise" (*SA* 2; 1: 141) or this "arising" of the divine which the divine as such is. If "man / Loves what is present" ("Menschen / Ist Gegenwärtiges lieb"; *SA* 2; 1: 140), he has to learn to endure its coming and its flight—the flight that goes hand in hand with its coming—the totality of presence as the point, the caesura of the measure which "each one has" (*SA* 2; 1: 148).

Calculation measures the instant, the presence, the brief scansion of life and sense, of the life of sense—and this is why, more than the poem, it also measures the poet himself, and in the poet, the individual above all: "The *apriority* of the individual / Over the whole" ("Die apriorität des Individuellen / über das Ganze"; *SA* 2; 1: 148). Here, the individual is not the sealed enclosure of an individualism. It is the indivisible *hic et nunc*, the precision of a measure which is one each time, whole in being the measure of all. Its apriority is neither a prevailing nor a primacy: it is, in strictest Kantian rigor, the condition of possibility for the experience of the whole. The whole can be present for all if it springs forth and fades away in a point that is each time unique.

> Alles ist innig
> Das scheidet
> So birgt der Dichter.
> (*SA* 2; 1: 321)

> Everything is inmost
> That separates
> And shelters the poet.

The poet has to "feel and appropriate the common soul, the soul common and proper to all" (*SA* 4; 1: 241; *Essays*, 62). And:

Die Dichter müssen auch
Die geistigen weltlich seyn.
(*SA* 2; 1: 156)

Poets must also, like
Spiritual ones, be worldly.

  Manche helfen
Dem Himmel. Diese siehet
Der Dichter. Gut ist es, an andern sich
Zu halten. Denn keiner trägt das Leben allein.
(*SA* 2; 1: 218)

  Many help
Heaven. The poet sees
Them. It is good to lean on
Others. For no one bears life alone.

Denn einsam kann
Vom Himmlischen den Reichtum tragen
Nicht eins.
(*SA* 2; 1: 244)

For no one can
Bear the wealth of the heavenly
Alone.

Thus the poet is the "individual," or individuated individuality as such: not someone separated in the arrogance of self-sufficiency or in the misery of isolation, but the very separation by which the whole shares itself with all, each time whole and each time one—very exactly the one as the measure of the whole insofar as the whole is itself shared in itself, insofar as it is itself nothing other than sharing: point, caesura, and passing, the clear remnant of the presence that presents itself thus.

This is why "original poetic individuality" has itself to be *aufgehoben* in the "boldest and ultimate attempt of the poetic spirit" (*SA* 4; 1: 252; *Essays*, 71) which, by its very "freedom," has to posit or present its own individuality outside it, before it, as an exterior figure, as the objectivity of its presence. This spirit cannot be "known" "in itself," but only outside itself. This "outside itself" of spirit—its body, its measure, its exactitude set down before it—is the present, impenetrable, punctual real.

Hölderlin's poetics is not a poetics of the possible and not, consequently, a poetics of the poem as a work, that is to say, as a possible world or world of possibilities, as another world imitating and sublimating this one. It is, on the contrary, a poetics of the real: not a "realism," but a poetics of the exteriority of every work, which it is the work's only task to inscribe or, rather, to exscribe. The exteriority of presence, the exteriority in which and according to whose measure alone is presence able to come, to come and go.

The real is always outside, always opposite, precisely where the "open look" ("dem offenen Blick") is fixed, the look for which "the radiance," the manifestness of "the heaven's bloom" opens: "Come! into the open, friend!" (*SA* 2; 1: 84) The real is that toward which one must go, infinitely, exactly, and poetic measure gives its distance and proximity, the distantiation and the imminence in the instant of its passing.

If man "dwells poetically"—"Rich in merit, but poetically, man dwells on this earth" ("Voll Verdienst, doch dichterisch, wohnet der Mensch auf dieser Erde"; *SA* 2; 1: 372)—it is not in the sense in which "poetics" would come to modify or, even less, embellish or sublimate "dwelling." Nor is it that, as Heidegger would have it, poetry makes of dwelling a dwelling. Rather, dwelling on this earth—being in the real, being toward the real—comprises the poetic: the earth, the outside, the irreducible, inappropriable presence, measures the divergence of whatever finds itself there, of whatever finds itself essentially there. Man does not dwell in the heavens, where there is no distance. But he has the heavens before him, and their manifestness measures him. What he risks poetically is turning toward the heavens with an exact, infinite look, at the right moment, at the correctly calculated moment: at every instant, providing it be the instant of such a calculation.

> Und siehe, wenn du kannst, zu diesem Licht,
> Dem Allesschauenden, empor!
> (*SA* 4; 1: 58)

> And lift, if you can, to this light,
> Your eyes, to the all-seeing.

The exact look does not take over his vision. It looks at what sees it, sees that it is seen from afar, from always further off in the unity of the whole. It touches on this dazzling, on its imminence, on its minute, ungraspable passing, never certain and yet so clear and so real.

Otherwise, people are becoming ever more dear to me, for I see in their small as in their large activities and in their characters one sole and same original character, one sole and same destiny. Indeed, it is this need to advance, to sacrifice an assured present to something uncertain, something different, better, and always better, that I consider the first cause of all the doings and gests of all the people who surround me. (*SA* 6; 1: 327)

Translated by Simon Sparks

Poetry's Courage

Philippe Lacoue-Labarthe

For Maurice Matet,
since it is a question of courage

That, after a brief period of fanatical activism in 1933–34, Heidegger's entire political preaching is to be sought in the discourse which he conducts on poetry—and on Hölderlin in particular—is not something we have invented. Heidegger says it himself.

At a transitional point in the course *Hölderlin's Hymns "Germania" and "The Rhine,"* given in the winter semester of 1934–35—the first, or very nearly the first, course he taught after his resignation from the Rectorship—one reads, as a reminder of the "task of the course" (*die Aufgabe der Vorlesung*), the following, which is perfectly clear:

The goal of this course remains to recreate finally in our historial Dasein a space and a place for what poetry is. That can happen only when we bring ourselves into the sphere of power [*Machtbereich*] of a true poetry and open ourselves to its effectivity. Why is Hölderlin's poetry chosen for this? This choice is no random selection from the poets available. It is a historial decision. Let us name three of its essential grounds: (1) Hölderlin is the poet of poets and of poetry; (2) From which it follows that Hölderlin is the poet of the Germans; (3) Since Hölderlin is this in a concealed and difficult way, poet of poets as poet of the Germans, he has not yet become a power in the history of our people. Since he is not yet that power, he must become so. To contribute to this is "politics" in the highest and most proper sense, so much so that whoever effects anything here has no need to talk of the "political."[1]

(I note in parenthesis: you will have recognized here the "unscrambled" transmission of the message that the Rome lecture of a couple of years later—"Hölderlin and the Essence of Poetry"—will go to great lengths to encode, at once very skillfully and very crudely, notably through the suppression of the signifier "the Germans." It is as though, far from home yet still on "allied" soil, Heidegger had obscurely prepared his future reception, his French reception in particular. And you will not be unaware of the consequences, including those for French poetry, that Henry Corbin's translation of this lecture has had since 1938.[2] But that is another question.)

That such a claim for a " 'politics' in the highest and most proper sense" (the word *politics*, we should note, is, even here, always in quotation marks), takes, beyond its manifest "nationalism," the exact form of the theological-political, is something Heidegger of course never mentioned—he would have challenged it most firmly. No one, with very few exceptions, has really noticed this. Yet it is no less clear.

I will take only one example, the one offered most immediately. It comes from the "Preliminary Remark" to the same course—perhaps (largely) written after the fact, though this is neither entirely certain nor of any great interest—soberly entitled: "Hölderlin." I quote:

> He must still stay long hidden, particularly now when "interest" in him is beginning to stir and "literary history" is in search of new "themes." One writes now on "Hölderlin and his gods." That is the most extreme misinterpretation, through which all possibility of action is forbidden to this poet, who still comes from the future of the Germans, under the pretext of finally doing him "justice." As if his work needed justice, especially from the poor judges who are today in circulation. One handles Hölderlin "historically" and fails to recognize this one essential thing, that his work, which is still without time and space, has already overcome [*überwunden*] our historicizing activity and founded [*gegründet*] the commencement of another history, the history which begins with the struggle [*Kampf*] where the fleeing or the coming of the god will be decided.[3]

(Allow me a second parenthetical remark here. Whenever this note may have been written—it sanctions, in any event, in a highly charged vocabulary, the careful and deliberate publication of a course from 1934 with which it is in perfect harmony—one thing is clear: up until the testamentary interview granted in 1966 to the editors of *Der Spiegel*, Heidegger will change practically nothing of the discourse which he will have started to conduct from the moment of his "withdrawal" from National Socialism. For the record, I recall the declaration of 1966: "Only a god can save us now. The only possibility that remains for us is to prepare, through thinking and poetry, a readiness for the appearance of the god or for the absence of the god in our decline; in the face of the absent god, we founder."[4] Let us pass over the very dated pathos of the decline; or, if you prefer, the haunting fear of nihilism. All the same, in a proposition of this sort what I do not think it exaggerated to call the theological-political finds itself confirmed. The theological-political is particularly tenacious Heidegger, and we must understand that if "real National Socialism" was, in his eyes, nothing but "politics," this was because it lacked not only an authentic sense of the *polis* (thought as the *Da* of *Sein*), but also a theology, and that is to say, as we have already seen, a notion of the essence of art and of the unique opportunity to have "grounded the beginning of an other history.")

I will not enter here into a consideration of the complex and cunning relations that Heidegger maintains with theology, a word which he never challenges. I take *theology* in its simplest sense: discourse on the divine. In order that things be totally clear, I will cite the passage that is probably the most explicit in this regard, the one where, in the "Letter on 'Humanism'" (1946), seeking to define an "ethics" before the ethical, that is to say, an "ethics" which responds or corresponds to what is no longer even called "ontology"—for this occasion, lacking anything better, I will call it an arche-ethics—Heidegger draws on the Hölderlinian motifs of the homeland (*die Heimat*) or of "German" (*das Deutsche*). Here is the passage—and do not forget that, at that time, this discourse also counts as a clearing of his name:

> "German" is not spoken to the world in order that the world might be recuperated through the German essence, but is spoken to the Germans in order that, from out of the destiny that links them to other peoples, they might become world-historical along with them. . . . The homeland of this historial dwelling is the proximity to being.
>
> In such proximity, if at all, it must be decided whether and how the god and the gods refuse themselves and the night remains, whether and how the holy day dawns, whether and how in this emergence of the holy an appearance of the god and the gods can begin anew. But the holy, which alone is the essential space of divinity, which in turn alone grants the dimension for the god and the gods, comes to appearance only when Being itself, beforehand and after long preparation, has shown itself and has been experienced in its truth. Only thus does the overcoming [*Überwindung*] of homelessness begin from being, in which not only men but the very essence of man lose their way.[5]

In 1934 Heidegger spoke to his students of the task of his teaching. There was an injunction in that. Such a task was already none other than the "task of thinking" insofar as, at the "end of philosophy," it chimes with the task it takes to have been that of poetry—Hölderlin's task: to prepare for the coming of (or announce the turning away from) god and the gods. We are face to face with a nihilism even more radical than that expressed in Nietzsche's "God is dead," which supported—this needs to be born in mind—the entire engagement of the Rectoral Address. The task of thinking, the task of poetry: this sort of arche-ethics would have ruled Heidegger's theological-political preaching.

If this basic hypothesis is right, I would like, in consequence, to put forward a few propositions—in the rather terse form of a simple reminder. There are five of them:

1. The theological-political is supported, quite obviously, by a theologico-poetics (once again, lacking anything better, I risk this phrase). Hölderlin's "preaching" authorizes Heidegger's own "political" preaching of the 1930's (and beyond), but only on the condition that we understand: that, on the one hand, this entire "preaching" (i.e. Hölderlin's—and I doubt that, in the end, it was one) thus has no authority since it both lacks authority and is authorized by nothing and no one (it is not a god who, as in the *incipit* of ancient poetry, dictates the poem; rather, it is the "failure of the god" that "helps" the poem, that, at the limit of the possible, renders it possible); and that, on the other hand, this authority is an authority only because it is, at the same time, the madness (the protective night of the gods) which has sanctioned the impossible possibility of the poem, a poem which has still not been heard (which is why, moreover, the Germans remain indebted to Hölderlin).

The ancient theological-political, as is confirmed by the *Introduction to Metaphysics* (1935), stems from Homer, under the orders of the Muse, giving "Greece its gods."[6] The modern theological-political suspends the poem of the annunciation—the gospel—of the coming or withdrawal of the god.

2. In the expression "theological-political," the political is what concerns the national, or, in Hölderlin's more precise locution, the "nationell."[7] It is, in other words, what concerns the people. Although in the "Letter on 'Humanism'" Heidegger credits Marx with having experienced the essential homelessness of modern man through the concept of alien-

ation (this insight needs to be born in mind: Marx is not so far removed from Hölderlin: the organized delocation of the masses by "industry is a fact, and it is a matter, in both Marx and Hölderlin, of an *alienation*), he challenges—and will always have challenged—internationalism, like nationalism, as a figure of unreserved subjecthood. By which one should understand their doctrines and, *a fortiori*, their ideologies. Just before the passage from the "Letter on 'Humanism' " which I recalled a moment ago—and after having remarked that Nietzsche, who had himself experienced this homelessness, "was unable to find any other way out" of it because he remained prisoner to the "reversal of metaphysics"—Heidegger says:

> When, however, Hölderlin composes "Homecoming," he is concerned that his "countrymen" access their essence within it. He in no way seeks this essence in an egoism of his people. He sees it, rather, in terms of belonging to the destiny of the West. But even the West is not thought regionally as the Occident in contrast to the Orient, nor merely as Europe, but as world-historical [*weltgeschichtlich*] from out of nearness to the origin.[8]

Be that as it may, during the 1930's this problematic of *Heimatlosigkeit* (of "uprootedness," as it was willingly called in the international of the conservative revolution) forced one to substitute for the Kantian question (in an indissociably *aufklärer* and metaphysical style) "What is man?" the question "Who is man?," a question which invariably finds its truth in the question: "Who are we?," which means, no less invariably, "Who are we, we Germans?," One possible response is, for example: "The philosophical people par excellence" or the "people of thinkers and of poets."[9]

3. The theological-political is, in turn, supported by an appeal or a call to myth. I have already tried to show this elsewhere, and so I will not insist upon it here.[10] This remark holds for the entire great (German) metaphysical tradition since its Romantic inception, with the Schlegels certainly, but above all with Schelling—a friend of Hölderlin, as we know. Heidegger's apprehension of poetry is overdetermined by speculative Romanticism. Which is why, moreover, poetry (*Dichtung*) is defined in its essence as language, *die Sprache*—or, what amounts to the same thing, language is defined as the original poetry (*Urdichtung*) of a people—and why it is in turn defined in its essence as *die Sage*: *ho muthos*. Not like *Heldensaga*, heroic legend, as Heidegger will say in the 1950's, but like the *muthein*, which, in its indistinct difference from *legein* (the gathering, as language, of the "there is"), is alone capable of pronouncing divine

places and names. In any case, *Dichtung*, like *logos* in its Aristotelian-phenomenological definition, is apophantic: *dichten*, by way of the high-German *thîton* and the Latin *dictare*, is *deiknumi*: to show, to designate, to make appear. Every sign (*Zeichen*) is a showing (*Zeigen*), which is to say, a naming (*Nennen*), on whose basis alone there is being. These correlations are well known. That Heidegger could, with regard to language, refer to von Humboldt is not unimportant.[11]

4. Politics, in the sense in which Heidegger understood it without ever wanting to hear it spoken of (always those quotation marks . . .), was nonetheless organically linked in the 1930's to National Socialism, to fascism, with all the compromises that one knows or can easily imagine. Despite his denials, his half-confessions, or his lies, Heidegger was never really able to defend against this. If one can thus speak of a *politics* of Heidegger during these years—and I maintain that there is one, inscribed, explicitly or not, well beyond the "ordinary" compromise, in the texts themselves—then I would willingly say that this politics stems from what I will call (once more for lack of anything better) an arche-fascism. Such an arche-fascism has nothing to do with the hyper-fascism with which Breton, around 1934, prided himself on having reproached Bataille. On the other hand, it has everything to do with the terrible sentence, all the more authoritatively pronounced for its timidity, passed by Benjamin when, at a meeting of the College of Sociology where Bataille was speaking about the sacred, he whispered in Klossowski's ear: "In the end, you are working for the fascists."[12] The logic of the *arche*, as Benjamin was well aware, is absolutely formidable. And in this context it is clear that, in reality, Heidegger's discourse against "real fascism" had no other ambition than to set free the *truth* of fascism.

5. Bound up, under these conditions, with the theological-poetical, the theological-political assigns to poetry a mission, in the most banal sense that this word has taken on since the Romantics. This mission is, as we have seen, a *struggle*. Now, if such a struggle is, in Heidegger's eyes, an absolute necessity, it is because there is danger. When, in the Rome lecture of 1936, he condenses his 1934 course to the thread of five *Leitmotive* lifted from the Hölderlinian corpus, Heidegger retains a prose fragment, opposed to the declaration regarding the "innocence" of poetry, in which language (*die Sprache*) is said to be "the most dangerous of all goods" ("der

Güter Gefährlichstes"; *SA* 2; 1: 147). He proposes this commentary, which, for its part, needs no commentary:

But in what way is language the "most dangerous of all goods"? It is the danger of all dangers because language initially creates the possibility of a danger. Danger is a threat to Being by beings. But it is first by virtue of language as such that man is exposed to the manifest, which, as beings, attacks and bombards him in his Dasein, and, as non-beings, betrays and disappoints him in his Dasein. Language first creates the manifest realm of the threat and the errancy of being, and thus the possibility of the loss of Being, that is—danger.[13]

And a few lines further on he adds, still with respect to language:

In language the most pure and the most concealed, as well as the confused and the common, can come to be expressed [*kann . . . zu Wort kommen*]. Indeed, the essential word [*das wesentliche Wort*], which has to be understood and so become a common property of all, must make itself common. Accordingly, in another fragment Hölderlin writes: "You spoke to the divinity, but you have all forgotten that the first born are never for mortals, that they belong to the gods. More common, more everyday must the fruit become; then it becomes the possession of mortals."[14]

Heidegger glosses this as follows: "The pure and the common both form in the same way a said."[15] *Ein Gesagtes, eine Sage*: a myth.

One can easily deduce at least three things from this text. First, in the face of danger itself, danger in its essence (a threat that weighs on Being), an arche-ethical quality is assumed and required of poetry: namely, courage. This is, moreover, probably the only arche-ethical quality that one can detect in the Heideggerian discourse of the 1930's. What I here try to call the arche-ethical consists purely and simply in the experience of courage. Since, second, every "ontological" danger is also and necessarily a historial danger—and the historial is, if you like, the political without quotation marks—a responsibility is given to poetry, as to the thought which responds to it. One could call this responsibility transcendental insofar as upon it alone depends the possibility of a people's history. (This, one could show, in a certain way amounts to saying that courage lies in the decision.) The entire question in the commentary on Hölderlin is one of knowing whether or not the Germans are capable of entering into history and of opening a history, of becoming Germans, just as the Greeks, with the unprecedented courage to which tragedy attests, became Greeks. Finally, the obvious consequence of this is that the poet is defined as a *hero*, in the sense given to this word in section 74 of *Being and Time*, where Heidegger says that historial Dasein (the people) must choose its heroes in the tradition. In direct lineage from the Nietzschean interpretation of history and from ancient agonistics, the poet is more than a model, an example. Or, according to the terminology of the moderns, as present in Heidegger as in Nietzsche, he is a figure: a *Gestalt*. Heidegger, who will violently refuse any figure stemming from Nietzscheanism—from Zarathustra to Rilke's Angel and even, certainly much more belatedly, to Jünger's Worker—will nonetheless accept, and more than accept, the figure of the hero of poetry, of the demigod, the mediator or intercessor between the gods and men, immortals and mortals: the one who, says Hölderlin (or at least a certain Hölderlin), seizes the signs of the divine at the pinnacle of danger—under the immediate threat of being struck by lightning—in order to transmit them to the people in a veiled form. (The second part of the course of 1934, devoted to the hymn "Germanien," develops this motif at length. But it runs throughout Heidegger.)

I resigned myself to this reminder only to introduce my real topic for this paper. Here, then, is its outline, and it goes without saying that it is only an outline. As you will see in a moment, a much fuller analysis would be necessary in order to do justice to the question I would like to raise.

It so happens that exactly twenty years before the beginning of Heidegger's commentary, Hölderlin and, in Hölderlin, the same motif of courage, were the occasion for an opposite, so to speak, interpretative gesture. But one not entirely opposite, which is, I believe, of tremendous consequence.

I allude here to Benjamin's famous essay, written in the winter of 1914–15, entitled "Two Poems by Friedrich Hölderlin."[16] As you know, this text remained unpublished until 1955, when Adorno and Scholem issued the first collection of Benjamin's scattered essays. Obviously, Heidegger could not have known it. Yet my intention is not to "compare" the two interpretations. That would be of only very limited interest. In turning to this text, it is much more a matter of taking the measure of an epoch (from which we are far from being free) and of the philosophical questioning which underpins it: our politics, and not only our politics, still depend on it.

The text, to which a brief introduction is needed here, is a study of two versions of a single poem—which is, in reality, made up of three versions, though Benjamin deliberately neglects the intermediary one—the second (and last) of which, most certainly dating from after his trip to France, attests to the work of "rewriting"—or rather *internal translation*—to which Hölderlin relentlessly devoted his so-called "final years." The first version, you will not be surprised to learn, carries the title "Dichtermut," "The Poet's Courage," according to the old meaning of *Mut* in German. The second version is entitled "Blödigkeit," a difficult word which is usually translated as "timidity" or "awkwardness."[17] The motif of the poem is the same in both versions. It is the motif of the vocation or the mission of the poet, a vocation or mission which requires—let us hang onto the word—the arche-ethical quality of courage. We are already on the same ground upon which Heidegger will situate himself.

It is not uninteresting to note, moreover, that Benjamin's essay responds to what he too defines as a task: *eine Aufgabe*. The task, this time, is not that of a teaching. Rather, it is that of critique, or of what Benjamin a little pompously calls "the aesthetics of poetic art."[18] And this task is organized by the search in the poem for what Benjamin names with a word that Heidegger will systematically employ: *das Gedichtete*—a word which neither invent (it is attested in Goethe) and which, as a concept, signals in both Heidegger and Benjamin toward the essence (or the Idea) of poetry. (The word is supposed to be untranslatable because *dichten* is untranslatable. Working on a suggestion by Beda Allemann, Maurice de Gandillac bases his French translation on the Latin etymon *dictare* and suggests "dictamen," in what he calls the "old" sense [though Rousseau and several others still take the term in this way] of "that which dictates conscience." An ethical sense, then. Faced with this difficulty, most of Heidegger's English translators elect "poetized." That is perfectly acceptable, but "hesitation" reveals all the more forcefully, if we can hold these two suggestions together, the connection that unites the problematic of poetry and that of ethics.)[19]

The question—the question which dictates, if I can put it like this, the task of critique—is thus the following: how are we to gain *access* to the *Gedichtete*, dictamen, or poetized? This is a question of method, as Benjamin will not cease to say, Benjamin, who will never trouble to distinguish himself from the vocabulary of philosophy, just as Heidegger will obstinately make use of it, first of all against all post-Cartesian metaphysics, at least up to Hegel, by reducing *methodos* to *odos*: *Weg*, *Wegmarken*, *Holz-*

wege, *Unterwegs zur Sprache*, etc. Despite the similarity of the "approaches" in many regards, the problematic of access (which is a sort of breaking in) is not that of wayfaring. Benjamin speaks neither in terms of essence nor even—although he will do so a bit later—of origin. The language he uses is the Kantian or post-Kantian one of the a priori. The principle reference which underpins his long "methodological" introduction is a fragment from Novalis saying: "Every work of art has in itself an a priori ideal, a necessity for being."[20] If there is an aesthetics here (an "aesthetics of poetic art"), it is, in a somewhat unexpected sense, transcendental. The dictamen, which is above all not the cause of the poem or that which would allow it to be "interpreted," and which in itself is in no way "poetic" (no more so than is the essence of *Dichtung* in Heidegger), is purely and simply the *condition of possibility* of the poem. Its prerequisite, says Benjamin.

In order to establish this, Benjamin founds his argument on two concepts.

First, the concept, borrowed from Goethe, of "inner form" (*innere Form*) or "content" (*Gehalt*)—a word to which, as one knows, Benjamin will remain faithful throughout his critical enterprise, as is evidenced, for example, by the opposition between *Sachverhalt* and *Wahrheitsgehalt* in the great essay on the *Elective Affinities*, a much later jewel.[21] Here is how he introduces this concept, and one will see once more that it is again, and not by chance, a question of task:

> The inner form, what Goethe described as content, will be shown in these poems. The poetic task, as the prerequisite of an assessment of the poem, is to be ascertained. The assessment cannot follow how the poet has solved his task; rather, the seriousness and greatness of the task determine the assessment, for the task is derived from the poem. It is to be understood as the prerequisite of the poem, as the spiritual-intuitive structure of the world to which the poem testifies. This task, this prerequisite, will here be understood as the ultimate ground open to analysis. Nothing will be ascertained about the process of lyric production, nothing about the person or worldview of the creator, but about the particular and unique sphere in which the task and the prerequisite of the poem lie.[22]

The poem's prerequisite (its condition of possibility) is thus the task, singular each time, of the poem—that is to say, for it amounts to the same thing, that to which the poem, each time, testifies. One will see in a moment that such a testimony is always a testimony to truth or, to the extent that it is singular and always singular, the attestation of a truth. Exactly twenty years later, Heidegger will also speak of testimony: in the Rome lecture which comes out of the winter semester course—I will stay with this single example—he will note that, when Hölderlin speaks of language as the "most dangerous of all goods" (one which consequently threatens truth itself), he also defines it as the gift made to man in order that he can "testify [*zeugen*] to what he is" (*SA* 2; 1: 147). It is here of little moment that Heidegger modifies the proposition and moves from *what* (*was*) man is to *who* (*wer*) he is, according to the political-ontological aim with which we are familiar. What *is* essential is that, in Heidegger as in Benjamin, whatever the difference in their interpretation of truth, poetry is defined as "telling truth" or as "speaking in the name of truth." If you like, poetry is the *martyr* of truth. (No doubt this is why—but I will come back to this—as soon as poetry is thus understood, the destiny of poets actually becomes, through a sort of metonymic sliding, that of martyrs.)

Task, testimony: in their unity (the task *is* testimony) these two thus constitute the poem's prerequisite, or its a priori, what one might call its "authorization" (an authorization very much prior to the authority of the poet, and doubly so insofar as, authorizing the poem, it makes the poet possible). Nonetheless, before access to the complete definition of *dictamen* can be entirely cleared, a second concept is missing. This concept—we have already heard the word in passing—is "figure" (*Gestalt*), which brings with it, in an absolutely necessary way, the concept of "myth." Here is how Benjamin introduces it: challenging all examination of the creative process and the creating subject, he devotes critique to the sole consideration of "the particular and unique sphere in which the task and the prerequisite of the poem lie"; and he adds:

> This sphere is at once the product [*Erzeugnis*] and object of the investigation. It can itself no longer be compared with the poem, but is the sole thing in the analysis that can be established. This sphere, which for every poetry has a special configuration, is described as the dictamen. In this sphere the characteristic realm containing the truth of the poem will be developed. This "truth," which the most serious artists so insistently claim for their creations, will be understood as the objectivity of their production, as the fulfillment of the artistic task in each case. . . . In its general form, the dictamen is the synthetic unity of the spiritual and intuitive orders. This unity gains its particular figure as the inner form of the particular creation.[23]

The dictamen is thus *Gestalt*, figure (I stick to this translation for reasons of economy). More precisely, for each poem the figure is the mode of the presentation *and* articulation of its inner form or its content. But why is it a matter of a figure? Why this word—and this concept—of *Gestalt*?

There are, it seems to me, two reasons for this.

The first is the one just advanced by Benjamin: in general, he says, the dictamen is the "synthetic unity of the spiritual and intuitive orders." This vocabulary is clearly derived from Kant's, along the lines of the derivations carried out by Jena Romanticism ("spirit" for "understanding," for example). Understood correctly, this proposition simply amounts to saying that the dictamen is the *transcendental schema* of the poem. The *Dichten*, the "poetizing," is referred to the transcendental imagination, that "art concealed in the depths of the human soul."[24] Or else, and this is a category I willingly use in recalling the opening Heidegger attempts to draw between *Kant and the Problem of Metaphysics* and the lecture series "The Origin of the Work of Art," to an originary *techne* which founds a world because it is the archaic or principial configuration (*Gestaltung*) of the world. In Benjamin's language: the dictamen is a "limit-concept," a limit between the poem itself and that to which it testifies, that to which it attests in its truth ("life", says Benjamin), being the condition of both, of one *as* the other. It guarantees the "transition" between two "functional unities," that of the poem (prior to the distinction between form and matter) and that of life (prior to the distinction between task and solution). It is simplest here to quote:

Through this relation to the intuitive and spiritual functional unity of the poem, the dictamen shows itself as a limit determination as regards the poem. At the same time, it is but a limit concept as regards another functional unity, since a limit concept is not possible except as the limit between two concepts. The other functional unity is now the idea of the task, corresponding to the idea of the solution which the poem is. (Task and solution are separable only *in abstracto*.) For the creator, this idea of the task is always life. In it lies the other extreme functional unity. The dictamen thus turns out to be the transition from the functional unity of life to that of the poem. In it, life determines itself through the poem, the task through the solution. The ground lies not in the individual life mood of the artist, but in a life correlation determined by art.[25]

The dictamen, as *Gestalt*, is a figure of existence. Or, what amounts to the same thing, life (existence)—insofar as the task of the poem is to testify to it—is itself, in its truth, poetic. A poem can say, in truth, that we live (exist) in truth. Which is to say poetically. ("Full of worth, but poetically, man dwells on this earth"; "Voll Verdienst, doch dichterisch, wohnet der Mensch auf dieser Erde"; *SA* 2; 1: 372), says Hölderlin in a line upon which Heidegger will later comment at length.)[26] Poetry is our destiny. Our dictamen, in fact.

Whence the second reason, I think, that Benjamin advances in order to justify the word—the concept—of *Gestalt* (a highly "charged" word—concept—at the time he uses it, from a philosophical point of view [less Hegelianism than Nietzscheanism] as well as from a political one: it will be one of the key words of the "conservative revolution" then beginning to dawn.) Benjamin remains cautious. He is speaking, he says, only in "general" terms.[27] Be that as it may: *Gestalt* refers to myth; or, if you prefer, every figure (but we could just as much say every model, or every example) is potentially mythic, not in an abstract or remote sense of myth, but in the sense that life (existence) itself is—I borrow the formulation of 1936 from Thomas Mann—"life in myth," which is to say, "quotation."[28] Here is what Benjamin says:

> The categories in which this sphere, the transitional sphere of both functional unities, is ascertainable are not yet preformed [*vorgebildet*], and perhaps their closest correlation is to the concept of myth. Precisely the weakest results of art refer to themselves through the immediate feeling of life, whereas the strongest, with respect to their truth, do so through a sphere akin to the mythic: the dictamen. Life is in general the dictamen of poems [*das Gedichtete der Gedichte*].[29]

The mythic, thus understood, is above all not the mythological, that is to say the stereotyped organization (or correlation) of mythemes, the weakening of their reciprocal tensions. The mythic—which *is* the dictamen or which the dictamen *is*—is existence itself in its configuration or its figurability. Which is why a poem is ultimately a gesture of existence—a gesture with a view to existence. The poem, let us respect Benjamin's vocabulary, is a figure of life. Which amounts to saying that life is poetic. Essentially. Not because it is "poetizable" (this would be mystification par excellence: mythology), but because the dictamen of poems, which is never this or that particular poem from which, nonetheless, it is indissociable, *dictates* life.

Here, the dictamen is *courage*. This is perhaps the dictamen of every dictation, the poematic of every poem. Of literature in general. Benjamin, in any case, thought that:

> The law according to which all apparent elements of sensibility and of Ideas show themselves as the epitomes of essential, in principle infinite functions, will be

called the law of identity. It describes the synthetic unity of functions. In each of its particular figures it is recognized as an a priori of the poem. . . . Until the applicability of this method to the aesthetics of the lyric as such, and perhaps of further realms, has been examined, additional exposition has to be ruled out. Only once this has taken place can the a priori of the individual poem, or that of the poem in general, or even that of other poetic types or of poetry in general, clearly come to light.[30]

Here, then, the dictamen is—let us for the moment hang onto this cautious localization—courage: "The Poet's Courage."

It is not my intention to reconstitute Benjamin's demonstration. I will simply and somewhat dryly give its principle. This principle is relatively simple, while the demonstration is, on the contrary, very complex.

From the outset, remarks Benjamin, "The Poet's Courage" is a mythological *topos*. And it is treated as such by Hölderlin in the first version of the poem. It is the *topos* of the properly heroic poet, the intercessor or mediator between the gods and men (other more or less contemporaneous poems speak of the poet as a demigod, precisely the motif which will hold Heidegger's attention), who consequently braves the greatest risks in order to complete his task or mission. In Benjamin's judgment, since critique is a matter of judgment, the simple repetition of the *topos* alone explains the weakness of the first version of the poem: neither does the relation of man to the divine (or of the poet to the god that he invokes as his model) escape convention, nor is the essence of fate (that is to say, the inscription of death in existence) truly grasped. The poem is powerless to give figure to a world:

Hölderlin's object in the first version of his poem is a fate: the death of the poet. He sings the praises of the sources of the courage to die this death. This death is the center from which the world of poetic dying must arise. Existence in this world would be the courage of the poet. But here only the most vigilant suspicion can feel a glimmer of this structure of laws from a world of the poet. Timidly at first the voice rises to sing a cosmos, whose own decline is signified by the death of the poet. But the myth is formed out of mythology. The sun god is the ancestor of the poet, and his death is the fate through which the death of the poet, at first only mirrored, becomes real. . . . The poem lives in the Greek world, animated by a beauty in line with the Greek and governed by the mythology of the Greeks. The particular principle of Greek figuration is not, however, fully unfolded.

Denn, seitdem der Gesang sterblichen Lippen sich
  Friedenatmend entwand, frommend in Leid und Glück

Unsre Weise der Menschen
  Herz erfreute . . .

For ever since the song from mortal lips escaped
  Breathing peace, availed of sorrow and happiness,
    Our melody has filled with joy
      The hearts of men . . .

These words only feebly contain the reverence that filled Pindar—and the late Hölderlin with him—before the figure of the poetic. Nor, considered thus, do the "bards of the people," "pleasing to all," serve to lay an intuitive world ground for this poem. The figure of the dying sun god testifies most clearly to an uncontrolled duality in all its elements. Idyllic nature still plays its particular role against the figure of the god. Beauty, in other words, has not yet become a completed figure.[31]

In the second version, on the other hand, the mythologeme of the setting sun, of the decline, of the twilight, etc., is abandoned. And with it, that of the hero mediator and the bards of the people. Benjamin's intuition here is that, by undoing the mythological, the second version paradoxically reinforces the mythical through which it seeks, as we have seen, to think the dictamen. The dictamen appears in its truth—the essence of courage attained—only with the collapse of the mythological, only at the precise point of its failure. Here, Benjamin speaks of the "deposition [*Umsetzung*] of the mythological."[32] Much later, still concerning Hölderlin and now directed against the Heideggerian commentary, Adorno will speak of "demythologization."[33] Obviously, such a failure of the mythological does not happen without a certain failure of the theological. In any case, not without a certain dissociation of the theological-political. I will come back to this.

The deposition of the mythological is a deliberate gesture: it moves toward a conquest of objectivity and of the concrete—that is to say, technically, if you like, toward a sort of literalization, toward a prosaicness, toward a phrasing which would conceal itself from the phrasing of eloquence and of antique *pathos*. Toward the abandonment of the stereotypes of sacralization. Let me give a single example, that of the revision of the first two lines—upon which, moreover, Benjamin is relatively uninsistent. The first version says:

Sind denn dir nicht verwandt alle Lebendigen?
  Nährt zum Dienste denn nicht selber die Parze dich?
(*SA* 2; 1: 62)

Are not all the living akin to you?
  Does not the Parca herself nourish you to serve her ends?

This is indeed the traditional *topos* of death: the Parca, absolute servitude before death, the community of all mortals ("all the living"). In the second version, however, one reads, in an unchanged rhythm:

Sind denn dir nicht bekannt viele Lebendigen?
  Geht auf Wahrem dein Fuß nicht, wie auf Teppichen?
(*SA* 2; 1: 66)

Are not many of the living known to you?
  Does not your foot stride upon the true, as upon carpets?

"Many of the living" is substituted for "all the living," "known" for "akin": this is a first step toward objectification, an initial renunciation of the abstraction of "mortals," an initial determination of the "living" (those who are "known"). Which immediately leads, in the second line, to the simultaneous abandonment of the nominal reference to the Parca and to the literalization of the mythologeme: from the threads of the "many" fates is woven something like a carpet of the dead, which is the truth of being mortal. Or of death not in its truth, but as truth itself. And it is because the second version attains this truth that the dictamen of courage can suddenly appear in all its force. At the end of his demonstration, Benjamin has no difficulty stating this new determination of courage—the most accurate there is, so long as we undertake to read Hölderlin through to the end:

The dictamen of the first version knows courage only really as a quality. Man and death stand opposed both motionless, having no intuitable world in common. Certainly, the discovery of a deep relation to death in the poet, in his divine-natural existence, had already been attempted, but only indirectly through the mediation of the god to whom death mythologically belonged and to whom the poet—again mythologically—drew closer. . . . The duality of man and death can only be based on a casual feeling of life. This duality ceases to exist as soon as the dictamen pulls itself into a more profound connection and shapes [*gestaltete*] for itself a spiritual principle—courage—from life. Courage is self-abandon to the danger that threatens the world. In it lies concealed a particular paradox, which for the first time allows the structure of the dictamen of both versions to be fully understood: danger exists for the courageous, and yet he does not respect it. He would be cowardly if he respected it, and if it did not exist for him he would not be courageous. This odd relation is resolved in that it is not the courageous himself that is threatened by danger, but the world. Courage is the life feeling of the man who abandons himself to danger, so that in his death he broadens it into a danger for the world and at the same time overcomes it.[34]

And a bit further on:

The world of the dead hero is a new mythic one, steeped in danger: this is the world of the second version of the poem. In it a spiritual principle has become thoroughly sovereign: the becoming-one of the heroic poet with the world. The poet has no need to fear death; he is a hero because he lives in the center of all relations. The principle of the dictamen as such is the supreme sovereignty of relation.[35]

The spiritual principle (the word is repeated twice), the sovereignty of relation: the dictamen of courage attests to the congruence—or the "fit": "But we ourselves / Bring fitting hands" ("Doch selber / Bringen schickliche Hände wir"), reads the end of the second version (*SA* 2; 1: 66)—between (poetic) heroism and the danger of the world. This is the point where the mythological, even the theological, fails, if both consist in the separation of "elements" and the refusal of "relation." Commenting *in fine* on the third stanza of the last version in order to note the "sublation of the order of mortals and the heavenly ones,"[36] Benjamin insists:

It is to be supposed that the words "a lonely deer" describes man, and this accords very well with the title of this poem. "Timidity" has now become the authentic stance of the poet. Transposed into the centre of life, nothing remains for him but the motionless existence, the complete passivity that is the essence of the courageous; nothing but to give himself over entirely to relation.[37]

As to the failing of the mythological (of the theological): I think one could speak even more precisely of a "deconstruction" in the sense that Heidegger had begun to give this term, though he—unlike Benjamin, who did not have the concept at his disposal—failed to see the extent to which it is at work in Hölderlin's last poems. That is to say, the extent to which it unworks them. In no uncertain manner, Benjamin finds the reason for this "deconstruction" in the principle of *sobriety* as it is developed by Hölderlin in the "Remarks" which accompany the translations of Sophocles, namely, as the guiding principle of modern poetry. As that which dictates, if you like, the task proper to it. At the end of his essay, he writes this, whose political—and more than political—import you will appreciate:

In the course of this investigation, the word *sobriety*, which might often have served as a characteristic, has deliberately been avoided. Only now must Hölderlin's phrase "sacredly sober" be named, its understanding having now been determined. It has been remarked that these words contain the tendency of his later creations. They arise from the inner certainty with which these creations stand in his own spiritual life, in which sobriety is now allowed, is commanded, because it is sacred in itself, standing beyond all elevation to the sublime. Is this life still that of the Greek? This is no more the case than that the life of any pure work of art as such could be that of a people, no more too than what we find in the dictamen could be the life of an individual and nothing else.[38]

And in order that things be entirely clear, I will cite this—the very last passage:

The contemplation of the dictamen, however, leads not to myth but—in the greatest creations—only to mythic attachments, which in the work of art are formed into a unique unmythological and unmythic figure that cannot be better conceived by us.

If there was a word with which to grasp the relation between that inner life, from which the later version of the poem arose, and myth, it would be Hölderlin's, from a period even later than that of this poem, "The legends which grow distant from the earth /. . ./ Return to mankind."[39]

Put differently, there is no way of hanging anything theological-political on this failing theological-poetical. The poem or the hymn has no historial mission. Sobriety, we know, is what Benjamin, a few years later in his thesis on Jena Romanticism, will identify with prose. He will say, in speculative terms borrowed from Fichte: "The Idea of poetry is prose."[40] And for this very reason he will make Hölderlin the secret—decentered—center of Romanticism. One could thus risk saying, somewhat tersely: sobriety is the courage of poetry. Or else: the courage of poetry is prose. Which does not, of course, exclude versification.

Poetry's courage: this can be understood or accentuated in two different ways.

Either it is a matter of a subjective genitive: in which case courage is the courage of poetry for itself, the courage of poetry for poetry (let us call this its stubbornness), and, in the discovery of the poematic or of the dictamen, it is a matter of something like the arche-ethical of poetizing as such. The poem is the arche-ethical act. We are here on the level of what the Romantics called "reflection," and even, up to a certain point, on the—Heideggerian—level of the "poetry of poetry." The regime is that of a pure intransitiveness. According to Benjamin's reading, poetry's courage means: the courage to leave the mythological, to break with it and to deconstruct it. It is the courage to invent poetry, to configure the poem as the testimony which it is. Thus the destiny of poetry, after what Jean-Christophe Bailly has called the "end of the Hymn,"[41] would in fact be prose as the "true-saying" of the poem about the poem.

Or else it is a matter of an objective genitive: courage is the courage poetry must have in its transitive (prophetic or angelic) function, through which it would confront a danger of the world and announce a task to be completed. The ethical act would then be less the poem itself than what the poem dictates as task. And perhaps one would no longer be on the level of the arche-ethical if, unlike the ethical, the arche-ethical is an ethics that does not know what the good is (I propose this distinction in the wake of Lacan's discourse in *The Ethics of Psychoanalysis*,[42] but also in the wake of that conducted by Heidegger in the "Letter on 'Humanism' "). Or if, one could put it thus, the obligation of the arche-ethical—perhaps impossible to meet, but that is its incommensurable responsibility—is to tear itself away from the (mimetic) ethics of the example, the most ancient and ruinous ethics we know: namely, the cult of the hero, the metonymic sliding from the *martyrdom* of poetry to the poet as *martyr*. Recall: (1) Hölderlin is the poet of the poet and of poetry; (2) Hölderlin is the poet of the poet insofar as he is the poet of the Germans. From one proposition to the other, "the poet of poetry" has been forgotten. A whole politics is decided there.

One must not jump to the conclusion that this difference in inflection or accentuation traces the dividing line between Benjamin and Heidegger, even if, on the ideological and political (which is also to say philosophical) register things could not be more clear nor the difference more cut-and-dried. In reality, in both Benjamin and Heidegger—although assuredly not in the same way—intransitiveness and transitiveness ceaselessly trespass upon one another. It is a matter, each time, of that to which poetry testifies in attesting as such, that is to say, in attesting in its relation to the true, in its *telling truth*. Is this the—modern—vocation of martyrdom? Courage itself? Yes, but only according to its failure. Yes, on the condition that we finally accept what is being testified to, the "lack of God," as Hölderlin put it, or—what amounts to the same thing—our *a-theistic* condition.

Let this be understood as homage to Benjamin. The most humble and grateful possible.

Whenever philosophy addresses itself to poetry, it takes on a responsibility—and, moreover, it demands that it do so: it responds, it says, to the responsibility with which poetry believes itself to be authorized.

Heidegger, a notable example, responds to Hölderlin in order to respond from him. It is a matter, he thinks, of courage as such, which is the courage of History.

This project, as I will show elsewhere, is theological-political. Formidable in that it reinforces and tries to *verify* fascism. I would want to cut in here, in a way which would no longer simply be that of an opposition, with a theological-poetical project. Such a project is Benjamin's: he would have the immense merit of recognizing the theological in the figure of its failure and of turning poetry—whence we came—toward prose, where we are.

Translated by Simon Sparks

'Winke'

DIVINE TOPOI IN HÖLDERLIN, HEIDEGGER, NANCY

Hent de Vries

. . . und Winke sind
Von Alters her die Sprache der Götter . . .

That Hölderlin's poetry revolves around certain notions of the divine, of gods, and the holy is almost a commonplace, though it does so in a manner very different from the discourses on the coming god familiar from the Romantics through the later Heidegger.[1] How, exactly, Hölderlin's reference to the divine is made is less clear, including what such reference means for our understanding of both Hölderlin and the traditions of interpreting the divine on which he reflects, from which he departs, or which he, in turn, has called forth. The notion of the divine is often introduced in Hölderlin in a language that invokes topological images, theophanic heavens, and semigodly rivers, each of which seem to mark a specific manifestation or presence that is at once ineluctable and elusive, inscribed on the face of the Earth and the sky but also "immediate" in an utterly undialectical and nonidealistic sense of this word. This immediacy does not let itself be reduced to that of mere empirical reference, biographical detail, or material inscription. Such refusal is obvious in that it finds no better expression than in the figure of the "nod," the signal or gesture of the *Winke.* Hölderlin says that this figure has from early on (*Vom alters her*) been the language of the gods (*die Sprache der Götter*). This phrase and the constellation it evokes cast surprising light on the problem of divine names—or the lack thereof—that pervades the history of Western philosophy, apophatic theology, and mysticism. Hölderlin's poetry, in ways all too often overlooked by many of its most insightful readers, in the first place Heidegger, liberates the question "What is God?" (*quid est deus*) from the dilemmas of linguistic ineffability, of the sayable and the unsayable, and reorients our thought—our gaze, to be precise—in more than one direction at once. In what follows, I will attempt to substantiate this hypothesis by establishing an interpretative context for a few relevant lines, taken from "Was ist Gott?" ("What is God?") (*SA* 2; 1: 210) and "In lieblicher Bläue blühet" ("In Lovely Blueness . . . blossoms") (*SA* 2; 2: 991–92), relatively short texts that have drawn considerable critical interest.

One of the most recent and most challenging interpretations of the figure of the *Winke* in relation to the question of the divine and of topology is Jean-Luc Nancy's *Des lieux divins*, "Of Divine Places," first published in a *Festschrift* entitled *Qu'est-ce que Dieu?* (*What is God?*).[2] Nancy engages the question of whether—and, if so, where—there is still a place for the divine, for its experience and its "presence." This question opens an inquiry into the concepts, the names, the tropes, or, more precisely, the topoi of the divine, one quite different from most other discourses about difference and the "Other," which often reduce this problem to that of linguistic insufficiency or ineffability. These apophatic discourses—ranging from Pseudo-Dionysius, to Eckhart and Angelus Silesius, to the atomistic premises which lead to the conclusion of Wittgenstein's *Tractatus Logico-Philosophicus* and beyond—presuppose another, more challenging suspension, retreat, or dispersion, one that Nancy pursues across his essay's fractured topology. In fact, "Of Divine Places" is nothing less than an attempt to elucidate and to situate the historial and epochal displacement which resembles a relentless *Dei paralysis progressiva*.[3]

It has often been claimed that in the modern epoch the question and the naming of the divine takes the form of an ongoing reduction, a continuous process of erosion, a *reductio perennis*. This process, Nancy writes, has irrevocably diffused the divine itself, well beyond the scattering of the deity and the gods as separate and significant—immortal, eminent, or infinite—entities:

"God," the motif or theme of God, the question of God, no longer means anything to us. Or else—as is all too obvious to an unbiased eye—what the theme of God might mean to us has already moved or been carried entirely outside of him. Is there any statement about the divine that can henceforth be distinguished, strictly speaking, from another about "the subject" (or its "absence"), "desire," "history," "the other" [*autrui*], "the Other," "being," "speech," "the sublime," "community," and so on and so forth? It is as if "God" were in fragments, an Osiris dismembered throughout all of our discourse (indeed there are those who will now continue to speak of the divine in terms of explosion, dispersal, suspension, etc.). As if the divine, God, or the gods formed the common name or place—common and as such erasable, insignificant—of every question, every exigency of thought: wherever thought comes up against the furthest extreme, the limit, against truth, or ordeal, in short wherever it thinks, it encounters something that once bore or seems to have borne, at one time or another, a divine name.[4]

Distanced from everything else, from beings and Being—even, as it were, from "itself"—the religious and theological notion of "God" loses all specificity and thereby all relevance:

Far from being rediscovered, God disappears even more surely and definitively through bearing all the names of a generalized and multiplied difference. Monotheism dissolves into polyatheism, and it is no good asserting that this polyatheism is the true word and the true presence of God in his distance from the supreme Being of metaphysics. For the infinitely absent god . . . should no longer be termed "God," nor be presented in any way as "God" or as divine. . . . There is no theology that does not turn out here to be either ontological or anthropological—saying *nothing about the god* that cannot immediately be said about "the event," about "love," about "poetry."[5]

On the one hand, this implies that the language of "modernity" subtracts the privilege of speaking of "the Other, the Infinitely-other, the Other-infinite" from the jurisdiction of so-called theology: the proper negativity of our present and future thoughts seems to enable us to address the "*absconditum*"[6] that was formerly reserved for esoteric discourse alone. And yet, on the other hand, this secularization or, rather, becoming-exoteric of the absolute places thought in a paradoxical situation. Nancy describes this aporia in no uncertain terms:

In baptizing our abysses with the name of God, we are guilty of at least two errors or incoherencies: we fill in the abysses by attributing a bottom to them, and we blaspheme . . . the name of God by making it the name of some*thing*. On the other hand, the most subtle—and most theological—error would doubtless consist in *believing* that the infinite cannot provide a bottom and that naming a person [for example, "God"] is not naming a sort of "thing."[7]

Not only does this aporia belong to the very nature of theological discourse, but its slippage and, indeed, blasphemy are as old as Western thought. Here, we encounter the unavoidable and, in a sense, transcendental illusion of all discourse, whether written or oral. The word and the notion "god" is always accompanied by a dual temptation: the seduction to baptize all experiences of the limit of our world with a divine name, and the desire to protect this name and the obscurity of our experience from idolatry and superstition. This double yet incoherent allegiance to two types of naming, of ascription and of *apophasis*, cannot be resolved as long as one continues to move within the horizon of reference, that is, of positive and negative attribution alone. If "god," as Nancy insists, is neither "a manner of speaking" nor some "ultimate truth of mankind,"[8] then the uses and abuses of this name—which is not a concept, a term, or a figure, properly speaking—are no longer governed by any semantics or metaphorics.

Given the inconsistencies of any discourse *on* or *about* God (*de Deo*), whether ontotheological or not, whether coming *from* God or not, it clearly no longer suffices to ask: "What is God, if He is?" The question *quid est deus?* could very well

turn out to be the surest means of falling short of the question (if indeed it is a question, if it still hides or still reveals a true question), for God has perhaps become everything (or nothing); perhaps he has become, potentially at least, every true question, exigency, or furthest extreme of thought.[9]

Instead, "Of Divine Places" emphasizes the question of *place* and *space*, with a remarkable turn of thought and an intensity that at once seems close to Heidegger's thought of the site (of the *Ort*, the *Erörterung*, as well as the *Situation* of which *Being and Time* speaks) and moves far beyond the confines of the thinking of (*Andenken*) as well as the prerogative of (the one) Being. Writing of divine places, Nancy interrogates the being-one, the being-indivisible, of the place and the determination of whatever place or space as a site of gathering Being and beings, divinities and mortals, heavens (the sky) and Earth. Thus, rather than asking whether or what "god" *is* or *names*, we should, Nancy claims, raise the more central question:

> *If there is a place* for god, if there is still room [*place*] for him: that is, a place where he does not become indistinguishable from something else, and where it is consequently still worth calling him by the name of *God* . . . could we then in fact be dealing with a question of place, of distinct location [*lieu*] and not with a question of being?[10]

The very title "Of Divine Places," thus stands for a fragmentary and partial, indeed *local* and *topical*, exploration

> of the gods and their places; of the places they have abandoned and of those where they hide; of gods without hearth or home, of nomadic gods; of the *here* where the gods are *also*; of the common places of God; of the gods common to all places, to some places, to no place; of God: in what way he is topos; topics and atopics of the divine; of gods and places: treatise on divine paronomasia [i.e., on the juxtaposition of divine names which all sound alike, but give the divine a different meaning]; where is God to be found? in what place?[11]

For all the subtlety of his rephrasing of a traditional question, Nancy retains, if not the question *quid est*, then at least the form of the question and of questioning as such. Here, the question "Where, if anywhere?" (or "Is there any place, somewhere, where . . . ?") is substituted for—takes the place of—the ontological interrogation "What is X in its very essence and existence?" Remarkably, Nancy refrains from putting into question the principle of identity in this context. The very exclusion of a third possibility beyond the alternative *X* or not *X* (*tertium non datur*) regulates the basic assumptions on which "Of Divine Places" is based. For the notion "god" to be meaningful and not a mere *flatus vocis*, an empty word or sound, it should, Nancy maintains, have some *differentia specifica*, some distinctive feature. This is not to say that the word "god" should have a stable semantic reference (it has none), nor that one ought to be able first to answer the question of God's existence and essence by proving that and how the substance called "divine" must of necessity have all perfect properties and can therefore, by implication, be said to exist eminently. Clinging to the principle of identity—that the notion and word "god" must refer to some "presence" *here and now* in order to make sense at all, that "god" is either "god" or "not-god," without there being any third possibility—Nancy rephrases the traditional question of whether "god" exists (*an sit*) and, if so, what He is in essence (*quod sit*) by introducing the more topical and timely question of *where* He is, not in fact but, again, in essence (*quo sit*). To ask where and *when* his essence can be said to exist comes down, Nancy asserts, to posing the question of "how he withdraws from existence, how he is not where we expect him to be."[12]

In the same vein, Nancy reiterates in his essay "The Calculation of the Poet" that: "The god is nothing but the place [*Le dieu n'est que le lieu*]; the place is the place of departure and of the return, of the coming that withdraws itself and that thereby makes sense."[13] Far from cultivating a romantic imagery, far from nostalgic and irresponsible desire for lost or coming gods, to read Hölderlin *here and now* means paying attention to the metric precision with which in his writing the without-measure is captured. Nancy formulates it almost apodictically: "Poetry: material calculation of the atheistic passage."[14] One might wonder, of course, why, if the name *God* can no longer be taken to refer to the highest or the supreme Being, the first cause of all as well as of Himself—*causa prima* and *causa sui*—this notion and the very "presence" that it evokes should be distinguishable from something else at all. Why should there be a specific difference where, precisely, a certain indelible sublime indifference vis-à-vis finite differentiations is at stake? How, in other words, could one not sacrifice the specificity of God, even or precisely while one claims to speak *to* Him, *about* Him, let alone *away from* Him and *without* Him? Is not this what the *à dieu*, the invocation or denial of an absolute witness as the condition of all claims to truth, to truthfulness—in short, of all utterances made in good *and* bad faith—expresses economically and most provocatively? Why should presence at or in a certain locus or locution provide the divine with a distinctive feature or a discriminatory marker? Would not a confusion and transgression that is always possible—an idolatry or blasphemy—constitute "God's" or "the god's" unpredictable occurrence, if and wherever He (or it) occurs? Is not every epiphany characterized by a transposition which in its very *transcendere*—in its movement upward *and* downward, in its *transascendance* and *transdescendance*, to cite Levinas, who in the opening pages of *Totalité et Infinité* cites Jean Wahl's *Traité de Métaphysique*—retains something (of the) undecidable? It is on both sides of the same line at once, as if this line, which is supposed to set apart, not only the finite and the infinite, but also two different forms or experiences of the infinite, were in itself infinitely divided. That is to say, it is divided, neither in any determinable measure nor randomly, but by being haunted at each given point by a necessary possibility of the intervention of chance.

Instead, in "Of Divine Places" Nancy recalls the Hölderlinean motif of the lack of holy or divine names, here and now, at this place, where we are, and opposes it to the at bottom metaphysical preoccupation with the inexpressible, the ineffable, the unsayable beyond any ontic or ontological determination, *epekeina tes ousias*, beyond essence. In our days, Nancy writes, "God" is not

> unnameable in the metaphysical sense of that being that is inaccessible to all names . . . including the name of being itself, according to an unbroken tradition that is the very tradition of ontotheology. God is not unnameable in that sense, because in that sense unnameability is the result of an overflowing of names and language, whereas the unnameability of the god to whom I address myself (if I can) results from *the lack of a name*. . . . There is no impotence on the part of names in general to express or refer to God (just as, conversely, the unnameable is neither necessarily nor exclusively divine. . . . In fact it could well be that the "unnameable" is never divine, and that the divine is always named—even if it is *for want of a name*.[15]

Nancy alludes to the last strophe of Hölderlin's elegy "Heimkunft" ("Homecoming")—"es fehlen heilige Namen," "holy names are lacking" or "there is a lack of holy names"—a classic topos for the retreat and want of the divine name. Heidegger also traverses this topos in his exegeses or elucidations (Erläuterungen) of Hölderlin's poetry, or rather—more emphatically—*Dichtung*. Neither Nancy nor Heidegger reads this lack against the background of the tradition of negative theology. The apophatic way, Nancy suggests somewhat surprisingly, was always more obsessed with the difficulty of using *concepts* rather than the singular designators called *names* when speaking of God, of His hyperessentiality, or, more precisely, His abode. He holds that the lack of which (or from which) Hölderlin speaks in "Heimkunft" cannot be interpreted as a merely empirical—historical, cultural, biographical, or even linguistic—absence, as if the names were lacking in the sense of not or no longer being ready at hand (or *vorhanden*).

> No doubts are cast, in Hölderlin, on the possibility of divine names. On the contrary, the assertion of a lack of sacred names implies that we know what such names are—names, as Heidegger's commentary puts it, "which are commensurate with the sacred (or the holy) and which themselves cast light upon it." These names are thus not only proper to the divine, they bring it to light, they make it known as the divine that it is. These names are the manifestation of the divine, they are thus perhaps not far from being the divine itself. It is simply . . . that these names, here and now, are lacking.[16]

I will argue that this lack is not a loss or negativity that would make itself available to a dialectics and thus—Nancy assumes, following a long tradition of engagement with Hegelianism—to metaphysical substantialization or ontologization. Hölderlin's thought here—as it lets itself be read through the prism of Nancy's *theotopography*—is not so much one of nostalgic, classicist, or romantic mourning, but a mourning that is impossible and, in that sense, a paradoxical affirmation. In order to demonstrate this claim, let me first sketch the horizon from (within) which Nancy departs.

In "Der Fehl Heiliger Namen" ("The Lack of Holy Names"), published in the 1974 *Denkerfahrungen*, Heidegger illustrates the Hölderlinean notion of the *Fehl*, or want, in heavily charged sentences that—at least typographically, in their very verse form—are reminiscent of poetic prose, of a writing that neither is philosophy strictly speaking nor, in its aspiration to further *thought*, claims to be *Dichtung* in the limited, let alone the Hölderlinean sense. If *Dichten* and *Denken* (and, let us not forget, thanking, *Danken*) go hand in hand, they nonetheless belong to distinguishable spheres and follow different regimens. According to Heidegger, a situation of need and distress, namely, the persistent deferral of the divine, solicits the poet's utterance: "What urges the poet into [his] Saying, is a distress. It hides itself in the deferral of the presence [or presencing] of the divine" ("Was den Dichter in das Sagen nötigt, ist eine Not. Sie verbirgt sich im Ausbleiben des Anwesens des Göttlichen."). This need insinuates itself throughout Hölderlin's mature and late poetry and is expressed in his at once "simple," "illuminating," and "mysterious" word ("in das einfache, alles klärende und gleichwohl geheimnisvolle Wort"). Heidegger holds this to be how "holy names are lacking." He hastens to point out, however, that only those who are granted an insight (*Einblick*) into the "experience" of the "origin" (*Herkunft*) of this lack can understand its singular meaning (*das Eigentümliche*). That origin lies concealed in a "reserve" of the "Holy" (*in einem Vorenthalt des Heiligen*). In Hölderlin's universe, Heidegger surmises, the "reserve" of the Holy and ultimate meaning of Being withholds itself and thus (but why and how, precisely?) holds us mortals back from access to the holy names or (which is hardly the same) from appropriately naming this divine.[17] For Heidegger, to make or write poetry—to poeticize, that is, *Dichten*—means to let oneself be addressed by the pure call of the presencing of Being, which gives itself only *in* and *as* its retreat and thus never without reservation:

To poeticize—this means here: to let oneself be said the pure call of presence [or presencing] as such, and be it only and first of all a presence [or presencing] of withdrawal and reservation. [Dichten—meint hier: sich sagen lassen den reinen Anruf des Anwesens als solchen, und sei dieses auch nur und gerade ein Anwesen des Entzuges und des Vorenthaltes.] [18]

In so doing, poetry is at odds with the primacy of "method" in the technological age. *Dichtung* forges a path, leads the way forever underway toward a "clearing." It proceeds, as it were, without "procedure" (*Verfahren*), without "proof" and without "mediation" (*Vermitteln*).[19] Only the thought which follows this path, which itself has the character rather than merely the form of the path (*in sich Wegcharakter hat*), can hope to prepare the "experience of the lack" (the *Fehl*) and thus help the poet—the sole figure who can and must put this lack into words through his *saying* (sagen)—to understand. Not that it helps him make the distress and the necessity—the *Not*—of the lack intelligible or understandable. Instead, it helps him to exist in or, rather, to *stand out* and *stand through* the *Not*. This happens in an original and originary sense: the poet's *ausstehen* of *die Not* concerns first and foremost the earliest and primal *Not*, that of the forgetfulness (*Seinsvergessenheit*) and the *self*-concealment (*Sichverbergen* or *Lethe*) of Being.[20] This self-concealment is the proper feature of Being's presence or essence (its essencing or *Anwesen*) and the "source" or "domain" from which the present *Not*, notably the *Not des Fehls "heiliger Namen"* stems. Originating in the forgetfulness of Being, the need for holy names *eventually* might also leap away from this source. The understanding of thought that could help the *Dichten* does not imply, then, making this *Not* "understandable" but persevering in it.

The forgetting in question, Heidegger adds, should not be identified with an all-too-familiar privative sense, as if we were dealing with a lack in the sense of *Mangel*, of something that should not have been, a fault that could have been prevented, a mistake, an omission, an *Unterlassung*. On the contrary, Heidegger continues, the word *Seinsvergessenheit* names the sending of Being in the very essence of its clearing, which can only manifest itself if this clearing—that is to say, if its truth, its *Aletheia*—retains or withholds itself, if it keeps to itself or keeps itself in reserve for thought, as it has done from (and, Heidegger writes, *as*) the very beginning of Western thought all the way through the different epochs of the history of Being up until now, that is to say, the present technological age.[21]

Yet as this forgetting has turned into the—forgotten—principle of thought, it has become increasingly difficult or even impossible to understand it properly. Consequently, the lack of divine (sacred or holy) names cannot be experienced in its own right ("Der Vorenthalt der Lichtung des Anwesens als solchen verwehrt es . . . , den Fehl 'heiliger Namen' als Fehl eigens zu erfahren"),[22] that is to say, in its origin or *Herkunft*. More than ever, we seem removed from the "possibility" (the *Möglichkeit*) of gaining insight (*Einblick*) into this situation and assessing the appropriate task of thought.

Heidegger leaves no doubt that this task consists in a turn, a turning away, from the dominance of method (the following of a path, *methodos*, of sorts) in favor of a different, less secure but more demanding path. Underway to whatever is given thought to think, we might come to learn (again?) that in the retreat and the reserve that characterize the dominion of *Lethe*, the epochal forgetfulness of Being, there reigns "a proper mode of presence (or presencing)," one that is *constitutive of*, that opens, enables, insight into the anxious situation of the "Fehl 'heiliger Namen.'" Only our staying (*Aufenthalt*) or dwelling in the openness of the primary domain (*Gegend*) guarantees the *possibility* (*die Möglichkeit*) of overcoming our present blindness and seeing what, here and now, somehow *is*, even though, or precisely insofar as, it is lacking.[23]

What lacks *is*, somehow, if not somewhere, if not at some given moment *in* time and *in* space. And this *is*, Heidegger implies, grounds the very possibility of an insight.

As so often in Heidegger, there is thus an emphasis on the ontological or epochal primacy of the possible, as well as on the grounding, the localization, of lack in the *Lethe*, that is to say, in the presencing, the *Anwesen*, of Being insofar as it absconds itself from the present. This preoccupation directs Nancy's reading. He asks whether or to what extent Heidegger's thought still reserves for itself the possibility of an ontological or dialectical "reserve," and, if so, whether the "Fehl heiliger Namen" becomes secondary to a supposed primacy of the *Lethe* at the bottom of *Aletheia*. Might this "lack" (*Fehl*) and the "reserve" (*Vorenthalt*) that Heidegger thinks it signals and presupposes be understood as the overture to a new and other beginning or to a return—a *re*-turn as much as a "re-run," a turning up again—in which the gods once more might make their appearance?[24] And would the poet—especially der *Dichter* par excellence, Hölderlin—have a decisive role to play here? Is that how Hölderlin understood his poetry and himself? Or, to avoid interpreting the poet *e mente auctoris*, is that how he is most fruitfully or plausibly read? Does Heidegger's later thought measure up to the difficulties of Hölderlin's text, or should we attempt a more complex reading, a *lectio difficilior*, that resists the possibilities of hermeneutic understanding and, indeed, the very premises of the possibilism on which such understanding ultimately rests?[25] If Heidegger's *Erläuterungen* somehow miss the mark, what help, if any, do Nancy's more elliptical observations offer?

A brief review of a few other commentators will help prepare the ground for an answer to these questions. Richard Sieburth, unafraid of quick identifications, writes in the introduction to his translation of *Hymns and Fragments* by Hölderlin:

> If the heavens are now desolate, it is because a double estrangement has taken place: just as man has forgotten his love for God, so God has ostracized man, leaving him to wander through the night like an orphan. And, by a paradox of negative theology, the disappearance of God may be his most mysterious gift: as Hölderlin tersely observes in his poem "The Poet's Vocation" ["Dichterberuf"], *Gottes Fehl hilft* ("God's lack helps"). In the new dispensation brought about by the death of God the poet's role is no longer to receive the fire from on high and offer it on in song to his fellow men. . . . Instead he has now become guardian of the empty intersection that defines the mutual infidelity of gods and mortals. In this silent locus his words now discover their source.
>
> But the very concealment or hiddenness (*Verborgenheit*) of the divine, as Heidegger's commentaries on Hölderlin darkly suggest, defines the ground for its disclosure. Radiant by his very occlusion, God reveils himself by taking veil. . . . The distance of God, then, is integral to his proximity; he is, in the celebrated opening phrase of "Patmos," at once "Near and / Hard to Grasp."[26]

"Patmos," Sieburth asserts, is Hölderlin's "greatest meditation on the *deus absconditus*" and, long before the death of God will be announced, testifies to the "terrifying withdrawal of divine presence as Christ, as the last of the ancient gods, abandons his disciples to a diaspora of darkness and devastation."[27] Sieburth's interpretation emphasizes that the poet's voyage upon being seized by a spirit is "at once spatial and temporal,"[28] pointing out that in Hölderlin's later hymns and fragments writing came to be associated with a "mapping" that is at once "mythic"[29] and decidedly modern. He suggests, moreover, that for Hölderlin the "distance" of the divine is "integral" to its "proximity" and that it is in his "absence" that God can be found to "establish his presence."[30] This, then, would be why Hölderlin makes so much of God and man being each a "sign" (*Zeichen*)—a sign premised upon a relative as well as absolute absence—giving each other signs, instituting their relation through signs that call for as many acts of "interpretation" (*deuten*). On this reading, which draws heavily on "Patmos," "Mnemosyne," and ". . . der Vatikan . . . ," the poet is positioned in the space between mortals and gods, and between the latter and God:

> His language institutes measure and establishes proportion, mediating between gods and mortals, bringing about a reconciliation of opposites in which, as Hölderlin learned from his study of Greek philosophy, part coincides with whole (*ἕν καὶ πᾶν*), (the one and all) and unity corresponds to diversity (*ἑν διαφέρον ἑαητῳ*), (the one differentiated in itself).
>
> The marriage of contraries, however, remains throughout Hölderlin's work a precarious, and utopian, synthesis. The apocalyptic conjuncture of heaven and earth is something that can only be experienced in memory or in anticipation, for it is an event no longer or not yet possible, located either in the mythical Golden Age of beginnings or in the Parousia that lies at time's end.[31]

In between, the poet's hymns—notably in the so-called *vaterländische Gesänge*—are at once an "act of praise" and a "public declaration":

> Drawing upon a tradition that reaches back to include the Homeric hymns and Pindar's odes, as well as the psalms and canticles of Protestant worship, Hölderlin accordingly defines the poet as the one who leads the polis (or the congregation) in choric celebration of its heroes and gods. . . . Occupying a privileged space between the fiery heavens and the earthly community, the poet not only solemnizes the manifestation of the divine but is also called upon to translate or mediate these to his fellow men (and hence to institute community) while articulating the necessary proportions that gods and mortals must respectively observe. Everything depends on balance: unmeasured praise is as dangerous as despair. Between these two poles—between the hymnic celebration of plenitude and the elegiac lamentation of loss—Hölderlin's *vaterländische Gesänge* maintain a precarious equilibrium.[32]

Whatever the merits of this interpretation, I would claim that it lacks some of the complexities and subtleties of the more elliptical reading proposed by "Of Divine Places."

By transcribing and displacing Heidegger's account of the topos *es fehlen heilige Namen*, Nancy to a large extent reassesses the situation as well as the situating of thought it evokes. He maintains that the lack is not tied to a cut, to a being cut off, marked by the caesura or circumcision of language which comes with baptism into a given community. The want of names is constituted, not by the lack of this or that name, but by the (temporary or perpetual) absence or cessation of the gesture of naming, that is, of appellatives, of invocation or prayer. This allocation has not merely been historically or historially displaced. The lack of names unsettles and uproots the address in a more radical sense than by simply transporting it into a different context. It decontextualizes or eradicates the very *hic et nunc* of saying or praying (or is it cursing?) *mon dieu!* This "apparent possessive" is the basis of the sublime destitution and abandonment that Nancy sees at the origin of the divine topoi.

This view might seem to allow one to hold at a distance, at least philosophically or with regard to the task of "thought" the all-too-familiar, naive, and dangerous attempts to explore and exploit a supposed renaissance, resurrection, or *réveil* of the religious. It would seem that the death of God is irrevocable, and that gods return only as idols and idle kitsch. Yet the presupposition underlying this confident distance, the assumption that for want of prayer there can only be a parody of re-citation, is vulnerable in terms of Nancy's own analysis and is, I would suggest, anything but convincing. Just as the validity of the concepts of the "idol" and "idolatry" are based, as Nancy rightly points out, on the silent presupposition of the *Idea* of which they are thought to reflect a merely distorted—anthropomorphic and profane—image, so also the suspicion of *mere* recitation makes sense only in view of the ideal of direct and pure address, of a full speech, as it were, that is unequivocal and present to itself, here and now. In its very purity, however, this address can never exist or be given as such. This is what resonates in Hölderlin's phrase about the lack of holy names. What matters most is neither nature nor culture but how, as Hölderlin puts it in the second letter to Böhlendorff, "the living relationship" and "destination" (*das lebendige Verhältnis und Geschick*), of which nature and culture are the "abstract elements,"[33] manifests or reveals itself only, *if at all*, in relation and *as* relation to its other, and thus, as we shall see, both *im-mediately* and *inflected*. For us moderns, it assumes the form of the immeasurable, of a sacred pathos or a fire from heaven. Far from mimicking the classicist—supposedly Greek—ideal, this is, in fact, its other.

If one holds to Nancy's terms, faith must be taken as "entirely an outward act of presence [*une comparution à l'extérieur*]," one, moreover, that resembles "turning one's face toward the manifest heavens."[34] This comparison should make us pause since it invokes a remarkable Hölderlinean topos which cannot be easily situated or elucidated (as Heidegger would say *erörtert* or *erläutert*). For while we seem to be dealing here with a *topos aistheton* rather than a *topos noeton*, this distinction inaugurates and constitutes the very project of a metaphysical cosmology that is no longer pertinent to understanding Hölderlin's hint. "Turning one's face toward the manifest heavens," Nancy holds, has much more to do with the singular "phenomenon" of an "idea" being inscribed on—and playing around—the face of an "idol."

Hölderlin, too, asks "What is God?," the question "Of Divine Places" starts out to discuss, if only to raise doubts about its topicality. The question figures prominently as the title and *incipit* of one of Hölderlin's poetic writings, taken as the point of departure for Heidegger's elucidations. In it Hölderlin poses and answers the question "What is God?" by pointing to the aspect or the face of the sky, the *Angesicht des Himmels*, its "lightnings," which strike mortals and signal a divine visitation, an appearance and a shining of the divine, a theophany. The poem, also cited in Heidegger's ". . . *dichterisch wohnet der Mensch* . . ." (". . . Poetically Man Dwells . . ."), reads as follows:

Was ist Gott?

Was ist Gott? unbekannt, dennoch
Voll Eigenschaften ist das Angesicht
Des Himmels von ihm. Die Blitze nämlich
Der Zorn sind eines Gottes. Je mehr ist eins
Unsichtbar, schicket es sich in Fremdes. Aber der Donner
Der Ruhm ist Gottes. Die Liebe zur Unsterblichkeit
Das Eigentum auch, wie das unsere,
ist eines Gottes.
(*SA* 2; 1: 210)

What is God?

What is God? Unknown, yet
Full of his qualities is the face
of the sky. For the lightnings
Are the wrath of a god. The more something is
Invisible, the more it destines itself to what's alien. But the thunder
The fame is God's own. Love for immortality
is the property, also, like ours,
of a god.

While Heidegger's commentary starts out from the question "What is God?," his reading takes its lead from a passage, in *In lieblicher Bläue blühet* ("In lovely blueness . . . blossoms"), whose authenticity has been disputed, but which for Heidegger, in its "substance" (*sachlich*) and its time, belongs to the same "ambience" (*Umkreis*)[35] of the poem "What is God?"

Ist unbekannt Gott? Ist er offenbar wie der Himmel? Dieses Glaub' ich eher. Des Menschen Maass ist's. Voll Verdienst, doch dichterisch, wohnet der Mensch auf dieser Erde. Doch reiner ist nicht der Schatten der Nacht mit den Sternen, wenn ich so sagen könnte, als der Mensch, der heisset ein Bild der Gottheit. Giebt es auf Erden ein Maass? Es giebt keines.

Is God unknown? Is he manifest like the sky? This rather I believe. It is the measure of man. Full of acquirements, but poetically, man dwells on this Earth. But the darkness of night with all the stars is not purer, if I could put it like that, than man, who is called the image of God. Is there a measure on Earth? There is none.

This late fragment confronts us with the difficult problem of the interplay of poetry or poeticizing and topology, topography, and theophany: a relation that is not one relation among others, but provides the measure for all these others. How, if at all, can it be thought? How, if at all, is it experienced?

The precise nature of religious topographical imagery—in a word, of *topo-theo-graphy*—has long been a central focus in Hölderlin scholarship. Richard Sieburth, for instance, suggests that

Hölderlin's ideal landscape is always theophanic, a scripture to be reverently read and interpreted, a radiant figure of divine design. Such visionary topography, however, does not exclude a precise attention to the minute particulars of the *genius loci*, whatever their local habitation or their name.[36]

Paul de Man addresses this relation at different places throughout his *Critical Writings*, nowhere clearer than in "Hölderlin's Riddle." Speaking of "Brod und Wein" ("Bread and Wine") and "Andenken" ("Remembrance"), de Man begins by observing what no attentive reader should ignore:

This is not descriptive poetry as we would find it in Wordsworth or Coleridge, nor is it the kind of reverie associated with a Rousseauistic response to natural settings. The landscapes are made up of an intricate network of forces whose relationships are strongly dramatized. As a result, despite the absence of explicit symbolism or allegory, one feels behind these landscapes a working principle that encompasses mind and nature within a larger element. Like landscapes in a dream, every detail seems to have a meaning, to refer back to a will, to a purpose, even if this purpose remains hidden. Hölderlin modulates almost without transition from nature descriptions to dramatic scenes describing the actions of entities endowed with more than human or natural status. The course of the Rhine becomes the bearing of a demigod; the fall of night over a city the way in which a god ambiguously manifests his presence in withdrawal; a sunrise in the Alps suggests the proper distance between god and man.

In a late eighteenth-century work, such sudden and apparently effortless transitions from nature to divine presences are by no means easy to understand. The word *god* in Hölderlin, in the singular or in the plural, does not have behind it the weight of doctrinal and literary tradition that gives its anagogic level of meaning as in Dante or Milton. Nor are we dealing with a humanized and secularized version of Hellenic or Christian symbolism, as when Shelley or Keats represents the historical destiny of humankind in mythological form. It would also be false to think of Hölderlin's poetry as a form of pantheism. He does not reach what he calls "the gods" through the mediation of nature; nature, in his work, is not closer to god than the thoughts and the deeds of man. Least of all does the theocentric vocabulary designate a religious experience in the traditional sense of the term: it refers to no dogma or act of faith.[37]

It is precisely this overflowing of religious and mythological figures, the fact that Hölderlin adopts a "more-than-human point of view"[38] throughout most of this writing that has led, de Man continues, to countless misreadings in the large body of textual criticism devoted to work. This pertains in particular to the understanding of "Nature" as an infinite, eternal, and ultimately all-encompassing notion—surpassing man and gods—a motif that can be found in the letter novel *Hyperion*, in the different versions of *The Death of Empedocles*, in the odes and elegies of the middle years no less than in the free-verse poetry of the later hymns. According to de Man, one nefarious "misrepresentation" stands out here and deserves further attention since it will enable us to circle back to the discussion of Nancy's "Of Divine Places" above. De Man formulates it as follows:

There has been a persistent tendency to treat Hölderlin as a prophetic and eschatological poet, the precursor of a new historical era that his work helps to prepare. The trend goes back to Stefan George and his circle, who were closely associated with the rediscovery of Hölderlin shortly before World War I. It prevails, in a subtler form, in some of Heidegger's commentaries on Hölderlin's poetry during the thirties, whose ideological and nationalist overtones others, at the same time, were stating much less obliquely. But even when it appears in a nonpolitical or politically acceptable form, the messianic schema that one tends to associate with Hölderlin's view of history distorts his actual statement.[39]

Jacques Derrida, in his *Mémoires: For Paul de Man*, reminds us of the stakes of this reading, which seems consistent throughout most of de Man's critical writing on the subject but finds its strongest statement in the essay "Heidegger's Exegeses of Hölderlin," in *Blindness and Insight*. Here, de Man addresses Heidegger's reading of Hölderlin's complex notion of "Nature" as distinct from the metaphysical understanding of *phusis* since the Greeks, including the pre-Socratics, and its disastrous distortion in modern, techno-touristic exploitation.[40]

De Man starts out by citing Hölderlin's "Wie wenn am Feiertage . . ." — "Denn sie, sie selbst, die älter denn die Zeiten / Und über die Götter des Abends und des Orients ist, / Die Natur" (For she, she herself, who is older than the ages / and above the gods of the Occident and the Orient, / Nature) — and then goes on to claim that Heidegger rightly reads these lines as stating the poet's major concern: the fact that nature, by being "marvelously omnipresent" (*wunderbar allegegenwärtig*), is not to be understood in any "pastoral" sense, but rather as "the immediate apprehension (presence) of that which serves as a support to all things, that which precedes and makes possible their representation to consciousness."[41] This possibilizing dimension, De Man goes on to suggest,

> is indeed that which in Heidegger's own terminology is generally called the presence of the presents, the common *Wesen* of all individual presents that makes for the all-presence of things. It is the immediate givenness of Being that, for Hegel, is "*just* Being" (*nur Sein*) as long as it has not been represented to consciousness.[42]

Much, of course, could be said about this all too hasty sketch of Heidegger's thought in terms of a thinking that — supposedly from the perspective of Hegelian dialectics — concerns "just Being" (*nur Sein*), and that as such, as de Man puts it, has "neither the possibility nor the necessity to constitute itself into logos."[43] None of this seems very precise or convincing. The relationship between Heidegger and Hegel, like that between Hölderlin and Hegel, is far more complicated. This triangular constellation sheds a peculiar light on de Man's confident assumption that in the passage cited above the poet, Hölderlin, gives voice to the "anguishing question" of "*how to speak of Being*" or, more precisely, how to "say Being itself."[44] Is this indeed the "subject" or the "experience" of this poem, of this poetry? For de Man there seems to be no question here: "Heidegger is right . . . to see in the poem a statement of the relation of poet to Being."[45]

De Man claims that only when Heidegger's commentary reaches the point of attempting to demonstrate that the poet, in fact or in essence, *names* "the presence of the present"[46] does it start to "distort" Hölderlin's meaning and make him say exactly the opposite of what he—in fact and in essence—says. This explains de Man's final—and finally deeply ambivalent—judgment:

> Whatever one may otherwise reproach in Heidegger's commentaries, their great merit remains to have brought out precisely the central "concern" of Hölderlin's work; and in this, they surpass other studies. *Nonetheless*, they reverse his thought.

Again, this judgment is based on a rather curious interpretation of the stakes of Heidegger's thought. Surely, no careful reader of the *Erläuterungen* would want to conclude that for Heidegger Hölderlin's poetic universe is a placeholder for Being, for its *Anwesen*, a presencing that sustains itself as time passes by. Yet de Man reduces Heidegger's exegeses to precisely this caricature:

> For the promise of Heidegger's ontology to be realized, Hölderlin must be Icarus returned from his flight: he must state directly and positively the presence of Being as well as the possibility of maintaining it in time. Heidegger has staked his entire "system" on the possibility of this experience. This may explain why, obeying a tactic perhaps not fully conscious, he feels the need to base himself upon the work which proclaims that it is this experience, among all others, that is totally forbidden to men.[47]

This analysis lays the ground for the debate that interests me here, speaking of topoi and topographies, of their *logos* or lack of *logos*. For Hölderlin, de Man concludes, the "eccentric road man travels to the primeval unity of the immediate"[48] is far from assured, certainly not a given. Hölderlin does not imply "that the poet dwells in the Parousia," only, de Man explains, that Parousia is "the principle of his becoming."[49] If, moreover, Hölderlin's word—de Man quotes the line "Und was ich sah, das Heilige sei mein Wort" (And what I saw, the Holy be my word)—states an "eternal poetic intention," then, de Man insists, it "can be no more than intention":

> It is not because he has seen Being that the poet is, therefore, capable of naming it; his word prays for the Parousia, it does not establish it. It cannot establish it for as soon as the word is uttered, it destroys the immediate and discovers that instead of stating Being, it can only state mediation.[50]

To the extent that re-citation constitutes the basis, if not "the heart," of all prayer, de Man is quite right to state that Heidegger cites Hölderlin "as a believer cites Holy Writ." But is "Hölderlin the only one whom Heidegger cites as a believer cites Holy Writ"?[51] If one abstracts from the polemical fervor with which this claim is made; if one recalls that, for Heidegger, the structure of language—in particular that of *Dichtung*, more precisely still, that of saying in relation to its said—is marked in advance by an affirmation which ties the possibility of the *Denken* to which it (*Sprache*) aspires to a *Danken* from which it comes, then Heidegger himself could not have hoped to cite, re-cite, and elucidate otherwise.

What is important here is perhaps not what de Man notes, namely, that Heidegger's commentaries "do no more than formulate his own thought and use Hölderlin as a pretext." Nor should it surprise anyone that the "exegeses"—a term Heidegger avoids with good reason—rarely, if ever, take the form of a "critical dialogue."[52] Whatever the philological impieties of his exegeses, whatever their enormous thematic schematizations, Heidegger's *Erläuterungen* of Hölderlin's poetry—and even their reduction of this poetry to *Dichtung*—obey a forceful repetition. They signal a distortion and a doubling at once. One might be tempted to ask whether de Man's text, in its turn, might not also double (and thus also distort) the reading propelled by Heidegger. The former is at once faithful and unjust to the latter, if only because de Man accounts much less for Heidegger's "insights" than for his major "blindness." De Man takes the proven misreading to be the sign of a total mismapping and mismeasurement of the Hölderlinean *phusis*. In doing so, he loses sight of a more crucial hint or *Wink* which Heidegger's text—here closer to the *Wortlaut* of Hölderlin's poetry than de Man suspects—also allows to be discerned (i.e., to be heard and to be seen).

For Derrida, it is with respect to the interpretation of the "law" that de Man "intends to rescue Hölderlin from appropriation-by-identification, from what might be called Heidegger's hermeneutic mourning."[53] Derrida cites *Blindness and Insight*, where de Man states:

When he states the law, the poet, the poet does not say Being, then, but rather, the impossibility of naming anything but an order that, in its essence, is distinct from immediate Being.[54]

Heidegger interprets the first lines taken from "In lieblicher Bläue blühet . . ." as saying that the divine, while retaining its invisibility and unknowability, is made manifest (*offenbar*) by the heavens, sending, dispatching, or destining itself in the appearing of the world, which nonetheless remains completely foreign to it. However, in doing so, he also takes the risk of reducing the modality of this manifestation to a logic of absence and presence, revelation and concealment. Here, in other words, the divine epiphany is still thought, Nancy suspects, in terms of an Idealist alternative, as the manifestation of a god who does or who does not wish to be "close to us."[55] Yet he urges us to ask, "Can we still be content to go on conceiving of God, with or against Hegel, with or against St. Augustine, as a form of extreme intimacy? Will a day not come when we shall have to confront a god outside, exposed in the open sky, nowhere hidden and internal to nothing?"[56]

Nancy proposes another, in my view far more provocative, reading of the lines cited above. The poem, he claims, maintains less that God is revealed "by means of the heavens" than that "the god" is only "as manifest as the heavens." "The face of God is as manifest as the *Angesicht* of the heavens."[57] This means that, for Hölderlin, the heavens are not the mere appearance, reflection, mirror image, or projection screen for a god whose existence (or presence) and essence ought to be located elsewhere, beyond this surface, or who would jealously retain some space, some reserve, some *Vorenthalt*, for himself. What alternative do we have for thinking divine manifestation as an occurrence that takes places *either* in interior (psychological or, rather, spiritual) enlightenment *or* that draws on the resources of an outer nature (from, say, "elements of the cosmos," *stoicheia tou kosmou*, through the created universe, to the Pascalian infinite voids of infinitely extended and infinitely divided spaces filled by divine grace alone): "The invisible divine lets itself be seen resting, itself, upon the face [the *Angesicht des Himmels*], or woven into it, sent or destined therein, but as another face that lets itself be seen *here, without 'here' serving as a mediation for it.*"[58] For lack of a better word, Nancy speaks here of the "im-mediacy" (*im-médiateté*) of the divine, the "*im-*" implying neither negation nor privation nor, for that matter, mere indifference, but rather a relation or a state which has absolved itself from that of both mediacy and immediacy, and, we should add, concerns man alone.

He discusses this relation, whose terms absolve themselves from every possible mediation—and thereby from the very mediation that is kept in reserve by every possible, whether transcendental, virtual or real—extensively in his *L'expérience de la liberté* (*The Experience of Freedom*).[59] There he notes that "immmediacy"—this time around, the immediacy of "freedom"—should not be confused with "sensuous immediacy":

Nor is it an absence of mediation in the intelligible. It is neither a sentiment nor an intellectual given. . . . This might resemble what we could call the specific pregnancy of the "feeling of reason," which for Kant is the *respect* for the law of freedom. . . . The immediacy of this experience [the "passion" of freedom] must . . . be understood as *the affective im-mediacy of freedom in existence insofar as freedom affects existence from an infinite distance.*[60]

One should not, therefore, confuse im-mediacy with "temptation itself, the cunning abdication of thought into [that other] immediate, into the 'lived,' into the ineffable, or into the praxis and art designated as the others of thought," as it can be found, for instance, in the well-known reference, in Hegel's *Phenomenology of Spirit,* to the image of the young girl who "presents the outstanding products of ancient art and the divine places that the gods have left."[61] Rather, Nancy insists, "im-mediacy," as that which enables and characterizes freedom, calls for a "returning of praxis to thought," and in this turn, he adds, a central element from Marx's writings "resonates" with "something from Heidegger."[62]

The "im-mediate" of which Nancy speaks in "Of Divine Places" gives a hint man can follow up on, or not. It is all too easy to miss the point, the pointer, which is nonetheless somehow *there,* out there, but "in," if one can still say so, a "there" that cannot be fixated—indeed, pinpointed—with the help of the usual coordinates. What the heavens and the god have "in common," is neither a common measure—"Giebt es auf Erden ein Maass? Es giebt keines"—nor an *analogia entis,* but "the sovereign interplay [*jeu*] of darkness and radiance [*l'éclat*], of radiance withdrawn into darkness and of darkness as manifest as radiance."[63] From this Nancy draws a dazzling conclusion. Seen in this light, against the background of this radiance, revelation can be thought neither as "presentation" nor as "representation," since it is not the conveying of some message, but merely as "the evidence of the possibility (never the necessity) of a *being-unto-god.*"[64] The "radiance," in its very indeterminacy, is itself as such the revelation, though it reveals nothing but a possible. It reveals a pure relation or relatedness from which the terms (addressor and addressee) have withdrawn or *absolved* themselves completely and irrevocably: "What there is revelation of is not 'God,' as if he were something that can be exhibited . . . , it is rather the unto-God [*l'à-dieu*] or being-unto-god. Or more exactly, it becomes manifest that such a being-unto-god is possible."[65]

These formulations expose the fragility of Heidegger's suggestion that the divine is made manifest *by* the heavens, that the presence-absence of the gods should be projected against the screen of Being's manifestation and concealment, that the Hölderlinean "time of need" (*dürftige Zeit*) should be taken as the time in which the gods have fled while the god is still to come. These are all premises of a reading that enables Heidegger *in advance* to "extract" a future "positive" out of a past and present, supposedly nonprivative and nondialectical, "negative."[66] Indeed many, if not all, of Heidegger's formulations which seek to determine the relation between Being and the highest Being, between the holy and the gods, between the gods or the holy and the Christian God (the others, we have seen, are either kept in reserve or not to be mentioned, let alone taken serious, at all), are deeply ambivalent in this respect. One example must suffice here to illustrate this claim. Heidegger writes, in *Erläuterungen zu Hölderlins Dichtung*:

> As Hölderlin founds the essence of poetry [*Dichtung*] anew, he determines . . . a new time. It is the time of the gods who have fled *and* that of the coming god. This is the time in *need*, since it stands in a doubled lack and not: in the no-longer of the gods who have fled and in the not-yet of the coming.[67]

Here, the doubled "not" has already the structure of a lack which calls for and calls forth a presence, if only by opening a space, by ending or, rather, closing an epoch, marked by the flight of the gods. These gods, in turn, serve, if not as the medium, then at least as the signs and the mediators of *the* god (*der Gott*): "Die Göttlichen sind die winkenden Boten der Gottheit. Aus dem heiligen Walten dieser erscheint der Gott in seine Gegenwart oder er entzieht sich in seine Verhüllung."[68]

How are we to understand these words? Do they confirm the basic tenets of Nancy's suspicion? Does not the thought of Being from the very outset prejudge the eventuality and the very modality of the coming of gods, by framing this coming or not coming in terms a of a possible and associating the present lack—the *Fehl "heiliger Namen"*—with a withholding or keeping itself in reserve, with a *Lethe* at the bottom of *Aletheia*? This question has far-reaching implications, not the least of which would be to problematize a common reading of the relation of Being and the divine topoi.[69]

Nancy's question posed to Heidegger's reading of the Hölderlinean manifest heavens, his rethinking of this figure as a mere "possible" or as the "possibility of a possibility," is itself open to question. This attempt to frame the relationship between Being and the heavens, gods and mortals, in terms of a—possibilizing—*dimensionality* remains therefore steeped in the ultimately metaphysical presupposition of the existence, the deontological force, and the aesthetic value, of dispositional capacities (i.e., the *Möglichkeit* identified with a *Vermögen*, a *dunamis*, a *potentia*). Such forms of thought can no longer satisfy, because in any thought of possibilization, what is supposedly made possible is *ipso facto* also *made* into what it is not. It is *screened* in all senses of the word: projected onto and limited by properties, and be it in light of a horizon deemed most proper to its, the openness of the *there* (*Da*) its ownmost being or the Being—again, the *Dimension* of Being—that opens it up while folding it back into the One, the very simplicity that for Heidegger, following a long tradition, remains the *index veri*. Indeed, Heidegger's attempt, in the *Beiträge zur Philosophie* and elsewhere, is, as has been noted by Courtine, to delineate a new post-metaphysical concept of and space for the transcendence or epiphany of the divine that departs from the "long christianization of God."[70]

Only when he reinscribes Heidegger's problematic formulations into the Hölderlinean topos of the *Winke* of the divine does Nancy efface the last remnants of the ontotheological legacy discernible in the restriction of thought to the openness to a possible being-unto-God. Reading Heidegger against Heidegger, without referring to other, perhaps more topical, discourses (ones developed by Levinas, Derrida, and Marion), Nancy proposes that a better name for the peculiar modality of the "presence" of "God," "god," "the god," and "the gods," as they are revealed and manifest *like* the heavens, is that of the *Wink*, the divine "nod" or "sign." The divine does not signal something; it signifies nothing in particular, nothing that exists, nothing phenomenal, not even the radiance of the heavens. But if the divine "glory" is said to be "open, offered, dazzling like that of the heavens and effaced like them,"[71] then this appears to mean, if anything (since the order of meaning and appearance is precisely what is displaced or erased here), that it has the "nature," the "mode," or the "quality" of a *Wink*, of a *Wink*, moreover, which in its elusiveness ambivalently evokes joy and fear at once,[72] thereby recalling the formal features of the sublime, not only in a Kantian sense. The god, Nancy writes, is only *im Vorbeigang*, "passing," "in passing,"[73] *de passage*, and, to the extent that all departure entails some sort of death (*partir c'est . . .*), has always already passed away.

Two observations should be made here. First, for all its continuity with Heidegger's interpretation of the *kairos*, in the early lectures on the phenomenology of religion, notably in a reading of St. Paul, the "passage" in question here has no longer the character of a Parousia.[74] As Courtine notes, we should seek the source of Heidegger's concern with the specific modality of the "passage" and, indeed, the *Wink* in the *Zweisprache* with Hölderlin. Here, Hölderlin's "Friedensfeier," "Der Einzige," and "Patmos" would be particularly revealing.[75] Second, it is important to remark that the divine *Winke* thus found is, in Heidegger's view—and here he may come closest to Hölderlin—intrinsically linked to an understanding of finitude which it radicalizes or pushes to its extreme.[76] Important differences, however, remain intact.

Citing Hölderlin's passage on divine *Winke*, Heidegger comments:

. . . und Winke sind
Von Alters her die Sprache der Götter . . .

The saying of the poet is the capturing of these nods, in order to gesture [*winken*] them further in his people. This capturing of the nods is a receiving and yet at the same time a new giving; for the poet sees [*erblickt*] in the "first sign" also already the accomplished and boldly puts this vision [*Erschaute*] in his word in order to predict the not-yet-accomplished. Thus, "fliegt, der kühnen Geist, wie Adler den Gewittern, weissagend seinen Kommenden Göttern voraus."[77]

Nancy emphasizes the double reading that these words allow. Hölderlin's verse does not presuppose that the realm of the profane has the ability to gesture toward the sacred. No analogical continuity surmounts the abyss between appearance and essence, the many and the One, the mortal and the divine. And yet, he adds, this does not exclude the possibility that "to 'give a sign' is perhaps always—divine."[78]

To be sure, the relation between the different modes of giving a sign while signifying nothing in particular, between the *Wink* that is a divine gesturing and gesturing that is a *Wink*, is hardly symmetrical, despite a secret correspondence. Not every giving of signs, calling, seducing, and so on, can pretend to present us ipso facto with the divine. For Hölderlin no less than for Heidegger, only poets function as the heralds—and, perhaps, the very *Winke*—of the *Winke* called "divine." Yet the privilege accorded their words can hardly be attributed to the semantic or metaphorical potential of poetic speech, just as the lack of divine proper names cannot simply be ascribed to a supposed "metaphysical surfeit of the thing over the sign, of the real over language."[79] On the contrary, if the lack of names is the absence of a *Wink* which "is" or evokes a gesture of naming, then the want of a name has nothing to do with some "signifying capacity"[80] or lack thereof. As a result, the lack of divine names "cannot be judged in relation to sense,"[81] and precisely this circumstance and not some "dispositional capacity" may enable all names, in principle and in fact, to address—or name—the divine.

In one of the texts published in *Denkerfahrungen* and dated 1941, Heidegger describes the *Winke*, his own "Winke,"[82] in careful words which deserve extensive quotation, not least because they circumscribe the status of the *Winke* while "forging a path" between the supposedly separate domains of poetry and thought, aesthetics and philosophy:

> The "nods" [again, Heidegger speaks of his own *Winke*] are not poems. Nor are they "philosophy" put into verse and rhyme. The "nods" are the words of a thought that in part needs this expression [*Aussagen*], but does not fulfill itself in it. This thought finds no anchor [*Anhalt*] in being[s], for it thinks *das Seyn*. This thought finds no example in being, since what is thought thinks being [*das Seiende*]. The saying of thought is other than the word of poetry without image [*bildlos*]. And where there seems to be an image, it is neither the poeticized of a poem nor what is intuited by a "sense" [or "meaning"], but only the desperate anchor of an attempted but failed abstinence from any image [*Bildlosigkeit*].[83]

Here and elsewhere, the word *Winke* retains an intimate relationship with the manifestation and the revealability of the essence of Being, that is to say, of Truth.[84] None of this, however, is assumed in Hölderlin's poetry or in the commentaries on which we have been focusing. On the contrary, the quintessence of the interrogation of the relationship between the poetic, on the one hand, and the thought of Being and (its) Truth, on the other, consists in questioning the primacy of the latter over and against the former.

How, then, can Hölderlin's "poeticizing"—the writing, reading, or re-citation of his "poem" as well as the simple *aisthesis* of the heavens of which it speaks—signal the turning of a possible being-unto-the-god, without at the same time returning this relation to the order of signification and of aesthetic representation? And where, if anywhere, does the *Wink* touch upon art? Nancy's demarcation between the two modes of manifestation seems to leave no room for confusion. Apodictically, "Of Divine Places" upholds a clear-cut distinction between the presence without (re)presentation of the divine, on the one hand, and the presentation of/as representation of art, on the other: "God . . . comes in the ruin of all appearing [*le paraître*]. Art, on the contrary, infinitely incises the edges of appearance [*l'apparaître*], but keeps it intact."[85]

At least three traditions seem to cross paths here, one of which is passed over in silence: the one which relies on the New Testament figure of the *hoos mè* (ὡς, as if not) that forms an important key to understanding Heidegger's "exegeses," not of Hölderlin, but of St. Paul. Nancy identifies the two remaining ones as two particular modes of the sublime, each of which seems to have left its imprint in the very texture of his essay:

There is the sublime in art, going from Kant to Benjamin and from there on to us. It signifies: to feel the fainting away of the sensible, to border on the furthest extreme of presentation. . . . And there is divine sublimity, that in terms of which Hegel seeks to characterize the Jewish moment in religion. . . . The coming of God reduces the phenomenon to nothing. Here the sublime is no longer to be found at the furthest extreme of presentation where presentation is transformed into offering. It is in a presence that ruins all presentation and all representation. It is no longer the gesture of offering, it is the imposition of glory. It is no longer the limit of forms and figures, it is the light that disperses the visible.[86]

Nancy tries precisely to capture the difference between these two sublime modes of signifying at the limit of and beyond the phenomenal world, at the limit of and beyond art, in the fragmentary—indeed paratactic[87]—form of presentation adopted in "Of Divine Places." Its most significant stylistic feature is that of "juxtaposition," without explanatory connectives, not so much of grammatical or syntactical elements (as in the parataxis of Hölderlin's late poetry), but, so to speak, of the structures of argumentation or, rather, elucidation. It takes, quite literally, the form of a fragmented topology or, rather, topography. It allows of no gathering, no *Versammlung*, no *logos*. It does not let itself be folded into any *Einfalt*, nor does it limit the domain (the *Gegend*) of thought to any fourfold (*Geviert*), folded, in its turn, into one.

"Of Divine Places" demonstrates—in its propositions as much as in their performative structure—that the sublime modes of the divine and the aesthetic are fundamentally *irreconcilable*. They cannot be translated into each other, for they have no common measure. And yet, Nancy writes, each mode can offer the other: each can offer itself *to* or *as* the other:

Between the "thing" of sublime art and the sublime "thing" of the divine, there can be said to be that infinitesimal (and in its turn sublime?) difference that lies between presentation at the limit and naked presence: it follows that each can offer the other, but also that it is impossible to confuse one with the other.[88]

One might argue that the reverse must be equally true. Given the "infinitesimal" difference and distance between the two sublimes, neither one of them can ever hope to offer the other, strictly speaking. Nonetheless, it is impossible *not* to take the one for the other. One cannot but keep them apart *and* confuse them. On both counts, the infinitesimally small but lasting difference or distance between the limit(ed) and the limitless regulates their impossible yet inevitable relation. Or so it seems. One hints at the other, as its other, its *Wink*.

Hölderlin's Christ

Jean-François Courtine

. . . Schön
Und lieblich ist es zu vergleichen . . .[1]

Confronting the question of Hölderlin's "christology" remains now, as previously, a risky task, despite Xavier Tilliette's convergent and ever more precise indications in *La christologie idéaliste*, *Christ de la philosophie*, and, recently, *La semaine sainte des philosophes*.[2]

If it is true, as Tilliette emphasizes, that Hölderlin's major "authorized" editors from Beißner to Sattler, like the commentators who more or less follow the Heideggerian reading of the poet, have in general tended to occult, or at any rate minimize, the christological dimension of Hölderlin's great poems, whether it be a question of the elegy "Bread and Wine" or of the final hymns, the resistance to this reading—which, if not neopagan, is at least resolutely Greek and syncretic—has nonetheless been notable. Suffice it to mention here Erich Przywara, Romano Guardini, Eduard Lachmann, or Heinrich Buhr, not to mention the discussions raised by the interpretation of the hymn "Friedensfeier" ("Celebration of Peace") after its discovery and publication in 1954 by Beißner.[3]

It is not my intention here to remake the history of this occultation or, inversely, to remake the history of that timid christological inflection in the reception of Hölderlin, both of these engaging in the interpretation of the famous "vaterländische Umkehr" ("Anmerkungen zur *Antigonae*"; ["Remarks to *Antigone*"] *SA* 5: 419). Is this "return to the homeland," which marks a clear rupture with the enthusiasm of Hölderlin's youth and his nostalgia for Greece and its fled gods, properly speaking an Occidental turning and even thereby the return of a christological figure? This question has been debated often since Wilhelm Michel's essay, which rather curiously associates the "Syrian" with the "Nordic," "Germanic" principle.[4]

It is important to emphasize, first, when we want to elicit the fundamental characteristics of Hölderlin's interpretation of the figure of Christ in the late works, up to the threshold of his madness, that if the "jewel of dwelling" is opposed to the Empedoclean principle, this is not simply to substitute it for the ancient gods or demigods by throwing them back into the enchanted realm of Fable. If for Hölderlin, as for Hegel and Schelling moreover, Christ indeed marks an end or a completion in the "Day" of the gods, this is also because he modifies the course of time in opening it to another scansion—because, despite his precipitated passage, he gathers the chorus of the gods around himself and even into himself through the gifts he leaves behind him. The mourning that characterizes the time of the fleeing of the gods is fraught with renunciation, but it also shelters within itself, without negation, the premonition of a reconciliation to come or of a Celebration of Peace, if only the "legend" (*Sage*) or the multivocal witnessing of the "Holy Scripture" are kept and cultivated with care.

The opening of the hymn "Germania" with its initial *nicht*, to which Heidegger drew attention,[5] far from marking the abandonment of a past that has lapsed irremediably, indicates just as well the expectation or premonition of the return:

> Nicht sie, die Seeligen, die erschienen sind,
> Die Götterbilder in dem alten Lande,
> Sie darf ich ja nicht rufen mehr, wenn aber
> Ihr heimatlichen Wasser! jezt mit euch
> Des Herzens Liebe klagt, was will er anders,
> Das Heiligtrauernde?
> (*SA* 2; 1: 149)

> Not them, the blessed, who appeared,
> The images of gods in the ancient land,
> I may no longer invoke them, but if
> You, native waters, now with you
> The love of my heart laments, what else does it wish,
> It that mourns the sacred?

And:

Entflohene Götter! auch ihr, ihr gegenwärtigen, damals
Wahrhaftiger, ihr hattet eure Zeiten!
Nichts läugnen will ich hier und nichts erbitten.
Denn wenn es aus ist, und der Tag erloschen . . .
(*SA* 2; 1: 149)

Fled gods! You also, you who are present, though formerly
More real, you had your time!
Nothing do I wish to deny here, and nothing demand,
For when it is over, and the day extinguished . . .

If, therefore, there is never negation, repudiation, or even renunciation other than through the singular memory of mourning, this is also because the awaited "return" is that of the Same, of that Sacred for which the names are lacking:

Die Alten, so die Erde neubesuchen.
Denn die da kommen sollen, drängen uns,
Und länger säumt von Göttermenschen
Die heilige Schaar nicht mehr im blauen Himmel.
(*SA* 2; 1: 150)

The Ancients, thus visit the Earth anew.
For those who must come press upon us,
And no longer will the sacred host
Of divine men linger in the blue heavens.

The second major point that deserves immediate emphasis is that this turning from Greece to the Hesperides takes place amid the guiding theme of a thorough meditation on Greek tragedy through translating Sophocles and elaborating the "Remarks" on *Oedipus* and *Antigone*.

The rigorous determination of Greek tragedy to which Hölderlin devoted himself since *Empedocles* and the "poetological" essays that accompany it, first among which is "Grund zum Empedokles" ("The Ground for Empedocles"), constitutes more than the backdrop, the horizon within which the singularity of the christological epiphany is sketched out. This is an epiphany of the god bound to the world (*weltlich*), taking human form, to be sure, and even assuming the condition of the "everyday" (ἐν μορφῇ δούλου), but most of all, that of the Conciliator whose premature recall[6] also attests to the concern for the measure that mortals can endure and whose departure is the evidence of eucharistic and/or scriptural presence.[7] In tragedy, this figure of the present-absent friend is opposed to the encounter, the unbearable collision of man and god. Just like Scripture (which is always to be interpreted, since it is above all important to attend to the letter), the veil or the clothing that God wears[8] contributes to the withdrawal or the "mediateness" by which alone human beings can resist the power of the tearing away of the divine, the violent drive in the direction of the aorgic that characterized the first versions of *Empedocles*. In this context, we would need to be able to follow in detail the revisions of the *Trauerspiel* and in particular the christological traits of the final *Empedocles* in the debate with Manes.[9]

In his "Remarks," Hölderlin noted about Tiresias, whose intervention marks the "caesura,"

> He enters the course of fate as the custodian of the natural power which, in a tragic manner, removes man from his own life-sphere, the center of his inner life into another world and tears into the ex-centric sphere of the dead. (*SA* 5: 197)[10]

Tiresias is not up to the Hölderlinian determination of the tragic—or better, of its presentation (*Darstellung*), which, beyond the reference, obviates against Aristotle's *Poetics* and the doctrine of κάθαρσις as the work proper to drama—which does not imply a reference to Christ, God made man. But in this unique instance of Incarnation, the coupling or the "copulation" God-Man, if they remain *ungeheuer*, unheard of, even monstrous (δεινόν), are precisely "bearable" and even "salutary," not simply because the god is here present in the figure of Life or of Love, but first of all because he is precisely present *in figure*, under the veil of the non-appearance of the "Everyday Man":

> The presentation of the tragic rests primarily on the unheard of—how the god and man mate and how natural force and man's innermost boundlessly unite in wrath—conceiving of itself, [rests] on the boundless union purifying itself through boundless separation. (*SA* 5: 201; *Essays*, 107)

The second "definition" of *tragische Darstellung*, this time as it figures in "Remarks on *Antigone*," in my opinion confirms the legitimacy of this connection, or better, of this rigorous differentiation of the epiphany of the Greek god at the center of tragedy and the fleshly manifestation of God as the Son of Man, since Hölderlin himself also evokes, in this key passage in the "Remarks," the god of the apostle—it being relatively secondary here, as far as the very principle of the contra-position is concerned, whether it is a question of the Pauline kenotic figure or the apocalyptic figure according to John:

As has been hinted at in the remarks on *Oedipus*, the tragic representation has as its premise that the immediate god is all at one with man (for the god of an apostle is more mediate, is highest intellect in highest spirit), that the *infinite* enthusiasm conceives of itself *infinitely*, that is, in consciousness which cancels consciousness, separating itself in sacred manner, and that the god is present in the figure of death. (*SA* 5: 269; *Essays*, 113)

This affirmation of "mediateness" as the essential and differential characteristic of the figure of Christ[11] would in itself suffice to make untenable the clear opposition that Allemann believes he can generalize between the royal and Empedoclean principle, which tends to unite with nature, the infinite All, to the point of dying for it, on the one hand, and on the other hand, the Hesperian principle[12] of differentiation, distinction, and "statute." But above all, it seems that this overly systematized opposition leads us completely into error when Allemann, from a reading of the late hymn "Der Einzige" ("The Only One"), interprets Christ as the "demigod who uproots and consumes,"[13] as he who returns abruptly to the Father according to a short route forbidden to the Hesperians. For the Hesperians, placing themselves under the authority of Zeus, the most proper, must rather accomplish the return toward this world far from the ex-centric sphere of the dead.

That is why I also can no longer follow Allemann when he interprets lines 48–50 and 83–86 of the hymn:

Ich weiß es aber, eigene Schuld
Ists! Denn zu sehr,
O Christus! häng' ich an dir.
(*SA* 2; 1: 154)

But I know it is my own fault!
For too much,
O Christ, am I attached to you.

And:

Es hänget aber an Einem
Die Liebe. Diesesmal
Ist nemlich vom eigenen Herzen
Zu sehr gegangen der Gesang.
(*SA* 2; 1: 155)

But my love clings
To only One. This time, yes,

Too much from my own heart
Has the song issued.

Allemann reads these lines as almost a guarding against all *imitatio Christi*, as though it were permitted to no one but Christ to take the shortest path to return to the One-All. For Allemann, Christ, in a direct prolongation of Empedocles (or more exactly, of the first version of *Empedocles*), would figure the "God at the beginning of all the gods who leave the world," while it falls to mortals, if they want to resist, as is necessary, the deadly call toward the One-All, to be "unfaithful."[14] If Christ is understood as the abducting and devouring demigod, this is the very possibility of the comparison—"Schön / Und lieblich ist es zu vergleichen" ("Beautiful / And good is it to compare") (*SA* 2; 1: 164)—between Christ and the "weltliche Männer" ("men of this world") that were Heracles and Evius. The comparison lapses just as the injunction addressed to the poets who "müssen auch / Die geistigen weltlich seyn" ("must also, / They who are spiritual, be of this world") (*SA* 2; 1: 156) becomes incomprehensible. To these difficulties, which we might be tempted to call "literal," is added the immense problem of the distinction and the complementarity between mediateness and withdrawal [*retrait*], a problem that by the same token is entirely eluded by Allemann's unilateral commentary.

I will not be able to treat the image of Christ in Hölderlin's oeuvre in all its complexity, and will limit my study, or better, my approach, to the three great hymns that can be considered, relatively cautiously, as "christological": the three versions of "The Only One," the hymn "Friedensfeier" and its preparatory version, "Versöhnender der du nimmergeglaubt . . ." ("Conciliator, You Who No Longer Believed . . .") and finally "Patmos,"[15] leaving aside the late fragments ("To Madonna," "What is God? . . . ," and "The Vatican").[16]

Far from being an uprooting demigod who would ignore the measure or the necessary distinction of the spheres,[17] Christ appears there, rather, as the Conciliator, the Reconciler, who surpasses the other gods or demigods (and the very ones to whom he is as close as a "brother") precisely because he leaves them a relative right and completely fulfills the work of the Father.

The elegy "Brod und Wein" ("Bread and Wine") already accentuated the proximity of Christ and Dionysus to the point that the two (equal) figures were superimposed on one another. Christ appeared there as that "silent genius" (*stiller Genius*) who announces the end of the day, the world's night that only the "holy memory" (*heilig Gedächtnis*) can endure (*SA* 2; 1: 94, 91). But the night is also the end of the world of the Greek gods, the end of the world of the presence and proximity of the visible gods. The end here is the fulfillment[18] that marks the term, the closing of the celebration of the heavens and the completion of time.[19]

To be sure, the death and disappearance of Christ constitute the opening of the long night of the world, of that time of distress given to melancholy and to mourning, "when the Father had turned his face from man":

> Aber Freund! wir kommen zu spät. Zwar leben die Götter,
> Aber über dem Haupt droben in anderer Welt.
>
> (*SA* 2; 1: 93)
>
> But my friend, we have come too late. The gods live, of course,
> But there, above our heads, in another world.

Neither nostalgia nor mourning are in place here, however. For in the solitude of the evening a "sign" remains, the announcement of the great return of the fled gods that is still to come, the sacrament of the Eucharist: "Thus we remember them in the heavens."[20]

Christ left us gifts as evidence of his return, the Bread and Wine that materialize the harmony between Earth and Sky and emphasize the proximity of Christ and Dionysus, the god of the Earth.

Hölderlin immediately associates the memory of the other divinities with the memory of Christ. The bread is the fruit of the Earth blessed by the Ether (*Vater Aether*). Wine also proceeds from Zeus, the thundering god, and is proper to Dionysus. Christ thereby finds his place in the "celestial chorus" once again, and the other demigods are already associated with his commemoration:

> Als der Vater gewandt sein Angesicht von den Menschen,
> Und da Trauern mit Recht über der Erde begann,
> Als erschienen zu lezt ein stiller Genius, himmlisch
> Tröstend, welcher des Tags Ende verkündet' und schwand,
> Ließ zum Zeichen, daß einst er da gewesen und wieder
> Käme, der himmlische Chor einige Gaaben zurük,
> Derer menschlich, wie sonst, wir uns zu freuen vermöchten,
> Denn zur Freude, mit Geist, wurde das Größre zu groß
> Unter den Menschen und noch, noch fehlen die Starken zu höchsten
> Freuden, aber es lebt stille noch einiger Dank.

Brod ist der Erde Frucht, doch ists vom Lichte geseegnet,
Und vom donnernden Gott kommet die Freude des Weins.
Darum denken wir auch dabei der Himmlischen, die sonst
Da gewesen und die kehren in richtiger Zeit.
(*SA* 2; 1: 94)

As the Father turned his face from man,
And, rightly, lamenting over the world began,
As, finally, a silent genius appeared, heavenly,
Consoling, and who announced the end of the day and disappeared,
As a sign that he was once there and would come back,
The heavenly choir left a few gifts behind,
In which, as before, we could take pleasure, as men,
For the great had become too great with spirit for joy
Amongst men, and yet, there is still a lack of those strong for
The highest Joy. But some thanks still live silently.
Bread is the fruit of the Earth; yet it is blessed by the light
And the joy of wine issues from the thundering god.
Therefore with it we think also of the Divine, who otherwise
Were there and will return when the time is right.

In the elegy "Bread and Wine," with which it is always fitting to begin in order to appreciate the relative novelty of the "christological" hymns, the final strophe evokes Dionysus in his proximity to the Crucified, of whom he is like a typological prefiguration. Dionysus also has the figure and function of a conciliator. He reestablishes the unity of what had been separated, torn so that the *Weltgang*, the course of the world, "has no gap," and so that Earth and Sky remain solidly bound in their differentiated unity. Dionysus is already he who grants Day and Night, Darkness and Clarity, together.[21] Through this work of conciliation, this "descent" into the obscure, Dionysus can legitimately be compared to the Son of the Most High, he who comes after the end of the Greek day as the torch bearer among the shadows:

. . . wir sind herzlos, Schatten, bis unser
Vater Aether erkannt jeden and allen gehört.
Aber indessen kommt als Fakelschwinger des Höchsten
Sohn, der Syrier, unter die Schatten herab.
(*SA* 2; 1: 95)

. . . we are heartless, shadows, until our
Father Aether is recognized by each of us and listened to by all.

But in the meantime, he, the Syrian, the son of the Highest,
Comes as a torch-bearer here below, amongst the shadows.

"Torch-bearer," *Fackelschwinger*: here, it is a matter of a classical epithet for Dionysus that is attested to by Sophocles (*Oedipus the King* 215–19) as by Euripides (*Bacchae* 145),[22] but that applies just as well to the "Syrian," that is, to Christ, without us necessarily having to decide between one identification or the other, in conformity with the general logic of the elegy (which was first entitled "Der Weingott" ["The God of Wine"]), whose entire economy tends to emphasize precisely the fraternal proximity of Christ and Dionysus,[23] to the point of the superimposition of the Eucharist and the fruits of the Earth and the Sky, bread and wine, that we have already noted.

From the second version of the fragment "Versöhnender der du nimmergeglaubt . . ." on, Christ appears linked to the Orient, notably through the evoking of his passage near Sichar, where he encounters the Samaritan (John 4: 5):

Und manchen möcht' ich laden, aber o du,
Der freundlich ernst den Menschen zugethan
Dort unter syrischer Palme
Wo nahe lag die Stadt am Brunnen gerne weiltest,
Das Kornfeld rauschte rings   still athmete die Kühlung
Vom Dunkel des geweiheten Gebirgs.
(*SA* 2; 1: 134)

And I would like to invite many, but O you,
Who are earnestly disposed to befriend men,
You there, under the Syrian palm tree,
Near where the city lay, liked to linger by the fountain,
The cornfield rustled around you   the coolness breathed silently
From the darkness of the hallowed mountains.

"The Only One" (1801–3) opens upon a clear opposition between Greece and the fatherland, or Hesperides.[24] In his too exclusive love for the ancient gods, the poet remains captive to his nostalgia for Greece:

Was ist es, das
An die alten seeligen Küsten
Mich fesselt, daß ich mehr noch
Sie liebe, als mein Vaterland?
(*SA* 2; 1: 153)

What is it that
To the ancient sacred coasts
Binds me, so that I
Love them still more than my homeland?

But if the religion of beauty, of the god who appears and gives himself to be seen in the image, responds as necessary to the flashing of the world into its light,[25] it also risks masking (*verbergen*) Christ, characterized not only as Lord and Master, but still as the "jewel of dwelling," the "last of your race," the very race from which the Greek gods stemmed. The Father is in fact he who has engendered and sent the gods, the heroes ("Der hohen Gedanken / Sind nemlich viel / Entsprungen des Vaters Haupt"; ["Sublime thoughts, / Many of them, / Have sprung from the Father's head"] *SA* 2; 1: 153), as well as the Son, the Unique. Thus, Christ is at once the last of the ancient gods, he who puts an end to the epiphany of the Greek gods, and he who keeps himself at a distance, withdrawing:

Mein Meister und Herr!
O du, mein Lehrer!
Was bist du ferne
Geblieben? und da
Ich sahe, mitten, unter den Geistern, den Alten
Die Helden und
Die Götter, warum bliebest
Du aus? . . .
("Der Einzige," *SA* 2; 1: 162)

My Master and Lord!
O you, my teacher!
Why have you remained
Distant? And when
I saw, in the middle of the Spirits, the ancients,
The heroes and
The gods, why did you remain
Apart? . . .[26]

But (such is the economy of the hymn) this recognition of Christ, beyond the dissimulation of the appearing and mundane gods, in turn risks excluding, because of an excessive love, all the others from the song and divine service of the poet. The central question of the hymn and Hölderlin's christology in general could therefore be formulated thus: How can

Christ be *der Einzige* without, for all that, pronouncing some exclusion of the other demigods? How are we to conceive such a nonexclusive unicity?

> . . . ist voll
> Von Trauern meine Seele
> Als eifertet, ihr Himmlischen, selbst,
> Daß dien ich einem, mir
> Das andere fehlet.
> (*SA* 2; 1: 162)

> My soul is full of sadness,
> As though, O you divine ones, you yourselves,
> If I serve one, must
> Lack the other.

But, in truth, a false alternative is at stake there. To characterize Christ as unique is not to exclude (in the too narrowly interpreted sense of Christian dogma) the Greek gods. Christ's unicity appears in truth when he himself reemerges from the milieu of the Greek gods as the last of the ancient gods—as "of the same race" as them.[27] Unicity characterizes Christ in that he has a *single*, *unique*, signification for this time between times (*dürftige Zeit*) that is the time of the fled gods, and during which the principle task of the poet is precisely to find the measure[28] that measure that founds and authorizes the comparison between Christ, the Only One, and the other gods or demigods. What is this "reserve" that in the first place holds the poet back from bringing together, in order to compare, Christ in his unicity and the Greek gods[29] Dionysus and Heracles, characterized as *weltliche Männer*, "men of this world"? Of what does the "daring," the "recklessness,"[30] of the profession of faith, of the "confession" that declares the following consist?

> . . . So sind jene sie gleich. Voll Freuden, reichlich. Herrlich grünet
> Ein Kleeblatt . . .
> (*SA* 2; 1: 163)

> . . . And so are they alike. Full of joy, plentiful. Glorious,
> A clover leaf becomes green . . .

Where is the mistake here? Of what does it consist?[31] Must we conclude from the fact that Dionysus and Heracles are characterized as *weltliche Männer* that Christ would be determined as more celestial, as the man of another world? The image of clover with its three leaves (Heracles, Dionysus, Christ) immediately forbids such an opposition. If the image of the three-leaf clover is daring, it is not so much because it would assimilate Christ, the most celestial, to the men of this world because he stems from another world or an other-world or from a different economy, that of the spirit. It is daring, rather, because the image risks "dissimulating" the radically "kenotic" character of the event of Christ, the humanity of his task (*Tageswerk*).

The true boldness in the knowledge of the poet lies here, in that, through the unity of the modes of appearance of the divine, all referring to the Father or the Most High, he recognizes the unicity of the reconciliation that is carried out by Christ. The comparison bears witness to the care the poet takes, conscious of his mission, to omit none of the cosmic theophany in order to greet the revelation of Christ.

If the poet evokes a potential mistake (*Schuld, Fehl*) with the reserve to which every decision to say and, especially, to name,[32] remains attached, this mistake is not connected to an unwarranted, reckless, or sacrilegious comparison, but to an excess of love or of attachment to the One that would not then be recognized as the mediate and mediating unity, which precisely emerges as such only from the comparison-distinction.[33] The mistake here is the excess of love:

> Gut machen will ich den Fehl,
> Wenn ich noch andere singe.
> (*SA* 2; 1: 155)

> I wish to make the mistake good,
> When I sing of still others.

The exclusive attachment is the mistake precisely because this attachment does not recognize the unicity of Christ. "Der Einzige":

> Es hänget aber an Einem
> Die Liebe. Diesesmal
> Ist nemlich vom eigenen Herzen
> Zu sehr gegangen der Gesang . . .
> (*SA* 2; 1: 155)

> But my love clings
> To only One. This time,
> Too much from my own heart
> Has the song issued.

In "Patmos," Hölderlin will return to the mistake of the excess of attachment. Approaching Christ, the disciples must be attentive not to stop at the first epiphanic moment of his economy of presence:

> Zu meiden aber ist viel. Zu viel aber
> Der Liebe, wo Anbetung ist,
> Ist gefahrreich, triffet am meisten. Jene wollten aber
> Vom Angesichte des Herrn
> Nicht lassen und der Heimath. Eingeboren
> Wie Feuer war in dem Eisen das, und ihnen
> Zur Seite gieng, wie eine Seuche, der Schatte des Lieben.
> (*SA* 2; 1: 182–83)

> But much is to be avoided. Too much love, however,
> Where this is adoration,
> Is dangerous, strikes most. But those did not want
> To leave the face of the Lord
> And their homeland. Inborn
> Like fire in iron, this, and walked
> Beside them, like a plague, the shadow of the beloved.

The true mistake that the connection—the parallel to be established between Christ and the other demigods—can lead to does not originate in some lack in Christ. Christ remains the Master, the Lord, the Council, he who, offended by the immediacy of the visible demigods, was sought for in vain for a long time. But if there is a mistake in the poem, it is in relation to the divine economy in general, in relation to the Father. However, it is only at the risk of that mistake, at the end of a struggle in the poet's spirit, that this new decisive perspective can be won, the perspective according to which the unicity of Christ emerges precisely only through its comparison with the other demigods. Christ's love must not distance the other demigods, as though they were jealous, since Christ is the Conciliator par excellence.[34]

Perhaps loving always implies loving excessively (*zu sehr*), but this excess, even if it is inscribed in the very essence of Christianity as the religion of love, must not, however, either exclude or misjudge the unity of and in differentiation. How is one to find the "measure" of love?

> Nie treff ich, wie ich wünsche,
> Das Maas. Ein Gott weiß aber
> Wenn kommet, was ich wünsche das Beste.
> (*SA* 2; 1: 155)

I never strike, as I would like to,
The measure. A god, however, knows
When what I wish for comes the best.[35]

Such is the question that is opened immediately at the resolution of "Der Einzige" (*SA* 2; 1: 155): "This mistake, I want to fix it if I sing again."

To find the measure, at any rate, is first to recognize that it is the measure of love and that the law is "das schönausgleichende Gesetz der Liebe" ("the beautifully balancing law of love"). *Eifersucht* (jealousy) would therefore not find its place in the sphere of the divine, which, rather, characterizes Conciliation, of which Christ is the exemplary figure:

Und schöner, wie sonst, o sei
Versöhnender nun versöhnt daß wir des Abends
Mit den Freunden dich nennen, und singen
Von den Hohen, und neben dir noch andere sein.

And more beautifully, as before, O be,
Conciliator, now reconciled, so that evenings,
With friends we might name you, and sing
From the heights, and beside you still others might be.[36]

It is the very unicity of Christ, the most beloved, sent by the Father, in the end, that assigns him an eminent consoling and, even more, reconciling function. The Father, here, must not be understood in the framework of the Trinitarian dogma as God the Father, he who sent his only son, *unigenitus*. The Father (Etherial-Father, the "most properly Zeus") designates that supreme instance, barely known and undoubtedly essentially anonymous, without name,[37] the instance that sent, that bestowed all the divine that has arisen in the Occident.

This Father, one and the same, "completely known"-unknown, is also he whose reign is always essentially differentiated: "For he never rules alone" ("Denn nimmer herrscht er allein") ("Der Einzige," *SA* 2; 1: 155). The difference, still unspoken in the first, incomplete version of "The Only One," will receive its most complete determination in the third version. It is mediateness, writing:

. . . erscheint Gott . . .
. . . Mittelbar
In heiligen Schriften.
(*SA* 2; 1: 163)

... appears God ...
... mediate
In sacred writings.

To be sure, Dionysus is also he who reconciles, and it is in that that he can appear as the figure or anticipation of the Crucified. He is *Gemeingeist*, as a famous late variant of "The Only One" indicates.[38] A common spirit, he is nonetheless opposed to all drunkenness, to all vertige. He is he who knows how to tame the "peoples' desire for death" ("Todeslust der Völker").[39] The god of institutions, of orientation, he indicates the sure paths (*die rechte Wege*). Finally, he reminds the people of "das Maas, daß einer Etwas für sich ist" ("the measure, that each is Something for himself"). In that, Dionysus is already he who teaches the *unterschiedene Einheit* (differentiated unity).

But Dionysus, as the worldly (*weltlich*) demigod (he who founds and dwells), far from being opposed to Christ, the "uprooting" god, even in being this worldly demigod gives the ground for the comparison, the analogy that connects him to Christ. Hölderlin's Christ, in fact, totally assumes the humanity and the anxiety of finitude. Even in the desert,[40] Christ refuses to use his divine power, and he takes on the whole human distress, even hunger, pain, and an ignominious death on the cross of thieves. But before the real and ultimate temptation, that of Jesus' prayer on the Mount of Olives, addressing his Father so that he spares him the cut of infamy, the desert, where the divine fire is near and devouring, is the place of temptation par excellence, that of the "Hinwegjauchzen von dieser Erde" ("Yes, always with a cry of joy the world withdraws from this Earth"), the temptation to leave the Earth precipitously. In resisting the temptation with all his might, Christ surmounted the *Todeslust* to which Empedocles succumbed. He took on the terrestrial condition fully. He separated, distinguished himself from the divine all-powerfulness, but at the same time he sublated his secret unity with the Father. For hence he can accept his task, his ultimate mission of Conciliation on this Earth, in taking to heart differentiated unity and thereby encountering Heracles and Dionysus.

So sind jene sich gleich. Voll Freuden, reichlich. Herrlich grünet
Ein Kleeblatt.

And so are they alike. Full of joy, plentiful. Glorious
A clover leaf becomes green.

The comparison of the three demigods has made what they have in common and also what remains unique about Christ emerge. The three gods are brothers: they are *Feldherrn, Heroen* ("field generals," "heroes"). And if mortals need them, therefore, it is because "ohne Halt verstandlos Gott ist" ("without end, God is incomprehensible") (*SA* 2; 1: 163).

Must we choose between the image of a profoundly "kenotic" Christ and the image, suggested by Wolfgang Binder, of a Christ without passion and without reconciliation (*Versöhnung*)?[41] If, in fact, Hölderlin never invokes Golgotha directly, must we conclude from this that he intends to stress Presence much more than expiatory sacrifice, *Opfertod*? I will concede to Binder that, notably in "Patmos," Hölderlin reflects first and foremost on divine "manifestation," on the measure of the appearance of the divine, of its epiphany in Christ.[42] But does this historical perspective make of Christ the final theophany, taking place in the series of the "cosmic divinities"? To what extent does Christ form a series with the other demigods?

Any response to these questions that would not take into account the extraordinary reserve of the poet, echoing John 16: 12, would risk amputating Hölderlin's christology of one of its essential dimensions. If we must guard against overinterpreting the recurrent motif ("Vieles wäre / Zu sagen davon"; "There would be / Much to say about it"), we would be equally wrong, however, to see in it merely some evasive strategy, as, it seems to me, is attested by the movement of the sixth strophe of "Patmos," which presents John, Christ's beloved disciple, in an extraordinary summary of the Last Supper and Golgotha:

> . . . und es sahe der achtsame Mann
> Das Angesicht des Gottes genau,
> Da, beim Geheimnisse des Weinstoks, sie
> Zusammensaßen, zu der Stunde des Gastmals,
> Und in der großen Seele, ruhigahnend den Tod
> Aussprach der Herr und die lezte Liebe, denn nie genug
> Hatt' er von Güte zu sagen
> Der Worte, damals, und zu erheitern, da
> Ers sahe, das Zürnen der Welt.
> Denn alles ist gut. Drauf starb er. Vieles wäre
> Zu sagen davon . . .
> (*SA* 2; 1: 167)

. . . and the attentive man saw
God's face exactly
As, by the secret of the vine, they
Sat together at the banquet hour
And in his great soul, calmly foreboding,
The Lord pronounced death and the final love, for he never
Had enough words to speak of goodness,
At that time, and to console, when
He saw it, the wrath of the world.
For all things are good. Thereupon he died. There would be much
To say about that . . .

The disciples believed they had found, in the immediacy of the divine, in Christ's face, the very object of their desire and their true homeland. But therein lay precisely the danger of too great a proximity and of the excess of measure. By way of his kenosis, by way of his departure, which is also the gift of his spirit, Christ saves:

. . . Drauf starb er.
Es sahen aber, gebükt, deß ungeachtet, vor Gott die Gestalt
Des Verläugnenden . . .
(*SA* 2; 1: 182)

. . . Thereupon he died.
Yet nonetheless they saw the figure inclined before God, otherwise
unnoticed,
Of he who renounces himself . . .

Humiliation, condescendence, and even making foreign, or exhaustion, could be understood in the nonappearance of the expression *sich bescheiden*, which Hölderlin uses in a late version of the hymn "The Only One":[43]

Christus aber bescheidet sich selbst.
Wie Fürsten ist Herkules. Gemeingeist Bacchus. Christus aber ist
Das Ende. Wohl ist er noch andrer Natur; erfüllet aber
Was noch an Gegenwart
Den Himmlischen gefehlet an den andern . . .
(*SA* 2; 2: 752–53)

Christ, however, resigns himself.
Like princes is Hercules. Communitary spirit, Bacchus. But
Christ is
The end. He is of yet another nature; but fills

What was still lacking in presence
For the heavenly beings to the others . . .

The image of the clover leaf is elucidated here only by the assimilation of Heracles to the hunter, of Dionysus, as the god of the Earth, to the laborer, and of Christ to the beggar (ἐν μορφῇ δούλου):

. . . Jene drei sind aber
Das, daß sie unter der Sonne
Wie Jäger der Jagd sind oder
Ein Akersmann, der athmend von der Arbeit
Sein Haupt entblößet, oder Bettler.
Nicht so sind andere Helden. Der Streit ist aber, der mich
Versuchet dieser, daß aus Noth als Söhne Gottes
Die Zeichen jene an sich haben. Denn es hat noch anders, räthlich,
Gesorgt der Donnerer. Christus aber bescheidet sich selbst.
(*SA* 2; 2: 752)

. . . But these three
Are this, that under the Sun
They are like hunters of the hunt, or
The laborer, who, breathing hard from his work,
Bares his head, or like a beggar.
Not so are the other heroes. But the conflict
That attracts me is this: necessarily, as the sons of God,
They carry the signs in themselves. For in yet another way,
opportune,
Has the Thunderer provided for. But Christ gives his own consent.

In following Detlev Lüders's invaluable indications[44] here, we understand this consent, this humiliation, as Christ's decision to take upon himself and voluntarily assume, right to the end, the human figure and its proper limitation. In this, Christ differs from the "men of this world," Hercules or Bacchus, hunter or laborer, who, conforming to the will of the Master of the storm (*Donnerer*), had to don the external signs of the terrestrial condition, while intending to remain "celestial." If, then, Christ is "of another nature," this is not, as Allemann would have it, because he would still belong to the beyond, but precisely because he divests himself of all divine "signs" and remains like a man who inhabits the world. The fact of "resigning himself," far from being the means by which Christ "can be compared to the two other demigods,"[45] on the contrary testifies to his "other nature," his "unicity," that very unicity that allows him to accomplish fully the mission assigned him by the Father. As Lüders writes nicely, "With Christ not only a historical point of view, the ancient day of the gods, is completed; in him the attainable measure of the incorporation, the incarnation, of the divine in the human also reaches its term in a vertical line. More resolutely and unconditionally than the other gods or demigods, he was a man. But it is thereby that, more than all the others, he can pull together, link Earth and sky. With him, the Father accomplishes his ultimate, his extreme operation."[46] I could not put it better.

Christ thus accomplishes what his predecessors still failed to do, his predecessors who thereby prefigure him—in the words of "Der Einzige":

> Dieweil sein Äußerstes that
> Der Vater und sein Bestes unter
> Den Menschen wirkete wirklich . . .
> (*SA* 2; 1: 156)

> Meanwhile the Father's utmost
> And best deed amongst
> Men really took form . . .

Christ's work will have been "real," "effective," and not only spiritual.[47] To be sure, for Hölderlin Christ's death coincides with the gift of spirit. "Patmos":

> . . . Drauf starb er. Vieles wäre
> Zu sagen davon. Und es sahn ihn, wie er siegend blikte
> Den Freudigsten die Freunde noch zulezt,
> Doch trauerten sie, da nun
> Es Abend worden, erstaunt,
> . . .
> . . . und lassen wollten sie nicht
> Vom Angesichte des Herrn
> Und der Heimath . . .
> . . .
> Drum sandt' er ihnen
> Den Geist . . .
> (*SA* 2; 1: 167–68)

> Thereupon he died. There would be much
> To say about that. And the friends saw him, as he triumphantly
> Looked up, the most joyful, at the very end,

> Yet they were sad, now that
> It had become evening, astonished,
> . . .
> . . . and they did not want to quit
> The face of the Lord
> And the homeland . . .
> . . .
> Therefore, he sent them
> The Spirit . . .

But Spirit is also what makes man capable of "dwelling," what gives him a heart that may well endure the categorical turning away of God such as it is performed in the kenosis and withdrawal of Christ.

Christ is not the *only* mediator, but he is the mediator who fulfills mediation to its most extreme limit. He is therefore the last mediator, for he does not simply present a *Bild Gottes* (image of God), but because in a sense with him God came upon this Earth. Christ is the sign of complete, perfected unity, because he carries to Earth the unity between Heaven and Earth, man and God. Thus, Christ is the last hero, the last god after whom no other god can appear, as the hymn "Patmos" equally teaches us: God (the Father?) can be grasped, apprehended ("close and difficult to understand, God"), only when the internal connection between all the peaks of time, no matter how distant from one another, is assured. Christ is *one* of the peaks of time, but also he who allows the harmonious reunification of all the others. That is why he can be named, in the fifth triad of the hymn, *Losungszeichen*, the "password," the "enigmatic word." If, then, "noch lebt Christus," as "Patmos" puts it, if Christ is still living, this is essentially as the mediator, present in the Holy Scriptures, of whom it falls to the poets—the "spiritual" beings who are also "of the world" (*weltlich*)—to keep the letter with care.

This care for the letter as divine service in expectation of what comes through the Day of Celebration, in which the spreading of the divine in its christological unity is gathered, is also what the prose outline of "Friedensfeier" magnificently expresses:

> Ein Chor nun sind wir. Drum soll alles
> Himmlische was gennant war, eine Zahl
> geschlossen, heilig, ausgehen rein aus unserem Munde.
> Denn sieh! es ist der Abend der Zeit, die Stunde
> wo die Wanderer lenken zu der Ruhstatt. Es kehrt bald

Ein Gott um den anderen ein, daß aber
ihr geliebtestes auch, an dem sie alle hängen, nicht
fehle, Und Eines all in dir sie all, sein,
und alle Sterblichen seien, die wir kennen bis hieher.

Darum sei gegenwärtig, Jüngling. Keiner, wie
du, gilt statt der übrigen alle. Darum haben
die denen du es gegeben, die Sprache alle geredet, und du
selber hast es gesagt, daß in Wahrheit wir auf
Höhen und geistig auch anbeten werden in Tem-
peln. Seelig warst du damals aber seeliger
jezt, wenn wir des Abends mit den Freunden
dich nennen und singen von den Hohen und rings
um dich die Deinigen all sind. Abgelegt
nun ist die Hülle. Bald wird auch noch anderes klar
seyn, und wir fürchten es nicht.
(*SA* 2; 2: 699)

We are now a choir. Therefore everything
Divine that was named, a number,
closed, sacred, must, leave pure from our mouth.

For look! It is the evening of time, the hour
in which the voyagers steer towards their place of rest. Soon
will enter
One god upon the other, but so that
Their most beloved, on whom they all hang, is also not
lacking, and so that all of them are One in you,
and all the mortals we have known until now.

Therefore be present, adolescent. No one counts
for all the others as you do. Therefore
those who you permitted spoke all languages, and you
yourself said that, in truth, we will worship
In the heights and spiritually in temples.
Blessed were you then, but more blessed
now that, evenings, with friends,
we name you and sing about those above, and all
around you are your own. The clothing
is now laid down. Soon still other things will be clear
and we do not fear that.

Translated by Jan Plug

Epistolary Writing, Fate, Language

HÖLDERLIN'S 'HYPERION'

Edgar Pankow

What is the relation between epistolary writing and fate? What is the relation between fate and language? Hölderlin pursued these questions at least since 1792.[1] In 1797, he published the first part of this interrogation; a second part followed in 1799. A terra incognita has been visited.[2] The name of the journey: *Hyperion, or the Hermit in Greece* (*SA* 3: 1–160).[3]

Hölderlin's novel was long neglected by the critics, although the situation changed to some degree after the publication of Lawrence Ryan's study in 1965.[4] However, even book-length studies barely pay attention to the novel's epistolary outline.[5] Setting out from Hölderlin's second letter to Casimir Ulrich Böhlendorff, this study will trace the constellation of epistolary writing, fate, and language in the work that constitutes Hölderlin's most extensive project in prose, *Hyperion*.

Toward the Other

In November 1802, Hölderlin wrote to Böhlendorff:

> If you would just write to me soon. I need your pure tones. The psyche among friends, the origination of thought in conversation and correspondence is necessary for artists. Otherwise we have none for ourselves; but it belongs to the holy image that we create. (*SA* 6; 1: 433; *Essays*, 153 [trans. modified])

Writing, among other things, about epistolary writing, Hölderlin addresses a personal need and an artistic necessity. The request for an early response to his letter aims at both. Hölderlin is in need of the "pure notes" of Böhlendorff, the friend and artist. Yet in a remarkable twist, Hölderlin proposes an epistolary exchange in which neither he nor Böhlendorff would figure as a direct addressee for each other. Rather than Böhlendorff, Hölderlin addresses what he calls, in an unusual phrase, "psyche among friends." Such "psyche among friends, the origin of ideas in conversation and letter is a necessity for artists." The address constitutes an incisive departure from conventional subject-centered notions of epistolary exchanges. Obviously, Hölderlin's projected epistolary commerce with Böhlendorff goes beyond the formal structure of a relationship which would locate the correspondence as a setting of individuals who refer to each other in their distinctiveness. Addressing the "psyche among friends," Hölderlin leaves behind an entire tradition of epistolary writing which understands the letter in the context of an exchange between independent subjects.

The syntax of the phrase "psyche among friends" is strikingly unconventional, and Adolf Beck, the editor of Hölderlin's letters, in fact does not hesitate to diagnose the stylistic particularity as a sign of Hölderlin's "weakening strength in the linguistic expression of deep insights" (*SA* 6; 2: 1086). Beck's commentary suggests a regrettable case of individual deficiency: Hölderlin, Beck claims, *no longer* is in command of his powers when he addresses himself to the "psyche among friends." Hölderlin's later fate—his "madness"—seems to justify this judgment. And yet, the design of the letter appears more complicated. Concerning anything as elusive as psyche, and in particular psyche itself, strength of expression is a highly problematic requirement. Hölderlin seems to aim at an entirely different kind of discourse here: namely, an open *epistolary* exchange that would respond to the artistic needs of two poets. The *psyche* "among friends" does not refer us back to Hölderlin's psyche, to a subject that speaks, but to the psyche of conversation and letter writing.

Psyche is a term rarely used in Hölderlin's writings. Moreover, it imports an element of classical Greek into a modern German context. The use of the Greek term in a letter to another poet—rather than the German *Seele*—reverberates with one of the most profound concerns of Hölderlin's poetic and theoretical work. Significantly enough, his first letter to Böhlendorff, December 4, 1801 (often referred to as *the* Böhlendorff letter), is almost exclusively devoted to the issue: here, Hölderlin outlines an integral part of his response to Greek antiquity. Given the context, the evocation of the Greek term in the second letter carries the full charge of Hölderlin's thinking about Greek classicism.

In the first letter to Böhlendorff, Hölderlin demands no less than "that, with the exception of that which has to be the highest for the Greeks and for us—namely the living relation and fate [nemlich dem lebendigen Verhältniß und Geschik]—we may not to have anything *in common* with them" (*SA* 6; 1: 426; *Essays*, 150 [trans. modified]). Read in light of these lines, the perspective on Hölderlin's use of the term *Psyche* in the second letter gains an additional dimension. Now it appears that concerning *Psyche*, too, since it speaks Greek and recalls Greek thinking about the soul, "we may not to have anything *in common*" with it, except "the living relation and fate." In the second letter to Böhlendorff the word *Psyche* belongs in an emphatic way to the realm of a foreign language. Yet the Greek term not only exposes "us" toward the other, the foreign, the unfamiliar, but also evokes what we have in common because it is "among" us, namely a certain *Geschik*, "fate," "skill," or "transference," that, Hölderlin hopes, will take the form of an epistolary exchange with Böhlendorff. It is noteworthy that *Psyche* introduces the sign of a foreign language at precisely the moment when Hölderlin proposes a new community and outlines the future correspondence with his friend. The use of the Greek word, in place of the German *Seele*, marks a movement toward the other *as* other, a transition or translation into an area that is unfamiliar. It is the Greek clue, situated "among" these two Germans, that evokes epistolary writing as a mutual exploration of otherness.

The use of *Psyche* in the letter of November 1802 denotes a shift from the German *seelenvoll* ("soulful"), which Hölderlin employed in the earlier letter to Böhlendorff in order to qualify Homer's encounter with the other. The "Greeks," Hölderlin wrote then, "are superior in the gift of representation from Homer on because this extraordinary man was soulful enough to capture for his Apollonian realm the Occidental *Junonian sobriety* and thereby truly to appropriate that which is foreign" (*SA* 6; 1: 426; *Essays*, 149–50 [trans. modified]). Here, Hölderlin presents Homer as someone who was "soulful enough" to appropriate that which is foreign. The soul brings something unfamiliar and foreign to Homer's mind (i.e., the "Occidental *Junonian sobriety*"); it captures that which is foreign *as* something foreign; and, most importantly, it alters its own mode of representation in the course of this experience. The soul induces Homer to speak in a new, hitherto unfamiliar tongue. Significantly, Hölderlin refers to Homer's appropriation of that which is foreign—which establishes a superior quality of representation—with a German adjective, *seelenvoll.* For Homer, writes Hölderlin, psyche spells soul.

In the address to both Böhlendorff and Homer, the terms *Psyche* and *Seele* speak a kind of foreign language that is not simply another language, a language of another country, of Greece or Germany, or of another time, but, in an essential way, what could be called the language of the other. Although it would certainly have been possible for Hölderlin to render *Psyche* as "soul" or *Seele*, and vice versa, translation cannot undo the linguistic rupture marked by the occurrence of the other. The essential problematic posed by the very existence of such a thing as "psyche" or "soul" still would linger on. The translator would still be met with an irreducible otherness, no matter in which national language he chooses to speak or, more precisely, no matter in which national language "psyche" chooses to speak the language of the other.

Hölderlin's reference to Homer's "soul" indicates that the encounter with "that which is foreign" does not happen outside the field of language. On the contrary, Homer's shift toward a new paradigm of representation demonstrates in an exemplary way that the other (in the case of Homer: the "Junonian sobriety") coincides with the structure of another, or foreign language. The soul or psyche, then, ties the speaking subject to a language that does not originate in the subject itself, and cannot be fully translated into the idiom of subjectivity. Hölderlin's two letters to Böhlendorff also demonstrate, however, that only the translation of the other (of Homer's "psyche" into "soul," of Hölderlin's "soul" into "psyche") can make evident what translation is *unable* to do: namely, to establish a direct contact—be it with Greek antiquity or an epistolary correspondent—without losing ourselves as artists and friends.

Hölderlin links the "psyche among friends" to "the origin of thought [das Entstehen des Gedankens] in conversation and correspondence." Like psyche, the origin of thought has no place other than in the in-between, the transsubjective difference. Hölderlin in fact does not ask Böhlendorff to make thinking the *topic* of their conversation or letter writing. He aims at the *origin* of ideas (*Gedanken*) as something that exceeds the thematic discourse about origins, but comes into being, as the gift of friendship, in conversation and letter writing—adding: "Otherwise we have none [i.e., no idea (*keinen Gedanken*)] for ourselves."

The second letter to Böhlendorff addresses the origin of thought as a formal detour, as dialogue or discourse. For Hölderlin, thinking necessitates that the artist opens up toward the other through conversation or letter writing. Unable to reach the originality of thought directly, and all by himself, the artist reaches out beyond the confines of subjectivity—toward an other which is not just another subject. It is this original moment of thought, arising in conversation or in an epistolary exchange, which situates the psyche "among friends."

Return and Repetition

Singulière fortune où le but se déplace,
Et, n'étant nulle part, peut être n'importe où!
Où l'Homme, dont jamais l'espérance
n'est lasse,
Pour trouver le repos court toujours comme
un fou!
—Charles Baudelaire[6]

In Hölderlin's novel, everything seems to begin with repetition. The first sentence of Hyperion's first letter to Bellarmin speaks about a return and about that which is happening again: "Der liebe Vaterlandsboden giebt mir wieder Freude und Laid" ("Once again the dear earth of my native country brings me joy and sorrow") (*Hyp*, 7; 2). No line is written, no letter sent until Hyperion returns to the fatherland and a certain joy and sorrow begin to unfold anew. The fact that the "hermit in Greece" should address the issue of his return first is particularly significant. The figure of return is the figure par excellence of a certain Greek classicism. Greek experience and thinking has been represented over and again as moving along a circular line. Thus Odysseus, the quintessential Greek hero, returns to Ithaca, and the Platonic soul returns, in an act of anamnesis, to the ideas it conceived before birth. In the horizon of this particular Greek interpretation, being is returning, and Hyperion, it appears, wants to return too—not to the modern Greece of his contemporary fatherland but to the experience and thinking of Greek antiquity. Hyperion wants to return to the possibility of returning. He wants to return again.

Significantly, it is the reading of one of the myths of return, where Plato speaks about "aging and rejuvenation,"[7] that stands at the beginning of Hyperion's relationship with Alabanda which will later culminate in the unsuccessful attempt to recapture the territory of Greek antiquity by means of force. In a letter to Bellarmin, Hyperion details the encounter with Alabanda, taking place in the timeless shadow of the "ever green laurel":

Ein Tag ist mir besonders gegenwärtig.

Wir waren zusammen auf's Feld gegangen, saßen vertraulich umschlungen im Dunkel des immergünen Lorbeers, und sahn zusammen in unsern Plato, wo er so wunderbar erhaben vom Altern und Verjüngen spricht. . . .

Wir sprachen darauf manches vom jezigen Griechenland, beede mit blutendem Herzen. (*Hyp*, 27–28)

One day is especially present to me.

We had gone to the country together and were sitting with our arms trustfully around each other in the dark shade of the ever green laurel, looking at our Plato—the passage where he speaks of aging and rejuvenation. . . .

We then spoke much of Greece today as it is today, both of us with bleeding hearts. (20)

In the first Böhlendorff letter, Hölderlin spells out an essentially modern intervention against the continuity of Greek classicism.[8] A similar concern permeates his epistolary novel. In one of its last letters—to Notara, enclosed in a letter to Bellarmin—Hyperion unmistakably states:

Aber es ist die vorige Welt nicht mehr, zu der ich wiederkehre. Ein Fremdling bin ich, . . . und wär' ich auch auf meiner heimatlichen Insel, in den Gärten meiner Jugend, die mein Vater mir verschließt, ach! dennoch, dennoch, wär' ich auf der Erd' ein Fremdling und kein Gott knüpft ans Vergangene mich mehr. (*Hyp*, 150)

But the former world is no longer to which I return. I am a stranger, . . . and if I were on my native island, in the gardens of my youth, which my father bars to me, ah!, even then, even then I should be a stranger on Earth, and no god would join me to the past again. (125; trans. modified)

For the general design of the novel, it is of utmost importance that Hölderlin has Hyperion not simply deny the possibility of a return to the gardens of his youth. Hyperion's is not an abstract negation, which could be phrased, for example, "I cannot return" or "There is no return." Signifying a move back to a place where Hyperion has never been before, the term "return" has a more intricate temporal structure. Hyperion's "former world is no longer," yet still he returns. He experiences homecoming as an encounter with foreign territory. Nonetheless he is returning, but this return displaces the opposition of fatherland and exile, things familiar and foreign, which—in the wake of the *Odyssey*—has governed Occidental imagination. Hyperion's return encloses the exile within itself; it is marked by the inaccessibility of its own origin. He returns to Greece, though it is no longer a Greek return that structures his journey.

The parameters of Hyperion's troubled homecoming give rise to a set of assumptions not easily reconcilable. If returning is possible for Hyperion, it is still impossible to return in the manner in which Greek antiquity seemed to conceive of returning. What gives Hyperion's declaration to Notara its paradoxical appearance is the fact that his notion of returning is not based on identity. Hyperion arrives at his fatherland, but the fatherland is no longer the same. He returns, but he misses the idea of identity for which he looked in his return. Neither Homer nor Plato—that is, the Homer and Plato who emerge according to Winckelmann's paradigms—would have been able to admit the particular quality of Hyperion experiences: that the essence of returning is not returning, that the essence of homecoming is not homecoming. Coming home, Hyperion continues to live on foreign territory and even declares that, "if I were on my native island, in the gardens of my youth, . . . even then, even then I should be a stranger on Earth, and no god would join me to the past again" (*Hyp*, 150; 125). In his return, Hyperion encounters an exile unknown to Greek antiquity.

Hyperion experiences at once a personal predicament and an epochal shift in the logic of repetition. The latter is all the more significant for its departure from the Romantic construction of Greek classicism. According to the idealized Greek model, the returning to (and of) the same is always possible, at least in principle, since the very memory that leads back to the true transcendental home of man is ultimately guided by the unchangeable character of timeless ideas. A returning of this kind is based on a concept of identity; it derives its authority from the assumption that the *act* of returning is immune to temporal displacement—in other words, that the act is, in fact, not acting. Whereas Platonic anamnesis imagines a return to the preexisting idea itself, Hyperion experiences returning as a displacement of the very place to which he intends to come back.

Untimely Measures

Hölderlin differentiates the outline of his novel from the romantic commonplace ("I cannot return") by a complex temporal dialectic in the course of which the possibility of the historical narrative itself is fundamentally challenged and reconsidered. Addressing Bellarmin, Hyperion writes:

Lieber! es war eine Zeit, da auch meine Brust an großen Hoffnungen sich sonnte, da auch mir die Freude der Unsterblichkeit in allen Pulsen schlug, da ich wandelt' unter herrlichen Entwürfen, wie in weiter Wäldernacht, da ich glücklich, wie die Fische des Oceans, in meiner uferlosen Zukunft weiter, ewig weiter drang.

Wie muthig, seelige Natur! entsprang der Jüngling deiner Wiege! wie freut' er sich in seiner unversuchten Rüstung! Sein Bogen war gespannt und seine Pfeile rauschten im Köcher, und die Unsterblichen, die hohen Geister des Altertums führten ihn an, und sein Adamas war mitten unter ihnen.

Wo ich gieng und stand, geleiteten mich die herrlichen Gestalten; wie Flammen, verloren sich in meinem Sinne die Thaten aller Zeiten in einander, und wie in Ein frolokend Gewitter die Riesenbilder, die Wolken des Himmels sich vereinten, so vereinen sich, so wurden Ein unendlicher Sieg in mir die hundertfältigen Siege der Olympiaden. (*Hyp*, 18)

Dear friend! there was a time when my heart, too, basked under the sun of great hopes, when the joy of immortality pulsed in my every vein, when I roved among beautiful projects as through the half light of a vast forest, when, happy as the fish in the sea in my boundless future, I pressed on, farther, ever farther.

How boldly, blessed Nature! did the youth leap from your cradle! how he rejoiced in his untried weapons! His bow was drawn, his arrows rattled in the quiver, and the immortals, the high spirits of Antiquity, led him on, and his Adamas was among them.

Wherever I went, wherever I stopped, their glorious presences were with me; in my thought the high deeds of all the ages were mingled together, and as those gigantic forms, the clouds of heaven, united in *one* exultant storm, so the hundredfold victories of the Olympiads were united in me, so did they become *one* never-ending victory. (11–12)

The passage presents a dilemma similar to the one concerning the spatial twists of Hyperion's return to a place no longer existing. The evocation of the "joy of immortality," the commerce with the "immortals," and the ecstatic fusion of the "deeds of all the ages" are nested within a grammar of temporality that seems unsuited to house its atemporal referent. Grammar and semantics are discontinuous. While the phrase "there was a time" seems to refer us back to a past moment within a stable historical continuum, the actual reference to the fusion of the "deeds of all time," cannot, by definition, be *part* of this continuum. The temporal ambiguity of the passage effectively merges two fundamentally different types of memory, one proceeding along the lines of a historical narrative, the other presenting itself as transhistorical event.

Throughout the entire text, the novel offers itself as an irregularly shifting network of these two forms of memory. It is evident that Hölderlin does not present us with a clear-cut alternative. Hyperion has no choice other than to use a grammar which affirms itself through a structured continuity, but necessarily begins to contradict itself once it is used to express a kind of experience which questions the idea of continuity in general. Similarly, Hyperion's transhistorial fusion of "the deeds of all the ages" is recalled by recourse to an idea of plurality which implies temporal and spatial differentiations within a historical grammar and its vocabulary.

In Hyperion's account, historical and transhistorical memory are counterparts; they harbor each other inside themselves as their nonthematic other. Historical narrative and transhistorical event inhabit each other by mutual implication. They are each other's interiorized, though severed memory. Both types of memory, one could say, are marked with the necessary forgetting that constitutes a psychic space which is proper and yet unavailable to themselves. Hölderlin's text is not concerned with these forms of memory on the literal level. They only merit interest because their way of being historical or transhistorical is *strategic*, that is, involved in a discontinuous exchange with the other. Hyperion's narrative is neither historical nor transhistorical, but the opening of an elusive space in-between that produces historical or transhistorical effects, a trace of psyche in-between history and forgetting.

Childhood and Metaphor

Concerning the important theme of childhood, Hölderlin writes early on in Hyperion's third letter to Bellarmin.

Wie der Arbeiter in den erquikenden Schlaf, sinkt oft mein angefochtenes Wesen in die Arme der unschuldigen Vergangenheit.

Ruhe der Kindheit! himmlische Ruhe! wie oft steh' ich stille vor dir in liebender Betrachtung, und möchte dich denken! Aber wir haben ja nur Begriffe von dem, was einmal schlecht gewesen und wieder gut gemacht ist; von Kindheit, Unschuld haben wir keine Begriffe. . . .

Ja! ein göttlich Wesen ist das Kind, solang es nicht in die Chamäleonsfarbe der Menschen getaucht ist.

Es ist ganz, was es ist, und darum ist es so schön.

Der Zwang des Gesetzes und des Schiksaals betastet es nicht; im Kind' ist Freiheit allein.

In ihm ist Frieden; es ist noch mit sich selber nicht zerfallen. Reichtum ist in ihm; es kennt sein Herz, die Dürftigkeit des Lebens nicht. (*Hyp*, 10)

As the laborer into refreshing sleep, so my beleaguered being often sinks into the arms of the innocent past.

Peace of childhood! heavenly peace! how often do I pause before you in loving contemplation, and try to conceive of you! But our concepts are only of what has degenerated and been repaired; of childhood, of innocence we have no concept. . . .

Yes, divine is the being of the child, so long as it has not been dipped in the chameleon colors of men.

The child is wholly what it is, and that is why it is so beautiful.

The pressure of Law and Fate touches it not; only in the child is freedom.

In the child is peace; it has not yet come at odds with itself. Wealth is in the child; it knows not its heart nor the inadequacy of life. (4–5)

At first, childhood appears in this passage as an original state of innocence, as "innocent past" and "heavenly repose." As such, childhood is related to various themes in Hölderlin's novel, including Hyperion's vision of classical Greece, as well as his relation to Adamas and Diotima. A more complex view of childhood emerges only a few lines later, when Hyperion confesses his proper ignorance: "of childhood, of innocence we have no concept." Hölderlin here departs from the notion of a "paradise lost," so often preoccupying the thinking about origins in myth, religion, and philosophy. Hyperion's childhood is not lost as much as *suspended* in the exchange of contradictory claims and counterclaims. He writes about childhood knowing he possesses no knowledge. It is this knowing nonknowledge, however, that marks the term "childhood" as a metaphor that covers up what appears for Hyperion as the original absence of childhood. For the remembering Hyperion there is no childhood to begin with, no historical presence of childhood which the work of memory could recall, again: "of childhood, innocence we have no knowledge." Hölderlin has Hyperion explicitly note that the significance of childhood has the same origin as its loss. There is no meaning of childhood for the child: "it knows not its heart nor the inadequacy of life." There is no way the child could know his or her proper—childlike—nonknowledge concerning his repose, divine being, freedom, richness, and immortality without ceasing to be a child, and thus appearing in the "chameleon colors of men." In these letters, childhood enters the realm of meaning only *as* childhood, at the moment at which a linguistic sign simultaneously refers to and distances a state of immediacy. Childhood is a metaphor that springs from an original delay of its proper signification.

Interpreters of Hölderlin's novel are often tempted to construct Hyperion's fate as a linear historical development according to the paradigm of the *Bildungsroman*.[9] Nothing could be further from the complexities of Hölderlin's novel than the historicist sequencing of temporal layers typical for this genre. When Hyperion writes about the past, and indeed about his current need for doing so, the act of writing originates a temporality more intricate than the one suggested by any linear historicist model. It is characteristic for this process that the term "past" operates also as a metaphor indicating Hyperion's current need to remember. For all the reasons that compel Hyperion to follow the traces of memory, the past is not simply past. The literal and metaphorical sense of temporal markers merge in the act of writing and remembering, so that the Hyperion who affirms the necessity to remember is also the Hyperion who acknowledges that the past is still with him.

Lawrence Ryan has pointed to the deep affinity between the concepts of the *Bildungsroman* and *Geistesgeschichte*,[10] and it is certainly not a coincidence that Wilhelm Dilthey's *Das Erlebnis und die Dichtung* (1905) had chosen Hölderlin's novel as the prime example of the genre more than half a century before.[11] However, the structure of a historical narrative, which these readings impose on the text of the novel, inevitably relies on an erroneous literalization of temporal metaphors. It is crucial to note that for Hölderlin a term like "childhood" does not function as a historical index that would mark the beginning of Hyperion's *Bildung*; and it should not be mistaken as having a literal meaning which would originate with the birth of the child itself. In the narrative, the originality of childhood originates *at the same time* as the loss of it. For Hyperion, the term is significant only as a figure that seeks to substitute the original lack of knowledge about childhood. His "loss of childhood," then, also figures as a movement of substitution which transports meaning where no meaning has yet been—a semiotic void which, retrospectively, is addressed *as* childhood. In Hyperion's account, there is no historical difference between child and man. Being ignorant about their origin, as Hölderlin reminds us, both begin to differentiate themselves only after the metaphorical substitution of their original ignorance has been affected. Hyperion's loss—or rather suspension—of innocence begins with the reading of childhood as metaphor, as an ever deferred promise of fulfillment rather than a fulfilled life in the past.

In certain important aspects, Hyperion's attitude towards childhood links itself with the intersubjective dynamic of his epistolary project. Writing to Bellarmin, he simultaneously remembers the past and anticipates the response of the other. Both acts originate at once, interfering with one another at the source. Metalepsis and prolepsis—the essential belatedness of the autobiographical writer and the no less essential prematurity of the one who imagines his future reader—are inextricably interwoven in the way Hölderlin's novel presents itself. There is no sequentiality of the two processes; both occur concurrently. Like the course of the Rhine river, in Hölderlin's eponymous poem, the line of writing in Hyperion's letters springs from a single, yet divided source or *Ur-sprung*.

Philosophical Shipwreck

The last letter of the first volume, comprising Hyperion's famous speech about Athens, has attracted more critical attention than any other letter of the novel. This preference among critics is not altogether surprising, since it concerns a text which in many respects presents itself as a theoretical discourse. Hyperion addresses the preeminence of Athens in Greek antiquity, and he seems to express himself in the most direct, unequivocal, and unmetaphorical manner possible.

His speech sets out with a refutation of various concepts that try to ground the "excellence of the ancient Athenians" ("Treflichkeit des alten Athenervolks") in the natural condition of the Greek climate—"das Klima hat es gemacht" ("the climate had produced it") (*Hyp*, 77; 3)—or in the advanced cultural forms and institutions of art, philosophy, religion, and government. Beißner has indicated that the refutation mainly takes aim at Winckelmann's conception of Greek antiquity as it is developed in his *Geschichte der Kunst des Altertums*.[12] In the course of Hyperion's speech, however, the function of this critique is only preparatory. It opens the field for the presentation of a cultural genealogy of an entirely different order. The critical preliminaries are quickly left behind: art, philosophy, religion, and government are said to be "Blüthen und Früchte des Baums, nicht Boden und Wurzel" ("flowers and fruits of the tree, not soil and root"); they are "Wirkungen" ("effects"), but not the "Ursache" ("cause") of Athenian splendor; and with respect to the supposed superiority of the Greek climate, it is added "daß auch wir darinn noch leben" ("that we still live in it ourselves") (*Hyp*, 77; 63). Hyperion's main focus, then, is beauty. He sees an innate sense of beauty at the origin of an entire cultural genealogy:

Das erste Kind der göttlichen Schönheit ist die Kunst. So war es bei den Athenern.

Der Schönheit zweite Tochter ist Religion. Religion ist Liebe der Schönheit. Der Weise liebt sie selbst, die Unendliche, die Allumfassende; das Volk liebt ihre Kinder, die Götter, die in mannigfaltigen Gestalten ihm erscheinen. Auch so war's bei den Athenern. (*Hyp*, 79–80)

The first child of divine Beauty is art. Thus it was among the Athenians.

Beauty's second daughter is religion. Religion is love of Beauty. The wise man loves beauty herself, eternal, all-embracing beauty; the people love her children, the gods, who appear to them in multifarious forms. So it was, too, among the Athenians. (65)

This concept of beauty offers an anthropomorphized ontology, in character not unlike many of the other myths populating Greek imagination. Under the parentage of beauty, art and religion come into being. Yet Hyperion's mythmaking is also a theory about origins. Beauty precedes art and religion in an essential way; it does not amount to merely one cultural phenomenon among others, like the gods who exist in the plural and are loved by the people. Eternal like the gods, beauty is unique and integrative, the love of the wise man and the true concern of philosophy.

Beauty also governs the essence of philosophical thinking. In the form of *Dichtung* (a term only inadequately rendered by "poetry"), it marks the principle and *telos* of philosophy itself. "Die Dichtung, sagt' ich, meiner Sache gewiß, ist der Anfang und das Ende dieser Wissenschaft" (" 'Poetry,' I answered, confident of my argument, 'is the beginning and the end of philosophical knowledge' ") (*Hyp*, 81; 66). Still, Hyperion does not suppose beauty to be a substance in the classical sense of the word. With reference to Heraclitus, he defines its essence as a differential event, as "the One differentiated in itself": "Das große Wort, das ἕν διαφέρον ἑαυτῷ (das Eine in sich selber unterschiedne) des Heraklit, das konnte nur ein Grieche finden, denn es ist das Wesen der Schönheit, und ehe das gefunden war, gabs keine Philosophie" ("The great saying, the ἕν διαφέρον ἑαυτῷ [the one differentiated in itself] of Heraclitus, could be found only by a Greek, for it is the very being of Beauty, and before that was found there was no philosophy") (*Hyp*, 81; 67). Philosophy, that is, not only depends on a sense of beauty in general, but originates with the definition of beauty. The articulation of the essence of beauty as such is the first act of philosophy—the founding act of philosophical thinking, the very act that brings philosophy into being.

The exalted degree of theoretical reasoning sets this letter clearly apart from the rest of Hölderlin's novel. Placed at the end of the first volume which Hölderlin published in 1797, two years before the second and last volume of the novel, the theme and design of the letter represent a watershed in the further development of the narrative. Critics like Ryan and Aspetsberger have taken the letter to represent the philosophical credo of the novel, and more recent studies, such as the one by Jean-Luc Nancy, largely discuss the letter as a philosophical text.[13]

The pointed philosophical perspective, however, may be misleading. Despite the seemingly direct theoretical character of Hyperion's speech, the novel certainly does not invite its readers to an exercise in philosophy. In fact, Hyperion's quite philosophical definition of the essence of beauty already points to the end and limits of philosophical thinking. As the narrative moves on, this sense of philosophical entropy also turns against Hyperion's speech. While Hyperion feels "certain" about the nature of beauty as long as the argumentation develops, an irony more profound than such certainty sets in. Tellingly, the discourse takes place on board a ship during a voyage to Athens; now, shortly after the end of Hyperion's philosophical musings, a new kind of decidedly unphilosophical astonishment arises: "Wie bin ich doch, rief ich, auf die troknen Berggipfel gerathen, worauf ihr mich saht?" (" 'How did I ever come,' I cried, 'to be on the arid mountain peek on which you saw me just now?' ") (*Hyp*, 84; 69). Obviously, Hyperion has been carried away over the course of his speech. The philosopher on the sea finds himself, retrospectively, on an arid mountain peak. Hyperion here realizes that the evocation of an explicit philosophy of beauty does not provide him with a reliable sense of place in the world, a quite familiar concern in Hölderlin's oeuvre.

There are many echoes of this sentiment in Hölderlin's letters, where the "airy spirits with the metaphysical wings" are often associated with a new onslaught of depression (*SA* 6; 1: 222). Hölderlin's letter to Neuffer of November 12, 1798, the year between the publication of the first and second volume of the novel, directly associates philosophy with the resignation of the poet: "O woe! the world has frightened my spirit back onto itself from early youth on, and I still suffer from it. There is certainly an asylum where every poet who is stranded like me can find a honorable refuge—philosophy. But I cannot abandon my first love, the hope of my youth, and I rather want to perish without merit than to separate myself from the sweet home of the muses from which only chance has driven me away" (*SA* 6; 1: 289).[14] Of course, none of this implies that Hyperion was wrong (or right) in his philosophical definition of beauty, but the very possibility of being right or wrong is fundamentally put into question. Hyperion does not simply err empirically about his chosen course, in a manner which, *in principle*, could be corrected. More precariously, it is philosophical discourse itself that produces the metaphorical displacement from the vessel on sea to the arid mountain top disorienting the philosopher.[15]

Hyperion's attempt to follow the itinerary of Greek thinking about beauty, as charted by Heraclitus, brings about a disfigurement of his sense of place. In a similar way, the naval voyage to Athens leads to the spectacle of a gigantic shipwreck. Approaching Athens, Hyperion exclaims:

> Ich sah, und hätte vergehen mögen vor dem allmächtigen Anblik.
>
> Wie ein unermeßlicher Schiffbruch, wenn die Orkane verstummt sind und die Schiffer entflohn, und der Leichnam der zerschmetterten Flotte unkenntlich auf der Sandbank liegt, so lag vor uns Athen, und die verwaisten Säulen standen vor uns, wie die nakten Stämme eines Walds, der am Abend noch grünte, und des Nachts darauf im Feuer aufgieng. (*Hyp*, 85)

> I looked, and could have fainted, so mighty was the spectacle.
>
> Like an immense shipwreck, when the gales have been hushed and the sailors have fled and the corpse of the shattered fleet lies on the sand bank unrecognizable, so before us lay Athens, and the forsaken pillars stood before us like bare tree trunks of a wood that at evening was still green and, the same night, went up in flames. (70)

Upon first seeing Athens, Hyperion again experiences a fundamental disorientation of his senses. In the image of the stranded city-ship, he visualizes that his contemporary "odyssey of the mind," in Schelling's phrase, has lost the means to return to the ways of Greek antiquity. The journey that was supposed to transport and carry over (μεταφερειν) comes up against the ruined metaphor of its raison d'être. What lies ahead for Hölderlin's hero is the suspicion that thinking about beauty is no longer an exclusively Greek enterprise—if it ever could have been conceived as such. Yet Hyperion's vision of Greek antiquity will never be simply an *illusion perdue*. For him, the dead flame of Greek spirit leaves behind the memento of an ashen trace.

Translating Bellarmin

Hyperion's memories take the form of letters addressed to Bellarmin. For Hyperion, as for the reader of Hölderlin's novel, Bellarmin figures as the absent but implied other without whom the existence of the letters could not be imagined. Paying tribute to Bellarmin at the beginning of the third letter, he extends his thanks: "Ich danke dir, daß du mich bittest, dir von mir zu erzählen, daß du die vorigen Zeiten mir in's Gedächtniß bringst" ("I thank you for asking me to tell you of myself, for making me remember earlier days") (*Hyp*, 10; 4). Writing these lines, Hyperion acknowledges a debt. He receives Bellarmin's request like a gift, as if the ability to narrate one's life has to come from the other. Indeed, nothing less is claimed in the second part of the sentence. Hyperion notices that it is Bellarmin who brings to mind the former times. Bellarmin, we learn, is not just the passive recipient of a life's confession, the uninvolved witness to the memory of his friend; rather, he endows Hyperion with the capacity to write and to form a memory where no memory existed before. Unlocking a process of retrospective meditations barred to the solitary musings of the hermit, Bellarmin elicits memories unavailable to Hyperion outside the dynamic of the epistolary exchange.

In short, Bellarmin's "making" Hyperion "remember earlier days" implies that Hyperion does not first remember and later communicates these memories in a second, independent step of letter writing. The addressee of Hyperion's letters is inscribed in his memory-work from the very beginning. Hyperion does not remember in the sense of re-producing or re-presenting events perished and forgotten. His letters are not representations of past events. In fact, the total recall of personal history is barred to him—as is, on a different level, the direct access to the experience of Greek antiquity. His writings are not mimetic. Rather than re-membering, Hyperion actively produces his memories. His epistolary construction of the past is a multitemporal hybrid, a memory compound closely entangled with current needs and belated illuminations.

The epistolary parameters of Hyperion's memory-work provide an additional dimension to these complexities. Their hermeneutical import relies on the fact that the addressee of his letters "brings to mind the former times" precisely because Hyperion addresses these letters to him. In the strict sense, it is hardly adequate to refer to these memories as exclusively originating in Hyperion. Rather they are the effect of an exchange of written communications, *epistolary* in their very origin, the product of a "psyche among friends." Throughout his writings, Hyperion's memory-work is marked by the dimension of the addressee, the anticipated horizon of the other. As such, the addressee is essential for Hyperion's epistolary construction of the past.

Yet the complexities do not stop here. This is largely due to the fact that the addressee is essential *and* absent, both fundamental for Hyperion's literary activity *and* removed from his direct access. Moreover, Bellarmin's absence is not just an empirical phenomenon. As the one who makes it possible for Hyperion to articulate his own experience of absence, otherness, and nonidentity, Bellarmin is the figure of absence par excellence. Hölderlin's resistance to fleshing out the life and times of Bellarmin's empirical character highlights this aspect: the essential absence of the other within the dynamic of the epistolary exchange finally enables Hyperion to acknowledge the absence of a Greek antiquity that seemed to exist beyond the experience of ontological incompleteness. In a certain sense, Bellarmin provides the white page that underlies and makes visible Hyperion's written productions. Like an empty screen that makes visible and separates, he figures as the literary blank before and within the epistolary exchange itself. He is the intangible organon of the memory-work that inhabits all of Hyperion's letters and provides the groundless ground from which the writing proceeds.

In a significant study devoted to the use of names in Hölderlin's oeuvre, Wolfgang Binder has proposed a more concrete interpretation of the figure of Bellarmin.

> Bellarmin means for Hölderlin: the beautiful German—beautiful in the sense of the novel's doctrine of beauty; German in the sense of the Germans' hero celebrated in the extensive Arminius literature of the eighteenth century.[16]

Binder is as concise as possible; the colons after the initial "Bellarmin means for Hölderlin" indicate that he aims at no less than a full equivalent of "Bellarmin," the name as conceived by Hölderlin. Focusing on what he has called, one line before, the "meaning of the name," "Bellarmin," writes Binder, "means for Hölderlin: the beautiful German."

Such transcription is by no means unproblematic; in particular, the more specific definitions of *schön* ("beautiful in the sense of the novel's doctrine of beauty") and *deutsch* ("German in the sense of the Germans' hero celebrated in the extensive Arminius literature of the eighteenth century") complicate Binder's reading. Not only is it highly questionable to assume the existence of an explicit "doctrine of beauty" (*Schönheitslehre*) within Hölderlin's novel, even more problematic is the way in which Binder merges the issue of aesthetics with the question of nationality.

It is well known that this "German question" is situated in a larger context and that this context is not simply a matter of aesthetics. Unfortunately, Binder raises the ambivalence of his procedure to ever higher levels as he suppresses the issue of aesthetic nationalism altogether and focuses exclusively on Arminius, Hyperion, and the "meaning" of things German. Barely twenty years after the concentration camps, he informs his readers about the very "*Deutschheit* which the name Arminius means"—as if the sense of *deutsch* remained untouched by the events which happened in the name of it.

One of the shortcomings of Binder's procedure stems from the fact that the process of *translation* (from the name *Bellarmin* to the supposed "meaning of the name," the beautiful German) is never acknowledged as such. This particular lack of recognition for what passes between different languages and cultures is not entirely accidental. Among other things, the German question has always been a question of translation concerning the ability—or the lack thereof—to understand what is different, and furthermore to understand it in its difference. Binder translates the name "Bellarmin," but he remains silent about the difference that translation makes. Somewhat removed from the immediate context of Binder's attempt, it may be easier today to note the bitter historical irony in his procedure. Yet there can be no doubt that, among German writers at the end of the eighteenth century, it was in particular Hölderlin for whom this kind of homogenization of linguistic differences seemed progressively more untenable. Aside from the novel, his work on Homer, Sophocles, and Pindar, testifies to the relentless effort to come to terms with the turbulences emerging in the process of translation between foreign languages and cultures.

The act of translation is by no means neutral; it marks a philological and political orientation within the said "meaning of the name." While

Bellarmin, following Binder, may be read literally as "the beautiful German," it does so speaking a language other than German. It can hardly be emphasized enough that Hölderlin's novel effectively bars any monolingual German speaker from reading what is inscribed in the name of Bellarmin. Anyone unable to reach beyond German linguistic territory misses nothing less than "the beautiful German." Bellarmin's name, then, can be read only by way of a multilingual and multicultural differentiation of the linguistic sign. Above all, Hölderlin seems to suggest that there is no exclusively German reading of "the beautiful German," no exclusively German way to name what is beautiful, and no exclusively German way to name what is German. Hölderlin approaches the German question via translation, a fact which also spells out an ethical and political predicament.

Naming and Being

The translation of a name always carries a certain impropriety. Names are not identical with meanings in any immediate sense. In the famous *Scheltrede* on the Germans, contained in the penultimate letter of the second volume, Hyperion laments the economically motivated identification of name and meaning. It is in this context that he locates the linguistic import of what seems wrong with Germany and the Germans:

> Ein jeder treibt das Seine, wirst du sagen, und ich sag' es auch. Nur muß er es mit ganzer Seele treiben, muß nicht jede Kraft in sich erstiken, wenn sie nicht gerade sich zu seinem Titel paßt, muß nicht mit dieser kargen Angst, buchstäblich heuchlerisch das, was er heißt, nur seyn, mit Ernst, mit Liebe muß er das seyn, was er ist, so lebt ein Geist in seinem Thun. . . . Deine Deutschen aber bleiben gerne beim Nothwendigsten. (*Hyp*, 153–54)

> Everyone follows his own trade, you will tell me, and I say the same. Only, he must follow it with his whole soul, must not stifle every power in him that does not precisely accord with his official designation, must not, with this niggardly anxiety, literally and hypocritically be only what he is called, let him be what he is, earnestly, lovingly, then a spirit lives in all that he does. . . . But your Germans choose not to go beyond necessities. (128)

Hyperion directly links the German malaise to the general economic stratification of everyday life. Indeed, some of the most ardent language is aimed at the social disintegration associated with the division of labor. Considering it linked to market economy, Hyperion notes the existence of a semiotic economy in which titles function as commodities. In a vicious circle, titles and "what one is called," that is, names, are dominated by the very market sphere which they helped to create in the first place.

Hyperion terms the anxious adherence to one's name "literally hypocritical": *buchstäblich heuchlerisch*. The fixation on one's name is "literal" because the level of the letter is never left, but it is also "hypocritical" because "one," "man," or "being" is seen as not being identical with its name. Hyperion's critique—organized according to the difference between "what one is called" and "what one is"—to some extent refers us back to the opening of the *Scheltrede* with its forceful indictment of the ubiquitous professionalization of everyday life: "Handwerker siehst du, aber keine Menschen." ("You see artisans, but no men.") (*Hyp*, 153; 128). Here, however, the critique takes a linguistic turn and is no longer socioeconomically restricted. Hyperion evokes a discontinuity radically different from the one between workers and men: namely, the discontinuity between naming and being. At this point, he no longer critiques the Germans in the name of another name—be it in the name of better working conditions or even in the name of man; rather, his critique is directed at the use of the name itself.

As a writer who addresses himself to Bellarmin, Hyperion, of course, cannot absolve himself from the necessity to use names and titles, too. While he does not launch his critique of modern Germany in the name of any positive knowledge, Hyperion still does what the Germans do: he writes and names. It is to this extent that Hyperion takes part in the very German malaise he critiques. Yet, different from those he designates as "literally hypocritical," he speaks in the name of that which has no name and deserves to be named and valued precisely because it is unnameable. While the Germans have become literal readers of their names and titles, Hyperion insists that the name is not identical with itself, but inhabited by a linguistic void, marked by a silence that is not the simple opposition of speech, but the inheritance of language in general.

Hyperion points to this prelinguistic abyss within language in a letter to Diotima, after having recorded the sensations of a happier time: "aber das sind Worte und die eigne Lust von solchem Leben erzählt sich nicht" ("but these are words, and the unique pleasure of such a life cannot be told") (*Hyp*, 112; 92). In his emphasis on a nameless experience that seemed to be exempt from the empirical constraints of temporality, Hölderlin has Hyperion speak like the Jean-Jacques of Rousseau's autobiography.[17] The nonidentity of sign and meaning, expressed in this passage, stresses the vital function of metaphor in Hölderlin's novel. Earlier, the dissociation between sign and referent proved to be the major obstacle on Hyperion's unsuccessful journey toward a truly integrated self-consciousness. Terms like "fatherland" and "childhood" were shown to be linguistic substitutions covering an original absence; the literal revealed itself as metaphorical. The use of figures and metaphors emerges in Hölderlin's novel as an ontological imperative.[18] Even before the days of exile and the failed attempts to regain an original state of being, Hyperion was linguistically expatriated, and it is to this extent that his actual fate enacts the linguistic predicament of the poet who mourns the absence of an original meaning he nonetheless attempts to describe.

Thus far, I have addressed two types of translation: one that appears to translate literally from one language into another (like Binder's rendering of the name "Bellarmin"); and another that metaphorically passes between the nameless and name. Both types of translation necessitate each other and call each other into being, coexisting without being fully mediated or reconciled. This double perspective on translation is clearly in effect when Hölderlin has Hyperion, the Greek, write his letters to his German correspondent. For one, Hyperion leaves the familiar linguistic climate of his mother tongue; more fundamentally, however, these letters know themselves estranged from the direct contact with an origin they still attempt to name. Underlying the literal shift from Greek to German (or from *Bellarmin* to its German transcription) is a metaphorical shift already operative in these languages themselves. Hölderlin moves, that is, from the problematic of literal translations between two or several languages to a metaphorical concept of *language as translation.* Nothing is lost in translation—or, at least, not because of it. On the contrary, *Hyperion* spells out translation as an epistemological, ethical, and poetic necessity.

In this context, it is particularly significant that Hölderlin makes use of a translation of a celebrated passage by Rousseau at precisely the moment when Hyperion develops his sharp criticism of Germany and the Germans:

Es ist ein hartes Wort und dennoch sag' ichs, weil es Wahrheit ist: ich kann kein Volk mir denken, das zerrißner wäre, wie die Deutschen. Handwerker siehst du, aber keine Menschen, Denker, aber keine Menschen, Priester, aber keine Menschen, Herrn und Knechte, Jungen und gesezte Leute, aber keine Menschen—ist das nicht, wie ein Schlachtfeld, wo Hände und Arme und alle Glieder zerstükelt untereinander liegen, indessen das vergoßne Lebensblut im Sande zerrinnt? (*Hyp*, 153)

It is a hard saying, and yet I speak it because it is the truth: I can think of no people more at odds with themselves than the Germans. You see artisans, but no men, thinkers, but no men, priests, but no men, masters and servants, but no men, minors and adults, but no men—is this not like a battlefield on which hacked-off hands and arms and every other member are scattered about, while the life-blood flows from them to vanish in the sand? (128)

These lines offer an almost literal transcription of a passage from the "Discours sur les sciences et les arts":

Nous avons des Physiciens, des Géometres, des Chymistes, des Astronomes, des Poëtes, des Musiciens, des Peintres; nous n'avons plus de citoyens; ou s'il nous en reste encore, dispersés dans nos campagnes abandonnées, ils y périssent indigens et méprisés.[19]

This adaptation of Rousseau dramatizes Hyperion's argument against linguistic nationalism: the language of the novel itself appears in a state of translation. The destructive effect of economic and semiotic identification, detailed in these sentences, finds its most direct testimony in the use of a language which can no longer lay claim to a national identity of its own.

Drawing attention to the interplay of these two passages, Karlheinz Stierle is certainly right in underscoring the "far reaching influence of Rousseau."[20] Yet the issues raised by the quotation of Rousseau may be more complex than is indicated by the question of influence. Stierle's contention that Hölderlin's notion of contemporary society in *Hyperion* is "not thinkable"[21] without Rousseau has the decided disadvantage of employing a historiographic scheme of continuous intellectual traditions whose very unreliability structures the narrative conflict of the novel itself. The translation of Rousseau is grafted on a more radical translation which also underlies Hyperion's critique of the compulsive identification of naming and being that he finds in Germany: the translation from the nameless to the name.

'Geschik'

Schiksaallos, wie der schlafende
    Säugling, athmen die Himmlischen;
. . .
Doch uns ist gegeben,
    Auf keiner Stätte zu ruhn . . .
—Hyperion's "Schiksaalslied" (*Hyp*, 143)

Outside of Fate, like the sleeping
    Babe, the Heavenly Ones breathe;
. . .
But to us it is given
    Nowhere to rest . . .
—Hyperion's "Song of Fate" (199)

The two modalities of translation are similar to the ones which, according to the outline sketched in the letters to Böhlendorff, characterize Hölderlin's relation to Greek antiquity. The question that plagued so many of Hölderlin's contemporaries—how to relate to Greek classicism in modern times—is subjected to a shift reminiscent of the treatment of translation in *Hyperion*. Crucial in the first letter is Hölderlin's recognition that Greek art—like Greek and German language in *Hyperion*—does not have its origin in itself. The Greeks, Hölderlin asserts, first needed "to appropriate that which is foreign [das Fremde]," that is, "Junonian sobriety," before achieving their particular "gift of representation [Darstellungsgaabe]" (*SA* 6; 1: 426; *Essays*, 149–50 [trans. modified]). Radically changing the paradigm of Greek antiquity as it was current at his time, Hölderlin designates the origin of Greek art as the locus of a cross-cultural transference that extends beyond the Greek sphere proper. More precisely, he departs from Winckelmann's aesthetics which glorifies Greek art as a homogeneous, original creation encompassing the totality of possible experience. For Winckelmann, the task of the modern artist still lies in the imitation of the original Greek model; the true artist is essentially Greek even if he speaks German and has to translate what was given by the Greeks into the national dialect of his time. Hölderlin's departure from the classicist's model of Greek art was far ahead of its time. His achievement is all the more remarkable when one considers that Winckelmann's conception has proven exceptionally resistant to critical revision. Winckelmann's notion of Greek classicism had a lasting impact on several generations of

German literary critics and in fact could maintain its hold even in more recent works of criticism.[22]

Among modern critics, Peter Szondi was first to point out that Hölderlin no longer perceived the origin of Greek art to be an exclusively Greek affair. Yet even Szondi's celebrated analysis of Hölderlin's first letter to Böhlendorff betrays serious shortcomings, in particular in regard to the constellation of nature and culture.[23] In a recent critique of his essay, Andrzej Warminski has charged Szondi with projecting the model of an ultimately Hegelian dialectic onto the Böhlendorff letter, enclosing it within the framework of a false mediation between Greece and Hesperia.[24] Despite their considerable differences, however, both critics agree on a point which merits special attention: Hölderlin's use of the word *Geschik*. The pertinent passage in the Böhlendorff letter, already quoted, reads: "I have labored long over this and know by now that, with the exception of that which has to be the highest for the Greeks and for us—namely, the living relation and *Geschik*—we may not have anything *in common* with them." For Szondi, the use of the term *Geschik* marks a crucial moment in Hölderlin's understanding of the Greeks: "The Böhlendorff letter as a whole appears to stand under the sign of this word."[25] Szondi insists that *Geschik* is to be read as *techne*, signifying the technical ability of the poet to lend a "living relation" to the poetic work. The Böhlendorff letter, writes Szondi, "is a letter from the workshop [Werkstatt]."[26] Warminski concurs: "Szondi quite correctly reads the *Geschick* . . . not as a common 'destiny' or 'fate' . . . but rather as 'skill,' 'ability': in short, as Greek *techne*."[27]

The letter to Böhlendorff, however, modifies such an interpretation in several respects. The letter, which starts out with praise for Böhlendorff's recently completed drama *Fernando*, explicitly links Hölderlin's and Böhlendorff's progress as poets in terms of a common "destiny" or *Schiksaal*: "Your *Fernando* has alleviated my bosom a good deal. The progress of my friends is such a good sign for me. We have one destiny [Wir haben ein Schiksaal]. If one is making progress, then the other will not be left behind either" (*SA* 6; 1: 425; *Essays*, 149). An exploration of the destiny or fate of the contemporary poet,[28] the letter also implies that Hölderlin's interrogation of poetical ability or skill takes place within the horizon of *Schiksaal*. If one chooses to read *Geschik* as τέχνη, as Szondi and Warminski both do,[29] it still would be imperative to note that Hölderlin takes this very *techne* to be a "sign" (*Zeichen*) of the destiny of the poet. It is evident, that is, that *Geschik* does not refer to the mere technical ability of the poet who self-sufficiently masters his poetry, or to the technical instrumentarium he might employ.[30] For Hölderlin, technical ability, or *techne*, entertains an essential link with *Schiksaal*.[31] *Techne* articulates what comes into being as the poet's destiny: the fate of writing, the fate of reading.

Both Szondi and Warminski base their reading on the highly questionable assumption that "living relation" and *Geschik* are indeed those things the Hesperian poets are to have in common with the Greeks. Szondi writes: "The Hesperian poet must have 'the living relation,' without which the poem is not a poem, 'in common' with the Greek—as is the case with the 'skill' (*Geschik*) as well."[32] Warminski, too, speaks about "the *Geschik* that we are to have in common with the Greeks."[33] The passage in question, however, seems rather to suggest that Hölderlin restricts what the Hesperian and Greek poets should have in common to "that which has to be the highest [das höchste]." The Hesperian and Greek poet, that is, are not supposed to have in common the particularities of their respective "living relation" and *Geschik*—but the epochal predicament that, for them, nothing else would precede or equal the "living relation" or *Geschik* in rank. The different kinds of artistic techniques available to the Greek and Hesperian poet are to articulate fate and living relation as the highest while for both the Greek and Hesperian poet the highest necessarily spells technical and artistic difference. Again: "I have labored long over this and know by now that, with the exception of that which has to be the highest for the Greeks and for us—namely the living relation and fate [*Geschik*]—we may not have anything in common with them."

Hölderlin develops this notion when he turns to Böhlendorff's drama *Fernando* once again. He praises Böhlendorff for having written an "authentic modern tragedy" (*SA* 6; 1: 426; *Essays*, 150) and especially emphasizes the invention of a new dramatic technique: the epical treatment of the tragic. The "highest" that both the Greek tragedy and Böhlendorff's modern tragedy have in common, Hölderlin makes out in the concentration on the movement of the "innermost soul," that is, in the "living relation." Still the "living relation" proceeds along quite different ways for the Greek and Hesperian poet and indeed necessitates, as Hölderlin clearly recognizes, a new concept of the tragic and a new dramatic treatment of tragedy in modern times. "For this is the tragic to us: that, packed up in any container, we very quietly move away from the realm of the living, [and] not that—consumed in flames—we expiate the flame which we could not tame" (*SA* 6; 1: 426; *Essays*, 150). A similar differentiation is at work when Hölderlin speaks about the role of fate on destiny (*Schiksaal*) in ancient and modern tragedy. Despite the radical differences between Greek and Hesperian fate, it is still the very portrait of their respective *Schiksaal* that Hölderlin wishes ancient and modern tragedy to have *in common*. It is this dramatization of fate in tragedy that evokes for both Greeks *and* Hesperians what has no destiny—Greek *or* Hesperian: namely, the image (*Gedanke*) of the god who breathes, as Hyperion's "Song of Destiny" ("Schiksaalslied") puts it, "Outside of Fate" (*Schiksaallos*). "The magnificent Jupiter, then, is after all the last thought during the death of a mortal being—whether he die according to our or ancient destiny—if the poet has depicted this dying, as he should" (*SA* 6; 1: 426–27; *Essays*, 150 [trans. modified]). *Schiksaal* or *Geschik* in the sense of fate, therefore, "must be the highest for the Greeks and for us" precisely because it points beyond itself and relates Greeks and Hesperians not only to the end of their respective individual or national destinies, but also to the end of destiny in general: to the gods.

Hölderlin's emphasis on *Schiksaal* also implies that the relationship between Hesperians and Greeks cannot fully be described in terms of a chiasmatical reversal. Certainly, the letter to Böhlendorff identifies "holy pathos" as Greek nature and Hesperian culture; it identifies, furthermore, "Junonian sobriety" as Greek culture and Hesperian nature; and it is to this extent that the Greek constellation of nature and culture appears to be "reversed" for the Hesperians (*SA* 6; 1: 426; *Essays*, 150). Yet the reversal itself is not Hölderlin's last word on the matter. He subjects this simple chiasmus to what he terms "the highest," that is, the "living relation" and *Geschik*, which in their turn displace the formalism inherent in the mere rhetorical figure. The use of the designation *das höchste* indicates that Hölderlin's initial setup of a "reversed" relation between Greek and Hesperian nature and culture has indeed to be read in the light of this "highest"—*Schiksaal*—and not the other way around, as some critics would have it.[34] The "living relation" and *Geschik* temporalize what otherwise would remain a purely formalistic conception of the relation between Greek and Hesperia.

Geschik, of course, may also be read as missive, send-off, or epistle. In the latter form it surfaces at the end of the second letter to Böhlendorff, where Hölderlin defines epistolary writing (along with conversation) as an artistic necessity. Like the first Böhlendorff letter, the second—though more difficult, intricate, and obscure—is an investigation of Greece, Hesperia, and the "mode of destiny" (*Schiksaalsweise*) (*SA* 6; 1: 433; *Essays*,

153), too. Yet in contrast to the first, the second letter also signals a telling transition from a discussion of the task of the poet in terms of *Geschik* and *Schiksaal* to the material form this discussion should take: namely, "conversation and letter" (*Gespräch und Brief*). The poet's letter, then, does not simply stand for one form of literature among others, but for the name of the destiny of literature in general. It calls into being what is necessary for the artist and what he cannot achieve alone, his proper *Geschik*: the origination of thought and the psyche among friends.

Diotima, Diotima

The mutual metaphorical interplay of *Geschik* and epistolary writing also structures Hyperion's relation to Diotima. Hölderlin presents the correspondence between them as part of the letters that Hyperion addresses to Bellarmin: copying his epistolary exchange with Diotima, Hyperion frames it with another correspondence of the same time—a constellation all the more crucial, since it places the figure of Diotima at the intersection of repetition, fate, and epistolary writing.

Significantly, Hyperion's first mentioning of Diotima's name—"O Diotima, Diotima, himmlisches Wesen!" ("O Diotima, Diotima, divine being!") (*Hyp*, 53; 42)—is marked by repetition. Recalling Diotima's name twice, he also recalls the phantasm of reduplication that had magnetized him with a certain image of Greek antiquity even before the two characters had met. For Hyperion, this double Diotima incorporates the promise to transcend the confines of temporality: "Laß uns vergessen, daß es eine Zeit giebt. . . . Was sind Jahrhunderte gegen den Augenblik, wo zwei Wesen so sich ahnen und nahn?" ("Let us forget that time exists. . . . What are centuries compared to the moment when two beings thus divine and approach each other?") (*Hyp*, 53; 42), he writes to Bellarmin immediately after the initial twofold evocation of Diotima's name. Diotima, then, is not *simply* divine for Hyperion; she is divine in an atemporal sense that spells out the possibility of return: Diotima, Diotima.

With Diotima, Hyperion hopes to recuperate the spirit of Greek antiquity in a Platonic manner: as an identical repetition of that which has been before.[35] Not only does he cite, with the name of his love, the name of the priestess in Plato's *Symposium*—a text which itself is presented as a series of quotations and copies, starting with Apollodorus, who quotes

Aristodemus, who quotes Socrates, who quotes Diotima, who quotes the ideas which eternally quote and copy themselves—he also turns her into a copy from which other copies should be made. On the eve of the bloody campaign around the Peloponnesus, Hyperion insists that the "world" he fights for and intends to rebuild is "Auch die deine, Diotima, denn sie ist die Kopie von dir. O du, mit deiner Elysiumsstille, könnten wir das schaffen, was du bist!" ("yours, too, Diotima, for it is copied after you. O you, with your Elysian quiet, could we but create that which you are!") (*Hyp*, 114; 94).

Hyperion's love is Platonic in the sense that it is structured by an idea of repetition that transcends temporal and spatial displacements. Not for long, however, since Hölderlin, of course, stresses that Hyperion is never more deeply in error than at the time, shortly before the campaign, when he imagines himself in a position to "strip chance of its power" and to "master destiny" ("Wir nehmen dem Zufall die Kraft, wir meistern das Schiksaal") (*Hyp*, 113; 93). Written with respect to the impending warfare, these words echo—once again—Hyperion's attitude to Diotima, an attitude overshadowed by death and forgetting. Already his first letter to her read, "Wie die seeligen Schatten am Lethe, lebt jezt meine Seele mit deiner in himmlischer Freiheit und das Schiksaal waltet über unsre Liebe nicht mehr" ("Like the blessed shades in the region of Lethe, my soul now lives with yours in heavenly freedom, and Fate has no more power over our love") (*Hyp*, 104; 85 [trans. modified]).

Soon, the "power of Fate" takes an explicitly epistolary turn; when letters come late or fail to arrive at all. Hyperion's attempt to reconcile himself with Diotima after the lost battle at the Peloponnesus is a case in point. The epistle intended to undo the effects of an earlier "Unglüksbrief" ("unhappy letter") (*Hyp*, 132; 109) arrives late and speeds up Diotima's death. Walter Silz's protestation that "faulty mail services should not be a factor in tragedy" misses the point.[36] It exiles fate from the mailbox. Even more important, Silz misses the metaphorical transfer from "die Todesgöttin, die Nahmenlose, die man Schiksaal nennt" ("the deadly Goddess, the Nameless One whom one calls fate" (*Hyp*, 106; 87 [trans. modified]) to literary form in general—or, more precisely, to the name of "fate" itself and the epistles which name it. For Hyperion, fate articulates itself *within* and *as* epistle. It is epistolary precisely because the movement from the nameless to the name is at the same time made possible and deferred by the very letters he writes.

Ashes

He conocido lo que ignoran los griegos: la incertidumbre.
—Jorge Luis Borges[37]

The predicament that prevents Hyperion's epistles from arriving at their destination is doubled by another one that concerns the act of writing. In a certain sense, all of Hyperion's letters are marked by a blind spot that is caused by the unavailability of the linguistic referent. Over and again, he necessarily fails to articulate an essentially nameless experience and produces a series of meditations interminable instead.

The account of Hyperion's celebrated vision at the end of the novel comes closer to this kind of disowned writing than any other constellation in the novel. Hyperion here introduces his response to what he takes to be the voice of the dead Diotima—"Und mir war, als hört' ich Diotimas Stimme" ("And it seemed to me that I heard Diotima's voice") (*Hyp*, 158; 132)—in the following manner:

Und Einmal sah' ich noch in die kalte Nacht der Menschen zurük und schauert' und weinte vor Freuden, daß ich so seelig war und Worte sprach ich, wie mir dünkt, aber sie waren, wie des Feuers Rauschen, wenn es auffliegt und die Asche hinter sich läßt—

"O du, so dacht' ich, mit deinen Göttern, Natur! ich hab ihn ausgeträumt, von Menschendingen den Traum und sage . . ." (*Hyp*, 159)

And *once* more I looked back into the cold night of men, and shuddered and wept for joy that I was so blessed, and I uttered words, I think, but they were like the roar of fire when it flies up and leave the ashes behind—

"O you," so I thought, "with your gods, Nature! I have dreamed it out, the dream of human things, and I say . . ." (132–33)

It is one of the most remarkable aspects of Hyperion's response that it does not simply take place within language: instead, it puts the possibility of linguistic response-ability itself into question. The phrase "I uttered words, I think" testifies to a fundamental insecurity about what one calls words and about the calling of words. Initially, Hyperion tries to bridge this insecurity with an analogy: the words, he remembers, "were like the roar of fire when it flies up and leave the ashes behind." But then, the analogy is just another word or *logos*; it provides words for the "words, I think," and thus consumes itself as part of the linguistic fire it describes, leaving nothing behind but ashes.

The last letter of *Hyperion* has proven to be especially troublesome for readers who insist on a homogeneous view of Hölderlin's novel. In order to make the case for an all-encompassing reconciliation, Ryan harks back to what he understands to be Diotima's "teaching of the reconcilability of all dissonances" which, he finds, is "still missing in [Hyperion's] last letter."[38] Dieter Henrich, too, interprets Hyperion's words as being spoken in a kind of "enthusiasm into which that insight has not yet been entered which can sum up a life at the end of its course."[39] The argument against the significance of Hyperion's last letter works like the *petitio principii* of the philosopher: since it does not confirm preestablished notions of reconciliation and insight, Hyperion's speech is judged "missing" the correct "teaching" or lacking the ability to "sum up a life." The novel, however, does not present Hyperion as being interested in the philosophical argument that the "roar of fire" may or may not provide. Rather, the rhetoric of fire bears the mark of what remains unarticulated by the process of articulation itself. As a writer rather than a philosopher, Hyperion addresses his experience as a textual predicament, composed of "words, I think," which in their own turn are "like" fire and ashes. The ashes left behind after the fires have died, are not "just ashes," as Henrich believes.[40] They do not signify the simple absence of insight, but are situated at the very origin of Hyperion's epistolary project itself.

What follows, in Hyperion's letter, is put in quotation marks. For one, these marks indicate that the ensuing discourse has already taken place and is now quoted all over again. Yet the possibility of such return to or of an original presence—be it that of Greece or of the word itself—has been put severely in question throughout the entire novel. The quotation marks indicate a rupture that irritates Hölderlin's novel from the very beginning: the uncertain state of meaning. Within these quotation marks, too, there is no linguistic refuge from the "roar of fire when it flies up" or the "cold night of men." Instead, all apparent certainties of Hyperion's vision are put into question again:

> "Auch wir, auch wir sind nicht geschieden, Diotima, und die Thränen um dich verstehen es nicht. Lebendige Töne sind wir, stimmen zusammen in deinem Wohllaut, Natur! wer reißt den? wer mag die Liebenden scheiden?
>
> O Seele! Seele! Schönheit der Welt! du unzerstörbare! du entzükende! mit

deiner ewigen Jugend! du bist; was ist denn der Tod und alles Wehe der Menschen?—" (*Hyp*, 159)

"We too, we too are not separated, Diotima, and tears for you understand it not. Living tones are we, sound together in thy harmony, Nature! which who can undo? who can part lovers?—

O Soul! Soul! Beauty of the World! indestructible, ravishing one! with your eternal youth! you are; what, then, is death and all the woe of men?—" (133 [trans. modified])

The fundamental linguistic perplexity of *Hyperion* continues here, since Hölderlin keeps the sentences suspended between affirmation and interrogation. Both paragraphs start out with a positive statement, both end with a question; and both exclamation mark and question mark cast a dubious spell on the entire sequence.

If one were to accept at face value the apparently positive phrasing at the beginning of each segment—"We too, we too, are not separated, Diotima" and "O Soul! Soul! Beauty of the World! indestructible, ravishing one! with your eternal youth! you are"—and read the questions rhetorically (that is, not as questions, but as affirmations), one would still be faced with a split between form and meaning, sign and referent. The wording of the question itself, that is, would go up in flames. Yet if one were to read the questions literally, as questions, they would, of course, undermine the certainties of the initial statements. Suspended between the literal and the rhetorical, at the very end of the novel, Hölderlin's Hyperion interrupts his quotation only to announce some more: "So dacht' ich. Nächstens mehr" ("So I thought. More soon.") (*Hyp*, 160; 133).

More fires, more ashes. More letters.

PART II

Figures of Duality

HÖLDERLIN AND GREEK TRAGEDY

Arnaud Villani

Near the opening of *The Birth of Tragedy*, Nietzsche writes:

> We shall have gained much for the science of aesthetics, once we perceive not merely by logical inference, but with the immediate certainty of vision, that the continuous development of art is bound up with the *Apollonian* and *Dionysian* duality—just as procreation depends on the duality of the sexes, involving perpetual strife with only periodically intervening reconciliations. . . . They continually incite each other to new and more powerful births, which perpetuate an antagonism only superficially reconciled by the common term "art"; till eventually . . . they appear coupled with each other, and through this coupling ultimately generate an equally Dionysian and Apollonian form of art—Attic tragedy.[1]

I would like to show that this discordant accord or harmony[2] is in fact the guiding theme of a conceptual reading of Aeschylus and Sophocles and that pulling apart precisely what appears to be an enigmatic intuition forcefully illuminates certain delicate passages in Hölderlin's essays, and especially his "Remarks" on the translation of Sophocles' tragedies.

The "More-Than-Two" of Greek Tragedy

The infrangible order of an assigned destiny cannot, in its inflexible application, produce the tragic. Destiny becomes tragic when it is transgressed. It is then the place of man's struggle to the death with that to which he must make himself the equal.[3] Angelus Silesius elicits a destiny when he says: "I must climb higher than God, into a desert."[4] This gain in human dimension, which installs the breaking of limits in violence and death, engages man on an inaugural and destinal path and makes a *deinon* of him. Inferior to everything, since he must die, he is superior to everything, since he does not know the limits of his power.[5] He is inferior to the gods and yet superior in one way, inferior to destiny while he surpasses it by sidestepping it, such that the fluctuations around the line of equilibrium are inessential, whereas the essential becomes the passage of the two elements toward each other in which each would be impure and asymmetrical. This is an equally efficient way of fleeing the immobile equilibrium of a false measure and the imbalance of an absolute turning away of the gods (their transcendence).

If tragic destiny is such that it includes its own transgression and the exchange of opposites, Greek tragedy would then be a huge variation on duality. This hypothesis requires confirmation. But let us not be impatient here: we are not looking for the two, but the two in the one, the *unitas multiplex*, the more-than-two, the duality that escapes the doubling of unity and would rather be the doubling of itself. Moreover, if this is a central preoccupation of Aeschylus and Sophocles, we can expect, from one to the other, a deepening of this concept, which is the stakes of a constitution of philosophy. Can we, in Aeschylus, demonstrate the existence of a model of this unitive duality, and, in Sophocles, measure the distance that has been traveled in this concept?[6]

This central presence of duality in tragedy is in general confirmed, first of all, by its "form." Hegel writes: "only those works of art whose content and form are perfectly identical are truly works of art." And, there is, he adds,"an absolute relation of content and form . . . a reversal of the one in the other, such that the content is nothing other than the reversal of the form in the content, and the form the reversal of the content in the form."[7] But the structural form of tragedies, such as Jacqueline de Romilly describes it,[8] is so manifestly dual that it is astonishing that her study did not insist upon the concomitant duality of content, linked only to a theory of "poetic genre" (Aristotelian *peripitia*, the contrasted duties, agon or the "tight and crackling contrasts," as Romilly puts it nicely without making any more of it). Equipped, on the contrary, with this principle of the reversibility of form and content, let us pay closer attention to the manifestations of a structural duality. Thus, the dithyrambic division between the chorus and the characters brings out, in the theatrical space, two places, the orchestra and the stage,[9] in the verbal space, two kinds of meter (lyric meter, iambic trimeter), in the vocal space, two tones, two modulations, in the choreographic space, two bearings, with the "strophes" and alternations of the evolution of the chorus. This marked division, which allows confrontation and "response,"[10] is not simply a reminder of the confrontation of man and god (such an agon with reversal exists in Euripides' *Bacchantes*), but seems to become meaning itself, the referent, the origin of tragedy. Division as the bilaceration-gathering of Pentheus's body, the image of the tragic body, consequently, is founded on a simple *allusion* to duality, a duality that must be felt and lived as suffering, the lips of a wound that must be sutured even in pulling them apart as far as can be tolerated. The body of the god-and-man, the concept of their "machine" or theatrical work of art, are the same, perpetuating the mystery of a Two. This dual machine, body or text, plays one against the other, arouses them to make the energy of their difference "grow in power." Force or strength is an alloy, the power of the gods its dosage as the "creating of tension." In this way, a physical model of the origin of energy[11] in this machine that endeavors, as Nietzsche shows, to think *and* to be, to think life according to the principles of life, is raised up (*monté*) at the same time that a means of struggling against the menacing rise (*montée*) of the secondary effects of positivity is discovered: dichotomous logic, the exclusion and valorization of the "negative," the exaltation of light, the displacement of the center of interest from the physical toward the metaphysical, "bad things."[12]

Now this duality is also what is at stake in philosophy: the "philosophical" draws back with a step that retrocedes from the secure base of the real and finds support in this drawing back. The "solid" then goes off toward the "unformed," where, in a vacancy of beings, questions can arise.[13] Beyond diversion, man thrown between two infinities, the Hölderlinian *Unförmliche* and the *Allzuförmliche*, the "choice of choice" of Kierkegaardian ethics, the *polemos* of Will and Representation, Force and Form, in Schopenhauer and in Nietzsche, *Kehre* and *Schritt zurück* in Heidegger, each express in their own way an archic duality. Arbitrating an incessant combat of the maximal opening and of the birth of forms, these philosophies, as the disciples of possibility, are the spirit of duality, not the "yes-no," the trap of a dichotomy that seems to permit choice, but the decisive instance of the "undecidable" of all real duality. If the *deinon* or the *thambos* of the Tragics also bring out a place from which the essential can take place, then a connivance of the Greek "symbol" and of philosophy is brought out in the universal problematic of the transcendental doubling of any question touching upon the foundation: *noesis noeseos*, the choice of choice, I know that I don't know anything, and so on. Or, to put it in the form of an example, suppose the couple man/god. There are two possibilities. Either man has something of a god in him, and the god something of man, *deinon* being what they have in common, and their indefinite battle is, through violence, transgression, unfaithful fidelity, and crossed loves, the guarantee of a tragic "rhythm" in which what is assailed is form in danger of being fossilized in the extenuation of force, and force in danger of being lost outside the forms it overflows; or man and god are rigorously antithetical, symmetrical, and opposite in a duality without "thickness" and are only "dead and non-opposed extremes"—they allow the case of the figure of an "unhappy Hespericity" to install itself. For the gods could well not be simply turned away, the height of infidelity, but radically fled (God is dead). Thus, through theatrical structure, the "more-than-two" of the two seems to be a dithyrambic commemoration, and, as we will see, the content of the tragedies seems to constitute a rampart for a unity of thought and Being.[14]

Aeschylus, Or, Doubled Duality

That duality is present in Aeschylus's oeuvre is obvious from the first reading. But the type of this duality must be drawn out carefully. *Suppliants* is the story of a dilemma: "A terrible war must be waged against one or the other [τοῖσιν ἢ τοῖς], and my hull is pegged, lashed tight, as though by cabestans. I see no way out now that is exempt from pain."[15] Two laws are present here: that of Zeus, who backs the cause of Io and the suppliants, and that of the sons of Aegyptos. This confrontation admits of no decisive solution inasmuch as another confrontation doubles the first: that of the males and the females. *The Persians* is the consequence of an unfortunate bridging.[16] Xerxes "crossed the marine capes connected by a bridge to the two continents" (*The Persians* 736). From that time on, the dream will be of the quarrel of two sisters (181, 185) coupled in the yoke of a chariot that they break. This broken yoke is the image of the *mechane* of the bridge suggested by a blinding Ate: "His expedition was double and presented two fronts" (ἀμφότερα, διπλοῦν, δυοῖν) (*The Persians* 720). Hence comes the sacrilege of a yoke imposed upon the sea, traditionally "without course" (ἄπειρον). Xerxes' real mistake, however, is in the bridging of the Earth to the Earth, which doubly neglects the natural agonistic couples Earth/Sea and Earth/Earth. In this way, Xerxes' bridge has an intention that is strictly opposed to that of Heidegger's.[17] The Heideggerian bridge enriches both banks in its respect for their difference, which has become "sensible" (in Hölderlin's sense), whereas Xerxes' bridge sticks to assimilating Greek culture to Persian civilization, such that Earth and Sea are in fact the same.[18]

With *Seven Against Thebes*, the central module is presented. The themes, the vocabulary (the "Hebraisms," as Pierre Fontanier puts it),[19] the syntax (oxymorons), the structure of the verse (chiasmus), the prosody (stichomythies), the architectonics (the two partially opposed demi-choruses), multiply the signification of a very particular Two: simultaneous fratricide. Catastrophes of one and of two that tip each other into their contrary, brothers of the same blood, they are One. But their One is still more intimate, coming from a paternal *and* fraternal seed. But their dispute is total; it is to the death. But it ends at the same death, at a single, unique destiny. But they are separated there by the mode of their burial: *nefas*, to which nothing is opposed, demanding a total sacrifice, annulling difference in a calm that has finally been restored. Open wound/suture. The multiplication of the Two: "Double is our anguish, double the suffering that has just been completed" (διπλαῖ . . . διδυμάνορα διμοῖρα) (*Seven Against Thebes* 849). Reduplications ("suffering follows suffering") that mime the "module" appear in line 509: ἐχθρὸς γὰρ ἀνὴρ ἀνδρὶ τῷ ξυστήσεται, "As enemies will the confrontation of man to man be produced." Asyndeton and doublets combine to build a literal "mimesis" of the module of the dual unity in a clear perception of the reversibility of form and content. The crucible of the tragic, as the pairing-tearing (the unbearable cut in its pure affect) flash, the pure form of the repetition of the same, is *terribly different from having to be different although it is the same*: brothers who are enemies. Ἄρχοντι τ᾽ ἄρχων καὶ κασιγνήτῳ κάσις, ἐχθρὸς σὺν ἐχθρῷ, "King against king, brother against brother, enemy against enemy" (*Seven Against Thebes* 674). The central duel is also mimed by the six other duels, among which is that between Hippomedon and Hyperbias, whose shields themselves are divinities at battle: Typhon breathing flames, Chthonian god against Zeus, the Ouranian. These are the effects of a *doubled duality* that culminates in lines 961–62.

> παισθεὶς ἔπαισας
> σὺ δ᾽ ἔθανες κατακτανών
> δορὶ δ᾽ ἔκανες δορὶ δ᾽ ἔθανες.
>
> Struck, you struck
> You were killed in killing
> In combat you killed, in combat you were killed.

Here, the plays of differential variation are limited to a radical (μελεόπονος / μελεοπαθής), a letter (ἔκανες / ἔθανες), when this is not the pure taking up of the word (ἠὲ) again. The conclusion is obvious: διπλᾶ λέγειν, διπλᾶ δ᾽ ὁρᾶν, doublets of the spectacle and of the word (*Seven Against Thebes* 972).

Prometheus Bound displays the compensatory violence of a god who is sensible to the dangers (is this why he withholds a secret from the gods?) of an imbalance in the fragile couple man/god. In giving knowledge to men, Prometheus opposes himself to the image of the gods as destructive (the same image as in Exodus 32): "I stopped the mortals torn to pieces from descending to Hades."[20] The intermediary between the two reigns, of which he constitutes the "caesura," he gives men self-knowledge. But in untangling mortals' thoughts, since before "they saw without seeing, listened without hearing, and haphazardly entangled everything throughout their lives,"[21] he does "too much" and delivers himself to Zeus's anger. This "too much" is too much love. But he holds onto a piece of the god in a secret, and, reversing positions, this secret gives the advantage back to Prometheus: "He will need me; he will break my chains and return to form an alliance [εἰς ἀριθμόν] and friendship" (*Prometheus Bound* 169). This is a beautiful illustration of the impure and asymmetrical couple, without victor, of the symbolic more-than-two in which the invisible thickness is the "exit door," the solution that can be found, the avoidance of unilaterals. Cunning (*mêtis*), as retreat, is the perception of the thickness and of the possible exit (*poros*) from the system: Prometheus is equally "δεινός εὑρεῖν κἀξ ἀμηχάνων πόρον," "skilled at discovering a path, even at a dead end" (*Prometheus Bound* 59). And it is this secret cunning that holds Zeus's unilateral order in check. Taking up Hegel on Antigone, Jacques Derrida understands this principle well: "These two laws do not confront each other as two solid volumes or surfaces, identical to themselves, homogeneous in themselves. Each law is fissured in its side, and already by the work of the other within it."[22] Beda Allemann speaks of the fraternal struggle between two "communicating principles" in a comparable way.[23] This is a paradox in the strongest sense: Prometheus is the figure of the innocent couple that was dear to Kafka, of Hölderlin's faithful traitor. Thus, the central module of a doubled agon is confirmed, an agon carried to his maximum power there, where metamorphoses and catastrophes do not turn the world upside down, but establish its correct equilibrium. Such is the source, such the re-sourcing, of the tragic. And it might even be that the allusion to Zeus's "thunder," the "two-pointed braid with a median handle," is but an approximate sketch of this module, whose decisive importance appears to us little by little.[24]

With *Agamemnon*, there is a return of the (deepened) dilemma, since the father must sacrifice his own daughter. The sacrificial relation cruelly doubles the relation of kinship, as it does for Abraham, whose episode is quadrupled significantly in *Fear and Trembling*: "My lot is cruel if I disobey. Cruel too if I must tear apart my child. . . . On both sides there is nothing but unhappiness" (*Agamemnon* 206). The structure is identical in *Libation-Bearers*. Orestes is caught between the vow to avenge his father and the terror of matricide: "Must I not fear killing my mother? It is better to make enemies of all on Earth than to risk the fury of the gods" (*Libation-Bearers* 899–904). "Watch out for the avenging dogs of your mother! But those of my father; where do I flee from them if I renounce?" (*Agamemnon* 925). Here again, the combat between men is mimicked by a combat between the gods, such that there are always at least four protagonists. Ares' double whip, which intervenes in the two plays *Libation-Bearers* and *Agamemnon*, recalls Prometheus's two-pointed braid (*Agamemnon* 642; *Libation-Bearers* 375). The doublets in *Libation-Bearers* echo those in *Seven Against Thebes*: Ἄρης Ἄρει, Δίκᾳ Δίκα, "Ares will battle against Ares, right against right"—the Greek is much more intense (*Seven Against Thebes* 461). Finally, a dilemma, a bifurcation opening onto two abysses, occupies the center of *Eumenides*: "Here is where I am: whether I let them stay or drive them away, in neither case could I escape the aporia" (ἀμηχάνως) (*Eumenides* 480). This doubling takes place by way of the intervention of the gods: Athena against the Chthonians/Erinyes and Apollo the "impure," the at once savage and luminous god, playing the intermediaries (*Eumenides* 750). This new occurrence of the module might support an interpretation that I will put forth cautiously: would not the agon, which redoubles duality and which is shown to be what is at stake in the tragic, be the struggle between the *apeiron* (the unformed of the night) and the *peras* (the forms carved from the day), put differently, already the struggle between Hölderlin's "too formal" (*Allzuförmliche*) and "informal" (*Unförmliche*)?

If, moreover, Aeschylus's tragedies make something of a survey of archaic Greek thought in marching out the conception of the equilibrium of a just middle ("illness is near great health," *Agamemnon*), the Anaximandrian principle of the debt of injustice ("there is an old saying . . . that perfect happiness becomes fertile and does not die without children," that is, without unhappiness) (*Eumenides* 750), the reciprocity of *time* ("honor") and *poine* ("castigation") articulated around *tisis* ("debt"),[25] and finally the refusal of accumulation ("if a wise fear, prudently maneuvering, can unload a bit of the acquired riches, the house will not sink despite its opulent charge" [*Libation-Bearers*]), is there not, in these principles that are still alive with the Sages and the first philosophers, the simple model of an agonistic duality lifting itself up on the homogeneous foundation of a support from "behind," whether it be the Parmenidean *suneches* or the Heraclitean *polemos*, each capable of taking in both possibilities without reconciling them in the dullness of a dialectical "sublation"?

From the dilemma of *Suppliants* to the torn-apart kinship of *Libation-Bearers*, there is but one Aeschylean stake, doubled duality.

Sophocles and the Hyperbolic Machine

Sophocles said somewhat severely of Aeschylus that he "does what is necessary without knowing that he does it." This was a good pinpointing of the distance and the complicity of their "break." A simple and allusive duality indeed appears here and there in Sophocles.[26] Yet for having encountered well constructed and otherwise subtle structures in his text, we will see Aeschylus' concept in Sophocles. And we will study three of his "machines": (re)doubling, reciprocal and asymetrical exchange, and symbolic oxymorons.

Doubling is first of all the ambiguity of the mantic word. Heracles indeed believed he was to be "released from the toils" laid upon him; he did not understand that his death was at issue (see *Oedipus the King* 1170–71). Oedipus proclaims his conviction of himself without knowing it when he announces: "On behalf of no far-off friend, no, but in mine own cause, shall I dispel this taint" (with a significant αὐτός αὐτοῦ) (see *Oedipus the King* 1170–71); "I may suffer the same things which even now I called down upon others" (*Oedipus the King* 251–52); "I will uphold this cause, even as the cause of my own sire" (*Trachiniae* 264–65). What tragic humor! How, after that, could Freud not be interested in this character?

Doubling is also the splitting in two of the system of truthful speech. We can, in effect, oppose Tiresias, the representative of the tradition of a magical-religious *Aletheia* (*Aletheia* that has already lost its automatic convincing force, its *Peitho*, however, since neither Oedipus nor Jocasta are convinced by it),[27] as *index sui* ("in me lives the force of truth"),[28] on the one hand, and the constitution of *Apate*, as ruse, autonomous illusion, on the other. Thus Clytemnestra, in *Electra*, saw salvation in the messenger's news and not a pretense to introduce his son the murderer into the palace. After a long battle, Ajax and Hector really believed they were offering each other gifts, but only offered mutual death. Deianeira, deceived by Hyllus's words, "did evil with good intent."[29] Oedipus's presumed parents, Polybus and Merope, hide the true nature of his family relations from him. And only this "supplementary" screen could set loose the series of catastrophes.

Bifurcation, for its part, recalls the Aeschylean dilemma. Two paths are opened. Which one are we to choose? Laius's narrative is composed of two versions, that of the bandits and that of the travelers. The first version abandoned, there remains the choice between a troop and a single passerby. There is the same problem with Neoptolemus at the crossroads of the moral: must he obey Ulysses' orders and respect the supreme interests of the Achaeans or rather admit everything in preserving at least his moral integrity (*Philoctetes*)? The very image of bifurcation, we will see, will have considerable importance in *Oedipus at Colonus*.[30]

The reciprocal exchange complicates this first level of the Sophoclean duality. We will read it in the crossed figures of *philia* and *pharmakon*.[31] The *philia* that henceforth links Hector and Ajax is insured ritually by an exchange of gifts. But these poisoned gifts, the trace of mana, deliver death.

> With the very girdle that had been given to him by Ajax, Hector was tied to the chariot-rail in such a way that it sawed his flesh, dragged him, tore him apart, without mercy, while 'twas from Hector that Ajax had this gift, and by this very dagger hath he perished in a deadly fall. (*Ajax* 1029–33 [trans. modified])

Now, seest thou how Hector, though dead, was to destroy thee at the last? (*Ajax* 1026–27)

The gifts of *philia* between these enemies (the Greek, significantly, conveys the duel, which, like in Sanskrit [*obha*], is the mark of an indissociable twin couple: τὴν τύχην δυοῖν βροτοῖν) held them in a chiasmus, each object representing in the other a part of the adversary, Ajax's shoulder-belt establishing in Hector and Hector's dagger establishing in Ajax a bridgehead for the interests of the antagonist. We again find in passing an Oriental principle: yin and yang, the two opposites interlock and cut into each other impurely and asymmetrically.

This poison present (*cadeau*) can also be interpreted as the "persistence of the gift [*don*]." Hence, its transmission in a chain. Oedipus's parents abandon him; a servant gathers what was gotten rid of to the benefit of the adoptive parents; out of piety, they withdraw and return to the point of departure with all the accumulated interest that will have to be paid at the highest price. The gift of death must be relayed by a woman who is in love in order for Heracles, who cannot be killed by anything, to be vanquished (*Trachiniae* 1207). Hector triumphs over Ajax "post mortem," through the relaying of his gift. Therein lies a figure forgotten by Hegel in his battle of consciousnesses, and it could be that the tragic is situated in the space of that forgetting.

This very primitive "contagion" of death designates the object as paradoxical: παλαιὸν δῶρον (*Trachiniae* 555). Just like the fatal handkerchief in *Othello* or the key in Hitchcock's *Dial "M" for Murder*, it never stops traveling. A too intense object that gives everything, including death, it is a two-faced *pharmakon*.[32] A gift carrying another gift, the opposite of Silenus (the Lamia), it doubles the gift in inverting it. Death takes hold of what is living, the object possesses the subject. It is the object that gives to the giver the subject of the gift, and not merely the giver who entrusts the subject with the object of the gift.[33] This is at once what Hegel saw in a certain type of "more"[34] and what Baudrillard detects in the Code, desiring in the place of the subject, whose desire no longer concerns anything but the intermediaries of intermediaries, and putting the subject in the position of the object, or in the "inexchangeable merchandise" whose connection with the Hölderlinian unbearable we would have to explore.[35]

That it is a question of a *pharmakon* is clear in the text: "And what Trachinian deals in spells so potent?" Heracles asks (*Trachiniae* 1140). His son is not a "murderer" but a "physician" (*Trachiniae* 1208–9). And it is a "magic charm" Deianeira believed she was passing along (*Trachiniae* 575). In Sophocles, then, the peace and prosperity that Oedipus gives to Thebes, the children he gives to Jocasta, the maledictions on the murderer, would be *pharmaka. Pharmaka* are his children, whom he must curse, *pharmakon* is he himself, the most dreadful man in Greece, "unwanted child, accursed in wedlock, accursed murderer" (*Oedipus the King* 1184–85). But having become φαρμακός (scapegoat), he will reverse his death in aid of Colonus and Athens. Love, the act of adoption, the helpful gesture, indomitable vigor, profound wisdom: everything has turned into its contrary, and this strophe reveals the two faces of the tragic hero that are so clear in the anti-Oedipus, who has "one eye too many," the eye of the secret of his other, his cursed part, his line of escape, both of them stemming from ἀμοιβός. Equally, then, in Sophocles the doubling becomes a crossing and tends toward a quadruple figure.

Still farther in the complexity of dual machines, the textual figure of the symbol, oxymoron, is presented. In an exaggeration of its principle, this figure concerns a familial relation that is disrupted in the Orestes and Oedipus cycles.[36] That father and son attach their desire to the same harbor, the same "soil," as the Greek puts it ("The same bounteous place of rest sufficed thee, as child and as sire also, that thou shouldst make thereon thy nuptial couch.") (*Oedipus the King* 1209–10), is a monstrous "double blow" whose duplication the text, to obtain the maximum tragic effect, need only mimic. The same pullulates—profound intuition; and it is thus that it is the most destructive. Put differently, Sophocles used oxymoron very differently from Heraclitus, for example. Opposite meanings are sought out in Heraclitus (ἕν πάντα; ἄνω κάτω; δίκην ἔριν),[37] whereas Sophocles looks for polyptotic repetition: ἄγαμον γάμον (*Oedipus the King* 1214).[38] There are indeed, in Heraclitus, oxymorons founded on words with the same stem: ἀφανὴς φανηρῆς ου ἀθάνατοι θυητοί.[39] But Sophocles concentrates the effect on the *parentela*: ἐξ ἀνδρὸς ἄνδρα (This has fascinating consequences, for the translation could be "man by husband" just as well as "husband by husband"—a new example of the "innocent guilty" party [*Oedipus the King* 1250].) And Sophocles simplifies the effect through a fixed movement: τέκνων τέκοι (*Oedipus the King* 1250). The succinct maxim, "she had borne . . . children by her child," summarizes the tragic in two words: it is to have subverted, through an overbidding, the directions of kinship, going from son to father and from brother to brother. . . .

Going further in kinship is the death-blow: πατέρας ἀδελφοὺς παῖδας . . . νύμφας ψυναῖκας μητέρας, with a ternary compound that is characteristic of turmoil. "Oh wedlock, you have begot fathers-sons-brothers . . . daughters-wives-mothers." And we must at all cost respect this writing by parataxis or asyndeton, just as one could resolve the debate opened by Heidegger over lines 360 and 370 of the chorus of *Antigone* by respecting the oxymorons through an asyndetonic juxtaposition.

This general turmoil in the family relations, which makes language vibrate against itself, is the reverse of the indefinite possibility of man: man is the terrifying, and it is "in an instant" (ἐξαίφνης) that Oedipus loses his wife, children, mother, honors, and power. Tiresias had seen this abyss into which Oedipus, the *hupsipolis* who is henceforth without state, fell. But as Hölderlin will know, Tiresias is the caesura of time (the suddenness of the god) and the abyss itself that he opens under the feet of the characters. This strophe of time makes us understand that oxymoron is but a textual figure for absolute speed, that of the exchanges in symbolic unity. Let us enter more precisely into this machine of tragedy.

That is, a throwing (*lancée*, βόλη, *physis* for Heidegger). It is exteriorization, a wager on time, an adventurous trace. It throws itself toward something. Let us posit it as unassimilable. A struggle without end is engaged, in which the two trajectories are secured. One fails [*défaille*] there and we come back to the exclusive One. The *symbolon* with which, among others, tragedy is occupied is the machine formed by the conjoined resistance of the two trajectories that exhaust themselves and renew their resources in their attacks—a unitary holding, a reciprocal exchange, the impossibility of eluding the rift [*faille*]. In this sense, the *symbolon* also functions as *diabole*. It bridges two elements by "throwing between them" but also constitutes the obstacle of their separation. It takes the space of the interval of a bridge and a rift, the Kafkaesque bridge. The *diabole* of the *symbolon* lies in that it joins and tears, just as the devil knows how to be slanderous (*diabolos*). Let us imagine, then, a perennial module in which it would be impossible for the *symbolon* to lose its equilibrium. That is possible only if we assume a basis for the double symbolic in a new double, such that the two symbols function perpendicularly (the fourfold) or in a chiasmus. *The absolute symbol finds its "resting point" in hyperbole.*

There is nothing abstract in these figures. They are often mimicked in small machines, in the wheels of a wagon, in vases, and in Chinese instruments used for prediction. They are the universal concept of symbolic thought. In Sophocles, a model of this thought attains its perfection: Antigone and Creon. But if Oedipus has "one eye too many perhaps" in Hölderlin,[40] it is because despite himself he is the parturition of that perfection. All over Oedipus's path one encounters these struggles and these couples, these bifurcations and these characteristic strophes. Oedipus is the part Laius and Jocasta have in common and in which they stand united at the heart of their greatest difference. His monstrosity is to redouble this part, for, being the seed of Laius in the flesh of his wife, he resows this flesh, all the while putting his parent to death: "Oh, how can the soil wherein thy father sowed, unhappy one, have suffered thee in silence so long?" (*Oedipus the King* 1211).

Everything happens, then, as if the natural connection of human groups in the *genos*, tightened by the symbolic struggles between clans (potlatch), was perverted and met its end in an excess. Oedipus's horrible gifts, the legacy of Jocasta or Laius, negative *megaloprepeia*, in the tragedy commemorate the painful end of the *genos*. And precisely because the thought that issues from a constitution may be mythic of a "religion of the family,"[41] but which, if we want to understand "political revolution" as the end of an anterior system, we are forced to postulate, this thinking of the secret, of the refusal of linear excess, of compensatory equilibrium, is close to giving way to the great movement of the "positivities" (money, philosophical and mathematical concepts, laws, the equality of citizens, . . .); it produces, on its extreme decline, a flowering of symbols of which tragedy seems to be a sophisticated expression. Thus, rather than staging a dithyrambic ritual, the key to which must have been lost to many of the Greeks, tragedy will tell of the death of the familial system in the throes of parricides and fratricides, as well as the struggle that was responsible for that death, that of *genos* and polis, of Antigone and Creon.

What has lost the *genos*, however, Sophocles seems to say, is a *too much*, the "too much" of the legendary conqueror of the Sphinx. His "too much" love for his mother. If Oedipus, Antigone, and, already, Orestes, Electra, and Clytemnestra sound the "toll" [*glas*] of this ancestral family,[42] it is for an excess of love that should have led the positive thought to meditation. Look at Oedipus the too good son who leaves his parents to spare them troubles; Jocasta, who loves her son like a lover; Clytemnestra, who is crazy about Iphigenia; Electra, the living memory of her father; and Antigone, who sacrifices everything for her brothers and her father. Look at Deianeira's too faithful love. The fortune of the *genos* seemed (perhaps in the imaginary) to be a re-enforcement of family bonds. In fact, it is what loses this bond. This is the hard lesson of the *pantoporos aporos.* At the limits or in the rift, the extremes are one.[43] But the polis, and Creon already shows this, was also doomed to exaggerate in a single direction, to linearize severely what had triumphed, and what worries us. Our Occidental age is the necessary thinking of this stake.

Oedipal hyperbole was the figure of the crossed relations of his two family relations with his own being double (which gives us, upon reflection, much more than one eye too many). Antigonean hyperbole goes further in perennializing the principle of *genos* in face of the polis or the law that seems to crush it. That is possible by way of a mutation of the excess of which Oedipus is the bearer. His transformation (terrifying for Theseus at the end of *Oedipus at Colonus*) is the preparation for Antigone's saving excess. It is here that we must take an extra step in order to follow the astonishing constitution of this symbolic logic to its end.

In fact, the displacement of the problem of the ritual confrontation of man and god to that of *genos* and polis does not erase the previous struggle in redistributing the "deal" of the Greek world. The lack of god (*Gottes Fehl*) in Hölderlin or Karl Reinhardt (tragedy, in *Sophocles,* is a "sign towards the enigma of the border between man and god"),[44] or in Hegel's statement, "It is the divine that constitutes the theme proper of primitive tragedy,"[45] expresses the importance of this theme. But the God has changed. He has doubled himself in order to keep the human instance itself bipolar. Therefore, on the one hand, a god of the city (support of human law and of religious formalism) responds to the Antigone-Creon couple, and, on the other, a god of the dead (who perennializes the *genos* and has the high hand over the forces below, the excessive forces that are always to be mastered). Antigone will be the revealer of their hyperbole, in which the harmony of a median point is this respect for the laws that can surpass itself out of respect for the foundation of these laws.[46]

It is thus not for her fraternal piety (even if her love is nearly carnal)[47] that Antigone dominates the tragedy by her stature, but for her just relation to the gods below. This just relation is understood after the fact as the endurance in which Oedipus's excessive ardor is reversed and as the "more" that lets us foresee his opening as a "broken" man, raised infinitely beyond himself. Not that a new hubris pushes him to take himself for a god, but he understood the lesson on respect espoused by the statuary god (he obeys the pious injunctions of the chorus) (see *Oedipus at Colonus* 190) and of the thundering god, the master of the *thauma*. This espousing and doubling lesson is named Antigone, and permitted Hölderlin to conceive of a nonchronological logic of the sequence *Oedipus the King/Antigone*.

Nor is it for having scorned his family ties (*Oedipus at Colonus* insists on his violence toward Oedipus and his daughters) that Creon is finally beaten down—without, for all that, as Hegel well knew, his political choice ceasing to represent a future (and in that he cannot be defeated), but only for his negligence of the eternal laws in which the very laws he defends take root. Sophocles' lesson is perhaps even more profound: whereas Antigone does not contest Creon's rights and understands these rights, Creon has the naivete of the clean conscience of the just, the prototype of a formidable political attitude that "knows only the True."[48] Antigone's law says yes twice; her *yes* is doubly symbolic. Creon's law says yes for the one, no for the other, without understanding any further. And Nietzsche and Hölderlin both grasped that in the tragedy.

Put briefly, Creon's hubris is the indefinite linearization of the power of positive laws, and Antigone's sacrifice is the instance of transgression required to reestablish an equilibrium with a curvature. A serious potentialization, in fact, can be obtained only through a return of the force that has reached its maximum, and thus offers itself a new field to pursue its trajectory.[49] The mathematical curve of the hyperbole or the theme, in these philosophers and in Hölderlin, of the hyperbolic and of the return expresses this curvature without regression that is balanced in the model of two inverse crossings of double compensatory curvature.

The latency between the two tragedies in which Oedipus is the protagonist, this slow maturation through suffering, plays the role of a "third" or of an "empty case," of the caesura constitutive of an unfolding of the god in two parts, which alone permits him not to decline. Man thereby also keeps his opening. The law is kept from becoming a rite without referent; the positive does not yet make the "desert multiply." Necessity in the tragedy of doubled dualities, of a "thickness" of the symbol, of the *kommos* in which the chorus and the characters are joined, thus becomes clear. By way of a machine that becomes more and more complex, tragedy mounts a strategic thought, a wisdom, in which the "more-than-two" as support is not, for all that, "conciliation."

The Instance of Greek Tragedy in Hölderlin's Poetic Achievement

Showing that the conceptual structure of Aeschylus' and Sophocles' tragedies (with Euripides "something changes")[50] is what Hölderlin understands as the "operation of the poetic spirit" (*SA* 4; 1: 251) will permit us to measure the strides the poet has achieved since the extreme trajectory of Greek thought. And in its reflection on the essence of poetry, Hölderlin's position is all the more original for admitting as a point of departure a conception that is very similar to that of Hegel, his fellow disciple and friend. They coincide on the definition of the lyric (the internal, the subjective) and of the epic (the external, objective)[51] and on the idea of an intermediary zone between internal form and exteriority ("It participates in the two extremes.")[52] or, once again, in the crossed exchange of the two: "That objectivity that has its source in the subject and that subjectivity presented in its reality and its objective signification form the totality of the spirit."[53] One will find, in Hegel, oxymorons, such as the "guilty-innocent."[54] But if Hegel saw well, notably for Antigone, the principle of having in oneself the reason for one's destruction and the guarantee of one's protection,[55] it remains true that the conception of tragedy as the calm conciliation that the "substantial truth" produces in "sublating" the unilaterality of the conflict, which is central in Hegel, is radically inoperative in Hölderlin's thought.

The detail of the essays and the "Remarks" will convince us of this. "Reflection" joins at once the contraries of temperance and enthusiasm: "Where temperance forsakes you, there are the boundaries of your enthusiasm."[56] And Hölderlin adds, in a sense that Nietzsche will not forget: "One can fall upwards just as well as downwards."[57] The idea is the one Sophocles spoke of: the rise in power (*Steigerung*) is not made linearly but in the curve that implies the sustaining of the equilibrium of the forces in tension:

> He must by no means think that he can only supersede himself in the *crescendo* from the weaker to the stronger; in that manner he will become inauthentic and too high-strung; he must feel that he is gaining in levity what he has lost in significance, that silence and the sensuous replace the vehemence and the impetus quite beautifully. (*SA* 4; 1: 234; *Essays*, 46)

How could one put this any better? Inversion, harmonic reciprocity, nonlinear excess, the least one can say is that the elements of Greek tragedy are not foreign to Hölderlin, even in their very concept.

Let us attempt, then, to enter into that formidable text that is "On the Operations of the Poetic Spirit" (*SA* 4; 1: 241; *Essays*, 62). Let us first stick to grasping the sense of this long, three-page, introductory sentence: "Wenn der Dichter einmal des Geistes . . ." The structure is strange: there is a sixty-two-line subordinate clause and a two-line main clause. If the principle of the reversibility of form and content is also verified in Hölderlin, such an obvious disproportion makes sense. In the first lines (lines 1–8 in the German), the poet must fulfill the condition of a grasping of the harmonious principles of alternation (*Wechsel*) and of progressive striving (*Fortstreben*), of the reproduction of the self outside itself and of the progress of the spirit (*Folgerungsweise*). But starting at line 10, the conflict (*Widerstreit*) intervenes that separates the synchrony (the simultaneity of the parts, the unity of the spirit) that is called spiritual content (*geistiger Gehalt*) and that is here designated by *a*, of diachrony (the reproduction of the self outside itself, progression, alternation) that is called spiritual form (*geistige Form*) and that is here indicated by *b* (lines 10–15). But *a* and *b* are themselves divided into two: in order to be grasped, spiritual content implies a difference in degree in sensible content (*sinnlicher Gehalt, Grade*) (= *a'*); spiritual form, for the same reason, implies an identity of sensible form (*sinnliche Form*) (= *b'*) (lines 15–26). The conflict can then develop a chiasmus that Hölderlin, without catching his breath, studies branch for branch. The poet, in fact, must understand that the conflict between *a* and *b* is resolved precisely by the fact that: (1) the identical sensible form (*b'*) compensates the alternation of parts (*b*) and satisfies the requirement of the identity of *a*; and (2) the differential sensible content (*a'*) compensates the need for unity (*a*) and satisfies the requirement of alternating progression (*b*). Hence, the first chiasmic figure (lines 26–43):

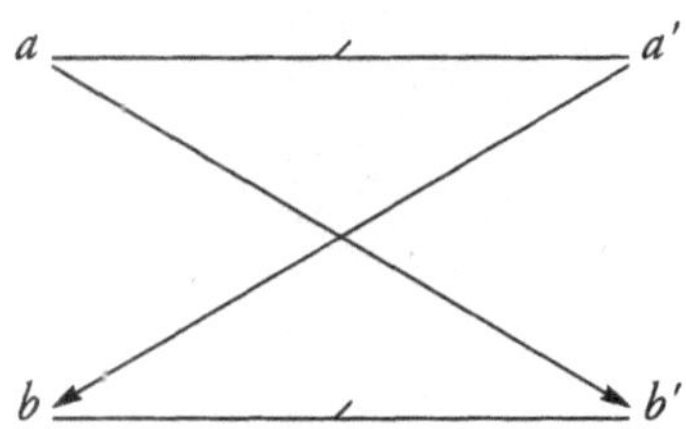

Scheme 1. —/— = compensates; —— = satisfies

The figure of the chiasm would not be complete, however, if the inverse movement were not verified. The antagonism between *a′* and *b′* must therefore be resolved by the fact that: (1) the spiritual content (*a*) compensates the loss of material identity in the sensible content (*a′*) and satisfies the material identity (*b′*); and (2) the alternating spiritual form (*b*) compensates the loss of material diversity due to the progression toward the culminating point of material identity (*b′*) and satisfies the sensible content (*a*). This is the second inverted chiasmus (lines 43–51):

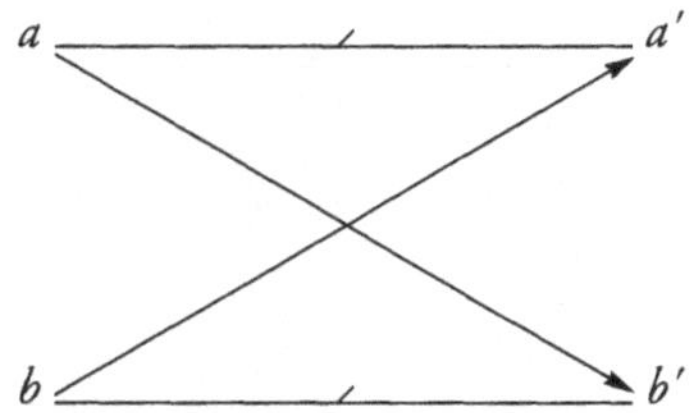

Scheme 2

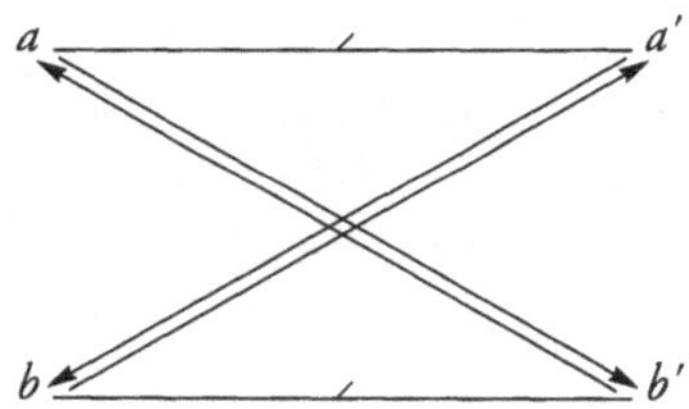

Scheme 3. A double chiasmus

It still remains for the poet to understand, if his head is not spinning too much, that it is inasmuch as they are irreconcilable (*unvereinbar*) that *a* and *b*, and *a′* and *b′* can become sensible (*fühlbar*) in their antagonism (lines 52–56). In this irreconcilability, each element can find its culminating point. But this is also its "point of rest" (*Ruhepunkt*), for the "identical progression" (*b′*) is the process toward the culminating point, and this point itself is double, satisfying at once stability and process as *akme*. This "culminating rest" realizes a "harmony" in which the contraries are balanced and maintain one another. We can thus bring out the complete scheme:

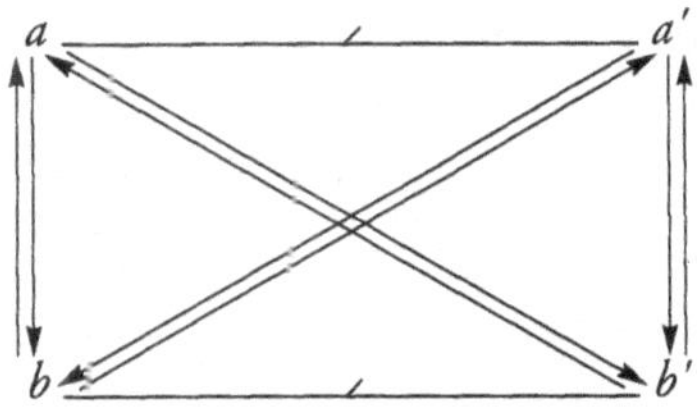

Scheme 4

Or, according to the hyperbole:

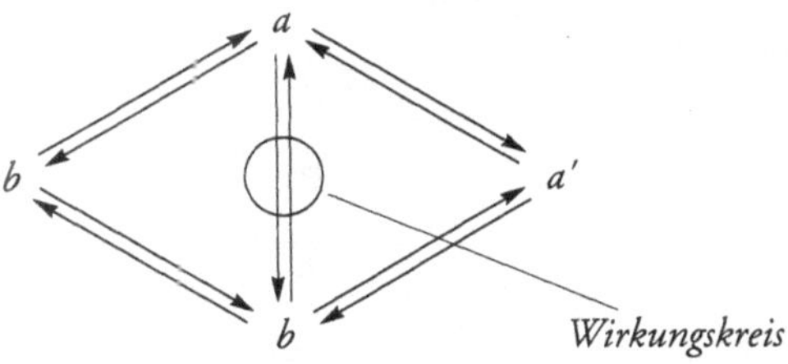

Scheme 5. Hyperbole

In this exceptionally rigorous research, Hölderlin without doubt wants to say that this torn apart, crossed, recurved operation must, however, impose itself in the evidence of one single "holding." Hence, the sentence in a single breath. The perfect facility of the poetic spirit thus appears at the highest point of difficulty: "Most near, but difficult to grasp, the god . . ." What is essential, we see, is not in the elements but in their fourfoldness as a sphere of influence (*Wirkungskreis*). (*SA* 4; 1: 245; *Essays*, 65). What is that exactly?

Between the "expression" (*Ausdruck*) of poetic matter and the "free elaboration" (*Behandlung*) of the spirit, it is foundation. Spirit-matter, it is spiritual reconciliation and material dissociation. "What distinguishes it is that it is also opposed to itself." The principle of change and of stability, it reverses values, separates the unified, liberates the fixed and does the reverse as well. "From that point on, signification reconciles even what is absolutely contradictory and is in all points *hyperbolic*." This is because the contents, having reached irreconcilability, are nonetheless reconcilable by the degree and the orientation of this opposition and are taken up in the *Allgemeinsten* of life, the contradictory union of all opposites as *harmonisch entgegengesetzt*. At this point, spirit becomes sensible as "infinite" and appears to itself in the "most sensible, most hyperbolic" manner in a "negative-positive" that seems to be a commentary on the epigraph to *Hyperion*. (*SA* 4; 1: 250; *Essays*, 70). Hölderlin can therefore write: "*The infinite unity* which is once the point of separation for the unified as such, but then again also the point of union for the unified as the opposed, finally is also both at once, so that what is harmoniously opposed within it is neither opposed as something unified nor unified as something opposed but as both in One, is felt as opposed in a unified manner as inseparable and is invented as something felt" (*SA* 4; 1: 251; *Essays*, 71). Present infinity, divine moment, completed art—this definition seems to us the montage of a simple machine of mystic mechanology. Chiasmus, inversions, symbol, excess, and hyperbole are the cogs of this "organ in itself," as Hölderlin names it. When this infinity must be expressed, the "Hebraic" reduplications reappear ("World in the world, voice of the Eternal addressed to the Eternal") in which, at the caesura, the divine breaths and the *kommos* become possible (*SA* 4; 1: 250; *Essays*, 70).

Hölderlin nonetheless attempts to go further than the tragedians in deploying a conceptual arsenal of definitions. The *harmoniously opposed* will be the element that changes place and qualities in an indefinite struggle. The *opposed harmoniousness* will be the "holding," the Greek ἁρμονία, in which the opposed must be understood as a Heraclitean ἀντιξοῦς.[58] *Alternation* or *change* is the reciprocity of the crossed movements. The *absolutely opposed* is the real irreconcilability of the real elements that do not struggle in a weak and insipid unity, but, at the end of their inexchangeability, find the strophe that joins them in reconciliation. The *caesura*, then, is the irreducible rift between the opposites, their architectonic rebalancing, the wound without healing, the Parmenidean γόμφος that, in the prologue of the *Poem*, forms a couple with περόνη: the corner that separates, the ankle that holds, Philoctetes' unbearable wound.

The same idea comes through in other texts. "On Religion" gives the recipe for an interlocking by way of a mortice and tenon joint: "In that each part goes a little further than is necessary, they will achieve their inseparability" (*SA* 4; 1: 281; *Essays*, 94). The enjambments in the poems and the characteristic *aber* without doubt issue from this surpassing. "Becoming in Dissolution" expresses the tragic as a chiasmus of the real infinity and the ideal finity (*SA* 4; 1: 282–87; *Essays*, 96–100). "The Ground for *Empedocles*" defines the "more-than-two": "When everything is entirely what it can be, and one combines with the other, compensates for the shortcomings of the other, which that one must necessarily have in order to be entirely that which it can be as a particular, then there exists perfection" (*SA* 4; 1: 152; *Essays*, 53). Then is born "the most sublime feeling man can experience, . . . when the two opposites . . . encounter each other" (*SA* 4; 1: 153; *Essays*, 53).[59] Hölderlin works more particularly on the concept of the "return." Nature is the aorgic (unlimited) and man the organic; each leaves its "native" place, tends toward its contrary, and then returns to the native. From that point, the two movements admit of two symmetrical points of encounter in which nature is more organic and man more aorgic at the same instant. At this point (ἐν μέσῳ), the overdetermined place of Greek thought, takes place,

> the moment [in fact, the subject of "takes place"] where . . . the organic that has turned aorgic appears to find and return to itself in that it keeps to the individuality of the aorgic, and where the object, the aorgic, appears to find itself in that at the very same moment that it assumes individuality it also finds the organic at the highest extreme of the aorgic, so that at this moment, *at this birth of the highest hostility, the highest reconciliation appears to be the case.* (*SA* 4; 1: 153–54; *Essays*, 54)

Such texts are radically unintelligible in an asymbolic logic in which neither Greek logic nor an Empedocles, as the living emblem of contradiction experienced at its most intense, are conceivable.

What, then, is tragedy? Let us try to potentialize the "Remarks" with the knowledge we have gained above.

> The presentation of the tragic rests primarily on the fact that the tremendous [*Ungeheuer*]—how the god and man mate [*sich paart*] and how natural force [*die Naturmacht*] and man's innermost boundlessly [*grenzenlos*] unite in wrath [*in Zorn Eins wird*]—is conceived by boundless union [*grenzenloses Einswerden*] purifying [*sich reiniget*] itself through boundless separation [*grenzenloses Scheiden*]. (*SA* 5: 201; *Essays*, 107)

The caesura (*Zäzur*) is nothing other than the "change of representations at its highest point" in a sense that is henceforth intelligible (*SA* 5: 196; *Essays*, 102). In the "calculable law" (*Gesetz*) of the work, the caesura balances the two parts of the tragedy. It is therefore the effective field, the suspension of forms, the tragic tearing from the sphere of life, the rift in which what is unbearable in an eccentric sphere of the dead takes place (*SA* 5: 196–97; *Essays*, 101–2). This *deinon* is also the original that springs forth when the sign equals "zero" and the antirhythmic suspension is produced in the vacancy of forms (*SA* 5: 196–97; *Essays*, 101–2). Man must not let himself be snatched up by this vacancy. It is the retreat of the being in his entirety that reveals the *peras* of a validating discrimination of forms, which, on the basis of their *apeiron*, can therefore make their "return," thus avoiding a new enrooting of the becoming-form in an unnecessary form. The caesura is therefore understood as the "walk in the unthinkable" of the Greeks and of the Hesperians.[60]

It is therefore from the very inside of the experience of aorgic fire that the retreat of forms toward the unformed validates the organicity of the forms, their harmony. Since this harmony is a crossing, the concept of tragedy will be the avoidance of a unilateral or noncrossed becoming-form. Thus, the gods punish the hubris that would consist in taking oneself for a god (Empedocles). Man must never be *without remainder*, which is what, however, he must strive for totally (the god). The unilateral, then, must be "purified" and pay its debt of injustice. (It is clear that Anaximander's word is a very profound understanding of the Greek and Hölderlinian tragic model.) It is therefore the ἄγνος rather than the καθαρός of an Aristotelian tragic model that *sich reiniget*, payment for "injustice," would signify. From that point, Suidas's citation would become: "The clerk of nature, letting its calamus macerate (the meaning of ἀποβρέχων) for it to become well prepared" (the meaning of εὐνοῦν in this position). And the citation, as certain commentators think, would concern Sophocles and, through him, Oedipus, who is made to macerate by a long suffering, just as the divinities who take him in turned from savagery to benevolence (Eumenides).[61]

An incalculable but conceivable law, the rift demands the confection of a machine comprised of an "against" (*gegenseitig*), excess (*allzukeusche, allzumechanische*), and exchange (*Ineinandergreifen*): "Hence the ever-contending dialogue, hence the chorus as contrast to the former. Hence the all too chaste, all too mechanically and correctly closing concatenation among the different parts. . . . Everything is speech against speech, one canceling the other."[62] But what does this excessive modesty (*allzukeusche*) signify?

The figure of Oedipus, lifting himself too briskly toward the celestial, keeps close to the *deinon* of becoming one with the god. In the too much of a logic of excess, he can only make an about-face (*kehret*), scorning Tiresias, the living figure of god. The potentialization of forces is preserved only at the price of the "categorical turning away," a double turning away in which the god and man "express themselves in the all-forgetting form of infidelity," the oxymoron of a "pious traitor" (*SA* 5: 202; *Essays*, 108).[63] But the god turns, for "at the utmost borders of the suffering, there exists nothing but the conditions of time and space" (*SA* 5: 202; *Essays*, 108) and the wholly mathematical figure of hyperbole. We know what we must understand here by 'mathematical". What is essential lies in the anti-symmetry or "impurity"—the fact that beginning and end no longer "fit" and that neither the god nor man can "equal the initial situation." In the *kategorischer Umkehr* (categorical turning), a mimetic nostalgia fascinated by identity gives place to what "follows" (*Folgerungsweise*, we have seen for the poetical) in a reciprocal relation in which fidelity is "too much" fidelity, respect "too much" respect (heretical), such that the two dangers of the relation are avoided: the status quo of inferiority and of superiority in which the positions of the god and of man are fixed; the definitive turning away into forgetful and radical exteriority ("bad conscience," and "God is dead").

A real and united opposition, crossed movements that are reversed,[64] the displacement of the crossing point, whether it is a question of the relation of the Greeks and the Hesperians, of the natural aorgic or human organic, of man and the god, these elements, which as principles can be called categorical, are always found again. But the doubled duality or the double chiasmus is not yet sufficient to make sense of this "living, . . . thousand-times-divided, inward whole" of poetry.[65] In fact, at the caesura, the crossing points are also points of hesitation at which the road can turn. Hesitation having reached its height, the potentiality of indefinite turns, is the "desert."

Oedipus, the *nefas* become *antitheos*, is driven by the categorical turning away into an "atopia" in which nothing (honors, family, power, time) any longer "makes a mark." This desert is the place of turns, the field of effectivity itself become sensible. Oedipus finds there, again, respect for the statutory god, and hence jumps to his loss, which assimilates him to being a god. Antigone follows the same trajectory, with but this exception, that it is her turning away from the merely statutory forms of divinity that sets loose a turning of the god doubling as a statutory and savage god. Antigone, from the desert of her tomb, jumps higher than the god of rites, toward the savageness of the dead. In both cases, at the point of encounter of the crossings between god and man, the hesitation or the latency of a walk in the desert, "under the unthinkable," is produced, and this latency is followed by an about-face in which man, instead of following a return to the homeland, sets about following an opposed path, toward something "higher than god." Every time Hölderlin knowingly modifies Sophocles' text, something capital is at stake, that "fundamental opening."[66] The semen that Danae hoards (ταμιεύσκε), in Sophocles,[67] becomes temporal becoming (*Vater der Zeit / die Stundenschläge*). Semen therefore becomes the universal spermatic of the conditions of space and time, in a time with neither protention nor memory. For the tragic to be born, it is not enough that man be torn from his sphere of interest. (This sphere is nowhere so appealing as in Haemon and Antigone's affection just before their wedding.) But a potentiality of man (his *deinon*) that is not unveiled must take over.[68] This nonmetaphysical epiphany also releases an epiphany of the god.[69] This is, then, to head toward a definition of the tragic as the "risk of a contour."

Lacking elasticity, fear becomes a mania for fixed forms and *monumenta*.[70] It forbids the "manifold development of strengths." On the contrary, tragedy and poetry, as risk, are the *trajectory that displace the suture point* of crossings. This is a leap over the abyss as the sons of the Alps alone know how to make it, and in that they know the proximity of the risk and of what saves (as in "Patmos"). But what does Hölderlin say of the risk in the work of art?

> The boldest moment in a day's course or in a work of art is when the spirit of time and nature, when the heavenly which takes hold of man and the object in which he is interested, oppose one another most ferociously, because the sensuous object extends only half of the way while the *spirit* awakens most powerfully where the *second half* sets in. At this moment, man has to sustain himself the most; hence, he also is most exposed fundamentally and takes on his proper outline or contour. (*SA* 5: 266; *Essays*, 110)[71]

The contour (*Fest* of "The Origin of the Work of Art" in Heidegger) is the miracle of a *dynamis* of form obtained through its passage into its absolute opposite, the "unformed" that "exposes him fundamentally." This is also what, in Sophocles already, outlines Haemon and Creon in playing upon the meaning of ἄρχη and σέβειν.[72] Ἄρχη means command and beginning; σέβειν means to respect and to be pious. Beyond the surface appearance of destiny that veins the intrigue, it is in the semantic interval that the decisive point is found, and yet an apparently minimal turn is enough to open the character. Thus, the *Mein* of *Mein Zeus* devours Antigone and the tragedy in "the spirit of the endlessly living, the unwritten savageness, the spirit of the world of the dead."[73] Hence, the scorn for the priest in the first version of *Death of Empodocles*, for he forbids that turn.

If the "Remarks on *Antigone*" is perhaps the most pregnant text in literature, it is because for the first time since the Greeks, the couple of man and god, the *spring of the tragic*, and that of the Greeks and the Hesperians, the *spring of the displacement of the tragic*, cross in a poetic form. "It proceeds from the Greek to the Hesperian" "in the middle" (ἐν μέσῳ) (*SA* 5: 267; *Essays*, 111). Tragedy forces us to think the "absence of rhyme," because something has changed since the principles that formed the foundation of the Greek spirit. Hence the paradoxes when the displacement of the suture point is the displacement of the barycenter, not only of a (Greek) thought, but also of two thoughts (Greek-Occidental). For example, Antigone repeats Empedocles' patience, Empedocles who in his joy could have rejoined Nature and the gods and who, not having done so in time, is condemned to a long suffering that is beneficial to man.[74] And there would be something to say about the graves (καταςκαφάς) that, in the waiting of a slow death of being buried alive, recall the wrapping up of our Hesperian death.[75] And in Sophocles, Antigone is also the firm determination of a Hesperian type of consciousness, in Hölderlin's eyes. However, it is indeed a Greek play in which the word directly kills bodies ("murderous mediation"), while in Oedipus or in the character of Creon, a more Hesperian word would speak, since both, having been saved, suffer the long wait of a word that is "immediately" (it affects the spirit without mediation) "wounding" (it does not destroy the body).

Therefore, we must come to a conception of a Greek-Hesperian whole functioning "impurely." The two opposed harmonics maintain one another through their lacunae and their reciprocal ties. The Greeks are less masters in sacred pantheism because it is native to them, and are excellent in moderate exhibition, which is what they have acquired. Each steals a part of himself from the other and fights the other on his own ground. With Sophocles' tragedies and their translation, with *Empedocles*, a central point is made precise for Hölderlin: the effectivity of the "infinite moment," which must have intoxicated him beyond all expression. But how could his time have understood Hölderlin's evolution (and have we understood it well?) from a nostalgia for Greece and for the One to that entirely machinelike and conceptual, antimetaphysical, conception in which what is essential is no longer a nontemporal aesthetic model, but "poetic dwelling," "philosophical enlightenment," the "contour" of a continuous crossing of the Greeks and the Moderns? And that Hölderlin's genius consisted in being placed at this point and in having begun to develop it poetically? "For it is always poetically that man inhabits this Earth."[76]

Whatever the meaning of *gegen*,[77] and whether or not Hölderlin orientalized in both cases or orientalized and simplified the Greek text in order to find a purity in it again, it remains the case that Hölderlin's step is not to separate the lot of the Greeks and of the Hesperians, preserving, in my opinion, a teaching of Greek tragedy itself, and not going to cut their unbreakable symbol. More profoundly, that means that Hölderlin knew that Sophocles' plays interest modernity by Oedipus's latency in which the knowledge of self that will be named "Antigone" is constituted. Let us not lose sight of the unwritten play about Oedipus's "walk in the desert," supported by his daughter. That is where Antigone, in silence, suffering from the horrible secret of her birth, impure as her father's daughter and sister, denied in her desire as a young woman, crushed by the task, forges for herself that consciousness that will transform time, the god, and the post-Schopenhauerian Occident into philosophy. This latency, at once Greek (the knowledge of self) and Hesperian (waiting), appears to us as the unsurpassable point of a final symbol, that of Sophocles and Hölderlin.

In the infinite space of the displacement of the crossing points that is named the future of the Occident, the stakes of modernity remain possible, in suspense, interrupted by a caesura.

Translated by Jan Plug

Monstrous History

HEIDEGGER READING HÖLDERLIN

Andrzej Warminski

Heidegger's lectures on Hölderlin's late hymns—his third and last lecture course on Hölderlin, given in the summer of 1942 and published in 1984 as volume 53 of the *Gesamtausgabe*—follow a path from and back to a commentary on Hölderlin's "Der Ister" (*SA* 2; 1: 190–92) by way of a long excursus on the Greek determination of man's essence in Sophocles' *Antigone*. This excursus to Greece—and hence Heidegger's entire interpretation of Hölderlin—turns, as always, on a translation from the Greek. Here it is the well-known second choral ode of *Antigone*, in particular one word in its opening, which Heidegger renders as follows:

> Vielfältig das Unheimliche, nichts doch
> über den Menschen hinaus Unheimlicheres ragend sich regt.

And which Ralph Manheim in turn translates as: "There is much that is strange, but nothing that surpasses man in strangeness."[1] This opening is something of a riddle—why is it, how is it, that man is stranger than strange, more uncanny than the uncanny?—and the lines that follow could hardly be taken as an answer: man goes out on the sea and on land, masters the Earth and the animals, teaches himself language and thought, cures illnesses, and yet comes to nothing, for he cannot escape death. Whatever the "answer" to the riddle of man, it has to do with what he can do and what he cannot do anything about, his living and his dying.

Heidegger's translation of the Greek words δεῖνόν and δεινότατον by *unheimlich* ("uncanny," say) is already an answer to the riddle: an account of man's living and dying, his always going out to that which is different and his always coming back to the same. But before going over to Heidegger's "answer"—to his determination of man's essence as the most uncanny of that which is uncanny—we should note that it is a little different from, not quite the same as, Hölderlin's own "answer." That is, Hölderlin translates the opening of the choral ode by rendering the Greek word not as *unheimlich* but as *ungeheuer*:

> Ungeheuer ist viel. Doch nichts
> Ungeheuerer als der Mensch.
> (*FA* 16: 299)

> Much is monstrous. But nothing
> more monstrous than man.

Heidegger is, of course, well aware of this difference and the apparent strangeness of interpreting Hölderlin's dialogue with Sophocles and yet not using Hölderlin's own translation of Sophocles: "Since Hölderlin himself translated the whole of Sophocles' *Antigone*, it would seem appropriate to listen to this choral ode in Hölderlin's own translation. Nevertheless, this translation (Übersetzung) is comprehensible only on the basis of the Hölderlinian translation (Übertragung) in its entirety and this in turn only in the immediate proximity of the original Greek word."[2] In other words, Heidegger does not quote Hölderlin's own translation because he does not want to quote it out of context. But the "context" of Hölderlin's translation in its entirety is not to be understood in any ordinary sense. As many scandalized philologists have pointed out, Heidegger has no trouble whatsoever quoting Hölderlin out of context—indeed, some would say that his whole project of interpreting Hölderlin rests on arbitrarily ripping lines out of context and making them mean something other than what they mean "in context." No, Heidegger means more than that by "only on the basis of Hölderlinian translation [Übertragung] in its entirety"—as the switch from *Übersetzung* to *Übertragung* suggests. He means that Hölderlin's translation of the choral ode would be understandable only on the basis of our already having understood Hölderlin's interpretation of the Greeks' historical specificity and our ("Hesperian") historical specificity in their sameness and their difference. In other words, the real reason for Heidegger's not quoting Hölderlin's translation here is the nature, the essence, of translation itself.

Translation, according to Heidegger's "Note on Translation" which he interposes directly after his quotation of the choral ode, is not the substitution of a word in one language by a word in another language as though one could coincide with the other. All translation has to be interpretation (*Auslegen*), not the preparatory step to interpretation but always the result of interpretation. "But the reverse also holds," continues Heidegger, "every interpretation, and everything that serves it, is a translation. For translation moves not only between two different languages, but rather there is a translation within the same language. The interpretation of Hölderlin's hymns is a translation internal to our German language."[3] Heidegger summarizes:

> All translation is interpretation. And all interpretation is translation. Insofar as it is necessary for us to interpret works of thought and of poetry of our own language, this indicates that every historical language in itself and for itself is in need of translation and not only in relation to a foreign language. This in turn indicates that a historical people is at home in its own language not of itself, that is, not without its contribution [its act in addition, *Zutun*]. Hence it can happen that we indeed speak "German" and yet talk in nothing but "American."[4]

If this is the case, if translation is not confined to what takes place *between* different languages but rather is what (always already) takes place *within* "one" and the same language, then it is no wonder that Heidegger, in his attempt to interpret the Greeks and Hölderlin, to interpret the dialogue of Hölderlin and the Greeks, does not (indeed cannot) use Hölderlin's "own" translation but rather has to retranslate both Sophocles' Greek and Hölderlin's German. For Hölderlin's German is not his own, is not truly German, *is* not in an authentically historical sense, except in dialogue with Greek—just as "a historical people *is* only on the basis of the dialogue of its language with foreign languages."[5] Hence when Heidegger retranslates the Greek word, the *x* of the Greeks' determination of man's essence, with his *unheimlich* ("uncanny") rather than Hölderlin's *ungeheuer*, it is no arbitrary substitution but an interpretation that would say the same thing as Hölderlin. For to say the same (*das Selbe*) is not to say the merely identical (*das Gleiche*): "The same is truly the same only in the differentiated," says Heidegger.[6]

In order to *think* the same of what Hölderlin says, it is necessary to say what he leaves unsaid, in other words, to say it differently, to say it otherwise. This is why Heidegger says *unheimlich* and not *ungeheuer* (i.e., *unheimlich* says the same thing as *ungeheuer* but [precisely *because*] it says it in its difference). As is clear, all the weight of Heidegger's thinking—of the same and the identical, of *Dichten* and *Denken*, of dialogue as *Auseinandersetzung*, and so on—and what Philippe Lacoue-Labarthe calls its "hyperbologic"[7]—in shorthand, the more it differs, the more it is the same—could be brought to bear in order to justify this retranslation. Nevertheless, questions remain. For one, it remains to be asked whether Heidegger's *unheimlich* preserves the internal difference proper to *ungeheuer*, whether his retranslation translates the word's self-translation. What happens when the *x* of the Greek, the *x* of the Greeks—their determination of the essential nature of man—is translated as *unheimlich* and not as *ungeheuer*? And since the question of the Greek *x* is precisely the question of translation—that is, going out to the different and returning to the same—this question also asks: what happens when the *x* of translation—the translation of translation, as it were—is translated as *unheimlich* and not as *ungeheuer*?

Heidegger would, of course, have no trouble solving the riddle: what happens is that we begin finally to understand Hölderlin. In other words, Heidegger's retranslation of the choral ode is thinking more Hölderlinian than Hölderlin—just as in order to understand the Greeks we have "to think more Greek than the Greeks themselves"[8]—and many commentators have pointed out that in his retranslation of the Greek Heidegger only follows a path already marked out for him by Hölderlin's own translation: that is, in the direction of a "pessimistic" reading of the choral ode—not as a hymn to the glory of man but as a putting into question of man as monstrous in his excesses as wielder of *Technik* when thought against the background of his essential finitude. How is *unheimlich* instead of *ungeheuer* more Hölderlinian than Hölderlin? Heidegger thinks man's essential "uncanniness" (*Unheimlichkeit*) on the basis of what he calls the law of history, or, better, "the law of historicality" (*das Gesetz der Geschichtlichkeit*). This law is to be thought as "the altercation of that which is foreign and that which is one's own" which is "the grounding truth of history."[9] If man is *unheimlich*—the most *unheimlich* of all that is *unheimlich*—it is because his essence consists in "coming to be at home" (*Heimischwerden*); and if his essence is "coming to be at home" (*Heimischwerden*), then this means that it is at the same time "not *being* at home" (*Unheimischsein*). "Coming to be at home" (*Heimischwerden*) and "not *being* at home" (*Un-

heimischsein) are mutually implicated, mutually determine the essence of man: if man has to *come to be* at home, then he *is* not at home; if man *is* not at home, then he has to *come to be* at home.

Coming to be at home in that which is one's own [Das Heimischwerden im Eigenen] is the only concern [or "care," *Sorge*] of the poetry of Hölderlin that has entered the form of the "hymn," whereby of course "hymn" means no fixed literary and poetic schema, but rather determines its essence on the basis of the saying of coming into that which is one's own. That which is one's own [das Eigene] is the native [das Vaterländische] of the German. That which is native itself is at home with the mother earth. This *coming to be* at home [Heimischwerden] in that which is one's own includes in itself that man first of all and for a long time and sometimes for always is not at home. And this in turn includes that man mistakes and denies and flees the at-home [das Heimische], perhaps even has to deny it. Thus coming to be at home [das Heimischwerden] is a passing through that which is foreign.[10]

This "law of historicality"—the mutual implication and intrication of not being at home (*Unheimischsein*) and coming to be at home (*Heimischwerden*)—is given an ontological interpretation by Heidegger—that is, in the terms of the fundamental ontological project of *Sein und Zeit* and the existential analytic of *Dasein*. That is, man's uncanniness (*Unheimlichkeit*) is grounded in his homelessness (*Unheimischkeit*) and this homelessness in turn has its hidden ground in man's relation (*Bezug*) to Being. In short: "The uncanniness of man has its essence in homelessness, but this homelessness is what it is only through this, that man is at home at all in Being."[11]

It is this reading of *Unheimlichkeit* on the basis of *Unheimischkeit* and in turn on the basis of man's relation to Being that allows Heidegger to interpret the choral ode in terms of the ontological difference between Being and beings, *Sein* and *Seiendes*.[12] That is, man goes out and seeks his home in beings (*Seiendes*), masters it through his technology, but is always called back to the same nothing by his essential finitude, his death, and this nothing is the nothing *of* Being—the only place he can *be* at home. In short, man's *Unheimlichkeit* is grounded in his essential *Unheimischsein* (not *being* at home) and *Heimischwerden* (*coming to be* at home), and Heidegger has little trouble interpreting Hölderlin's *ungeheuer* as meaning the same as *unheimisch*: "The monstrous is at the same time and properly speaking the unfamiliar. The familiar is the intimate, the at home. The monstrous is the not at home" ("Das Ungeheure ist zugleich und eigentlich das Nicht-Geheure. Das Geheure ist das Vertraute, Heimische. Das Ungeheure ist das Un-heimische").[13] If *unheimlich* means *unheimisch* and *ungeheuer* means *unheimisch*, then *ungeheuer* means *unheimlich*. Hence Heidegger can justify his retranslation: "In that we translate *deinon* with *unheimlich*, we think in the direction of the not familiar" ("Indem wir das *deinon* mit 'unheimlich' übersetzen, denken wir in die Richtung des Nicht-geheuren").[14] In short, Heidegger translates Hölderlin's going over and coming back on the basis of the essence of going over and coming back and this means by himself going over and coming back: from Hölderlin to the Greek(s) and back. Hence it is no wonder that Heidegger also has no trouble in interpreting the enigmatic lines of "Der Ister":

> Der scheinet aber fast
> Rükwärts zu gehen und
> Ich mein, er müsse kommen
> Von Osten.
> Vieles wäre
> Zu sagen davon.
> (*SA* 2; 1: 191)

> But it seems almost
> To go backwards and
> I mean, it must come
> From the East.
> Much could
> Be said about this.

If the Ister—the Greek name for the Danube (*Istros*)—seems to hesitate at its source and origin in Germany before resuming its west to east itinerary and thus seems almost to go backwards (i.e., east to west, as though it had come from the East), it is not just on account of the difficulty (for a demigod) of forgetting the source. Rather it is also because the river Ister fulfills the law of historicality—in short, coming to be at home by going out to and back from that which is foreign, the alteration of *Unheimischsein* and *Heimischwerden*—*has* always already fulfilled the law of history at its source, at its origin, by having invited the Greek Hercules as a guest (*den Herkules zu Gaste geladen*), as the second strophe puts it, when he came from the hot Isthmus (*vom heißen Isthmos*) to the shady source of the Ister looking for the shady olive tree to plant it in the shadeless festival arena of the Olympic games (*SA* 2; 1: 191). "The Ister," writes Heidegger, "is the river, for which that which is foreign is a guest and present already at the source, the river in whose flowing the dialogue of that which is one's own and that which is foreign always speaks."[15] If the Ister at its source seems almost to go backward, as though it had come from the East, it is because its foreignness is *at* the source, it has always already at its source gone out to the foreign and come back to the fatherland. The Ister does not want to go east because it has always already at its source gone to the East and come back to the West—thus fulfilling its authentically historical destiny, its essence as *Halbgott*, and hence as essentially poetic (*dichterisch*), and so on.

This brief and insufficient sketch of Heidegger's interpretation *of* going over, and coming back *by* going over and coming back should nevertheless indicate the "law" governing his interpretation of Hölderlin. That "law" is the law of history itself—"according to which that which is one's own is the most distant and the path to that which is most one's own is the longest and most difficult"[16]—and it is grounded in an ontological interpretation of historical man's essence. This is why it is futile to object to Heidegger's interpretation on the basis of any prematurely "philological" grounds: if Heidegger interprets Hölderlin "out of context," he nevertheless thinks that poetry within a more rigorous "context"—that of fundamental ontology and its account of the relation between apparent and nonapparent meaning, between said and unsaid, thought and unthought—than any mere philology can come up with (because *as* mere philology it cannot think its own essence and therefore always has to interpret on the basis of an un-thought [i.e., metaphysical] interpretation of Being). Nevertheless, a question can still be asked. An oblique way to formulate it would be to ask whether Heidegger's interpretation of Hölderlin's *ungeheuer* as *nicht geheuer* (and this in turn as *unheimisch*) leaves anything over, a self-translating difference from itself that is not quite gotten across by changing *un-* to *nicht*. On a purely "verbal" level—when we translate *ungeheuer* into English or French and not into German—it is clear what does not make it across: the monstrous and the monster, the monstrous of man and the monster that is man. (That Hölderlin does, to some extent, mean monster is corroborated by his famous statement of "the tragic" in the "Notes to *Oedipus*": "The presentation of the tragic rests principally on the notion that the monstrous [das Ungeheure], the way in which god and man mate [or "couple," *sich paart*], in which the natural force and the most inner part of man become one, boundless, in rage, is understood through the purification, by a limitless scission, of the boundless act of becoming-one.") [17] A less oblique way to ask the question of this excess monstrosity (of translation) [18] would be in terms of Heidegger's "law of history": that is, if in thinking the same as Hölderlin, Heidegger remembers to say it differently, to say it in its difference from itself—in short, remembers that Hölderlin has "his own" language only in dialogue with the foreign language of the Greeks—does he also remember to think the same of the Greeks and of the Greek in its internal difference from itself? In short, does he remember that the Greeks also had "their own" language only in dialogue with a foreign language? In other words, we are asking about Hölderlin's famous first letter to Böhlendorff, dated December 4, 1801, and its interpretation of the conditions of poetry writing for the Greeks and for us "Hesperians" in terms of differing relations between that which is one's own and that which is foreign, *das Eigene* and *das Fremde*, for the Greeks and for us (*SA* 6; 1: 426).

Heidegger not only refers to the letter and its scheme throughout the two hundred pages of his lectures—for instance in his abbreviated version of the "law of history": "that which is one's own the most distant—the path to that which is most one's own the most difficult" ("das Eigene das Fernste—der Weg zum Eigensten der schwerste") [19]—but also gives it a more elaborate explicit interpretation than anywhere else in his work. In brief, the scheme runs as follows: that which is national, natural, their own (*das Eigene*) for the Greeks, what they are born to, is what Hölderlin characterizes as the "fire from heaven" and "holy pathos"; whereas that which is foreign and needs to be appropriated for the Greeks is what he characterizes as "the clarity of representation" and "Junonian sobriety" (*SA* 6; 1: 426). For us Hesperians, it is the reverse: our nature, what we are born with, that which is our own (*das Eigene*), is precisely "the clarity of representation" and "Junonian sobriety"; whereas that which is foreign and in need of appropriation for us is the "fire from heaven" and "holy pathos." What is natural and *das Eigene* for the Greeks is foreign (*das Fremde*) for us; and what is foreign (*das Fremde*) for the Greeks is natural and *das Eigene* for us. There is a chiasmatical reversal in the relations of *das Eigene* and *das Fremde* for the Greeks and for us, and the reason is easy to see: what we are born to, what is natural, national, and proper for us is precisely that which the Greeks—whose nature was different from ours—appropriated: that is, their culture. Our nature is Greek culture. The ramifications of this reversal are far-reaching: for one thing, it means that we Hesperian artists cannot simply imitate Greek art—treat it as though it were nature—because that art is the response to a different nature from ours. In short—as Peter Szondi puts it—it means a wholesale rejection of Winckelmannian classicism.[20] On the other hand, it does not mean a wholesale turning away from the Greeks, for that which is one's own, says Hölderlin, has to be learned just as much as that which is foreign; and because it *is* that which is one's own—natural, national, that which we are born with—it is the most difficult to use freely and this is why we will never surpass the Greeks in the clarity of representation and Junonian sobriety that we are born with and that for them was foreign (and therefore easier to use freely). In any case, much could be said about this.[21] Here we are more interested in what Heidegger does—and what is done to Heidegger—when he comes to interpret this letter and its historical scheme. Although he follows the letter faithfully in his interpretation of it, Heidegger nevertheless makes a slight shift when he applies the letter's scheme of *das Eigene* and *das Fremde* ("that which is one's own" and "that which is foreign") to interpret the "Ister" and the Greek determination of man's essence.

In short, whereas Hölderlin's scheme maintains an internal doubleness of that which is one's own and that which is foreign, *das Eigene* and *das Fremde*, both for the Greeks and for us, Heidegger identifies Hesperia (in these lectures, simply "Germany" or the "Germans") with that which is one's own, *das Eigene*, and Greece with that which is foreign, *das Fremde*. In doing so, he not only changes Hölderlin's bipolar scheme of a relation (by inversion) of bipolar terms into a simply bipolar scheme of us (Hesperians, Germans) and the Greeks, that which is one's own and that which is foreign, *das Eigene* and *das Fremde*, that is to say, of simple terms—and thereby reduces Hölderlin's version of the relation between the Greeks and us to one of the terms of that relation. No, more important than what this shift does to Hölderlin's scheme is what it does to Hölderlin's *Greeks*. That is, whereas Heidegger is able to preserve a certain doubleness of, a certain internal difference to, us Hesperians, Germans—insofar as we are ourselves, *das Eigene*, only in dialogue with the Greeks, *das Fremde*, and so on—in identifying the Greeks with the simply foreign, *das Fremde*, he nevertheless is not able to preserve the Greeks' difference from themselves, their own "dialogue" between that which was their own, *das Eigene*, and that which was foreign for them, *das Fremde*. In other words, by calling the Greeks simply *fremd*, "foreign," Heidegger collapses that which is their own and that which is foreign, *das Eigene* and *das Fremde*, for them. This is evident throughout his interpretation in his constant identification of the foreign, *das Fremde*, with the Greeks and the Greeks with the East and the South, the fire from heaven. In short, Heidegger—in calling the Greeks foreign, *das Fremde*, for us—quite simply reverses Hölderlin's terms and calls that which for Hölderlin is our own, *das Eigene*—the clarity of representation, Junonian sobriety, that is, Greek culture, say—foreign, *das Fremde*. And in doing so, he also renders that which is radically foreign for us—that is, the Greeks' nature, that which is natural and their own, *das Eigene*, for the Greeks: the fire from heaven, holy pathos—our own, that is to say, *Greek*. That this shift effaces the Greeks' internal difference to themselves becomes clearer if we identify the Greeks' *nature*, that which is their own, *das Eigene*—that is, the fire from heaven, holy pathos: namely, as the East, the Orient, or, in some of Hölderlin's texts (for instance, *Hyperion* or the third version of *Empedocles*), Egypt and the Egyptians. That is to say, the Greeks' nature, that which is their own, *das Eigene*, is somebody else's culture: the Oriental fire from heaven and holy pathos (just as our Hesperian nature, that which is our own, *das Eigene*, for us is also somebody else's culture: the Greek clarity of representation, Junonian sobriety). If it is legitimate to read Hölderlin's scheme in this way, then what happens in Heidegger's identification of the Greeks as foreign, *das Fremde*, for us—as the fire from heaven and holy pathos, as the South and the East—is his turning of what is a threefold historical scheme of the Orient, Greece, and Hesperia into a twofold scheme of Greece and Hesperia: in other words, Heidegger turns a scheme of us (Hesperians) and them (Greeks) and *their* them (the Orient) into a scheme of us and them, Hesperia (or "Germany") and Greece.

To turn the Greeks into that which is simply foreign for us (when for Hölderlin it is that which is our own), then, means to collapse the Greeks' nature and the Greeks' culture, *das Eigene* and *das Fremde*, and thereby, quite simply, to suppress the radical difference of the nature of the Greeks: that is, to suppress the Orient, the East, Egypt, and so on. This suppression of the Orient is legible throughout Heidegger's interpretation of Hölderlin and the Greeks in his consistent reduction to a twofold scheme—Greece and Hesperia—of what is well marked everywhere in Hölderlin as a threefold scheme: Orient, Greece, Hesperia.[22] In the "Ister," for instance, the threefold is marked by the names of three rivers: one Oriental (the Indus), one Greek (the Alpheus), and one Hesperian (the Ister):

Wir singen aber vom Indus her
Fernangekommen und
Vom Alpheus, lange haben
Das Schikliche wir gesucht,
Nicht ohne Schwingen mag
Zum Nächsten einer greifen
Geradezu
Und kommen auf die andere Seite.
Hier aber wollen wir bauen.
(*SA* 2; 1: 190)

But we sing from the Indus,
Having come from afar, and
From the Alpheus, long have
We sought that which is fitting,
Not without wings may one
Seize what is nearest
Straightway
And reach the other side.
But here we want to build.

If we come *here*, to the West, to Hesperia, to the Ister—"Man nennet aber diesen den Ister. / Schön wohnt er" ("But one calls this one the Ister. / Beautifully he dwells")—from the East and the Orient *by way of Greece*, then what this means, at the very least, is that our origin and source cannot simply be a matter of us and the Greeks—of the Ister and his Greek guest (Hercules)—cannot be a simply Greco-German origin. Rather the origin is, as it were, a Greco-Oriental (or Greco-Egyptian) origin, and this would offer a different way to read the lines about the Ister's seeming to go backwards:

Der scheinet aber fast
Rükwärts zu gehen und
Ich mein, er müsse kommen
Von Osten. Vieles wäre
Zu sagen davon.

But it seems almost
To go backwards and
I mean, it must come
From the East. Much could
Be said about this.

That is, the Ister seems almost to go backwards not in its hesitating at the source and seeming almost to go east to west but rather in its natural itinerary, in its going from west to east—as though its historical origin were in Germany and the West, and not in the East—whence (from the Indus through the Alpheus) we have historically come—and whence, historically, *it* must come: "*und / Ich mein, er müsse kommen / Von Osten*" ("and, / I mean, it *must* come / from the East"). In short, rather than *going* backwards—west to east—the Ister, as truly historical, must *come* from the East (and the antithesis between "going" [*gehen*] and "coming" [*kommen*] would support such a reading).

This reading of Hölderlin's historical scheme cannot help but have implications for Heidegger's "law of history"—in shorthand, again, "that which is one's own the most distant"—and the "hyperbologic" of not *being* at home and *coming to be* at home, *Unheimischsein* and *Heimischwerden.* In other words, that which is the furthest, the most distant, the most foreign, for us is not Greece and the Greeks—strictly speaking, that is, Hölderlin speaking, this is our own, *das Eigene*—but rather the East and the Orient. And it is most foreign for us because we are separated from it by our *Greek* nature—our Greek metaphysics, ontology, epistemology, aesthetics, and so on. If we are not at home, exiled, we are *not* at home in relation to, or exiled from, Greece and the Greeks, but rather "in relation to," and exiled from, the East, the Orient. And because the relation of nature and culture, that which is one's own and that which is foreign, *das Eigene* and *das Fremde*—that is, the clarity of representation and Junonian sobriety on the one hand and the fire from heaven and holy pathos on the other—for the Orient was exactly the same as it is for us (how else could the Oriental fire from heaven have been natural, their own, *das Eigene,* for the Greeks?),[23] to be not at home in relation to, or exiled from, the Orient means to be not at home in relation to, or exiled from, ourselves: that is, not the Greco-Hesperians or Greco-Germans but the Hespero-Orientals, Germano-Orientals. In short, we are not at home not because we are exiled *from* Greece but, rather, because we are exiled *by Greece* from ourselves: the Orient, the East, Egypt, and so on. Again: it is not just that we are not at home, but rather that we are not at home in relation to not being at home; or, better, we are not just exiled (from Greece, say) but rather exiled *from* exile (exiled from the Greeks' exile, say), as Blanchot puts it in writing on Kafka.[24] From such an exile (from exile), there is no return—not even a return to statements (or "laws") of history like "Hesperia is the Orient" or "We are the Orientals (or the Egyptians)." For such statements would identify that which cannot be identified—that which makes all (self-)identification impossible—in the terms of any ontology, no matter how fundamental. In other words, there is a radical disjunction between us Hesperians and the Orient, and Greece *is* it. To "identify" ourselves as Orientals would amount once again to reducing that which is radically foreign for us to that which is our own, the Oriental to the Greek. It would amount to the same thing as saying "We are dead" and meaning the (monumentalized) death of the Greeks—that which is natural and our own, a death in which we can (elegiacally) recognize and identify ourselves. To say "Hesperia is the Orient" or "We are the Orientals" is to say "We are dead" and to mean the death of the Orient—that which is radically foreign for us, a death in which we cannot identify ourselves because it is not *our own* death (i.e., the death of the Greeks), not being dead for ourselves, but an other death: being dead for somebody else, a death without death. This would be one way to read Hölderlin's determination of man's historical essence as monstrous: the monstrosity of history, history as the monster.[25]

A perhaps more straightforward way to put this would be by way of two quotations from Paul de Man's "The Riddle of Hölderlin"—a review essay in the *New York Review of Books* that dates from 1970 and that remains one of the best essays on Hölderlin. The first is an explanation of the attraction of Hölderlin's scheme for critics with an utopian or apocalyptic bent:

> Thus the transposition of Hölderlin's philo-Hellenism into a literal historical scheme yields an interpretation of the present that is, to some critics, reassuring; during a period of history that is part of our civilization, men could think of the gods as actual presences from which they were not separated by transcendental distances. If this was possible for a consciousness not essentially different from our own, it follows that the absence of gods, painfully experienced as everyday reality, may be only a passing dark phase between two stages of unity, one past but another still to come.[26]

Although this statement does not apply to Heidegger's interpretation immediately—not without some translation to make it fit (for example, of the terms "consciousness" and of "essentially different")—it is close enough to his law of history, being not at home and coming to be at home, and so on. De Man's—that is, Hölderlin's—answer is uncompromising:

True wisdom begins in the knowledge of its own historical ineffectiveness. When Hölderlin evokes the possibility of future moments of historical splendor, comparable to what Greece used to be in the past, such evocations are accompanied by the foreknowledge that people will be conscious of the achievement of these periods when they have ceased to be and have become in turn parts of the past. Nothing could be more remote from schemes that conceive of history as either apocalyptic failure or salvation.[27]

To transpose this into the terms of our reading of Hölderlin's scheme: if we as Hesperians can be the nature for an other Greece—other Greeks and other Greek gods to come after us—it is only as *dead*, as the Orient or the Egyptians: that is, as dead not for ourselves, but as dead for somebody else, an other. The upshot would be that "Greece" and "the Greeks" is something we invent in order not to face this other death: an other death that reminds us that there never were any Greeks in the first place—and therefore no Orient or Hesperia (or Germany) in the first (or last) place either—but only a wearing away and wearing down, the ceaseless erosion of monstrous material history.[28]

Disowning Contingencies in Hölderlin's 'Empedocles'

Stanley Corngold

> For everything outside the phenomenal world, language can be used only allusively [andeutungsweise], but never even approximately in a comparative or analogical way [vergleichsweise], since, corresponding as it does to the phenomenal world, it is concerned only with property and its relations.[1]

Hölderlin wanted to complete and stage his tragedy *Empedocles*: it mattered enormously to him to do so, but he never finished writing it. One sort of difficulty arose from his loyalty to "the tragic form"—to its "proud refusal" to portray contingencies however "brilliantly or finely strung."[2] In practice, this ascesis leads to doubt about the communicative possibilities of the *Bild* ("image," "metaphor"), an aporia that frustrates dramatic representation.

In examining this obstacle, one wants to do more than add to the literature on the death of *Empedocles*; one wants to say "something new" about it. But the intractability of this premise will become clear after a moment's thought. To say something new still means only *adding* something onto something—in the best case, onto a "tradition of the new." In the recent history of writing about *Empedocles*, two events stand out as signposting roads of inquiry that have not yet been much taken: the first is D. E. Sattler's reconstruction of Hölderlin's manuscripts (his volumes on *Empedocles* appeared in 1985 as numbers 12 and 13 of the *Frankfurter Aus-*

gabe); the second is Andrzej Warminski's deconstructive reading in 1987 of portions of the third version of the drama—a reading that does not, however, acknowledge Sattler's changes.[3] In fact, the two lines of this tradition cross with interesting results.

While working on *Empedocles* in the years 1798–1800, Hölderlin composed a variety of plans and theoretical essays to articulate the aims of his project. They include the Frankfurt plan for *Empedokles: Ein Trauerspiel in fünf Acten*, which was never written; the first version, untitled; the second version, titled *Der Tod des Empedokles: Ein Trauerspiel in fünf Acten*, never completed; The tragic ode . . . ; *Allgemeiner Grund* ("General Basis"); *Grund zum Empedokles*; plan of the third version; *Szenisches Konzept*; the third version, never finished; and then, finally, the essay "Das Werden im Vergehen" ("Becoming in Disintegration")—constituting what Jürgen Söring calls the *corpus empedocleum*.[4] Immersing oneself in this body of work is to come away marveling at the rigor of Hölderlin's will to precise expression (it can be said about him what Doctor Faustus said about himself: "I sought the hard and labored might and main");[5] marveling, too, at the boldness of Hölderlin's philosophical imagination. The subject of *Empedocles* is none other than the possibility that one perfectly lived life might have a salutary, an enlivening effect on an entire civilization—might engender, upon the hero's death, a new historical moment.

This ramified body of work is partly the result of a wound, an injury at the core. Determined to absorb even "destructive things" so as to enable the "fullest representation of [his] inmost [life]" (*FA* 12: 14), Hölderlin was unable to produce the tone that struck him as most needed. In the draft of a letter to Schiller, he wrote: "I am trying as seriously as possible to develop the tone closest to my natural, undistorted sensibility. . . . I believed I would be able to achieve that tone most completely and naturally in the tragic form and have ventured on a tragedy, *The Death of Empedocles*" (*FA* 12: 19). This enterprise led Hölderlin continually to revise *Empedocles* in light of the exigency of tragedy that he calls its "stolze Verläugnung alles Accidentellen," "proud denial of all accidental features" (*FA* 12: 18).

The word *verläugnen*, to "deny" or "disown," has opposite connotations in Hölderlin's German—of denial and negation and also of only seeming to negate and deny, of negating and denying despite the covert will to affirm the very thing at stake.[6] Indeed for Hölderlin, *Verläugnung* is more than a concept: it is a field of relations situating the interinvolvements of his empirical identity, the "alien analogical material" of his poetic topoi, his states of so-called divine possession, and the fate of this "excess of inner intensity" in the tragic work. All these terms are caught up in a turmoil of doublings and denials—vivid, for example, in the "General Basis," whose theme is the *Verläugnung* of the poet's personal intensity:

Thus, in the tragic-dramatic poem, too, the sacred [das Göttliche], which the poet feels and experiences in his world, also expresses itself; . . . however, as this image or metaphor [Bild] of inwardness disowns and has to disown everywhere its ultimate basis precisely to the extent that . . . the inwardness is more infinite, more ineffable, closer to the *nefas* . . . , the more severely and coldly the image has to distinguish man from his feeling world [sein empfundenes Element], so as to hold the feeling firmly in its bounds; all the less can the image express the feeling immediately, it has to disown it as much in respect of its form as its matter; the matter must be a bolder more alien figure [Gleichnis] and example of it. . . . The alien forms must be all the livelier, the more alien they are; and the less the visible matter of the poem resembles the matter that is at the basis, the sensibility [Gemüth] and the world of the poet, all the less may the spirit, the sacred, as the poet experienced it in his world, be disowned in the artificial alien matter. . . .

For the most inwardly intense feeling becomes vulnerable to passing away [Vergänglichkeit] in precisely that degree to which it fails to disown the true temporal and sensory connections. . . . Precisely for this reason the tragic poet, because he expresses the deepest inward intensity, wholly disowns his person, his subjectivity, as well as the object present to him; he carries it over, metaphorizes it into alien, foreign personality, into alien objectivity. (*FA* 13: 869–70)

This litany of negations—especially of the immediate "physical and intellectual context" (*FA* 13: 870) of the empirical person—jeopardizes a certain assured empirical visibility, "the visible matter at the basis . . . , the world of the poet."[7] The tragic denial of empirical visibility aims principally at the poet's "person" and "the object present to him"—his "true temporal and sensory connections"; yet this *Verläugnung* takes place under a condition that in fact jeopardizes its work of negation, for "every poem, the tragic as well, must have come forth . . . from the poet's own world and soul, because otherwise the right truth is everywhere lacking, and nothing at all can be understood or given life, if we are not able to carry over our own sensibility [Gemüth] and our own experience into an alien, a foreign analogical material" (*FA* 13: 869). How could "our own experience" be preserved when our own "true temporal and sensory connections" are annihilated? Hölderlin constructs an aporia, whose thrust, on the one hand, is to keep intact the "characteristic inward intensity," "the intensely inward relationship of the figure [Gleichnis] with the matter" (*FA* 13: 869), and on the other to represent the distinctive experience of an empirical personality in the work of images so "cold and severe" that they must border on the unintelligibly (in)human and hence invisible.

As a result, the question arises for readers of Hölderlin in the wake of its arising again and again for him: What could the form of a tragedy be that rigorously disowned all contingencies? This sort of tragic poem becomes identical with a report of the reflexive consciousness of an absolute subject, and nothing would remain of "die großen / Gestalten dieser Welt" ("the great / shapes of this world") (*FA* 13: 943) as they appear. From the standpoint of the Idealist subject congruent with this aim, would not other persons in their empirical spontaneity—persons who share the city where the tragic hero dwells—also come under the head of accidentals?[8] The difficulty is compounded by the fable of Hölderlin's *Empedocles*, according to which Empedocles' world is torn by contingencies (in the form of the hyperpolitical agitation of the Agrigentians, for one).[9] How could such fateful, hence necessary disorder be conveyed without the theatrical representation of inessentials? Furthermore, there is the question of the historical epoch itself, the new age, which Empedocles' death is meant to empower and enliven. How could it have an essential identity all the time his own world is only inessentially constituted by persons and institutions in which being has been forgotten or instrumentalized? This binary model conveys a relation that is alien to the major strain of Hölderlin's thought, for it posits essence entirely at one place at one time; yet "only that is the truest truth, in which error too, because it places it in the whole of its system, in its time and place, becomes truth" (*FA* 15: 59). At this point, the sustaining myth of *Empedocles* will seem factitious—namely, that the individual hero in whom alone a general truth is concentrated must shatter so that this truth can be returned to a disintegrating whole.

Finally, when all accidentals have been denied, would not the empirical being of the hero, the finite, historical, bodily entity, also be instantly drawn up in this vortex of *Verläugnungen*? If the denial of the body—of the all too visible reminder of the sensory root of experience, of "the matter that is at the basis, the sensibility and the world of the poet"—is the result of a rigorous elimination of contingencies, then, of course, it is perfectly appropriate that the play *Empedocles* should end with a suicide. But by this logic it should begin with a suicide. Nonetheless, is such a play a tragedy, even where this suicide is, as Empedocles will claim, "nach göttlichem Geseze" ("divinely ordained") (*FA* 13: 945)?

Tragedy could still be defined, as Hölderlin himself defined it, *in theory*, as an event occurring between a human being—the tragic figure—and nature—or a god. Thus in a fragment on "the meaning of tragedies," "Die Bedeutung der Tragödien," written around 1801, Hölderlin poses the paradox:

In the tragic the sign [of the original principle] is in itself insignificant, without effect, but the original principle just comes forth. . . . Insofar as . . . the sign in itself is supposed as insignificant = 0, the original principle, the hidden ground of that nature, can also be represented. If in reality nature represents itself in its weakest potency, the sign when nature represents itself in its strongest potency = 0. (*FA* 15: 379)

The sign, "the visible matter," must be emptied out, it must be *verläugnet*. But given only an empty sign, how then could Nature "*represent* itself"? Again, in the "Anmerkungen zu *Ödipus*" of 1803, "The representation of the tragic is based above all on this: that the monstrous [Ungeheure], the way god and man are paired and the force of nature and the innermost of man in a fury become unlimitedly One, understands itself thus: the unlimited Becoming One is purged through unlimited separation" (*FA* 16: 257). We have seen this formula before in the "General Basis," where the agency of the "unlimited separation" of "man and his feeling world" was, precisely, "the image or the metaphor"—the *Bild* grown "bolder and more alien"—a *Bild*, however, which to be adequate to its task of separation, must not be accidental but essential; but which, as the materially visible, borders essentially on the contingent.

And here the reader may already think ahead to the many connections, both in the *Grund zum Empedokles* and at the close of the third version, of "Bild" *as such* with *Trugbild*—"phantasm" or "spook." To the astonishment of longtime readers of the *Grund* in the Beißner edition, the key moment—"the unifying moment" (*FA* 13: 871), that fuses, after a series of intricate chiastic crossings-over, Nature and Art, but then dissolves like a *Trugbild*—turns out, in Sattler's reading, to be only *ein Trug*, "a delusion," as if to write "Trug*bild*" were only to write *Trug* twice; hence, *Bild* need not be specified. All the same, Empedocles is called, by Manes, *Trugbild*, to intensify the sense of his passing away—this, shortly after Empedocles himself has rejected the festival of earthly life, of earthly imagery itself, as

> der Erde trunkenes Bild . . .
> . . .
> Der geschäfftge Reigentanz, womit den Geist
> Die Sterblichen . . . feiern
>
> the Earth's ecstatic image . . .
> . . .
> The busy round dance, with which mortals
> Celebrate the Spirit
> (*FA* 13: 936)[10]

The fact that Hölderlin cannot get around, however, is that tragedy needs to empower, at least provisionally, precisely this "busy round dance" of the mortals; it is the field—or *Trauer-Spielplatz*—of essential conflict, of the tragic agon.[11]

I shall mention in passing Pierre Bertaux's wonderful, if unprovable, assumption that underscores this point.[12] Hölderlin originally conceived of *Empedocles* as a state play, to be performed in honor of the longed for but never realized Swabian Republic. He did so with hopes born of the success of a comparable venture in Paris. Marie-Joseph Chénier, the brother of the better known André, became the official dramaturge of the new Republic. But the notorious decree of General Jourdan on March 16, 1799, made it clear that there would never be a Swabian Republic. And abandoning this hope of *Representanz*, of *Darstellbarkeit*, of aptness for representation in one strong sense[13]—which would, moreover, have led to his securing a bourgeois existence—Hölderlin would have abandoned his play, for the leading motive for its completion would have fallen away. This major thesis is entertaining but precarious; there is better evidence of Hölderlin's struggle with the factor of visibility in the tragic drama. This is found in the manuscript itself, in a note in the text of the first version that turns literally on the problem of representation.

Hölderlin writes in the manuscript for the beginning of the second act: "Here the sufferings and abuse he endured must be so represented that for him it becomes impossible ever to turn back, and his decision to go to the gods appears more nearly obtained by compulsion than voluntarily. That furthermore his reconciliation with the Agrigentians is represented as the highest generosity" (*FA* 13: 731). In this passage the term "represent" (*darstellen*) changes its value from the first sentence to the last.

The passage is so interesting because in it Hölderlin struggles to bring his drama into explicit conformity with a conception of tragedy as the visible agon of the heroic individual and the embodied will of his adversary—the mob, which has initially rejected him. At first, one might be inclined to read the passage as saying that the seemingly voluntary inner life of Empedocles is in actual fact coerced by his public fate. That this coercion might be an objective fact—objective through its "lively representation" (*FA* 13: 710)—is supported by the final sentence, which speaks of the grandeur of his reconciliation with the people. This largeness of public spirit is forecast by an earlier note: "Keinen Fluch!" ("No curse!"; i.e., Empedocles must not curse the priest and the archon who reject him). "He has to love, to an infinite extent. . . . He must consume, so to speak, the last of his powers of reconciliation, which without this might perhaps have helped him to return again to his former holy, cheerful life" (*FA* 13: 719). There is no basis, then, for denying the truth of his reconciliation shown by a scenic representation. But then what is the status of the events "represented" (*dargestellt*) at the beginning of the note—the pains and coercion he has suffered at the hands of the others? In fact, these alleged pains only contribute to an illusion, indeed, to a deception; and now what was represented only seems to be the case. For it is not true, it only seems true, that his decision to go to the gods is obtained from him by coercion. Assuming that the picture of Empedocles' character we have in the third version holds true from the first one on, his adversary Manes must not be proved right, and Empedocles must not be shown choosing to die because he has been "provoked" (*gereizt*) (*FA* 13: 943).

Furthermore, how are we to read the words *for him* in this passage? Someone believes it will not be possible for Empedocles to turn his feelings of outrage into feelings of charity; is this "someone" the audience? Is the audience supposed to know something about Empedocles' beliefs that Empedocles does not know himself? This is an unlikely disadvantage to assign to a demigod. Hence, this someone who alleges the impossibility of his freely reversing his feelings and of his regaining a sense of autonomy must be Empedocles himself. But since this assumption is wrong, *for him* makes Empedocles himself in this note the deluded center of consciousness.

This reading is fatal in its implication, because it means that Empedocles is an unreliable perceiver and enactor of his own motives. In thus becoming the mere actor, or *Darsteller*, of his inner life, he also becomes audience to his own play. (This is not the first or strongest hint of this idea to emerge from the drama.) Moreover, if his suffering and reconciliation take place as an only theatrical performance, then it follows that nothing he thereafter states and thinks has anything but a scenic truth: it has entirely the status, in words from *Hyperion*, of "bloße Phänomene des menschlichen Gemüts" ("mere phenomena of the human spirit") (*FA* 11: 589). Yet even without the surmise that Empedocles is the center of consciousness in this passage, it can seem that whatever (in this Novalisian moment) is "represented" (*dargestellt*) deserves at once to be "denied" (*verläugnet*).

Read in its entirety, this note confuses the distinction between Empedocles, as an independent agent, and the coercive public world, between what he is and what he seems, between enactments of his self and their merely theatrical representation. The effect of this erasure of difference is to isolate him, in Peter Szondi's words, to romanticize him as "the isolated self thrown back upon itself."[14] Such a self, by a dialectic made familiar by Adorno writing on Kierkegaard and Kafka, necessarily loses contour and distinction; its claim to authority verges on vacuousness.

It appears that, as Hölderlin proceeds from the first to the third version, a radical will to the truth of reflection makes itself felt, and the aptness of his drama to imaged representation becomes less and less a consideration. Equally, its departure from the order of the image can be thought of as thematizing the problematic loss of the image as a kind of death. Empedocles becomes the vehicle of an adventure of Hölderlin to realize the claim of a subject to its autonomy—an adventure of poetic thought that will collide with the tragedian's ambition. He will take the risk; he will radicalize Empedocles' errant way of self-deification without concern for its suitability as theater and at the same time register the resistance to this impulse. What is now at stake is not that Empedocles be visible but that Hölderlin, his poetic analogy—who increasingly becomes less his analogy, his *hypocrite lecteur*, than his double, indeed his self—that Hölderlin be poetically rigorous and yet not be required to follow his hero, but to *live*.

Hölderlin's doubt of the image is more nearly ethical than epistemological, and the term *autonomy* for him is not merely or chiefly a poetological category. The point is that the only justification for a literature of intensity is the representation of inwardness, and inwardness has or might have no correlatives free of the contamination of contingent experience.

The longing for *another* imagery of inwardness: this is so-called "poetic language."

One can observe in Sattler's text with fine precision the operations of Hölderlin's developing intent to detach Empedocles from accidentals. We see Hölderlin at work enlarging Empedocles' capacity for conscious experience; lines 70–76 from the second version cost him immense trouble. In this scene Hermocrates is speaking of Empedocles' crime: he has cast his good words before an ignorant and undeserving people. His reward for doing so has been mockery. Through version after version (as the manuscripts show), Hölderlin conceived the matrix of the scorned preached word as one of *correspondence* between an external entity, the mob, and a subject-agent, Empedocles—correspondence being the principle of representability as such. To put this more precisely: the matrix is one of systematically distorted correspondence. This relation is conveyed by a parody of the time-honored Classical-Romantic trope of the echo—indeed, by a doubling, an echoing of the figure of the echo in the figure of mockery (a distorted echo *of* an echo). This pattern grows more and more intricate in Hölderlin's variations; as we shall see, the factor of distorted correspondence between these echoes is displaced onto the subject alone, with the result that the logic of representation is skewed.[15]

The first draft reads:

> HERMOKRATES. Weist *du* es
> *Siehst denn nicht?* ihn *haben*
> *es*
> *Den hohen Geist die Geistesarmen*
> *Geirrt, die Blinden den Verführer.*
> Die Seele warf er vor das Volk,
> Verrieth der Götter Gunst
> Gutmüthig den Gemeinen,
> Und rächend Äfft der T
> todte Wiederhall
> Der Wiederhall den Seher.
> (*FA* 13: 764)

> HERMOCRATES. Do you know it
> Do you not see? Him have
> it
> The high Spirit those poor in spirit
> Misled, the blind the seducer.
> His soul he threw before the people,

Betrayed the grace of the gods
Equably to the commoners,
And taking revenge apes the d
dead echo
The echo the seer.

The action of the mob, in its deadly way of repeating back to Empedocles and to one another Empedocles' original word, mocks Empedocles' original preaching. The "dead echo" (*todte Wiederhall*) apes Empedocles in an objective sense—in a public scene that can be represented by means of a correspondence perceptible to both the speakers and Empedocles himself.

But Hölderlin must revise this construction. He continues to compose this passage, adding on some dozen lines and then stopping in order to return to line 73 and revise, writing, struggling:

Die Seele warf er vor das Volk,
verrieth
er *warf* D*ie Seele*
Gutmüthig den Gemeinen,
Verrieth der Götter Gunst[,]
Und rächend äfft
Drum e
'auch *aus leerer Brust*
äfft' auch rächend
Doch rächend äfft aus le
stu
lebensloser Brust
Der leere Wiederhall den warmen *Seher*[.]
(*FA* 13: 765)

His soul he threw before the people,
betrayed
he threw His soul
Equably to the commoners,
Betrayed the grace of the gods[,]
And taking revenge apes
Thereupon h
too from his empty breast
taking revenge too apes from his li
mu
lifeless breast
The empty echo the warm seer[.]

These lines point up a key transformation. The field of the mockery, the distorted echo (*of* the empty echo), is now no longer the public scene but the "empty" (*leere*) breast and then, more tellingly, the "lifeless" (*lebenslose*) breast of the subject-agent Empedocles. The once "dead" echo of his preaching becomes merely "empty"; now it is the breast of Empedocles himself that is dead. Death is no longer a public, representable feature of the mob's vile repetition of Empedocles' speech; death is not first introduced as the death of the good word on the lips of the rabble. Death is a dimension of the subject, who in a sense has already died for having uttered his speech. Even in the perspective of Hermocrates, Empedocles is shown suffering the burden of his guilt, a kind of death, for having spoken publicly.

But this transformation does not go far enough in the direction in which Hölderlin's conception of his work is taking him. He continues: "But taking revenge apes from the lifeless / dead breast" ("Doch rächend äfft aus lebensloser / todter Brust") until there is the flash of a breakthrough. The phrase bursts into the manuscript with an excitement that we can feel: "as empty echoing" (*leeren Wiederhalls*). The new phrase transforms the case of the agent that reciprocates; the nominative is replaced by the impersonal genitive, freeing the subject position for *another* concept. The "completed" text reads:

> Siehst du denn nicht? es haben
> Den hohen Geist die Geistesarmen
> Geirrt, die Blinden den Verführer.
> Die Seele warf er vor das Volk, verrieth
> Der Götter Gunst gutmüthig den Gemeinen,
> Doch rächend äffte leeren Wiederhalls
> Genug denn auch aus todter Brust den Thoren.
> (*FA* 13: 819)

> Don't you see? The weak in spirit
> Have misled the lofty spirit,
> The blind their seducer.
> His soul he threw before the people, betrayed
> The grace of the gods equably to the commoners,
> Yet taking revenge, of empty echoing
> [Having had] enough, then aped from the dead breast the
> foolish one.

I read the subject of the act of aping as the "he" of line four.[16] The agent producing the correspondence of mockery is not now the echo or any abundance of echoing but Empedocles himself. Empedocles is "aper" of himself, "the fool"—at once himself and his own degraded facsimile. The subject, both agent and impotent persona of itself, stages its own mocking echo. The new text displaces the locus of the echo from the field of self and other in the simplest mode of interpersonal exchange, from imitation, from the mirror stage, onto a position of symbolic interiority—the "breast"—which is also not a position, being "dead."

What is most telling is that the act of aping is now seen to avenge the sickening abundance of echoes—it is an acting out of an outraged sense of truth. One sort of echoing (self-mockery, self-condemnation) aims now to annihilate the consciousness of an echoing of a simpler kind, which proposed a correspondence between the subject and the public world. The once real, objective correspondence producing the echo vanishes into a remote obscurity, into a remote etiology: it is an event in the past he is now only sick of, and nothing "representable" remains.

This passage—the turn to interiority and to another halted temporality—is driven on by the relentless will to transform the world into an experience of it already suffered through. It is true that the experience remains suspended in a play of oppositions, but the impulse is plain: Empedocles himself becomes the author of his own fate but in the same instant *Darsteller*, "aper" of himself. This demonstration of the will to poetic autonomy forces into the open a doubt as to its representability at the same time that it splits the inner world into a stage and an audience, with negative implications. Throughout its various versions, *Empedocles* enacts experimentally the Idealist program, with a characteristic result: in disowning the contingent image, the work impedes representation (as visibility) at the same time that it transforms consciousness into a scene of representation, into a "stage" marked by the very mimetic falsity that the Idealist project seeks to surpass. Thus, the play can be thought of as ending in suicide and, hence, in the hopelessness of the Idealist program; or, the play being incomplete, and the suicide undepicted, it can be thought of as leaving the program undecided.

A look at a pair of famous cruxes in the third version will heighten our consciousness of this parlous denial of all accidentals—of the inward turn, incorporating contingencies; a theatrically impeded dialectic; aberrant temporalization—and of Empedocles' persistent, if questionable, will to autonomy. We are concerned, now, with the famous Manes scene (Manes, who is said to be dead, represents one sort of metaphysical and poetic solution to the requirement of a strange enough representability.), in which Empedocles tells the story of his life to Manes in a form that fits Manes' prescription for the "neue Retter" ("new savior"), the one who is qualified by his suicide to bring about a rejuvenation of the historical moment. The stories coincide; the result is that Empedocles, and Manes, too, conclude that Empedocles is indeed the new savior.

In the first of these passages, Empedocles focuses on a moment of the acquisition of self-consciousness in relation to the presences of nature.

> Da ward in mir Gesang und helle ward
> Mein dämmernd Herz im dichtenden Gebet;
> Wenn ich die Fremdlinge die gegenwärt'gen
> Die Götter der Natur mit Nahmen nannt' . . .
> (*FA* 13: 943)

> Then song was in me and bright
> Became my dawning heart in poemlike prayer;
> When I the Strangers the Present Ones
> The Gods of Nature named with names . . .

The passage is immediately remarkable for the plenitude from which it speaks: if the *poesis* is described as supplicative (it is a "prayer"), it is also productive (it is "poetizing"): it mingles lack and plenty at its source. More, in this matter of poverty *as* abundance: these lines prove the famous claim about the iconicity to which Romantic nature poetry, especially, is held to aspire—it proceeds simultaneously to perform the thing it also only describes as being done. The passage, in other words, proves Adorno's definition of "form" in Hölderlin as "sedimented content."[17] In the second half of this verse the familiarizing sequence—"the Strangers the Present Ones the Gods"—enacts literally the process by which, in song/poetry/prayer, the dim wonder becomes clear.

The passage continues with an account of what is at work when self-consciousness is achieved through poetic activity:

> Und mir der Geist im Wort, im Bilden sich
> Im seeligen, des Lebens Räthsel löste.
> (*FA* 13: 943)

> And for me the Spirit in the word, in forming
> In the blessed [act], resolved for itself the enigma of life.

These lines, which describe Empedocles' acquisition of poetic language, could seem orthodox Idealist; Söring terms them "the poetic reflection of the genesis of the 'Idealist' word."[18] They become a bone of contention, however, in the strong if amazing reading proposed by Andrzej Warminski. Though he published his book in 1987, two years after the first appearance of Sattler's *Empedocles* in the Roter Stern edition, Warminski uses Beißner's text without discussion:

> Und mir der Geist im Wort, im Bilde sich
> Im seeligen, des Lebens Räthsel löste.

> And for me the Spirit in the word, in the image [figure]
> In the blessed [act], resolved for itself the enigma of life.

Warminski appears, however, to misconstrue this sentence, making it say that "life's enigma . . . resolved itself."[19] In fact the passage says, importantly, that the so-called Spirit resolved life's enigma for itself. Why bother, then, with a reading based on a text and translation so plainly unreliable? Because Warminski's reading of *im Bilde* as "figuratively" will allow him to read another crux in Empedocles' speech interestingly—indeed, it will allow him, it seems, to solve an enigma. At the end of this speech, Empedocles, confident of his rectitude as the new savior, describes the bliss of imminent suicide:

> Nun nicht im Bilde mehr, und nicht, wie sonst,
> Bei Sterblichen, im kurzen Glük,
> Im Tode find ich den Lebendigen
> (*FA* 13: 944)

> Now no longer in the image, and not, as otherwise,
> Among mortals, in brief happiness,
> In death I find the one who lives

About "Now no longer in the image," Warminski writes: "Whereas in youth life's enigma (*des Lebens Rätsel*) resolved itself for him in word and figure (*im Wort, im Bilde*), now he will find that which is living in death and no longer in image or figure (*nicht im Bilde mehr*)"—no longer figuratively, no longer merely *im Wort und Bilde*. Here, for Warminski, Empedocles makes himself liable to the error of wanting life's truth "literally"; yet such truth, Warminski maintains, can never be experienced in any mode other than the figurative, other than fiction. Empedocles' patent error establishes Hölderlin's controlled superiority to his hero: Hölderlin would then be ahead of him, let alone the literal-minded reader, every step of the way.

Warminski's reading institutes a solid difference between the literal and figurative senses of *Bild*—between *Bild* as *Wort*, and hence as figurative, and *Bild* as the literal absence (however delusory) of the verbal factor. For Warminski, truth goes to *Bild* as *Wort*; *Bild* as immediate life is the literalist error. Yet, in reality, Hölderlin has instituted no such difference here.[20]

Given Sattler's emendation of *Bilde* to *Bilden*, no distinction can be made in the first crux between literal and figurative meanings in favor of the figurative (*Wort* as *Bild*). In fact, just the opposite is true: the lines speak of the most immediate incorporation of truth in language, in the spirit of Walter Benjamin's "(probably general) law of lyric poetry: words achieve their perceptual meaning [*anschaulichen Sinn*] in the poem without sacrificing the figurative."[21] The terms *Wort* and *Bild* are not parallel—first, because the word *Bild* does not occur in the first citation—and second, because the sense of *Wort* in its first appearance ("Und mir der Geist im Wort, im Bilden"; "And for me the Spirit in the word, in forming") has next to nothing to do with the category *Bild* in the line *Nun nicht im Bilde mehr*.

What, then, is the force of this first passage? It is a play chiefly of fusions and of fine distinctions. In the way of fusions, it means to contain in a single action the contrary motions of giving shape or form (*Bilden*) to the alien beings of the world ("die Gestalten dieser Welt" [line 398], "die Fremdlinge" [line 405]), and also of solving, resolving, or dissolving the enigma of life, which is its result. More importantly, it claims an audacious intimacy between Empedocles' poetic consciousness and what is called "the Spirit." It is true that the word *Wort* comes marginally between them and hence distinguishes the subject (the poetic self, *mir*) from the absolute (*der Geist*). But by the same token the word *word* brings them as close together as is thinkable and requires one to take seriously this claim to the not-merely-fictive character of poetic language.

These extreme tensions of unification and distinction have their effect on the self-reflexive aspects of the passage, in particular the construction involving *mir* ("for me") and *sich* ("for itself"). In the act of giving strange presences gods' names and thus making gods of them, the so-called Spirit solves life's riddle "for itself" (*sich*), for its own sake, but also "*for* the self" (*mir*). What in the previous sentence was described as the naming activity of the self (*ich . . . nannte*) is recast as the activity of the absolute spirit performed in lieu of the self. With this change a more intricate stage of consciousness is realized by the poem. Hölderlin revises the claim of the self to name the gods as in fact a productive error properly recast as a claim made on behalf of the so-called Spirit. The self that alleges it knows divinity through its activity is in reality claiming something else: that the absolute spirit performs its work (of giving form) for the benefit of this self.

The so-called Spirit makes use of the speaker, a space in him not obviously congruent with the space occupied by the subject of consciousness (where "helle ward / mein dämmernd Herz"). In some other place visible to the self—for the self and yet "in" the self—the absolute spirit performs its work of giving names. This activity of problem solving is blissful: the Spirit is at work with the gaiety of self-making—*for* Empedocles; its exhilaration is exemplary. This conclusion reads the relation of authority to its subject as one of benevolent tutelage. But there is an opposite and in the long run more important implication at work. This other place or scene of revelation suggests a *Schauplatz*. The self provides the space in which the so-called Spirit performs its work for the self, which in this way becomes the theater of its own *Bildung*. Indeed, in line 443, we hear of the "Geist, der uns *erzog*" (*FA* 13: 944). Yet this theatricalization of consciousness jeopardizes the truthful character of the *Bildung*.[22]

What then of the second, later crux—*Nun nicht im Bilde mehr* ("Now no longer in the image")? Can this line be read without Warminski's help—and incidentally without Söring's, too, who, like Warminski, refers this passage to the previous one, seeing in it "the old enthusiasm for death, . . . no longer, however, in the 'Ideal' word?"[23] To what other likely textual place could one connect the phrase "Now no longer in the image"? An important candidate is the passage where Empedocles likens the earthly round (*den geschäfftigen Reigentanz*, "the busy round dance") to an image (*der Erde trunkenes Bild*, "the Earth's ecstatic image"), while emphasizing its precarious spatial properties as an object attractive to perception (lines 171–73, unlike line 407):

Es glänzt . . .
. . . dir gegenwärtig . . .
Es rauscht und regt durch alle Lande sich . . .

There gleams . . .
. . . present before you . . .
It rushes and stirs through all lands . . .

The spatial character of this visible image is significant. For in favoring the encounter provided by death, Empedocles has rejected both of the following terms: (1) the living spirit as it might be glimpsed in the image; and (2) as it might be found "in brief happiness" (*im kurzen Glück*). Not in the brevity of empirical bliss and not in the space of perception. It is consistent with the rigor of detail in Hölderlin's construction that he would want to oppose the envisioned permanence of the god in death with the perishability of his form in time (being subject to the "Lord of Time") as well as in the space of fascinated perception.

At the same time, however, other clues continue to recommend the previous crux (in which Empedocles names the gods) as the contact passage, even if it no longer anticipates the phrase *im Bilde*. The basic passage opens by conjuring the very divinities whom Empedocles earlier named:

> O Geist, der uns erzog, der du geheim
> Am hellen Tag und in der Wolke waltest
> Und du o Licht! und du, du Mutter Erde!

> O Spirit, who formed us, who secretly
> Reigns in the bright day and in the clouds
> And you o Light! and you, you Mother Earth!

More importantly: it repeats the two-sided construction *mir und sich*, "for me and for himself," that determined the logic of the previous passage. The present crux concludes:

> Und heute noch begegn' ich ihm, denn heute
> Bereitet er, der Herr der Zeit, zur Feier
> Zum Zeichen ein Gewitter mir und sich.
> (*FA* 13: 944)

> And even today I shall meet him, for today
> He, the Lord of Time, is preparing in celebration
> As a sign a storm for me and for himself.

Recalling the previous construction—"Und *mir* der Geist im Wort, im Bilden *sich*/Im seeligen, des Lebens Räthsel löste" (lines 407–8)—the later passage (lines 451–53) reproduces the same situation of Empedocles as spectator at his own drama of a sought-after intimacy of self and so-called Spirit. And at the very moment that this passage envisions unity, it enacts unbridgeable division, for the moment in which Empedocles *finds* "the living one" and *encounters* the "Lord of Time," he must lose his inmost self in the image. The self cannot represent in an image the self that sees itself; the seen self is the *Darstellung* (representation as well as performance), yet never the seer.

Empedocles envisions death as a kind of scene, a staging, in which the constraining environment of the sensory image is cast off ("Now no longer in the image"). The envisaging of death in this passage thus becomes a sort of negative ekphrasis, in which the text denotes the imageless work of art that it itself would be. In this passage—in a death without a sting, a death that allows the consciousness of the vanishing body to flourish—the text envisions utopianly the very play, *Empedokles auf Ätna*, which sought to stage in a condition of the most ascetically reduced, the coldest and most alien imagery a world *without* contingencies, *without* "die wahren sinnlichen Beziehungen" ("the true sensory relations"), without "den physischen . . . Zusammenhang" ("the physical . . . context") (*FA* 13: 870). And indeed from the moment on in which Empedocles says, "Nun nicht im Bilde mehr" ("Now no longer in the image") (line 448) until with a thrilling freshness the phrase "dieser grünen guten Erde" ("this green good Earth") strikes in line 486, he does not, with one exception, conjure a single image as a concrete particular (all the other seeming images are figures of speech or allegory). The entire action of the play becomes the theater of the separated soul—its own actor and spectator. The scene tries not to register the inevitably phenomenological splitting of the subject into seer and seen and its captivation by the image, but it cannot evade for long the revenge of the material image. For here too a *sign* (*Zeichen*) is given to the dying man to gaze at—the scenic sign of his death, the envisaged sign of the storm (*das Gewitter*), which has been prepared for him. Only a god, not Empedocles, could enjoy, by intellectual intuition, imagery in which there was not a jot of contingent opacity. In this way the materiality of the letter, the sign—which does not appear in the earlier crux, in Empedocles' acts of naming and in his experiencing the solution to life's enigma in the word—reappears despite its implied cancellation, in the natural sign, "the storm." The sign-character of the word, which went unspecified in the earlier passage, is here explicit as the natural sign that, according to Empedocles, will be transparent to its meaning as an expression of the benevolent, authorizing will of the Lord of Time. This is mythical thinking; its tendency continues in Empedocles' effort to redeem the natural sign, which surfaces in the valorization and equivalence of the terms "green good Earth" and his wish to remember his past—his friends. Memory will be a kind of transparent naturalism—a thinking in images purged of their contingency in the way that the storm would be transparent to the will of Jupiter and the green good Earth transparent to the memory of friends. This myth prepares the way for Hölderlin's late poetry of memory.[24]

If *Empedocles—pace* Warminski—is not a tragedy about the deluded reader who reads images literally, what sort of tragedy is it? It could only be the tragedy of an incomplete, a stumbling self-deification, a tragedy of hubris at its far extreme. Hölderlin ponders the self-deification that remains only hubris; the hero calls down punishment not for his attempt at divinity (an imageless autonomy) but for his failure to achieve it. Failure is not in the distance with which he separates himself from the measure of the human; his failure is in the distance still separating him from the divinity that he leaves untraversed—a failure consisting specifically in his not detaching himself from the empirical image: to be "no longer in the image" is the optimal condition. Put inversely, in the perspective of the achieved god: to want to be wholly image is the most radical hubris. A remark of Walter Benjamin in another context is apposite: "Here before us is a highest expression of identity: the Greek god is entirely given over to his own principle, the plastic form [Gestalt]. The highest sacrilege is understood as hubris, which, attainable only for a god, transforms him into a dead form. To give plastic form to oneself, that is the definition of hubris."[25]

For Empedocles, death does not appear to be his way; his punishment is at once not to die *and* not to find another way. Before *The Death of Empedocles*, tragedy is the tragedy of an unwanted death; this tragedy is of the impossibility of the wanted death—a representation of death-in-life against the will that this death be something more. At the end of the play, Empedocles is not dead; he is seen turning back, at least in thought, to the polis to remember friends. This act of remembering evokes the sheer recalcitrance, the obstacle of attachment that bars his passage.

This obstacle is the material image in its insistent theatricality—its capacity (being never severe or cold, strange or artificial enough) always to insinuate itself into the warm consciousness, to split the subject into actor and spectator, with effects, as we have seen, of delusiveness, self-mockery, or bravado. The ethical correlative of this epistemological loss is the fall back into the misery of the empirical. The penultimate lines wishfully connect remembering with a happy spatial perception:

> Von dieser grünen guten Erde soll
> Mein Auge mir nicht ohne Freude gehen,
> Und denken möcht' ich noch vergangner Zeit . . .
> (*FA* 13: 945)

> From this green good Earth
> My eye shall not go from me without joy
> And I should like to think on times past . . .

But they are followed by the speech of the chorus that defines perception with terrifying finality as mere delusion:

> . . . und es hängt, ein ehern Gewölbe
> der Himmel über uns, es lähmt Fluch
> die Glieder den Menschen, und ihre stärkenden, die erfreuenden
> Gaaben der Erde sind, wie Spreu, es
> spottet unser, mit ihren Geschenken die Mutter
> und alles ist Schein—
> (lines 495–500)

> . . . and there hangs, a vault of ore
> the heavens over us, a curse paralyzes
> The limbs of men, and her strengthening, the delightful
> Gifts of the Earth are, like chaff
> with her gifts our Mother mocks us
> and everything is seeming—

But while this aporia goes some way toward answering the question of why Hölderlin did not finish his play as a tragedy, it does not answer the question of why he did not finish the play *tout court*. The answer could have to do with the excessive intimacy between Hölderlin the writer and his "alien, analogous material figure" Empedocles. Tellingly, in the first version, Empedocles addresses a "phantom" (*Schattenbild*) of himself as "poor Tantalus" (*FA* 13: 706). Later, in the Böhlendorff letter of December 1801, Hölderlin fears "lest it come to pass in the end with me as it did with old Tantalus, who had from the gods more than his fill" (*SA* 6: 428). If both Hölderlin and Empedocles are Tantalus, then the requirement is violated according to which the figure of the other must be alien (*fremd*). Sattler has noted the heightened identification of author and persona: "In the identity of the poetic agency with the poetic form, the historical difference disappears. The protective metaphor becomes too transparent" (*KT* 6: 290).

The transparency of the metaphor can be seen as resulting from Empedocles' thrust toward self-transparency, in the sense of a mind from which all contingencies have been burned away, all complaints and all attachments to sense. The tendency of the poet writing a drama in which all accidentals are also disavowed will be to identify himself with such a hero-poet. The resurgence, however, of the contingent image at the mimetic order works against this transparency at the poetic order, at the same time that it disqualifies Empedocles as a figure good enough to sacrifice—even, perhaps, as a figure good enough to avow. The dramaturge's recoil from intimacy with his hero is shocking. Hölderlin is left with a hero whom the good "coldness" of the poetics has not been able to alienate for tragic ends, and worse—a hero contaminated, perhaps, by the "coldness of the common day" ("die eiskalte Geschichte des Tages") (*SA* 6: 290). An exemplary alienness that would ban the familiar image is spellbound by the coldness of the ordinary: such is the economy of denegation.

Gerhard Kurz subordinates this loss of poetic distance to a more generally tragic movement by which differences are collapsed.[26] At the same time, however, the virtual identity of poetic subject and figure should not be conceived as a mistake or indeed as an accident of Hölderlin's narcissism. It is his ethos to imagine as deeply as possible the condition of his poetic persona, and so he goes Empedocles' way with him—a way that could mean going to his death. He takes him to the point of his death and then turns back.

The potentially killing point of intimacy is the discovery that Hölderlin elaborates through Empedocles' tragedy. The tragic theater that would renounce the empirical image proves the persistence of the renounced. The image resurges where it has been most discarded: remembered, it reinserts itself into the heroic consciousness to make a scene. Here the subject splits into a being seen (linked to the image) and into a seer deluded as to his autonomy. The logic of the image jeopardizes, though it does not finally destroy the Idealist project of subjective autonomy.

Through Empedocles, Hölderlin takes himself to the point of his own death and then turns back; he will not let him die by his own hand. The abandonment of *Empedocles* is the price of an insight into the insistence of the image in the myth of a hero transparent to the autonomous self of poetic agency. The two moments that conclude the third version diagram Hölderlin's concerns thereafter. History becomes memory, a language full of discrete images: like Empedocles, Hölderlin will remember his friends (one thinks of the ode "Andenken"). History becomes a world-time without sacred renewal, a godless time: the speech of the chorus ending *Empedocles* states the feared conclusion:

> . . . die erfreuenden
> Gaaben der Erde sind, wie Spreu, . . .
> Und alles ist Schein . . .
> (lines 497–500)

> . . . the delightful
> Gifts of the Earth, are like chaff, . . .
> And everything is seeming . . .

Memory attempts to endure the absence of divinity, bound to contingent images, without pretension to truth: that is Hölderlin's vision to follow, but it required his renunciation of Empedocles' tragedy, who intended to die like a god. In the words of "Stimme des Volkes,"

> . . . er selbst
> Der Mensch, mit eigner Hand zerbrach, die
> Hohen zu ehren, sein Werk der Künstler.
> (*FA* 5: 594)

> . . . he himself
> The man with his own hand shattered, to honor
> The lofty ones, his work, the artist.

Reading the 'Poetics' After the "Remarks"

Christopher Fynsk

In the third section of his "Remarks on *Oedipus*," Hölderlin offers a definition of the object or affair (*die Sache*) of tragic representation. It is a kind of catharsis, the purifying separation, as Hölderlin puts it, of a monstrous union of the divine and the human. The "separation" may be read as a movement of absolute reflection, but it is no less a catharsis, as Hölderlin indicates when he interrupts his exposition with a reference (in Greek) to Aristotle:

> The presentation of the tragic rests principally upon this, that the monstrous, how God and Man couple, and limitlessly the natural force and the innermost of man become one in anger, is conceived in that the limitless becoming-one purifies itself through limitless separation. *Tes physeos grammateus en ton kalamon apobrechon eunoun.*[1]

There are other echoes of the *Poetics* in Hölderlin's "Remarks,"[2] enough to suggest that Hölderlin is engaging in his own manner in the kind of speculative appropriation of Aristotle that Schelling pursued in the tenth of the *Letters on Dogmatism and Criticism*. But the citation from Suidas indicates that the *Poetics* has offered more to Hölderlin's reflections than what could be simply at that time a set of conventional tropes for discussions of tragedy (or, as Peter Szondi would say, the tragic).[3] The astounding and almost brutal alternation of tones that occurs as Hölderlin passes into Greek sets Aristotle off in almost quaint relief. And yet Hölderlin seems also to be telling us that Aristotle was writing as a Greek what Hölderlin has just written (I refer to the definition of tragedy) as a modern; or, to put this in more compact terms, that the *Poetics* rest in their own (Greek) manner on such a "conception" of the tragic. Aristotle, "nature's scribe," would thus have been articulating the structure of tragic *Darstellung* as a presentation of the tragic essence of Being—inscribing "naively," perhaps, but in generous understanding, the mimesis of *physis* itself.

But is there actually a *reading* of Aristotle implied here—a reading in the strong sense of a confrontation with the text? Or is Hölderlin simply remarking, magnificently, the historial difference that *requires* a new reflection on the tragic, a reflection that will account for and answer to the *vaterländische Umkehr*? A satisfying answer to this question would require a thorough reading of the "Remarks" and a consideration of Hölderlin's relation to Aristotle within the framework of his understanding at this time (the time of the translations of Sophocles) of the relations between the Greeks and the "Hesperians."[4] For this occasion, I will try merely to suggest the possible reach of the question by sketching briefly something of what happens if we *return* to the *Poetics* in the light of the "Remarks." For if we bear in mind Hölderlin's description of the manner in which the tragic transport articulates itself in a rhythmic structure,[5] we are offered an intriguing solution to what has traditionally been the fundamental problem in the reception of the *Poetics*: Aristotle's understanding of *catharsis*.

Catharsis is not defined in the *Poetics*, and is referred to at only one point (though a significant one, in conjunction with the very definition of tragedy in Chapter 6: ". . . in a dramatic, not in a narrative form; with incidents arousing pity and fear, wherewith to accomplish its catharsis of such emotions").[6] Aristotle tells us at various moments in his discussion of plot what structures are most pitiable or frightful and best produce (*dia mimeseos*) the proper effect of tragedy (1453b12). But he does not really tell us how this effect is produced by these structures, and he never defines the effect itself.

The only other significant reference to catharsis in Aristotle is to be found in Book 8 of the *Politics*, where he takes up the place of music in education. The discussion is brief, and Aristotle promises us further development "when hereafter we speak of poetry" (1341b38). That lost book again . . . But he tells us nevertheless that music has an effect on character because its melodies and rhythms involve a mimesis of qualities of character and provoke a kind of sympathetic movement in the listener.[7] He recognizes also a cathartic effect that is not due to an "intellection" regarding such mimesis (an "aesthetic" contemplation of form), but rather to something *like* a medical treatment (*hosper iatreias . . . kai katharseos* [1342a15] that *also* provides an innocent pleasure.[8]

This is hardly the place to review the history of the application of these remarks from the *Politics* to the interpretation of catharsis as a homeopathic process. But I will note simply that recent readings of the *Poetics* informed by formalist principles—and many of the best readings since Gerald Else's may be characterized this way[9]—have disregarded the discussion of catharsis in the *Politics*, and this despite the fact that Aristotle offers a genealogy of tragedy in the *Poetics* that links it to forms of song and dance in which, as Aristotle says in Chapter 1, characters, actions, and emotions were represented by rhythm given form, *dia skhematizomenon rhuthmon* (1447a27). Instead they have turned to his statements about the cognitive function of poetic *mimesis* and have emphasized in particular Aristotle's statement in Chapter 4 regarding the pleasure we take in the *graphic* depiction of form.

The lines in question from Chapter 4 follow Aristotle's assertion that the origin of poetic art lies in the natural human inclination to mimesis, an inclination that takes two forms—the first deriving from the fact that humans, as the most imitative of animals, learn from mimesis, the second being that humans naturally take pleasure in mimesis (1448b6–9). He continues,

> The truth of this second point is shown by experience: though the objects themselves may be painful to see, we delight to view the most realistic representations of them in art, the forms, for example, of the lowest animals and of dead bodies. The explanation is to be found in a further fact: to be learning something is the greatest of pleasures not only to the philosopher but also to the rest of mankind, however small their capacity for it; the reason for the delight in seeing the picture is that one is at the same time learning—gathering the meaning of things, e.g., that the man there is so and so; for if one has not seen the thing before, one's pleasure will not be in the picture as an imitation of it, but will be due to the execution or coloring, or some similar cause. (1448b9–19)

The pleasure we take in mimesis, then, is essentially a "philosophical" pleasure of recognition. An image presents to us the *form* of an object in such a way that we can recognize that object in its nature, and indeed *learn* to recognize that object. It is the recognition of form that gives us the principle pleasure in mimesis, not the sensible qualities of the presentation (a fundamental point for the *Poetics*, as we see in the almost constant derogation of spectacle [cf. 1450b17–20]). Moreover, this is a pleasure that transmutes a kind of pain—the pain we have, Aristotle says (the example is not without interest in the context of a discussion of tragedy), in seeing ignoble animals and cadavers.

Roselyne DuPont-Roc and Jean Lallot, who offer in their critical edition of the *Poetics* what I take to be the most thoroughgoing and careful version of what I have called the formalist argument, conclude from this passage in Chapter 4 that the catharsis of pity and fear effected by tragic mimesis is to be understood from the basis of an analogy between the distillation of forms that occurs in painting and the synthesis of facts and events that occurs in the formal composition of the tragic *muthos*.[10] For tragedy, as Aristotle argues, also presents *forms* of action. This, of course, is why he says that tragedy is more philosophical than the chronicle. Its action obeys the logical criteria of the probable and the necessary. And for the purpose of producing its proper effect (which is the pleasurable catharsis of pity and fear), it offers something like the *form* of the pitiable and the fearful—a sequence of events that moves us by the way in which it concerns us (for it occurs to someone more or less like us) but that transforms the emotions its stirs through its ordering of these events, however terrible that order might be.[11] The pitiable and the fearful are inscribed, we might say, in the "syntax" of the facts that constitute the *muthos*—a system of facts, as Aristotle puts it in Chapter 14 (*systaseos ton pragmaton*) that follows the schema outlined in Chapter 13, that is to say, the passage from fortune to affliction of a man who is not eminently noble, but better rather than worse. This tragic *metabole*, entailing in the best of cases a reversal (*peripeteia*) and a recognition, is a form of the pitiable and the frightening. Our apprehension of it, or simply our conception of it (because in a good tragedy we should need no more than to hear the facts of the *muthos* [1453b1–7; cf. 1462a13]), effects catharsis—a purification that is in this reading more of an ordering of feelings through intellectual satisfaction and aesthetic gratification than a release offered by a homeopathic process.

From the basis of the analogy between the apprehension of form in a painting and the apprehension of the logical form of the tragic action, DuPont Roc and Lallot are inclined to see the tragic sequence as forming a kind of tableau for an aesthetic vision.[12] They recognize that a certain participation in the tragic movement is required of the spectator (their acknowledgment that the spectator must be able to identify with the protagonist, their emphasis given to the notion of surprise and their convincing articulation of this notion with Aristotle's remarks on the complex structure, finally their reading of the description of spectacle in Chapter 17, all point in this direction). But they nonetheless finally privilege the analogy with painting by emphasizing the spectator's aesthetic experience of tragedy. They are certainly not violating Aristotle's text in doing so. The remarks on organic form and the criteria for its beauty in Chapter 7 obviously lend themselves to such an interpretation. But I would suggest that there are strong reasons for suspecting that this formalist interpretation of the tragic effect cannot suffice.

My doubts here are the same ones I have regarding the interpretation offered by DuPont-Roc and Lallot of the cathartic effect of music—an interpretation that seems to me insufficiently dynamic.[13] Aristotle is insisting too heavily on the engagement of the spectator in the tragic action for the aesthetic interpretation of catharsis to be quite adequate. Two of the points I mentioned as being acknowledged by DuPont-Roc and Lallot are worth emphasizing here: the motif of surprise (*to thaumazein*) in the tragic action, and the problem of spectacle. I will go quickly over the first and then pass to the second, since it is a long-standing source of difficulty for the interpretation of the *Poetics*.

DuPont-Roc and Lallot argue quite convincingly (both on philological and interpretative grounds) that Aristotle's remarks on the effect of surprise in Chapter 9 are to be linked to his remarks on the structure of the complex plot in Chapters 10 and 11. Aristotle states in the last paragraph of Chapter 9 that pity and fear are most strongly evoked when a sequence of events linked causally occurs against all expectation and when we perceive in an astounding and seemingly chance occurrence a certain design. We might stress the fact that the perception of order in the most extreme circumstances is what intrigues Aristotle: pity and fear will be most stirred when the intellect is pushed to its limits and is offered a glimpse of order in the most startling chance event. Pleasure would come from the tension between chance and necessity and its resolution in the perception of an orderly design. But I would suggest that, insofar as Aristotle links surprise with the elements of peripeteia and anagnorisis—that is, insofar as he seeks to inscribe the surprising event in the ironic turn that occurs in a complex plot structure wherein a causal sequence suddenly turns about—he is also taking into account the spectator's engagement in the action. If the complex structure is best suited to producing the catharsis of pity and fear, and if the element of surprise that occurs as a "causal sequence of events comes about against all expectation" is an integral part of this structure, then the effect of surprise must come from more than a kind of intellectual astonishment. There must be something like a movement in a causal sequence that provokes us to shudder at the course of events (even if we only hear the tale told) and wonder at their design.

Again, my suggestion, here, is that catharsis must proceed from a certain movement—an engagement in the tragic action (and particularly its metabole, its turn) that involves something more than contemplation. We must be *taken* by the tragic turn of events. When Aristotle comes to describe the poet's attention to the tragic spectacle in Chapter 17, he is apparently thinking of precisely such an engagement, and this is why it is necessary to look closely at his discussion of this point.

The topic of the spectacle is another classic problem for interpretations of the *Poetics*. At least twice (in Chapters 6 and 14), Aristotle suggests that spectacle, though the most seductive part of the presentation, has little to do with art—and these statements are quite consonant with his emphasis on the structure of the *muthos* and his subordination of all other elements to it, including character. Yet in Chapter 26 (1462a15), he appears to contradict himself by describing the resources of the spectacle as one of the elements that make tragedy superior to epic, and in Chapter 17 he advises the poet to place the action before his eyes so that he can guarantee its coherence for the spectator. Here, consideration of the spectacle becomes a determining element in the very process of composition. The first paragraph of Chapter 17 reads:

> At the time when he is constructing his plots and engaged on the diction in which they are worked out, the poet should remember to put the actual scenes as far as possible before his eyes. In this way, seeing everything with the vividness of an eye-witness as it were, he will devise most effectively what is appropriate and avoid incongruities. [Aristotle goes on to provide an example of precisely such an unhappy incongruity and then continues.] As far as may be, too, the poet should even act his story with the very gestures [skhemasin] of his personages. Given the same natural qualifications, he who feels the emotions will be the most convincing: distress and anger, for instance, are portrayed most truthfully by one who is feeling them at the moment. Hence it is that poetry demands a man with a special gift for it, or else one with a touch of madness in him; the former can easily assume the required mood and the latter may be actually beside himself with emotion. (1455a22–33)

So Aristotle tells us here first that the poet must visualize the scene in order to avoid contradictions and to give the action a maximum of coherency.

The visualization he recommends has a kind of controlling function. But Aristotle also says very precisely that he who sees the action as though he himself were a spectator will find or invent with the most efficacy (*energestata*) what is most fitting—by which Aristotle probably means, what is most fitting for the diction. The poet who sees most precisely will be most inspired, we might say, or the most energetic in finding the fitting words. The recommendation, then, has more than a controlling function.

Aristotle then makes a second recommendation. To give a finished form to the tragedy, the poet must not only *see* the action, he must also have recourse to gestures that will evoke in him the emotions he is attempting to convey in his writing. Aristotle recommends such gestures in the *Rhetoric*, suggesting how the orator must interpret the text in order to give it a vividness that will provoke pity (2.1386a32). But in the *Poetics*, Aristotle is actually recommending to the poet a kind of theatrical exercise designed to awaken in him the passions he is attempting to portray. A technique of *mimesis* (in the sense of this term that so worries Plato in the opening books of the *Republic*) is recommended to the poet so that he may work himself into a state of possession that will make possible the effect of truth in the verbal representation of passions.

But how are we to conceive the relation between the poet's mimetic activity and his writing? Dupont-Roc and Lallot suggest that this relation is in fact already implied in the word *skhemata*, for in the Greek aesthetic vocabulary, the word designates both corporeal and linguistic forms: figures of dance (as noted, the word is used this way in Chapter 1) and figures of expression. Throughout the Greco-Latin rhetorical tradition, they remark, there is a strict affinity implied in the word *skhemata* between gesture and speech. "Before being a codified and fixed 'figure,'" Dupont-Roc and Lallot write, "the *skhema* is to language what the gesture is to the body—movement and rhythm." They cite Cicero's reference, in this regard, to the Greek notion of *skhemata* as "gestures of discourse.'"[14]

So there is a link between gesture and expression, and Aristotle would appear to be proposing to the poet a physical technique of autosuggestion (the rehearsal of gestural *skhemata*) in order to promote an effective *lexis*—one that produces the effect of truth on the listeners and brings them to share the passions portrayed. As Dupont-Roc and Lallot put it, "To give the work its accomplished expression . . . is to imprint in the text itself . . . the expressive forms, the 'figures,' which, in transposing the emotive movements from which they have issued, will lend themselves in turn to the vocal and gestural retranslation that defines the interpretive function of the actor. . . . Whether it is grasped in the moment of invention, at the moment of writing, or in that of interpretation, the *skhema* appears always as a dynamic form, a movement and rhythm of the body . . . the text . . . and elocution."[15]

So we are back to rhythm, at least as concerns the diction—a secondary element of composition vis à vis the invention of the *muthos*, but nevertheless that element of the mimesis that produces an effect of truth and engages the auditor by *communicating* to them the represented passions.

Still, the good tragedy, Aristotle suggests, should be able to produce its proper effect even without a well composed diction, let alone spectacle. We need only hear the general line of the story, the general schema, we might say. In the paragraph immediately following, Aristotle makes a third recommendation corresponding to the fundamental thesis of his argument: "His story, again, whether already made or of his own making, he should first simplify and reduce to a general schema, before proceeding to lengthen it out by the insertion of episodes" (1455b1–3).

In translating *ektithestai katholou* as "general schema," I am following the French version of Dupont-Roc and Lallot (McKeon gives "universal form"). Strangely enough, the French authors never justify their use of the term "schema" here (though they normally justify *everything*) and never note the possible connection between the *skhema* in the sense of the rhythmic figure and the "general schema" they have Aristotle propose to the poet. The apparent oversight is remarkable in the light of the extensive and even daring commentary they devote to the word *skhema* in the commentary of the preceding paragraph. Why translate *ektithestai katholou* as "general schema" in the first place, and why not remark upon (if only to dismiss) the relation between gestural schemata and plot schemata? I want to suggest that we are witnessing a kind of return of the repressed here, and that Dupont-Roc and Lallot are unwittingly pointing to a further dimension of the source of catharsis produced by the tragic *muthos*. For I would argue that in defining the "general view" as a schema, they are reintroducing the question of rhythm at the level of plot. They implicitly suggest that the "mainspring" of the play, which, as we have seen, is a *form* of the pitiable and the frightening, a kind of logical schema, is comparable to a *skhema* in the sense of a gesture or expression, and thus that listening to the main lines of the plot will be equivalent to engaging with a kind of rhythmic movement in discourse.

If the general movement of the play is comparable to a rhythmic figure, would it not be reasonable to suppose that tragic catharsis is effected at least in part through participation in a rhythmic movement? No doubt there is an important dimension of intellection and a certain aesthetic experience accompanying the experienced order of movement presented in the tragedy. Yet the tragic experience offers more than a theoretical or aesthetic contemplation. As Hölderlin observes, there is also the experience of a rhythm.

Did Hölderlin observe this in Aristotle? The reference to Aristotle in the "Remarks" appears immediately, as I have noted, after an allusion to the notion of catharsis. "Nature's scribe, dipping his benevolent reed." I cannot help recalling here Benveniste's remarks on the word *rhusmos*. *Rhusmos*, or *rhuthmos*, he argues, does not originally connote rhythm in the modern sense, but connotes rather form.[16] The word is used by Democritus and Leucippus in a technical sense to designate one of the ways in which bodies distinguish themselves by their form or configuration. It is Aristotle who identified and preserved this usage in his *Metaphysics*, explaining its meaning with an example from Leucippus. The letters *A* and *N*, he says, differ from one another by their *skhema*, which Aristotle equates with the word *rhusmos*. Everywhere in Greek prose up to the fifth century, and even in Plato and Aristotle, we find *rhusmos* designating a notion of *form*. But the root *reo*, which means "to flow," and originally prompted the spurious interpretation of *rhusmos* as deriving its meaning from the regular movement of the sea's waves, seems to have lent the word to designations of fluid and transient forms. Plato was then the first to give it the meaning of rhythm in the sense of a recurrent form that is measurable or countable, as in dance or music.

"Nature's scribe, dipping his benevolent reed." Might the conjunction of letters and a certain fluidity (there is ink, of course, but perhaps even an association of "reed" and water) point to *rhusmos*? This is not a connection to force. No more than I would want to suggest that we might easily return to Hölderlin's "Remarks" and seize the essence of what he was doing with his own notions of catharsis and rhythm from the basis of what we have seen in Aristotle. We might well say that for Aristotle what requires purification is a kind of monstrous confusion (the preeminent tragic event is the murder of a family member). But he certainly does not think of catharsis in terms of a separation of the human and the panic power of nature—a separation that for Hölderlin takes the form of time. Still, for

Hölderlin, the presentation of this separation occurs in a rhythmic structure that follows (but also defines) something like a calculus of the rhythm of what Aristotle would have called the soul. Catharsis, for Hölderlin, appears to involve a *measuring* of the diverse human faculties by which the human being (in the person of the hero) learns to "count time."[17] I suspect that Aristotle in his own way thought of catharsis as a kind of measuring. But as a Greek of the fifth century B.C., he could not think time as a transcendental condition of human sensibility and obviously could not think the veering of time that required the *vaterländische Umkehr* in the modes of representation. Only a modern could propose a definition of the tragic (and its *Darstellung*) wherein beginning and end would no longer rhyme.

Ancient Sports and Modern Transports

HÖLDERLIN'S TRAGIC BODIES

Rainer Nägele

Земля гудит метафорой[1]

The coincidence of the word play in the title is produced by a body of texts that seems infinitely far removed from the frivolities of puns. Whereas the sports stadium as a model for the theater seems appropriate for the provocations of the Brechtian epic theater, Hölderlin's classification of Greek tragedies in terms of different sport competitions in the late remarks on the Sophoclean tragedies is startling: "The grouping of such persons can be compared, as in *Antigone*, to a competition of runners where the first one to catch his breath heavily and to jostle the opponent has lost, whereas one can compare the struggling in *Oedipus* to a boxing match, that in *Ajax* to a fencing match." (*FA* 16: 421)[2] These *Kampfspiele* combine the deadly seriousness of the battle, the agon (*Kampf*) with the playfulness of tragedy as a game (*Spiele*). While the figural grouping on the theater is described in sportive terms, the temporal sequence appears as the "tragic transport" in Hölderlin's remarks on *Oedipus* (*FA* 16: 250).

Hölderlin writes the word *transport*, but he does not use the word *sport*, which entered the German language only in the mid-nineteenth century. The configuration of the words *sport* and *transport* is a linguistic after-effect.

There are puns in Hölderlin's poetry. But if they are acknowledged at all, the anxiety of the critics quickly searches for "deeper" meaning and legitimization. The genealogical trees of etymology must spread their branches to provide a support over the abyss of literal contingencies.[3] Yet the branches wear thin when, in the transport of poetic invocation, *Vater Aether!* is echoed by *Vater! heiter!* in a serene abandonment to the phonetic correspondence between the Greek word *(h)aitèr* ("Ether," "air") and the German *heiter* ("serene").[4]

This phonetic echo resounds at a crucial moment in Hölderlin's elegy "Bread and Wine," invoking the violence and power of the sleeping word ("es wächst schlafend des Wortes Gewalt") that prepares the coming of the gods "in truth." As in many of Hölderlin's texts, the passage stages the question of the truth in language as a question of the poetic ground and the poetic effect. The *and* between ground and effect signals the abyss of a linguistic causality that turns from the substantiality of a legitimating ground to the airy or arabesque flightiness of linguistic and scriptural effects.[5]

The ideological petrifications of academic and public critical scenarios make it difficult to address the question of truth and language outside the dichotomies of rhetorical proclamations and moralistic resentments that dominate the various institutions of literary criticism. Ever since Nietzsche's text "On Truth and Lie in the Extra-moral Sense"[6] became the focus of critical attention,[7] Nietzsche's provocative answer to the loaded question "What, then, is truth?" has stirred critical aggressions that testify to profound anxieties. Nietzsche's answer to his question:

> A mobile army of metaphors, metonymies, anthropomorphisms, in short, a sum of human relations that were poetically and rhetorically intensified, translated, ornated and that, after long use, seem firm, canonical, and binding to a people: truths are illusions of which one has forgotten that they are illusions, metaphors that are worn out and have lost their sensual impact, coins that have lost their image and have become mere metal, useless as coins.[8]

No wonder, then, that Nietzsche has been declared the gateway to postmodernism,[9] that shares with other posts the charge of linguistic and rhetorical reductionism. But these historical displacements of major texts of Modernism to postmodernity (including Kafka, Brecht, Joyce, and many others) may indicate a refusal to read the texts of Modernism at their most disquieting points.

Most disquieting for many literary critics is the confrontation with that which constitutes the subject of the discipline. The phantasmatic instrumentalization of language as a "mere" vehicle, "mere" formality produces as its complement the phantasm of "real presences" as the true and legitimate ground of literature. The less epistemological and analytical attention this dichotomy can sustain, the more rhetorical and moral pathos it will invest in its proclamation. This moral pathos apparently will stop at nothing; it even calls upon the Holocaust as its instrument.[10]

The pathos of suffering, death, and despair supposedly silences the babble of mere words. Tragedy might then signal the limits of language. But it does so with and within language, more precisely in the tragic or sad display of the wounded body and being that speaks. The difference in the relationship of body and language[11] constitutes for both Hölderlin and Benjamin the historical difference between antiquity and modernity.

Hölderlin's tragedy project takes a curious course: it begins with the failed attempt of writing a tragedy (*Empedocles*), and then moves through various articulations of a theory of tragedy to the translation of two Sophoclean tragedies (*Oedipus* and *Antigone*). In the hierarchy of the aesthetics of the period, this course seems to invert artistic authenticity, as it moves from creative spontaneity to ever more mediation and secondariness. The inversion signals a poetological revolution. As if to stage this turn even in the scriptural constellation, one of Hölderlin's major (unfinished) poetological essays ("Wenn der Dichter einmal des Geistes mächtig . . ."), is written in inverted order from the back of Hölderlin's notebook toward the front, and toward the unfinished text of *Empedocles*.[12] The textual constellation indicates that we are not dealing with a simple and linear movement from tragedy to theory, but that there is also an inverted movement to a point of encounter and nonencounter: to the point of an intersection and a cut. At the place of that intersection, there will be neither theory nor tragedy, but translation and commentary.

The scriptural figuration of the text as the (non-)encounter of tragedy and theory places Hölderlin's tragedy project firmly in the center of the philosophical-aesthetic revolutions after Kant in the late eighteenth and early nineteenth centuries. Tragedy, or rather the philosophy of the tragic, provides a major scenario for an attempt at a philosophical-aesthetic healing of the wounds left by Kant's rigorous critical-analytical cuts. The first sentence of Peter Szondi's *Essay on the Tragic* succinctly points to this situation: "Since Aristotle there has been a poetics of tragedy, only since Schelling there is a philosophy of the tragic."[13]

The move from poetics to philosophy is demarcated by a linguistic shift from *tragedy* to *the tragic* (*das Tragische*), signaling a subtle and yet fundamental transformation of the subject matter. The philosophy of the tragic is no longer concerned with the descriptive and prescriptive rules for making specific literary texts such as tragedies or elegies, for example, but addresses instead the essential qualities that constitute genre differences as the lyrical, the epic, the tragic. In his poetological writings, Hölderlin also tends to use the adjectival forms, differentiating the tone qualities and laws of the lyrical, the epic, and the tragic poem.[14]

What Szondi describes as a move from the poetological description and prescription of tragedy to a philosophy of the tragic touches on a historical constellation in which aesthetics become the scene of a particular encounter of philosophy and art. It is itself a scene of the tragic in the sense that the tragic assumes in this encounter: a deadly struggle of two opposites that are yet one. The philosophy of the tragic is also the tragic of philosophy: not only art is irreducibly "contaminated" by philosophy to the point where philosophy declares the end of art, but, no less, philosophy is "contaminated" by art to the point that the "oldest system program of German Idealism" declares aesthetics the legitimating ground and goal of philosophy. It is this latter contamination that marks every major philosophical text of Modernism as a decentering center.[15] But neither the philosophical declaration of victory nor the aesthetic dialectical reconciliation of the opposites in the beautiful can prevail on the tragic scene. What emerges from the tragic antagonisms and struggles is the structure of an otherness that does not allow for any victories or reconciliations other than local and momentary fragile arrangements. As this other structure emerges, the tragic scene is transformed and opens up to a sad play and a display of sadness in the *Trauerspiel*. The philosophy of the tragic must yield to the poetics of the caesura.[16]

What we called Hölderlin's tragedy project is a working through the scenarios of the tragic constellations to the point of their radical transformation. From the beginning, philosophy imposes itself on the writer as a saving and threatening companion: "I have much leisure now for my own work, and philosophy is once more almost my only occupation," Hölderlin writes on February 24, 1796, to his philosopher friend Immanuel Niethammer (*SA* 6: 202). But only a few lines later, this seemingly happy cohabitation with philosophy appears in a different light: "Philosophy is a tyrant [eine Tyrannin], and I tolerate her coercion more than I subject myself voluntarily [freiwillig] to it" (*SA* 6: 203). Philosophy is an agent of division and separation, less because it produces division then because its force becomes compelling when division and separation have torn the subject apart, when the *one* who writes is no longer one ("wenn ich mit mir uneins wurde").

In this split of the one there is a will for philosophy, although the writer does not subject himself to it through free will (*freiwillig*): "In the philosophical letters, I want [will ich] to find the principle that explains to me [das mir . . . erklärt] the separations in which we think and exist, but that is also capable to make vanish the conflict [Widerstreit], the conflict between subject and object, between our self and the world, and even between reason and revelation . . ." (*SA* 6:203). The split is staged grammatically between I and me ("*ich* mit *mir* unseins—will *ich* das Prinzip finden, das *mir* . . . erklärt"). Yet both times, the *me* split off from the *I* does not appear in the position of the direct accusative object, but in the somewhat displaced indirect position of the dative.[17]

The terms of this separation (subject/object, self/world, reason/revelation) as well as the dialectical goal of making the oppositions vanish, place Hölderlin's poetic project at the center of the emerging philosophical aesthetics and aesthetic philosophy. The grammatical indirection introduces a very slight linguistic displacement into the dialectics of subject and object.

The "separations" (*Trennungen*) that Hölderlin wants to explore in a series of philosophical letters also constitute the ground of the tragic in the form of an original, primal separation of the highest unity. Philosophy as the tyrant of the split "I" and the tragic as the unfolding of a primal separation thus meet and intersect in separation. It seems that Hölderlin, at least initially, envisioned an ideal collaboration in which philosophy would explain the principle of separation while tragedy would work through it on the level of experience, of *Fühlbarkeit*.[18] The separations are already doubled and joined in the double mode of the subject as one who thinks and exists ("Trennungen, in denen wir denken und existieren"). The *and* between thinking and existing marks the desired unity and the irreducible split that separates the subject of enunciation from the enunciated subject and projects that split in the subject's existence.

It might seem then that the "real" split is the existential split that finds its privileged articulation in the dialectic of subject and object as "central theme" and "ground-difference," at least during a certain period of Hölderlin's work.[19] But this articulation has already entered the scene of language when the paradigm of original separation appears in the wordplay of *Urteil* ("judgment"): "*Judgment* is, in the highest and strictest sense, the original separation of object and subject which are most intimately united in the intellectual intuition, that separation through which in the first place object and subject become possible, the arch-separation" ("*Urtheil.* ist im höchsten und strengsten Sinne die ursprüngliche Trennung des in der intellectualen Anschauung innigst vereinigten Objects und Subjects, diejenige Trennung, wodurch erst Object und Subject möglich wird, die Ur=Theilung") (*FA* 17: 156).[20]

The dialectic of subject-object as the dialectic of self-consciousness provides a structural paradigm both for the post-Kantian speculative philosophy and its enactment in the philosophy of the tragic. At the moment when Hölderlin articulates this dialectic in the terms of his philosophical context, he also sets up the terms of its displacement from tragedy to poetology and translation. A scenario of theater and gesture, of allegory as the *Trauerspiel* of *Darstellung* and of play as display emerges.[21]

Hölderlin defines the tragic constellation as the necessary separation of an original unity. The definition still fits well within the dialectical concept of the tragic. Its displacement begins at the moment when the separation is specified as subjection to the law of succession: that everything, in order to be presented, needs to unfold in the temporal and spatial display of one after the other. It is the law that Empedocles does not want to accept.[22] It is the law of *Darstellung.*

Darstellung is the formative law of poetry and art, its *Gestalt.* In the aesthetic vocabulary of the time, the word *Gestalt* has a special emphasis as a concept that indicates organic wholeness and inner formation more than external form and shape. But when Hölderlin translates the Pindaric phrase *epeôn de thesin* ("the setting [positioning] of words") in the third Olympic ode as "der Worte Gestalt," he recalls the derivation of the word *Gestalt* from the verb *stellen* ("to posit, to put something in a place"). By translating Pindar's vertiginous syntactical virtuosity word by word, he performs *Gestalt* as syntax in a translation that obeys in the most radical way the law of the words' position. The result is a poetic style that violently unravels the *Gestalt* of German syntax to present a *Darstellung* of *Darstellung* as succession and as the arrangement of words.

The intimate connection between Hölderlin's tragedy project and this translation of the Romantic *Gestalt* into the mechanics of *Darstellung* are presented in the fragmentary notations of an ode under the title "Gestalt und Geist":

Gestalt und Geist

Alles ist innig

Das scheidet

So birgt der Dichter
Verwegner möch[ ]test von Angesicht zu Angesicht
Die Seele sehn

Du gestest in Flammen unter[23]

The words scattered over approximately a third of the page note almost mechanically the first three movements of the tragic tones. From an original, all too intimate (*innig*) closeness and unity, separation and division result. The antagonistic violence then moves protectively into a subdued, enclosed space of a more sensual phenomenality far removed from the burning spirituality of the original unity. This third move is presented in the notations for the poem as *bergen*: as hiding, veiling, and protecting. It is the moment that the continuation of the ode was supposed to unfold in a further phenomenalization of the title constellation of *Geist und Gestalt* as *Seele* and (we might assume) *Gestalt*. The *Gestalt* of the soul finds its most common phenomenalization in the body. The beautiful body, represented ideally in Greek sculpture, was the Classical-Romantic aesthetic paradigm for the delicately balanced reconciliation of inside and outside, of spirit and materiality, and therefore the most perfect symbol of artistic expression.

Hölderlin's sketch of the poem marks a moment of a wish for the immediacy of the soul and warns of its consequences. The wish wants to erase the *Gestalt* of the body, but inscribes it in the negation itself: *von Angesicht zu Angesicht* ("from face to face"). The wish for an unmediated vision (*Sicht*) and view (*Ansicht*) is phenomenalized in the metaphorical appearance of immediacy as a face (*Angesicht*) that points metonymically at the erased body. But the metonymy has dissolved the wholeness of the body's *Gestalt* in order to inscribe it in the arrangement of parts, the order of words (*epeôn de thesin*) in the process of *Darstellung*.

The poetic sketch sets up a double bind with deadly consequences: either the subject bursts metaphorically into flames in the desire for immediacy (as did Semele, when she demanded to see the lover-god face to face) or is torn apart in the articulation and dismemberment of *Darstellung*. An apparent slip of the pen curiously refigures the (spiritual-metaphorical) death in flames. Hölderlin writes: *Du gestest in Flammen unter*. The editors usually emend the *gestest* as *gehest*. The neologistic slip, if it is a slip at all, writes into the deadly face of the spirit (*Geist*) both an old form of its name (*gêst* in old Saxonian) and its re-emergence in the theatricality of the gesture (*Geste*). The going under in the face to face with spirit and soul is performed as a gesture: *du gestest*.

The play of the pen and the words enacts Hölderlin's poetological theory. Hölderlin's poetology of *Wechsel der Töne*, "alternation of tones," constitutes a general theory of writing and the vicissitudes of *Darstellung*. One of the most elaborate texts on the alternation of tones—in Beißner's edition contained as "Über den Unterschied der Dichtarten" ("On the Difference of Poetic Genres")—begins with a succinct definition of three poetic forms: the lyric, the epic, and the tragic poem.[24] Hölderlin's poetics is built on a musical variation of three "tones" which he identifies as naive, heroic, and ideal. Each genre is constituted by a specific constellation of these tones and their specific permutation according to a "calculable law" (*gesezlicher Kalkul*).[25] The relation of the tones is again defined by three terms, as in the example of the tragic poem: "The tragic poem, heroic in appearance [dem Schein nach], is idealistic in its significance [in seiner Bedeutung]. It is the metaphor of an intellectual intuition [einer intellectuellen Anschauung]" (*FA* 14: 369).

Leaving aside for the moment the curious multiplication of triadic divisions that repeat themselves throughout Hölderlin's poetic and theoretical texts, we will focus first on the three terms that define the relations of the tones: appearance (*Schein*), significance (*Bedeutung*), and metaphor. They define the text in a familiar vocabulary of metaphysics and poetics in which a latent, invisible ground or meaning is phenomenalized in a manifest appearance through a metaphorical transfer or translation. Each text is thus split at every moment into a latent ground and a manifest appearance. Each text, however, is also a temporal unfolding and a succession of alternating tones. The beginning tragic tone constellation of *ideal/Heroic*[26] (i.e., a ground of intellectual intuition and speculative ideal unity manifested in heroic antagonistic oppositions and struggle) must unfold through a series of seven major permutations in which each of the three tones enters into a constellation with the others.[27] On the most general level, then, Hölderlin's poetic text theory accounts for two axes: a metaphorical axis of representational translation and a metonymic axis where each element is followed by another and points at another.

The transfer from significance to appearance or the reading of the significance in the appearance is called *metaphor*. The term of metaphor seems to function here in its literal meaning as a move (a transport) from one to the other.[28] It designates the work of translation itself, not the resulting meaning or figure. By analogy to Freud's "labor of the dream" (*Traumarbeit*), we might call it the labor of the text. In his remarks on *Oedipus*, Hölderlin will use the latinized form of metaphor: *transport*. But now metaphor seems to have been transported itself to the other axis, as it describes the successive torrent of "representations" (*Vorstellungen*) (*FA* 16: 250).

The familiar vocabulary of appearance and significance is shifted to other terms in the continued elaboration of the text. Significance becomes first *Grundstimmung* ("ground mood," "ground tuning") and then *Grundton* ("ground tone"). Appearance becomes *Kunstkarakter* ("art character"). The ground as tone and tuning has a musical quality, while the artistic form or appearance as "character" has a more outlined, defined, and graphic quality. In this qualitative differentiation of textual ground and its phenomenality, Hölderlin develops a poetological theory that will be echoed in later aesthetic and textual theories. Nietzsche's differentiation of the Dionysian and Apollonian as the constitutive forces of art marks the difference in the opposition of musicality and image or sculpture. Analogous categories of differentiation seem to be at work in Kristeva's textual theory in the constellation of the semiotic *chora* as a kind of rhythmical semiotic predisposition and the thetic, positing force that articulates the phenomenal text.

In the aesthetics of the late eighteenth century, the truth of poetry and art generally finds its legitimation and ground in the lived experience, in the continuity of life and art as *Erlebnis und Dichtung*. Hölderlin does not negate the urgency of this question, but he shifts its terms, and most of all the term of experience to a scene which Benjamin, not without recourse to Hölderlin, calls *Erfahrung*. Hölderlin's tragic constellation as the constellation of *Darstellung* is the unfolding of *Erfahrung und Dichtung*.

The privileged position of the tragic constellation in Hölderlin's poetics is based in "intellectual intuition" (*intellectuale Anschauung*) as the highest (or deepest) mode of being and cognition. Indeed it is the speculative unity of being and cognition, the fullness of presence. "All works of this kind [i.e., the tragic]," Hölderlin writes, "must have an intellectual intuition as their ground" (*FA* 14: 370). Intellectual intuition constitutes the ground as a "unity with everything that lives" (*FA* 14: 370). It provides the deepest foundation of speech and expression, but it cannot be expressed as such. It must exit from itself ("weil es aus sich herausgehen müsse"); but even this necessity is only a necessity of speech, because it can be pronounced only in this exit and enunciation ("und am leichtesten sich ausspricht, daß man sagt"). The expression (*Ausdruck*) constitutes the ground as its effect in a kind of after-pressure (*Nachdruck*).[29] This is the point of rupture between Hölderlin's poetics and most of contemporary aesthetics. There is no continuity between the legitimating ground and its expression, only a complex economy of exchange.

The exchange affects the most valorized terms: life, intimate interiority (*Innigkeit*),[30] and fullness. It takes place as an exchange between totality (*das Ganze*) and particularity (*Theile*). The process of this exchange is presented as a mutual gain: totality gains content (*Inhalt*), the parts gain intimate interiority; totality gains life, the parts gain liveliness; totality gains feeling, the parts gain fullness.

If the gains are complements of a lack on each side, we must think of totality (and thus of the deepest ground) in its pure state as empty and without life (because it gains content and life only through dissemination in the parts). When Hölderlin continues to formulate the "eternal law" of this economy, however, he refers to totality as *gehaltreich*, which Pfau translates sensibly as "substantial." But the translation erases the close relationship between *Inhalt* and *Gehalt*, both of which can be translated as "content." Totality, then, is both empty and rich in content. The difference between *Inhalt* and *Gehalt* corresponds to the difference between *innen* (interior) and *innig* (intimate). As the *innig* transfers the spatiality of interiority to its condensed limit of nonextension and metaphorical transport, *Gehalt* transfers the spatiality of content to a point of substantial compactness where it is ready for the metaphorical transport into pure significance. *Gehalt* (as a literary critical term) is significant content, no longer the literary or artistic material, but its transformation under the mark of the signifier.[31]

Hölderlin claims that the intellectual intuition can be recognized by the spirit, but that it cannot be felt by the more limited "inner senses" or "disposition" (*Gemüth*). It is, then, the task of the poetic operation to produce *Fühlbarkeit* ("sensibility," "tangibility") for the intangible intellectual intuition according to a "poetic logic" (*FA* 16: 411) that treats all faculties of "the perceptive system" (*Empfindungssystem*) that we call "human being" (*der ganze Mensch*) (*FA* 16: 250). The human being as a *system* of sensations is subject to a calculable law, and as a system of *sensations* and *perceptions* is subject to the organization of the perceptive organs that constitute the body. The body is the phenomenal appearance and materialization of the system: it is the medium of the tangible that can be touched, hit, wounded, and killed. It represents the space of vulnerability. It can be touched and affected—not least by words. This affectability by words removes the body from any simple graspable referentiality. As the tragic body it is a special body: an intersection of language and body.

Pfau's translation of *fühlbar* as "tangible" points at the tactile connotation of feeling.[32] It is the emphatic *noli me tangere* in the gesture of many Hölderlinian texts that testifies to the extreme vulnerability of a subject deciphering the painful inscriptions of the traces of experience. To be touched or not to be touched is a question of poetic authenticity for Hölderlin, but it is also a question of vulnerability. The life of poetry demands contact with the world, but the contact with the world threatens as a deadly impact, "because I am more easily destroyed than others" (*SA* 6: 290). The "warm inner life" ("das warme Leben in mir"), the "truest life" is threatened by the "icecold (hi)story of the day" which it yet needs for the liveliness of its *Darstellung* (*SA* 6: 290). Put to the test (and to the text), Hölderlin's poetics of the tragic (where the whole gains life, and the parts gain liveliness) encounter the bar of the *Fühl*bar*keit* as a rupture in the transport from life to "life."

Hölderlin's emphasis on "life" as the authenticating ground and animating content of poetry could be part of the aesthetics of expression that replaced poetics and rhetoric in the course of the eighteenth century. But while the rhetoric of aesthetics claims art as an expression of life, as a continuous flow of an overflow, Hölderlin's idiosyncratic terror of life in its (hi)storial daily presentation marks a caesura that constitutes the constellation of a poetics of modernity as a poetics of shock. Benjamin later deciphered this poetics of shock via Freud's "Beyond the Pleasure Principle" in Baudelaire.

Hölderlin's definition of *der ganze Mensch*, the whole human being, as an *Empfindungssystem* foreshadows in more than one way Freud's psychic apparatus, most of all in the shifting accent of its functions. While Hölderlin's system of sensations with its dynamics of representation (*Vorstellung*), sensation (*Empfindung*), and reasoning (*Räsonnement*) is conceived, like Freud's apparatus, as a perceptional machine and a graphematic plate, the emphasis shifts increasingly to its protective and screening function against the violent impact of what Hölderlin calls "the element" and Freud *Reiz* ("stimulus," but also "charm").

Instead of the continuous flow of expression there is contact as well as impact. The violence of this moment is proportional to the sensitivity of the registering organ. Freud describes consciousness, generally considered the organ of perception par excellence, as an organ of defense, a shield against the impact of perceptible stimuli. Hölderlin's letters are permeated by gestures of defense. He defends the "virginity of the spirit" as a necessary phase of protection (*SA* 6: 249) and he invokes his poetics as a system of protection that subordinates the potential damaging impacts in the system of "alternating tones" (*SA* 6: 290), but the system of integration and digestion itself has its limits: "Formerly I could jubilate about a new truth, a better view of that which is above us and around us; now I fear I could end up like old Tantalus who received more from the gods than he could digest" (*SA* 6: 427). The autobiographical notations of the letters are themselves already poetological readings of the autographical marks that Hölderlin's poetics and poetry attempt to decipher. The marks of impact provide the alphabet of a writing that, in its flight from the impact, is propelled by it. *Fühlbarkeit* is also *Berührbarkeit*. The degree of *Berührbarkeit* is one of the marks of difference between antiquity and modernity. Certain words are less dangerous to us than to the ancients because "we are not *affected* by them," Hölderlin writes in a note to his *Empedocles* (*FA* 13: 702 [emphasis added]).

"Life in poetry now occupies most my thoughts and senses. I feel deeply how far away I still am from the goal of hitting it [es zu treffen]" (*SA* 6: 289). *Treffen*—to hit something or the mark—is the word that marks for Hölderlin the lucky and dangerous moment of a constitutive impact. It is a lucky moment because it is the moment of the right constellation; it is dangerous because *treffen* is always also *getroffen sein*: to be hit, to have received the impact. The poet, who wants to hit the mark, must be hit by Apollo.[33] One of the Homeric epithets for Apollo is *hekatèbolos* (the one who hits everything from afar, the one who does not miss).[34] Pindar refers to the words of his poetry as arrows that will hit their aim.[35] The opposite term in Greek is *hamartia*: the fault that misses the mark and characterizes the tragic hero. For Hölderlin, it seems the *Treffen*, not the missing, that accounts for the tragic constellation, to the degree that the tragic constellation is the constellation of genuine *Darstellung* itself. *Treffen* as the moment of an impact erases the difference between active and passive, between hitting and being hit. If *treffen* is the precondition for life in poetry, the impact also threatens the life that sustains the poetic word: in order to protect his "truest life," the poet shrinks from the "vulgarity of real life" (*SA* 6: 289). But without it there is no poetic word. Its nobility depends on the vulgar. It must be colored by the "color of fate" (*SA* 6: 290) as soon as it is expressed and presented. Hölderlin's much laughed-at literal translation of the Greek *kalchainein* in *Antigone* (line 20) as "red coloring"—"Du scheinst ein rotes Wort zu färben"—gives the testimony of truth to Antigone's word.

The word is true only insofar as it is colored and contaminated in the materialization of its appearance. Hölderlin translates the Greek *dèlois* as *Du scheinst* ("you seem to color a red word"); the Greek can also mean "to reveal, to manifest." But the revelation, like the procreation by the father, is always uncertain. Only in the figural move from *zeugen* as procreation to *zeugen* as giving testimony, witnessing, the word can have a claim as living word. The transport from *zeugen* to *"zeugen"* is effected by a rupture, by the caesura that Hölderlin locates in the appearance of Tiresias in both *Oedipus* and *Antigone*. In *Oedipus*, Tiresias enters as the one who refuses to speak. Oedipus therefore calls him pitiless, or, literally, "hard, unsoftened" (*ategktos*), which Hölderlin translates as "farblos" ("colorless," line 340). He is outside of the coloring of fate and *Darstellung*; he is the figure of the inexpressive that cuts, as the caesura, into the stream of expression and brings it to its truth. The "living" word is the interrupted word as witness rather than a "creative" procreation. Hölderlin's *Oedipus* translation marks this interruption most radically at a point when the Greek *ektribèsetai* ("is annihilated") is given as "gezeugt seyn" ("is procreated") in line 435.

The figural transport is the transport of *Darstellung* itself. It involves a sacrifice: "Thus we must make a sacrifice from time to time to the godhead, that is between me and you; the light, the pure sacrifice that consists in speaking to each other of the godhead," Hölderlin writes to his brother on November 28, 1798 (*SA* 6: 293). To speak of the divine ground that constitutes the true relation is a necessary sacrifice. Hölderlin seems to take literally the German *Opfer*, derived form the Latin *offere*, "to offer," "to present." *Opfer* is presentation (*Darbietung, Darstellung* in German) and presentation is *Opfer*.[36]

Sacrifice is the tragic figure, the third term in the alternative between the (poetic) word and the (political) action. While the ode "An Eduard," written at the time of the *Empedocles* project, is set in the constellation of an alternative between poetry and action, Hölderlin's reflections in the *Grund zum Empedokles* posits the tragic figure as a figure of sacrifice. Neither poetic word nor immediate action can reconcile the extreme oppositions of the time. This reconciliation can occur only in the "sacrifice where the whole human being [is exposed], where that wherein the fate of his time seems to dissolve becomes real and visible . . ." (*FA* 13: 873). Since the "whole human being" is defined by Hölderlin as an *Empfindungssystem* of perception and representation, the tragic sacrifice is nothing but the exposure and presentation of (re)presentation.

In the sacrificial offering and presentation, something invisible becomes visible, but something visible also must disappear. The colorless participates in the sacrificial coloring as the necessity of an intervention of the inexpressive in the expression. The inexpressive appears as an act of will only in the "divine seer" (*theion mantin*), who thinks both the said and the unsaid ("Gesagtes, ungesagtes," in Hölderlin's translation [line 304]), that counters the blind will of Oedipus who wants to say everything and who wants everything to be said: "sagen must dus mir!" ("you must say it to me!") (line 347); "ich werde nichts, wie auch der Zorn seyn mag, / Weglassen" ("I will leave out nothing without regard for anger") (lines 349–50). Nothing is said where everything is said, is Hölderlin's experience, already indicated in the letter to his brother of November 1798: "I am not saying that our letters are something extraordinary in terms of their thought and wit and manifold concepts and matters; but there is something in them that one may call the sign of all living utterances, namely that they say more than is apparent, because a heart stirs in them that in principle can never in life say all that it would like to say" (*SA* 6: 293).

That words cannot say what the heart wants to say seems to be merely a Romantic topos. A worn figure of speech appears as *the sign* of the living word. The sign reveals a curious economy of surplus and lack. The living utterance says more than is apparent. But if this surplus does not appear, how can we read it? What kind of sign is this sign that does not appear? It seems precisely the absence of something, the constitutive, principal impossibility to say all "at all in life" ("überhaupt im Leben") that is the sign of the living word ("lebendiges Wort," "lebendigen Äußerungen"). The constitutive sign is the topos as atopos: it is a sign, not a presence or a thing. it is the most worn-out figure of speech, the worn coin that is marked and stamped by the absence of something. It is the moment of incalculability in the calculable law of presentation: "der lebendige Sinn, der nicht berechnet werden kann" (*FA* 16: 250). It rests on an irreducible silence.

In the sacrificial offering and presentation of the tragic, something invisible becomes visible, but something visible also must disappear. The pure ground must be contaminated in order to enter into *Darstellung*. It must assume the colors of fate and historical circumstances. The phenomenalization of the ground is supposed to be more than a fleeting appearance, however, the figure of its appearance must disappear. This is the tragic paradox of Hölderlin's *Empedocles* and the *Grund zum Empedokles*. The more Hölderlin works through the alternations of the visible and invisible, the more a certain stubborn resistance seems to arise from the phenomenalized elements. The transport of the tragic bodies turns into a major problem. The question of what is to be done with the corpses is not only the tragic motif of *Antigone*, but it is also the motif of Hölderlin's letter to Böhlendorff in which he differentiates modernity from antiquity according to the way in which they deal with the corpse.[37]

Empedocles hardly seems to have a body. The figure of the body seems to be consumed in the embodiment of the deepest tragic conflict of unity and separation, dissolved and concentrated at the same time into a pure sensibility where the slightest impact is felt as an earthshaking tremor. This kind of body is untouchable: "Mich ängstiget die Gegenwart des Reinen / Des Unberührbaren" ("The presence of the pure one, of the untouchable one frightens me"), says Panthea (*FA* 13: 757). Before Empedocles is visible on the stage, he is presented as an effect in speech, first by the two young women Panthea and Rhea (later renamed Delia), then by his opponents Critias and Hermocrates.[38] From the first lines, this presence in the speech of others is strongly emphasized: "Sie sprachen damals viel von ihm, und immer / ist sein Nahme mir geblieben" ("They spoke much of him then, and always his name remained with me") (lines 11–12). At one point, Panthea first introduces a speech about Empedocles with "I think" which Hölderlin then corrects into "One says" (*FA* 12: 45–46). This presence in speech is also questioned. It is in competition with the visual (re)presentation. Panthea interrupts her own speech: "Doch / Was sagts? Du must ihn selber sehn!" ("But what does it say? You must see him yourself [or: himself]!") (lines 21–22).[39] Rhea once saw Empedocles as an athlete in the Olympic games (lines 8–10). In a brief glimpse, the tragic hero appears in the figure of the athletic body. As if this were too much embodiment, Panthea immediately invokes another visibility: "You must see him now! now!" (line 13). Her emphatic insistence on the here and now, accompanied by strong deictic gestures—"This is his garden" (line 1); "There he stood" (line 2)—invokes the visibility of a presence that is still absent.

Seeing seems to function as the ground for the authentic word: only through seeing does the word receive its testimonial power. Yet the following dialogue between Critias and Hermocrates turns this hierarchy around. Empedocles has withdrawn, and Critias, the commonsensical ruler, hopes that his invisibility will weaken his power over the people. But Hermocrates, the priest, knows a little more about the power of the symbolic that resides in the word. Nothing could be worse than Empedocles' invisibility; they must see him for the power to weaken (lines 178–79 and 240–41); only when they see him with their own eyes (260–61), can the priest hope to break the spell of his power. His disappearance, (or invisibility) gives the ultimate power to the word. Disappearance becomes the marking of the word. In his translation of *Oedipus,* Hölderlin boldly translates the anaphorically repeated Greek *phthinousa* ("dwindling," "fading") as a mark: "Sie merkt den Tod in Bechern der fruchtbarn Erd' " ("She notices [marks] death in the cups of the fruitful Earth").[40]

The (re)marking in dwindling and dissolution makes tragedy the scene of a rejuvenation and of a revolution of language. It is the priest, the representative of positive law and order, who most clearly sees the dissolving, Dionysian function of Empedocles: the man is drunk (line 183), and the people are drunk through him (line 185); he recognizes no law, no necessity, and no judge (lines 186–87); the cultural customs turn into an incomprehensive roaring and rushing ("Von unverständlichem Gebrause") (lines 187–88). The word *rauschen* ("rushing," "rustling") is not only one of the key words of Empedocles—"da rauscht' es anders" (line 377); "und rauschten sie mir / All deine geflügelten Melodien" (lines 398–99)—but also of Hölderlin's poetry. It signals a moment of disarticulation and inebriation (*Rausch*). Its disarticulation affects the three terms of cultural stability: "Gesez und Kunst und heilge Sage" ("posited law and art and holy saying") (line 227). In his farewell speech to the citizens, Empedocles formulates this cultural revolution as a total revolution.

The paradox of tragedy is the representation of the unrepresentable. More specifically, it is the representation of the moment of its own constitution as the theatrical visibility (*Vorstellung* in German is also a theatrical presentation) of the invisible. As the figure of the caesura that is also the figure of the *appearance* of *Vorstellung* itself (*FA* 16: 250), Tiresias appears in Hölderlin's translation in the paradox of a colorless appearance: "Erscheinst so farblos du" (*FA* 16: 115).

The tragic transport toward the constitutive moment of language, (re)presentation, and experience is common to antiquity and modernity. It produces what Hölderlin calls "the living relation" (*SA* 6: 426) and the "living sense that cannot be calculated" (*FA* 16: 250). The direction of the transport, however, is different according to Hölderlin's first letter to Böhlendorff. The apparent oppositional symmetry between antiquity and modernity closely resembles Hegel's differentiation in the *Phenomenology of the Spirit*. The formation of "natural consciousness," the appropriation of nature in language and in concepts, is, according to Hegel, the task of antiquity;[41] the formation of the "fire from heaven" into the "clarity of representation" (*Klarheit der Darstellung*), of "sacred pathos" into "occidental Junonian sobriety" is the achievement of Greek art, especially of the "Homeric presence of spirit" (*Geistesgegenwart*) and "gift of representation" (*Darstellungsgabe*) according to Hölderlin. The situation of modernity is different: we are already born into a world of abstract forms and concepts according to Hegel; *Darstellung* and sobriety are our natural property according to Hölderlin. Our task is to work back toward the moment of the experiential impact, toward the fire from heaven, the sacred pathos, to liquefy the petrified forms and concepts.[42] To give form to "things unknown" and "to airy nothing a local habitation and a name" is the task and achievement of antiquity, but no longer of the modern poet's pen. When habitation and name have become habit, the poetic task is de-habitation, defamiliarization. The second stasimon of *Antigone* is the moment when this dehabitation becomes word in Hölderlin's translation of *ta deina* as "Ungeheuer" ("monstrous," "uncanny"), a word etymologically related to *huir/huis* and thus literally the negation of habitation.

The skill (*Geschik*) required for such de-habitation imposes a paradoxical task on modern poetics. If poetics is the calculable law of *Darstellung*, the skill to give form to the shapeless, the position of modern fate (*Geschik*) requires the undoing of *Darstellung*. Poetics must work against itself. *Darstellung* becomes *Entstellung* (displacement, deformation). There is, then, a radical asymmetry between antiquity and modernity. While the poetics of the Greeks achieve their goal in the "Homeric presence of spirit" as poetic presentation (*Darstellung*) and thus for Hölderlin as *art*, modern poetics is compelled to undo the essence of art, which is *Darstellung*. But it cannot undo it without undoing itself as art and poetics, without giving up the presence of the spirit which is precisely not the presence of the pure spirit, but the achievement of successful *Darstellung.* Thus modern poetics cannot in a simple, symmetrical reversal of movement regress to the poetic ground, to the living sense. It can only loosen up or, more violently, shake and rattle the firm forms of the given, as Hölderlin did, when he subjected the German syntax to the violence of Pindar's Greek syntax whose *epeôn thesin*—the setting of the words—Hölderlin turns into a *Darstellung* as the radical *Entstellung* of German syntax. Instead of the metaphorical transformation of the original fire in the embodiment of *Darstellung,* modern poetics can only metonymically point at the violent origin. Thus, whereas the Greek body is consumed by the fire it embodies, the modern body can only be put aside as an embarrassing corpse—"very quietly packed up in some container," Hölderlin writes to Böhlendorff in December 1801 (*SA* 6: 426).

When Hölderlin writes again to Böhlendorff, almost a year later, after his return from Bordeaux, the diction has changed in a striking manner that cannot easily be described. The relation of antiquity and modernity still seems to be a central question, but compared to the earlier letter to Böhlendorff, with its intricate philosophic—historical argument, Hölderlin's diction now seems more descriptive and narrative. In a sequence of rather short paragraphs, the letter describes certain aspects of the southern French landscape around Bordeaux and narrates its effects on the letter writer and his perception of the ancients and of his present environment. More precisely, it is a descriptive account of a certain gaze, of what "[I] have seen." The bracketed *I* points at the enigma of this letter, which begins with the word *Ich* and a negation and then drops the pronoun: "Ich habe Dir lange nicht geschrieben, bin indeß in Frankreich gewesen und habe die traurige einsame Erde gesehn" ("I have not written you for a long time, have been meanwhile in France and have seen the sad, solitary Earth"). Grammatically, the dropping of the *I* is not unusual, but in the context of this letter, its manifest absence seems overdetermined. *Ich* appears on the scene as one who has *not* written, has been elsewhere, has seen. In seeing, however, the subject of seeing has shifted to the position of the object that has been captured and seized by what it has seen: "The tremendous element, the fire of the sky and the stillness of the people . . . has continually seized me." Only after this shift, an *Ich* re-emerges as the grammatical subject of speech: "And as one says of heroes [und wie man Helden nachspricht], I can well say that Apollo has struck me." The "I" that emerges from the abyss of its absence in the *indeß* ("meanwhile") is presented as the product of a shock. It speaks from the position of a radical aftermath; it speaks literally "after," as one speaks of heroes: *wie man Helden* nach*spricht.* In the belatedness of this speech of a subject that is the object of an Apollonian blow, tragedy's privileged ground, the *intellectuale Anschauung* ("the intellectual intuition or vision") seems to come to a realization that is curiously displaced.

Intellectual vision is based on the ideal of a fundamental unity and suspension of differences. Tragedy is the necessity of its *Darstellung* in violent differentiations. We have seen how this process is presented in a vocabulary of purity and contamination: the pure ground that must assume the colorations and contaminations of fate. But if it seems that purity is on the side of unity and contamination in separation, the later description of ultimate unity as "the monstrosity of the way that god and man are copulating" raises some doubts about the clear distribution of the difference between pure and impure.

Perhaps we have reached here a differentiation that is the pivot of all cultural differentiations. René Girard asserts as much: "The sacrificial distinction, the distinction between the pure and the impure, cannot be obliterated without obliterating all other differences."[43] Nietzsche locates in the opposition of pure/impure not only the genealogical origin of the moralizing transformation of good/bad into good/evil, but also the interiorizing linguistic transport from literal to metaphoric meaning.[44] The privileged scene of the differentiation of purity and impurity is the human body, which is also the scene of the accompanying differentiation of exteriority and interiority, and thus the figure of figuration itself.

The landscape and figures appear in an enigmatic linguistic twilight of physical literality and eerie spirituality in a temporal constellation of postrevolutionary presence and the physiognomic traces of antiquity. The "anxiety of patriotic doubt and hunger" marks the first set of human figures—"shepherds of southern France and scattered beauties, men and women"—encountered in this landscape. The historical presence is immersed in an element that the first letter to Böhlendorff attributes to the natural endowment of the Greeks, "fire from the sky" or "heaven" (*SA* 6: 426), now reappearing as "the tremendous element, the fire of the sky" or "heaven" (*SA* 6: 426). The slight shift from "fire *from*" to "fire *of*" indicates a shift in the relationship of this fire to the world: "fire from the sky" suggests that it is coming from there to the people; "fire of" simply attributes its origin and belonging to another sphere. Indeed, it stands in marked contrast to "the stillness of the people" who are set off in protective limitation, enclosure, and contentedness. The only one who seems to be struck by this fire is the unprotected figure of the writer.

Struck by the fire of the sky, the writer begins to read the physiognomic traces left by that other fire from the sky. It is not certain that this reading can still be read. A heroic landscape appears, indistinct from its figures: "In the regions which border the Vendée, the wild and martial [character] interested me, the purely masculine where the life-light becomes immediate in eyes and limbs and which feels itself in the feeling of death as in a virtuosity and fulfills its thirst to know." The blurring of the landscape and its figures in the qualities of the wild, martial, and masculine, manifested in eyes and limbs, seems to point at a scene of ancient tragedy, to the monstrous coupling of god and man, of life-light manifesting itself unmediated in heroic discord and deadly virtuosity.

The letter seems to differentiate between the southern French region in general ("the shepherds of southern France . . . the athletic [character] of the southern people") and the regions bordering on the Vendée in particular. The hunger of the men and women of southern France and their confined contentedness stands in marked contrast to the wild and purely masculine fulfillment of the thirst to know in a deadly virtuosity through which the light of life has taken possession of eyes and limbs. The immediacy of this life-manifestation resists *Darstellung*. It is less a Greek tragedy than its failure, or its displacement from the scene of *Darstellung* to a deadly impact of life. Yet the regions bordering on the Vendée are part of southern France, and the purely masculine ferocity points at the "athletic [character] of the southern people, in the ruins of the ancient spirit." In the present athletic bodies that move like ghosts through the ruins of the ancient spirit, the writer reconstructs the athletic bodies of the ancients, who, in contrast to the wild masculinity of the present, had the wisdom and the skill to protect their genius in and through their bodies against the violence of the elements.

The act of reading in this letter, its clairvoyant penetration of landscape and bodies, seems to emerge from a visionary transport that is the radical opposite of the message deciphered in the athletic bodies.[45] The writer of the letter seems to come close to a fulfillment of his thirst to know. What he learns is the disquieting knowledge of a heroic Greek body shaped by an economy of protection against the violence of an excessive transport. And yet it is only in the visionary transport of a writer struck by Apollo that the Greek tragedies become readable as sports events.

The rest of the letter is marked by protective gestures against the violence of the transport of its reading. At the end it is signed only by the initial of the name: "your H." It is a mere breath. According to Hölderlin's notes, a tragedy like *Antigone* can be compared to "a competition of runners where the first one to catch his breath heavily and to jostle the opponent has lost."

Color Read

HÖLDERLIN AND TRANSLATION

Aris Fioretos

. . . (O, transport!) . . .[1]

Every translation, however felicitous or free, must rely on some form —and thus suggest some notion—of transport. Regardless of the assumed proximity or presumed distance to the original, in order to be an example of the textual practice it claims to instantiate, translation has to entail not only the transmission of semantic goods from one order of signification to another, but also, whether in a lax or strict manner, the conveyance of figurative patterns of organization. As the only guarantee that such a translation—rather than mere restatement or citation—does indeed take place, transport necessitates at once the extension and termination of the original. Surviving only as the phantom of itself, the original meets its end through the distending severance occurring in and as the transition from one language to another. Once rendered in a foreign tongue, the original cannot be translated back into its proper language, at least not without demonstrating an inherent instability which would make it not only different from its former self, but also, as a consequence, neither definitive nor original. Yet on the other hand, it cannot be translated further *as* translation, lest it lose entirely any affinity with primariness. Transport, then, both canonizes the original and renders dubious its inherent claim to prevalence.[2]

In order that the termination of the structures of articulation singular to the original language not simply become a dead end—or, in fact, be the end—theoretical assumptions concerning transport must presuppose that language and thought may be dissociated, and hence rely on the priority of the signified over the signifier. Accordingly, the content of thought is understood to remain intact in the interlinguistic transferal of semantic property, implying that translation only clothes in new words the convictions and beliefs contained in the original. Whether fitting or unfettered, figuration remains a matter merely accidental, a tinge coloring a body of thought which has been removed successfully from one linguistic setting and introduced into another. Presupposing the presence as well as the persistence of semantic transparency—and, by implication, the possibility of articulating the same content in different verbal guises—repeatability, or the-same-put-differently, provides the basis for every version of interlinguistic transport.

Although the figurative characteristics singular to the target language were by no means ignored in early modern attempts to make accessible the texts of antiquity, it was not until the late eighteenth century that these became a matter of consistent critical reflection. When Goethe, himself a noted translator, maintained that, "in translating, one must approach the untranslatable," since only thus is it possible to become "aware of the foreign nation and foreign tongue,"[3] he named what was soon recognized, and consequently defined, as the crux implicit in any attempt to conceptualize the relation between antiquity and modernity: untranslatability. What was truly foreign in the texts of the former was, strictly speaking, impossible to render in the language of the latter, and hence could never become an object of transport. The perimeters of modernity were defined by what would not be carried across them. To the extent, therefore, that a translation was desired to be modern, a manner had to be conceived of in which what remained outside the precincts of the linguistically familiar might nonetheless be accounted for, since "the untranslatable" did not only mark the borders separating modernity from antiquity, but also determined what could be said about the difference as such.

Discussing the varieties of translation in his notes "to a better understanding" of the *West-Eastern Divan*, Goethe suggested three distinct means of coming to terms with the particular foreignness of antiquity. The first method, also termed "prosaic" translation, "makes us acquainted with the foreign according to our own sense." Such translation effaces alien idiosyncracies and permits us to perceive what is in reality foreign as something inherently familiar. Luther's Bible translation providing the strongest example of this particular practice of translation, its virtue resides in the ability to expunge both cross-cultural and interlinguistic difference, stressing the ground common to all cultures and languages—or at the very least those of Greece and Germany—as implicitly epitomized in the target language.[4] The second form of translation, termed "parodistic," occurs "when one attempts to transpose oneself into the conditions of the foreign land, but is in fact concerned only with appropriating foreign sense and representing it with one's own." While historically, Goethe admits, such translatory practice may have elicited exciting results, it is nonetheless premised on the questionable assumption that whatever is foreign can and must be replaced with a "surrogate." Rather than transposing the foreign into the familiar, the familiar is transposed into the foreign, exploiting it on terms nonetheless entirely its own. Notoriously, this is the case with "the French" (Wieland's translations provides a German counterpart), whose underlying premise is the necessity of adaption.[5]

Whereas, in the case of prosaic translation, transport serves to underscore cross-cultural mutuality, in the case of parodistic rendition, it merely aids an ideologically misguided attempt at appropriation. The third kind of translation proposed by Goethe, however, exemplified only in rare, but not random, cases (all German), amounts to "the highest and ultimate" form of relating to foreignness. Here, the purpose is to "make the translation identical with the original," but "in a manner such that the one does not pertain instead of the other, but rather in its place."[6] If prosaic translation underscores a correlation between first and target languages based on a mutually espoused content, and parodistic translation considers everything expressed in the first language to be replaceable with the surrogate figurations of the target language—both thus coming into being at the expense, or *instead*, of the original—the third form of translation, never named by Goethe, identifies the two without permitting the latter to cancel out either the significance or the validity of the former. A phantom proxy, it does not replace but rather substitutes. The particular danger involved in this method, not present in the two earlier ones, is a tendency toward literalization which threatens the translation with irreconcilable alienation. "A translation which aims to identify itself with the original is," Goethe remarks, "ultimately inclined to approach an interlinear version." Thus, "the translator who follows his original closely" is forced "more or less" to give up "the originality of his nation, whereby something third emerges for which the taste of the masses has yet to arise." Using rhetorical and metric means peculiar to the first language, and not refraining from chaste literalizations of its idiomatic expressions, this ultimate translatory practice, acquainted with foreignness like no other, makes of Tasso, Shakespeare, and Calderon "germanized foreigners" or *eingedeutschte Fremde*.[7]

Regardless of how radical Goethe's ultimate form of translation may be, and how much the masses apostrophized may still have been in need of education, the classicist doctrine of translation underpinning this—as well as most other—attempts to render the texts of antiquity into German during the years before and after 1800 differed nonetheless merely in degree and distinction from those of the scorned French (exemplified, for example, by Racine and Voltaire). While the principle of beautiful infidelity of the latter was replaced by the former's insistence on felicitous accuracy, hence substituting congenital inventiveness with decorous—if literalistic, or even interlinear—loyalty, for all extent and purposes the interrelationship between first language and target language remained identical. No mediating amelioration was allowed, nor any attempt at a recontextualizing adaption that would cloak the idiomatisms and patterns of thought peculiar to antiquity in those of modernity,[8] yet the languages of both original and translation continued to be considered inter- as well as correlational.

Invoking, in the preface to his 1816 rendition of *Agamemnon*, the topological distinction between the inherent and extraneous suggested by Goethe, and arguing that a translation should appear "foreign" without creating the sensation of "foreignness," Wilhelm von Humboldt emphasized a similarily conceived cultural continuity and epistemic persistence, attributing its only potential ramification to the alienating abilities of temporal distance:

> Fidelity must pertain to the true character of the original and not . . . to its accidentiality, in the same way that all good translations generally are borne by a simple and unpretentious love for the original. . . . Necessarily related to such a view is, of course, the fact that the translation must carry a certain color of foreignness, but the limit at which this turns into an undeniable error is very easy to see here. As long as not foreignness, but that which is foreign is recognized, translation has accomplished its function; where foreignness itself appears, however, and perhaps even obscures what remains foreign, the translator betrays the fact that he does not master his original.[9]

Allowing for the presence of what is *fremd* though not of *Fremdheit* itself, hence admitting attribution but not essence, Humboldt's distinction between translatory fidelity and betrayal remained firmly articulated within the epistemic boundaries set by classicism. Relying on a view of "the true character of the original" which permitted the conscientious translator to both recognize and respect what was foreign in the Greek, while at the same time exercising enough of his loving erudition to transform its foreignness into contemporary relevance and parity, Humboldt's concept of turning the alien into the familiar remained based on the belief that transport, always occurring in time and as temporal distention, was possible as well as secure. As long as the particular repetition caused by translation appeared tinged by "a certain color of foreignness," hence amounting to attributive addition or figurative vivification, it paid due respect to the unalterable truth of the original. Impregnating the text entirely, however, thus mutating it into bare coloring, translation would become gratuitously rhetorical (that is, French), and thus instance an unwarranted *Fremdheit,* rather than being simply and fittingly *fremd.* The original had to be clothed in new words, yet an entirely new verbal costume would slight its linguistic customs and pay no heed to the figure it cut. The text would be entirely transported, as it were, becoming exclusively metaphorical. Accordingly, it would no longer exhibit either relation or relevance. Stripping and abandoning the text of the original in this fashion, translation would in fact do away with it. Only the presence of unquestionable familiarity with the contemporary pertinence of the classics—and consequently, an acknowledgment not only of modernity's indebtedness to antiquity, but also of a temporal distance between the two bridgeable by propitious transport—would protect the original from becoming "colorless and frigid and dead."[10]

By making translation a matter of showing color, however, the topological demarcation traced by Humboldt's distinction between "foreign" and "foreignness" introduced a difference no longer easily mastered by the classicist theory of linguistic transfer which was espoused. Although claiming that to draw a "true line of demarcation" was never difficult, Humboldt nonetheless perceived it as a task to be assigned only to the "unimpressed reader."[11] Not only a matter of philological exegesis or cultural hermeneutics, the particular coloring which occurred in and as the transport implicit in all translation required the critical capacity of reading. Respecting the foreignness of the original, the *uneingenommener Leser* imagined by Humboldt would be prone neither to disregard it by demanding that translation turn the original into something thoroughly familiar, nor to become so carried away as to lose her sense of what was proper. Literally a question of "not being taken in," the suggested reading is marked by resistance—above all else to appropriation. While aware of her own difference from the original, Humboldt's imagined reader nonetheless preserves a position from which she will not be taken in by the adept ministrations of translation. Able to distinguish between "foreignness" and being "foreign" in what remains somehow familiar, she perceives, in reading, that "certain color" which gives to translation a character no longer possible to appropriate in terms purely its own. In contrast to interpretation, which would focus on matters hermeneutically conceivable, and thus primarily on questions of semantics raised in and by translation, such readerly activity would attend to a dimension of the text which would be brought out only in and through the transport between languages, and moreover could not be reduced to a matter of signification. Pertaining neither strictly to the original nor exclusively to the translation, but to their relationship to each other as texts, such "color of foreignness" belongs neither to any particular language, nor to the interrelationship between particular languages, but rather, is proper to language as that which remains foreign to it.

The young Friedrich Nietzsche was one reader of such color proper only to linguistic difference. In October 1861, he wrote a letter to a fictitious friend in which he discoursed on why the imaginary recipient had to read his favorite author, the poet and translator Friedrich Hölderlin. Criticizing the addressee's overly facile and undifferentiated refutation of Hölderlin's allegedly "obscure discourse; his thoughts articulated at times as if by someone mentally insane; violent outbursts against Germany; the deification of the heathen world, in turn naturalistic, pantheistic, and polytheistic . . . an utter chaos," the letter went on to address the point of divergence which most radically accounted for the difference between Hölderlin's conception of modernity's relationship to the classical heritage and that of his literary or philosophical contemporaries, including Goethe and Humboldt. In the epistolary novel *Hyperion*, the seventeen-year-old high school student argued, his *Lieblingsdichter* indeed "launches sharp and cutting words against German 'barbarianism.' Yet this disgust in face of reality can be combined with the greatest patriotic love."[12] Facing pitiful reality, it is suggested, *Hyperion* turns the German tongue against itself. Working within the language of *Barbarentum*, that is, in the parlance of those who speak incomprehensibly, and therefore as foreigners, the novel is nonetheless able to express the strongest affection for the *Vaterland*. By exploring an inherent disjunction within German—one which, while defining it as German, makes it subtly though firmly alien to itself—*Hyperion* offers a linguistic example of authentic patriotism. Sharp and cutting, the text's expressions of discomfort and disdain restore to dis-like its genuine meaning.

Articulated as a difference-within-the-same, the trope organizing this by no means particularly unusual understanding of *Hyperion* treats the contrariety between foreignness and familiarity as essentially linguistic in nature. Implicating it in the very definition of a native tongue, Nietzsche's observation does not so much restate the norm of a *mutatis mutandis* prevailing within classicism, as suggesting a principle of inherent dissimilarity. Such ἕν διαφέρον ἑαυτῷ, of which *Hyperion* speaks, quoting Heraclitus in the original,[13] informs the most decisive aspects of Hölderlin's conception of the relationship between Hellas and Hesperia, and nowhere is its validity at stake as evidently as in the practical confrontation with Greek texts. Based on a notion of transport as *Übergang* rather than *Übertragung*, his translations offer the most dramatic possibility of accounting for "how it proceeds from the Greek to the Hesperian," in the famed prescription proposed by the adjacent remarks on the 1804 *Antigone* translation.[14] Less exemplifying the particular textual activity by which *Bildung* is achieved—passing through the foreign so as to acquire, consolidate, and enrich its true, though not yet manifest, identity—the notion of translation emerging from Hölderlin's interlinguistic activity proposes a different, more eccentric and more disjunctive path. Delineating the interrelationship between Greece and Hesperia as a process as well as a procedural problem, Hölderlin's resolution no longer assumes either a continuous course or uninterrupted succession, however tainted by what may remain alien, or by any identical or identifiably similar set of concerns that would allow for the establishment of an inherent teleology. On the contrary, as the first letter to Böhlendorff cautions, "it is . . . dangerous to deduce the rules of art for oneself exclusively from Greek excellence." Written on December 4, 1801, shortly before Hölderlin's fated trip to France, the missive continues by prefiguring the Sophocles translations on which he would begin work soon thereafter: "I have labored long over this and know by now that, with the exception of what must be the highest for the Greeks and for us—namely, the living relationship and destiny—we must not share anything identical with them" (*SA* 6; 1: 426; *Essays*, 150).

In the context of a direct *Auseinandersetzung* with the Greek texts, such contrastive dissimilarity both suggests and necessitates a different notion of translation than the one inherited from and instantiated by, say, Goethe's *Iphigenia,* in which epistemic congruity and figurative fidelity remain prerequisites, however much difference may be paid attention to and accounted for.[15] When the adolescent Nietzsche commented on *Der Tod des Empedokles,* considering it a text in which "the melancholy tones of the future of the unhappy poet, of the tomb of a madness spanning numerous years, are permeated . . . by the purest Sophoclean language and in an infinite supply of profound thoughts,"[16] he also signaled why Hölderlin's only—if repeated—attempt at a Hesperian tragedy had to fail. Speaking *as* Sophocles, in a manner proper to the Greek and according to the modes of procedure singular to its language, the only tragedy available was that of its own death and ruin. After three attempts, all failed, the *Empedocles* project was aborted, and the particular "tomb" to which Nietzsche alluded—encompassing thirty-seven or so years of *Umnachtung*—was effectively sealed by the Sophocles translations.

Although the "Sophoclean language" addressed by Nietzsche may have augured the translator's own imminent "benightedness," Hölderlin's actual renditions of Sophocles' language exhibited few signs of irrationality, and even less trace of the inanity lamented by his contemporaries. Writing in December 1808 about the idiosyncrasies evidenced by his translations, Karl August Varnhagen von Ense related their peculiarities not only to the particular mode of transport they employed, but also, allegorically, to their prefiguration of the translator's own transition from mental sobriety to psychic derangement. This *Übergang* from sanity to insanity, manifested in the "madness" instanced and embodied by the transition from Greek to German known as *Ödipus der Tyrann* and *Antigonä,* marked for a long time the understanding of the particular textual labor undertaken in these texts. Before going literally mad, Varnhagen von Ense proposed, Hölderlin "handed in a Sophocles translation for publication, which seemed rather insane, but only literarily so." The "noble poet of *Hyperion,*" he continued, "wanted to bury his pain [after losing Susette Gontard] in work, and thus translated Sophocles; the publisher, who printed and published the first part, did not sense that some traces of the transport, which regrettably all too soon became evident in the author, already may be found in the volume" (*SA* 7; 2: 371). Confounding life and letter in a manner not limited to the contemporary reception of the translations, but encompassing also the particular *Trauerarbeit* thematized in Sophocles' play, this conception of "transport" is characteristic for the distinctive form of linguistic transaction in which Hölderlin's renditions were considered to engage. Prefigured in monstrous literalizations and bizarre colloquialisms akin to the "parodistic" translation lamented by Goethe, the poet's cerebral excitation, caused by the loss of his own Diotima, distanced him progressively from any means of communication collectively shared. As the record of his severance from linguistic convention, Hölderlin's versions become the melancholy tomb housing a lost spirit.

Despite the fact that Friedrich Wilmans, the publisher responsible for these alleged examples of both literal and literary madness, may only have been well-intentioned when he announced them in the *Hallesche Allgemeine Literatur-Zeitung* in July 1805—speaking of a "classical translation" in which "the philologist, who has compared it to the original, will find everywhere that faithfulness, precision, and the spirit of the German language have been observed" (*SA* 7; 2: 288)—he inscribed Hölderlin's project into the very view of the relation of translation to modernity, normative at the time, from which the renditions attempted to depart. Interpreting the ἕν διαφέρον ἑαυτῷ, fundamental to this enterprise, according to classicist doctrine, Wilmans did not so much observe Humboldt's "true line of demarcation," as he displaced it onto a terrain where it no longer made its particular form of difference. Although decidedly less appreciative in their judgment, other considerations of the Sophocles translations did not differ significantly in their use of categories of evaluation. Regarding "the spirit of the German language," referred to by Wilmans, as alien to what was commonly conceived to be known as German, Hölderlin's Sophocles appeared not merely *fremd*, but was considered *Fremdheit* embodied. Gerhard Merkel, for example, writing about the translations in *Der Freimüthige* in January 1805, asserted that these German renditions were "one of the most pretentious examples of pedantry," and added that if "Sophocles had spoken as rigidly and stiltedly, and in as *un-Greek* a manner as this translation is *un-German*, his audience would have left the theater in a hurry."[17] Another reviewer, perceiving a similar monstrosity, went as far as to claim that the notoriously difficult remarks to the plays surpassed the actual translations in "nonsense."[18] And Schelling, Hölderlin's old classmate, delivered the psychologizing verdict in a letter to Hegel in July 1804, which for considerable time was to remain the final word on the matter:

Hölderlin's "Sophocles translation," he asserted, "clearly expresses his dilapidated state of mind."[19]

Among contemporary reviewers, Johann Heinrich Voß, Jr., may have possessed the greatest erudition, or at any rate the most practical experience. Writing at length about the translations in the *Jenaische Allgemeine Literatur-Zeitung* in October 1804, he came to a halt midway through his account of the philological errors riddling the texts, noting with exasperated resignation: "We cannot continue in our exposition without throwing a glance at two problems—Mr. H's knowledge of Greek and his view on Sophocles. We charge that he has not looked at his original with foreign eyes, but with his own."[20] Employing neither earlier translations or commentaries, nor dictionaries or Latin versions of the texts, Hölderlin was taken to task for not having engaged and respected the ocular abilities or penetrating insights of others. The perceptions of the Greek language and of Sophocles which can be gleaned from his translations were not merely "foreign," but indicated a foreignness constitutive of the language proper to the renditions themselves. By looking at the Greek with his "own" eyes, Hölderlin had alienated not only Sophocles, but also the German into which his texts had been transferred. Whatever means of transport he may have used, it brought neither him nor Sophocles into a realm of articulation familiar to the reviewer and his contemporaries. Referring to the radicalization of the task of the translator not particularly implicit in this verbal undertaking, Voß then went on to politely demand: "We would very much want to know more about the conditions under which such an original work could emerge"[21]—while continuing on a franker note in a letter to a friend, which since has become a staple reference in accounts of Hölderlin the translator:

> What do you make of Hölderlin's Sophocles? Is the man mad or does he just pretend, and is his Sophocles a veiled satire on bad translators? Recently one evening, as I had dinner with Schiller at Goethe's, I entertained the two poets. Reading the fourth chorus of *Antigonä*—you should have seen Schiller laugh; or *Antigonä*, line 21: "Was ist's? du scheinst ein rothes Wort zu färben." I suggested this passage to Goethe as a contribution to his optics.[22]

Although diverse—and at times more, rather than less nuanced—the general reception of Hölderlin's translations has remained confined to this view that either there is little more to them than meets the eye, or else they offer a bewildering repartee ridiculing inferior translations by exaggerating their lexical laxness and idiomatic inadequacy. While later critics, including Lesky, Reinhardt, and Schadewaldt, have contributed significantly to the re-evaluation of Hölderlin's Sophocles renditions once initiated by Nietzsche, by considering them as originals in their own right, though marred by scant philological acuity and insistent literalizations of figurative or idiomatic expressions, they, too, have inscribed the translations into the framework of the very dichotomy with which the texts struggle metacritically.[23] Paralleling the polarity between foreignness and sameness to that of original and copy, such appraisements may no longer conceive of the translations as blindfolded perversions of the Greek, which leave untouched few assumptions of translatory fidelity to the letter and much else to be desired, yet, nonetheless, they assign to them the compelling status of original only *as* copy. While earlier critics determined Hölderlin's renditions as subversions of primariness, monstrously attempting to make Sophocles readable in a language as idiosyncratic as misrepresentative, monstrosity is now considered an emblem of peculiar originality. Regardless of valorization and preference, philological subtlety and hermeneutical refinement, *mutatis mutandis*, the status of the relationship between Greek and German texts thus remains not merely identical, but also confined to an understanding whose analytical categories are unable to account for Hölderlin's particular—eccentric—example of transport.

Writing to Leo von Seckendorff in March 1804, in part in order to alert him to the translations which were to appear a month later, Hölderlin emphasized that what preoccupies him the most is "the national insofar as it differs from the Greek" (*SA* 6; 1: 437). Although his remarks on the plays may still be read (and have, indeed, mainly been read) according to the conventions of their time—that is, as attempts to spell out certain formal "principles and limits," how they conform with, as well as differ from, the craft of the Greeks, and in what ways "this greatly changes the essential and patriotic representations" of the Hesperians (*FA* 16: 418; *Essays*, 113)—his actual renditions of Sophocles offer acts of linguistic transferal based on a consideration not of the language of the original as much as that of the translation. Rather than articulating themselves in accordance with assumptions concerning the customs and beliefs of Sophoclean Greece, Hölderlin's German renditions are determined by the transpository character of their own language. The difference mentioned in the letter to von Seckendorff is not simply that of another culture or system of beliefs in need of fitting articulation—it neither implies nor organizes a symmetrical opposition between "Greeks" and "Hesperians"—but pertains rather to a contrariety structuring their respective differences from themselves. If what is most Greek to the Greeks is their particular ἕν διαφέρον ἑαυτῷ, such difference cannot be identical to—nor indeed be rendered into the verbal arrangement of—the one peculiar to Hesperians. The difference between the two cultures and languages concerns, rather, the difference of difference. While structurally similar, though neither identical nor interchangeable, these mutually distinct relations between familiarity and foreignness remain not only unlike each other, but dissymmetrical in a manner impossible to reconcile with the teleological models of understanding that had informed most theories of translation up to this point.[24]

The problems that such difference of difference poses for the concept and practice of translation are both considerable and inevitable. Neither aspiring to the traditional goal of classical translation, nor intending to efface the foreign character of the original, the translation must aim to articulate the first language's own difference in the target language, with the result that *its* character as translation is made evident, while at the same time not confusing such difference with that of the target language's own.[25] In order to observe the principle governing any act of modern translation—"we must share nothing identical with them"—that which is most Greek in the Greeks, their ἕν διαφέρον ἑαυτῷ, should be rendered in a manner which neither appropriates it, nor rarefies it into foreignness proper. Accordingly, if this selfsame difference is thought of as a tension between trope and meaning, modern translation is forced to convey the particular transport characteristic of antiquity in a manner which neither effaces it, nor converts it to the transport distinctive of modernity. By implication, the transaction presupposed in the transfer from one language to another pertains to a transport which cannot be conveyed, thus is untranslatable, because it remains peculiar to one of the two languages. That is, the latter-day transport effected by Hesperian translation must stay empty (in a sense still to be qualified). As Hölderlin's theoretical observations suggest, any modern translation of tragedy always also seems to imply a tragedy proper to translation.

Of the two Sophocles renditions Hölderlin saw through publication—a third attempt, *Ajax*, was to remain a fragment—*Antigonä* offers the most telling example of the stakes involved here. Thematizing the very transport on which interlinguistic rendition is predicated, the play is, in large measure, about different conceptions of transportation, most notably that of Polyneices' body.[26] As such, it allegorizes the conditions of translation, offering an indication of Hölderlin's radicalized notion of the tragic which implies less of a view than a reading of the Greeks. When Humboldt spoke of how a certain color of foreignness characterizes all true translation, and Voß lamented line 21 of *Antigonä*, both relied on a perception of the Greek original which left little room for the unimpressed reader, who would resist the immediate, phenomenalizing correlation of one language to another without residue or difference. Any interpretation of the interrelationship between *Antigone* and *Antigonä*, however, or so Hölderlin's translatory practice seems to indicate, must not only consider and inspect the disjunction at work between the plays, but also read the one operating within each one of them. In the first letter to Böhlendorff, explicitly concerned with the difference between Greek "fire of the sky" and "Western *Junonian sobriety*" (*SA* 6; 1: 426; *Essays*, 149), Hölderlin at one point mentions that, "among everything" he "can see of God," there is one "sign" which has become for him the "chosen one." The letter then proceeds to quote, freely, from Goethe's poem on the "limits of man"—"The old holy father pours down lightenings with calm hand out of red clouds" (*SA* 6; 1: 427; *Essays*, 151)—making clear that, while there may indeed be things to "see" about God, the signs chosen are always read.[27]

Having spoken of "the color of fate" in a letter two years earlier (*SA* 6; 1: 290),[28] Hölderlin's conception of the relationship between antiquity and modernity appears guided by a notion of translation which makes "certain and characteristic principles and limits" of the target language, which supposedly has only "the living relationship and destiny" in common with the first language, a matter of coloration (*FA* 16: 249; *Essays*, 101). Predicated on the manner in which due observation of Greek difference "greatly changes the essential and patriotic representations" of the Hesperian "mode of procedure" (*FA* 16: 418; *Essays*, 113), the *Verfahrungsart* thus assumed is exemplified by Hölderlin himself with reference to the Phrygian Niobe, "the image of early genius" (*FA* 16: 415; *Essays*, 112). Underscoring the importance of bringing the Greek "closer to our mode of representation," he quotes his own rendition—

> Sie zählete dem Vater der Zeit
> Die Stundenschläge, die goldnen.
>
> She counts to her father Time
> The striking of the hours, the golden ones.
> (*FA* 16: 415; *Essays*, 112 [trans. modified])

—and suggests that such translation is appropriate, rather than the more fluent "verwaltete dem Zeus das goldenströmende Werden" ("administered for Zeus the golden stream of becoming") (*FA* 16: 415; *Essays*, 112). Measure, then—and, with it, discretion and parataxis—is preferable to the continuous growth and syntactic seamlessness intimated by a more opportune rendition of the Greek. The "golden stream of becoming" is turned into the paratactic and mechanically measurable "striking of the hours, the golden ones." Time no longer appears in a constant flow of becoming, as it would be if the lines were translated from a perspective similar to a Greek worldview (as well as commensurate with the classicist fidelity according to which appropriate translatory measures ought to be taken); instead it is submitted to "striking" and being counted, that is, to a lawful calculation particular to Hesperian culture and articulated in verses rendered in a meter proper to its specific mode of procedure.

In order to trace the consequences of the difference thus implied, one which is intimately linked to the question of color, both of time itself and of the translation required by time's changing, it is necessary to turn to Hölderlin's attempt to render precise the notion of "transport" proposed in the adjacent remarks. Offering a designation whereby the measure of the difference between *Hellas* and *Hesperien* may be taken, he explicitly relates transport to a term derived from the theory of meter: the caesura (*FA* 16: 250; *Essays*, 102). Also referred to as a "counter-rhythmic rupture," this metrical disjunction is constitutive, since without it there would be no "rhythmic sequence of . . . representations wherein [which] transport could present itself" (*FA* 16: 250; *Essays*, 102 [trans. modified]). The caesura accounts, that is, for the possibility of tragic presentation as such. Itself expressing nothing, neither an intention nor its representation, it effects no more and no less than the form of representing. It is a "pure word," according to Hölderlin, neither a motion nor an interval, in which "nothing exists but the conditions of time or space" (*FA* 16: 250 and 258; *Essays*, 102 and 108 [trans. modified]).[29] As such, the caesura intimates a site which precedes even figuration, providing the condition of possibility for transport, rather than amounting to one of its—however necessary—effects. Identified formally, in the case of both *Ödipus* and *Antigonä*, as the "speeches of Tiresias" (*FA* 16: 251; *Essays*, 102), hence understood in discursive rather than dramaturgic terms, and linked to moments of vatic translation of material events into celestial meaning, the caesura appears to be not only the sine qua non of the particular transport on which these dramas rely formally, but, by implication, also the greatest challenge to its translation from Greek into Hesperian circumstances.

In large measure a drama about language, *Antigone* is concerned not only with its ways of functioning and manners of gaining validity, but also with the emergence of different, conflicting interpretations of its social, political, ideological, and religious definition.[30] Against Creon, king of Thebes, claiming to speak the language of the law and thus that of the community, stands Antigone, whose position is that of the solitary individual without linguistic authority. The tragic conflict of the play occurs when the latter defies the explicit decrees of the former by referring to the "unwritten laws" of the sky. Attempting to bury Polyneices, her brother, Antigone commits a transgression of the written laws applicable to all. Because her rebellious burial attempt threatens to tarnish the city, she must be exiled, and Creon argues for her isolation in order for the city to be cleansed of "pollution" (line 805; 776). It soon becomes evident, however, that this *μίασμα* consists less in Antigone's ability to infect the bodies of the city, than in her capacity to contaminate public opinion. Opposing Creon's interdiction against burial, Antigone attempts to move the dead body from the realm of the human and material, in which it is visible, to that of the divine and spiritual, in which it is not. As a schism originating in the corpse of Polyneices, the conflict between Creon and Antigone thus explicitly concerns the premises of transport.

Despite the difference in the plays' respective "rhythmic sequence of the representations wherein transport presents itself," in both *Ödipus* and *Antigonä*, the caesura occurs in the "counter-rhythmic rupture" provided by Tiresias's interventions. Taking place in the second scene of the fourth act of the latter text, it is located by Hölderlin in a passage primarily concerned with the vatic act of interpretation and, concomitantly, with the particular transport by which the signs to which it is devoted may link literal event to figurative meaning. As prophet, hence as someone with the ability to connect secular signifiers with celestial signifieds, Tiresias's function is double: to offer counsel to the king and to prophecy about the future to which he alone has hermeneutical access. An interpretive pro, his art consists of understanding signs (*τέχνης σημεῖα*) (line 1035; 998). Yet on this occasion, the prophet's hermeneutical skill seems not to work. With the help of his assistant, he witnesses the flight of birds and how they attack each other "murderously" (φωναῖς), gouging eyes (line 1041; 1000). While "the whirring of the birds" itself is not "incomprehensible"—"nicht unverständlich war / Der Flügel Sausen" (lines 1041–42; 1001)—the sacrificial fire Tiresias attempts to ignite in response to this avian auto-destruction falters nonetheless. In the ashes of failed fire, fat fizzles without flame. At this point of ornithoscopic collapse, the prophet, observing the shanks visible through the sputtering fat, speaks of "frustrate signs of oracles grown dumb" (ἀσήμων ὀργίων μαντεύματα) (line 1051; 1010). No longer wielding any proleptic authority, his auguric attempts break down because the only signs made visible in the fizzling ashes are explicitly ἄσημα, "signless," and where there are no signs, there is nothing to read. The indications needed in order to interpret the situation correctly, to correlate material signifiers with spiritual signifieds, and thus to administer proper counsel, yield neither meaning nor indication; prophecy predicated on asemantic sacrifices remains φθινόντ', "wasted" or "useless" (line 1010).

The German rendition of this passage, however, by Hölderlin designated the constitutive disjunction of the play, offers a more radical interpretation when it translates the Greek as "der zeichenlosen Orgien tödliche Erklärung"—literally, "the deadly explanation of signless sacrifices." Instead of resigning as the foundation for his "art of signs" dwindles, Hölderlin's Tiresias interprets the asemantic quality of the birds' bloody, empty omen: "prophecies" becomes "explanation," the plural μαντεύματα is turned into the singular *Erklärung*, and the frustrated Greek auguries are replaced by a German token of exegesis. Instead of the plenitude of uninterpretable sacrifices in the Greek, there is one single and deadly explanation in the German. The pure signlessness which Tiresias interprets in the caesura of the play seems, that is, in what the remarks on the play term "our time and mode of representation," to solicit clarification. Hölderlin's *Antigonä* offers such *Erklärung* when it interprets the signlessness as such, turning the absence of multiple signs into the single matter interpretable. In the original, the birds tear each other apart, yet Tiresias is unable to connect the bloody event with prophetic meaning; it remains strictly asemantic. By contrast, in the Hesperian version, the animal bloodshed is given an explanation grounded in its very signlessness. In the Greek, the seer does not succeed in translating a material occurrence into a spiritual meaning, whereas in the German, such transport is achieved by substituting the fiery and exalted μαντεύματα for the cooler and more sober *Erklärung*. What had been a flight of birds yielding no proleptic transparency, has become, in Hölderlin, a hermeneutic obscurity providing "explanation."

Reflecting the difference between "the fire of the sky" and the "Western *Junonian sobriety*," of which the first Böhlendorff letter speaks, the implications of this difference are evidently significant, since in the transition from Greek to German transport is not accomplished. Hölderlin's translation fails, in a sense, as it finds a "deadly explanation" for the signlessness of the birds, by which an event in the secular world is joined with a spiritual meaning in the sacred. His *Antigonä* veils, that is, the transport which never occurs in the original—providing a repetition of what occurred in the original without ever taking place.[31] In exemplary miniature, this may be the tragedy of translation as presented by *Antigonä*: in the caesura of the play—that structural gap which, according to Hölderlin, constitutes its core, the form of presentation but not the presentations themselves, and which is also referred to as an "empty" transport (*FA* 16: 250; *Essays*, 101)—a pure signlessness is uncovered which the original could not interpret, but which the translation exhausts as "explanation." Hölderlin's text gives semantic emptiness a meaning and hence covers up what never occurred in the original. According to his observations in the letter to Böhlendorff, however, this may be what a Hesperian translation must do, since it belongs to a world whose tragedy consists in the necessity to transport the dead hidden in containers and thus is unable to know whether transport truly takes place. Only in this way can the original's truth of transition be maintained, hidden in the emptiness which the German, with acute accuracy, defines as "deadly." What remained untranslatable in the Greek is translated as something translatable in the German, thus in turn observing a difference between the two languages which itself remains untranslatable. With greatest rigor, Hölderlin's *Antigonä* seems to follow the dictum: "Our poetry must be patriotic so that its themes are selected according to our world-view and their representations patriotic" (*FA* 16: 419; *Essays*, 113). Only thus may Hellas and Hesperia be related to one another—"through boundless separation" (*FA* 16: 257; *Essays*, 107).

In the conception of the tragic thus exemplified, tragedy becomes the particular linguistic form which renders evident the confrontation of man and god. Yet in Hölderlin's translation of Sophocles, the gods change appearances and the men are motivated by different political persuasions. Hence tragedy also appears to provide the truest articulation of historical change. Antigone is an *Antitheos* with republican pathos, and in her actions Hölderlin locates the first attempts at that which spells revolution. "The course of events," he claims in the adjacent notes, "is that of a rebellion where, to the extent that it is a patriotic course, it is important that everything senses itself as being seized by infinite reversal and, being deeply affected, senses itself in an infinite form within which it is affected." To this he adds: "The form of reason which takes shape here as a tragic one is political, that is, republican." This new, Hesperian form of reason termed republican is both tragic and in a state of becoming. When the remarks on *Antigonä* speak of the form of the play, they point out that it is "relevant precisely because the infinite, like the spirit of the states and the world, cannot be grasped other than from a skewed perspective" (*FA* 16: 419, 421, and 421; *Essays*, 114, 115, and 116 [trans. modified]).

Weimar lacked an eye for such *linkisch* perspective, succumbing instead to the linguistic chaos known as laughter. Translating the Greek δηλοῖς γάρ τι καλχαίνουσ' ἔπος as *du scheinst ein rothes Wort zu färben?* (*FA* 16: 266–67), Hölderlin's German rendition of *Antigone* was met only with incredulousness. Yet it may be asked if his rendition of καλχαίνουσ' as "red word" was not, in fact, a consequence of the theory of translation outlined in his adjacent remarks. By transcribing the chromatic connotation implicit in a figure (the Greek καλχαίνουσ', derived from καλχαίνω, literally means "to make red" but figuratively, "to brood" or "to harbor dark thoughts"),[32] Hölderlin's translation renders evident the tinge which to an extent greater than any other colors the events chronicled in Sophocles' play—from the blood flowing in the streets of Thebes and covering Tiresias's whirling birds, to its virtual signaling of the heroine's familial, and thus blood-related ethos. Nonetheless, and more decisive, his *Antigonä* would not be Hesperian if it did not do so by bringing out a color already present—figuratively, metaphorically, in transported form—in the original. Hölderlin's practice of translation is concerned not merely with content, that is, nor just with form, but rather with the manner in which formal issues are already thematized in the original, requiring translatory strategies of distinct complexity. *Antigonä* may clothe the thoughts of Sophocles in other words, giving them a different coat of signification, but it does not limit itself to such task, nor does it restrict itself to making evident a dye which runs from the revolutionary word of Antigone, over the dead body of Polyneices, to the bloody birds of Tiresias—all of which would still belong to a practice of translation possible, in principle, to subsume into one of the methods of translation proposed by Goethe. Rather, the taint Hölderlin's rendition makes readable amounts to something already present in the original, as figure, but which is brought out only in translation—solely as a matter of course, that is, in the course of time. Suspended in the temporal relationship of two texts to each other but belonging strictly to neither, only futurity is able to convey meaning to such coloration; only the promise of history, seen, as it were, from a *linkisch* perspective, appears able to render it legible. As in the case of the entombed Antigone, the redness of *καλχαίνουσ'* is put in suspension, visible only when being repeated reliably, according to a lawful calculation similar to that of Niobe in the desert, whose time was said to be "more calculable" through the golden rays of the Sun. Its closest form of expression may, in fact, be that of arrested annunciation.

While the transition from the literal to the figurative is accomplished in Tiresias's attempt to explain the signlessness of the birds, such mediation does not take place in the corresponding transport from Greek to German. Amounting not only to a tragedy translated, *Antigonä* makes evident the necessary tragedy of Hesperian translation. The depth of this other tragedy can be measured only linguistically. Hence it may well be an indication of modernity's historical truth that it remains *ἄσημα*. If so, Hölderlin's theorization of the relationship between Hellas and Hesperia as a linguistic relationship would also seem to undermine the notion of historical continuity governing the thoughts of his colleagues in Weimar. His *Antigonä* is not, as has been proposed, "a homecoming (tragically frustrated) to oneness between consciousness and the world."[33] On the contrary, the text reflects incessantly on its own mode of presentation—a metacritical awareness which seems to prevent it from being neutralized by any reading aiming at a telos at the end of time. By way of the tragedy of translation itself, implying a radical disjunction labeled an "empty" transport, Hölderlin's *Antigonä* instead becomes a text which is less prophetic than apocalyptic—but apocalyptic in a sense devoid of standard theological implications; it feels, that is, to quote from the passage on Niobe, "*with* the change of time . . . without the intellect anticipating the future based on the present" (*FA* 16: 416; *Essays*, 112 [emphasis added]). Its textual mode is not *futurus* as much as *adventus*: less consisting in the movement of the past into that present which is its specific as well as projected future (*Antigone* figured, sublated, and completed in *Antigonä*), it is given *with* time as something which can only come from something yet to come—monstrously.[34] If at all projected, it seems to be toward this unknown, radical reversal yet to come, that Hölderlin's disjunctive practice of translation aspires.

In the remarks on *Antigonä*, it is stated that the Hesperian word "is more immediate by taking hold of the more spiritual body," and furthermore: it is "more murderous-factual than deadly-factual" (*FA* 16: 417; *Essays*, 113). It kills, that is, as language. If this is the case, it is not surprising that red should be the color rendered readable by Hölderlin's translation, that doubly tragic drama of transport which occurs "whenever time is measured in suffering" (*FA* 16: 415; *Essays*, 112).[35] Neither is it surprising that the *traur'ge Arbeit* of which Antigone speaks in the opening lines of the play (line 5), but which lacks a counterpart in the Greek original, could be understood as the tragic suffering occurring when a transport cannot (yet) take place. Hölderlin's Sophocles translation, the lawfully calculated labor of an *uneingenommener Leser*, may be treated as such a "mournful work," one in which *Trauer* becomes an attribute to "labor" (*traur'ge*), and hence a quality bestowed onto the "work" in its temporal unfolding, rather than an inherent trait characteristic of an effective labor of suffering (or *Trauerarbeit*). Considered as a "mournful" work of linguistic transmutation, Hölderlin's *Antigonä* thus also conveys, in reading, what Benjamin once referred to as "die Wehen des eigenen"—the suffering of language itself.[36]

PART III

The Philosophy of Poetic Form

HÖLDERLIN'S THEORY OF POETRY AND THE CLASSICAL GERMAN ELEGY

Cyrus Hamlin

> So, ihr Lieben! auch mir, so will es scheinen, und niemand
> Kann von der Stirne mir nehmen den traurigen Traum?[1]

Much attention has been given in the past several decades to the importance of Hölderlin for the history of Idealist philosophy in Germany. This concern began quite early in the reception of the poet during the present century, notably in an important essay by Ernst Cassirer,[2] and assumed a particular urgency following the claim made by Martin Heidegger in his *Erläuterungen zu Hölderlins Dichtung* (*Elucidations of Hölderlin's Poetry*) that Hölderlin's later poetry constitutes an authentic and originary form of language, which extends far beyond the limits of Idealist philosophy. Heidegger's claim has been challenged recently by the research of Dieter Henrich into the historical origins of such philosophy during the mid-1790's at the University of Jena, where evidence has been summoned to suggest that Hölderlin played a crucial role in the transition between Fichte's *Wissenschaftslehre* (*Theory of Knowledge*) (1795) and the emergence of Hegel's dialectical thinking in his *Phänomenologie des Geistes* (*Phenomenology of Spirit*) (1807). Attention to such questions from the history of philosophy has also resulted in a neglect of more difficult and more important questions concerning the nature of Hölderlin's poetry as poetry, particularly where it may be argued that the later poems put into conscious practice a philosophical theory of poetic composition.

Two assumptions above all motivate the choice of the elegy "Menons Klagen um Diotima" ("Meno's Lamentation for Diotima") as the focus of discussion in this essay. First, the context of the classical elegy in German poetry, as it developed during the 1790's through the efforts of Goethe in his "Römischen Elegien" ("Roman Elegies") and Schiller through his theoretical discussion of the genre in his essay "Über Naive und Sentimentalische Dichtung" ("On Naive and Sentimental Poetry") and in his own representative elegy "Der Spaziergang" ("The Walk"), all published in Schiller's journal *Die Horen* ("The Graces") in 1795–96, provides the immediate background for and direct influence on Hölderlin in his subsequent work in this genre. The poet's "Lamentation" may be read as a conscious and programmatic response to the pioneer work of his mentors in the genre of German elegy. A survey of this generic history is provided in Theodore Ziolkowski's *The Classical German Elegy 1795–1950*,[3] an important work of criticism known primarily to scholars who work in this somewhat esoteric field. Second, the elegy of Hölderlin, it may be and has been argued, applies an elaborate and particularly obscure theory of poetic composition, consisting of a sequence of varying poetic tones, to the organization of the poem. This claim has been made, above all in a groundbreaking study by Lawrence Ryan,[4] with reference to the theory of tonal variation developed in Hölderlin's fragmentary essays on poetic composition.

"Menons Klagen um Diotima," insofar as critics of Hölderlin have given any attention to the poem, has generally been read with reference to the poet's biography, in particular his misfortunate affair with the mother of his pupil in Frankfurt, Susette Gontard, where he had been hired as a private tutor. Particularly important in this regard is the interpretive essay by Karl Viëtor, "Hölderlins Liebeselegie."[5] My argument goes contrary to such biographical assumptions. I suggest that the elegy is a complex reworking of the conclusion to Hölderlin's novel *Hyperion, oder der Eremit in Griechenland* (1797–99), where the hero has a quasi-mystical experience of communion with the spirit of his dead beloved Diotima.[6] So far as I am aware, no critic of Hölderlin has yet noted the possible connection of the elegy to the novel, which was probably completed only a short time prior to the initial composition of the poem, which, like the first version of Schiller's "Der Spaziergang," has the simple generic title "Elegie." Also of interest is the probable influence on Hölderlin's "Lamentation" of

Goethe's important elegy, "Euphrosyne," which involves a confrontation high in the Alps between the poet and the spirit of a young actress from the Weimar theater, who had recently died. Goethe's poem was published along with several shorter odes by Hölderlin in Schiller's *Musenalmanach für das Jahr 1799* ("Muses' almanac for the year 1799"), shortly before Hölderlin's elegy was first conceived.

The central claim of my argument, which cannot be fully developed here for limitations of space, is that Hölderlin's theory of tonal variation in poetry, somewhat akin to the inner form of the classical sonata in music as perfected by Beethoven at approximately the same time, seeks to define and control a dynamic process of thought and experience, which the language of the poem is intended to embody through a sequence of tones which move through stages of development, reversal, and retrospective self-reflection. The central hermeneutical task of the elegy is to communicate this process of thought to the reader, as the *Wirkung* ("effect") of the poem, through the complex metaphorical dynamics of its poetic form. Thus several specific features of the final text of "Menons Klagen um Diotima" are singled out as representative of Hölderlin's poetic technique in this extremely difficult communicative strategy. The elegy turns out not to commemorate only the loss of the beloved, as indicated by the symbolic names of the title, but in fact to articulate the poetic process through which such lamentation takes place as a quest journey of the isolated self in its attempt to reflect upon its situation and to comprehend it within the tenets of what the German Idealists called *Selbstbewußtsein* ("self-consciousness").

1

Hölderlin has been rediscovered twice in this century: first as a poet and now as a philosopher. His rediscovery as a poet in the initial decade by Norbert von Hellingrath in conjunction with the George-Kreis (the group associated with the poet Stefan George) has been often acknowledged and thoroughly researched. His rediscovery as a philosopher has occurred only recently, through the study of German Idealism in its historical development at the University of Jena during the 1790's. More than anyone else, Dieter Henrich may now claim a position alongside Hellingrath in the history of Hölderlin's reception, having devoted heroic labor to the authoritative documentation of Hölderlin's central role in this history. Henrich's recent publication, *Der Grund im Bewußtsein: Untersuchungen zu Hölderlins Denken (1794–1795)*,[7] a monumental volume of more than 850 pages devoted essentially to the brief period of less than a year when Hölderlin was at the University of Jena in close proximity with Schiller and Fichte, provides a powerful challenge to earlier views of the poet's development.

Two particular directions of inquiry seem likely to emerge from Henrich's work for the critical study of Hölderlin's poetry. We must shift our attention, first, backward from the dominant preoccupation among critics with the late hymns and fragments, those various poetic works composed after 1800 in the years just prior to the poet's final breakdown. This preoccupation derives, of course, from Hellingrath himself, who in his edition of the late poems labeled this late work as the *Herz, Kern und Gipfel* ("heart, core and summit") of Hölderlin's poetry.[8] But there has also been a pernicious tendency, at least since the essays devoted to these poems by Martin Heidegger from the mid-1930s on and continuing through the reception of Heidegger's questionable readings by various radical critics under the heading of poststructuralism and deconstruction, especially in France and America, to divorce these complex poetic texts from any consideration of either history or biography, as if they existed in a hermeneutical vacuum free of all contingency.

Equally important, secondly, will be the question of a larger context in literary history and the history of thought, within which the career of Hölderlin's writing may now be reevaluated. Henrich's book offers a challenge to literary critics as much as to the historians of philosophy. Where does Hölderlin stand as a poet within his own time? How do the so-called late poems—the *Spätwerk* of a writer (we should remind ourselves) in his early thirties, composed during a period of intensive isolation in the span of no more than two or three years—relate to the developmental history which Henrich has traced for Idealism and which for historians of European literature might equally well be called Romanticism? These questions need to be carefully reconsidered, assuming younger scholars may be found who are able and willing to undertake such labor in the wake of Henrich's research.

But there is a further challenge to criticism, which Henrich's book implies, even if it remains thus far unspoken. If Hölderlin is so important for the history of philosophy in his philosophical thinking—a claim which I, for one, am not prepared to challenge—then what difference does it make for our reading of Hölderlin's poems? How may our understanding of the poetry be affected by our awareness of the philosophical thought which stands behind it? These are questions which may lead in several directions.

One further twist to the complex history of reception for this uniquely challenging poet should be mentioned. When I first began my own study of Hölderlin as a student in the late fifties and early sixties, no other issue for interpretation seemed more controversial and more urgent than the question of Hölderlin's own poetics, the theory of poetic composition which he himself had drafted in the fragmentary essays written in Homburg just before the turn of the century, which had come to be known as *Hölderlins Lehre vom Wechsel der Töne* ("Hölderlin's theory of tonal variation"). It is one of the ironies in the study of Hölderlin that virtually nothing has been added to the pursuit of this question to supersede the work of Ryan, apart from several essays and monographs which address exclusively the challenge of explicating the argument of one or another of these essay fragments.[9] Above all, virtually nothing more has been written about the most controversial claim of Ryan's thesis, that the theory of "tonal variation" might help us to understand Hölderlin's later poetry.

What should be made of this failure to respond to the challenge of Ryan's work (a question which may well have been put by Ryan to himself)? Is the answer perhaps to be found in the absence of any adequate understanding of the broader historical and intellectual context in which Hölderlin's theory developed? And if that is so, are we not in a position now, thanks to the research of Henrich, to return to the question of "tonal variation" as a structural principle in the composition of Hölderlin's own poetry? I wish to appeal for just such a reconsideration based upon a fuller sense of the poet's development within the larger context of Idealist philosophy during the latter years of the 1790's than would have been possible at the time when Ryan first addressed the question in the mid-1950's.

Due to limitations of space, my thoughts will be addressed exclusively to the question of poetic structure in the specific genre of the elegy as Hölderlin conceived it at the time his theoretical essays were drafted and immediately thereafter. What I have in mind is the question of "tonal variation" as it operates specifically in the elegy "Menons Klagen um Diotima," which in its earlier version, with the generic title "Elegie," was almost certainly drafted in Homburg during the year 1799. This question is not new in the history of discussing this particular poem. Indeed, Ryan himself has some helpful remarks to offer on it in his study.[10] We are now in a position, I believe, primarily because of Henrich's work, to attempt a reconsideration of what Hölderlin's poetics may imply for a critical reading of such a poem.[11]

2

We may begin with the familiar threefold designation of the tones, which Hölderlin uses in the various essay fragments and tables: *Naive*, *Heroic*, and *Ideal* (*naiv*, *heroisch*, *idealisch*), which are also occasionally associated with emotional or psychological qualities: *Empfindung*, *Leidenschaft*, *Phantasie* (see, e.g., *FA* 14: 325–26). The question of sources for such terminology immediately arises, though I shall not pursue this here. More fundamental is the issue of how any poetic text could exhibit a tone, and how a reader might determine what that tone is. Hölderlin clearly associates a quality of language with specific states of mind, attitude, or mood. Equally problematic is the attempt to identify corresponding kinds or styles of poetry (*Stilarten*) under somewhat more conventional generic headings, namely lyrical, epic, and tragic. At the same time, this triad of tones may apply to the content or substance (*Stoff*) of a poem, as indicated in the following descriptive statement near the beginning of the longest of the fragments (traditionally entitled "Über die Verfahrungsweise des poetischen Geistes"), though the names of the three tones do not actually appear:

> Der Stoff ist entweder eine Reihe von Begebenheiten oder Anschauungen, Wirklichkeiten, subjektiv oder objektiv zu beschreiben, zu mahlen, oder er ist eine Reihe von Bestrebungen, Vorstellungen, Gedanken oder Leidenschaften, Notwendigkeiten, subjektiv oder objektiv zu bezeichnen, oder eine Reihe von Phantasien, Möglichkeiten, subjektiv oder objektiv zu bilden. (*FA* 14: 305 [punctuation emended])

> The subject matter is either a series of occurrences or views describing or painting realities subjectively or objectively, or it is a series of efforts, ideas, thoughts, or passions denoting necessities subjectively or objectively, or a series of fantasies forming possibilities subjectively or objectively.[12]

That each of these three alternatives relates to one of the three tones—the first to the Naive, the second to the Heroic, the third to the Ideal—is apparent from a footnote added to this statement by Hölderlin:

Ist die Empfindung Bedeutung, so ist die Darstellung bildlich, und die geistige Behandlung zeigt sich—episodisch.

Ist die Intellektuale Anschauung Bedeutung, so ist der Ausdruk das materielle leidenschaftlich, die geistige Behandlung zeigt sich mehr im Styl.

Ist die Bedeutung ein eigentlicherer Zwek, so ist der Ausdruk sinnlich, die freie Behandlung metaphorisch. (*FA* 14: 305)

If the meaning is feeling, then the presentation is pictorial, and the spiritual treatment reveals itself episodically.

If the meaning is an intellectual intuition, then the expression, the material is passionate, the spiritual treatment reveals itself more in the style.

If the meaning is more a particular purpose, then the expression is sensuous, the free treatment metaphorical. (trans. C. Hamlin)

These passages, and others which could be cited, indicate the degree to which Hölderlin thinks about poetic kinds, styles, and tones in groups of three. Precisely how he came to formulate such triadic relationships as he did might occupy considerable speculation. One crucial source for such thinking, however, if not indeed the exclusive origin of Hölderlin's procedure, is found, as is generally recognized, in Schiller's essay "Über naive und sentimentalische Dichtung," which appeared in three installments in the journal *Die Horen* at the end of 1795 and the beginning of 1796. There can be no doubt that Hölderlin studied this text closely during his initial weeks in the house of Gontard in Frankfurt.

Important above all for my purpose are several footnotes introduced by Schiller at the beginning, respectively, of the second and third installments of the essay, in which he attempts to clarify his definition for the three poetic genres of sentimental poetry, namely, satire, elegy and pastoral, on the exclusive basis of what he calls *Empfindungsweisen* ("modes of feeling"). The second and longest of these footnotes is the most important for defining what Schiller intended the sentimental elegy to be. I quote the middle paragraph of the footnote:

Die sentimentalische Dichtung nämlich unterscheidet sich dadurch von der naiven, daß sie den wirklichen Zustand, bei dem die letztere stehen bleibt, auf Ideen bezieht und Ideen auf die Wirklichkeit anwendet. Sie hat es daher immer . . . mit zwei streitenden Objekten, mit dem Ideale nämlich und mit der Erfahrung, zugleich zu tun, zwischen welchen sich weder mehr noch weniger als gerade die drei folgenden Verhältnisse denken lassen. Entweder ist es der *Widerspruch* des wirklichen Zustandes, oder es ist die *Übereinstimmung* desselben mit dem Ideal, welche vorzugsweise das Gemüt beschäftigt; oder dieses ist zwischen beiden geteilt. In dem ersten Falle wird es durch die Kraft des innern Streits, *durch die energische Bewegung*, in dem andern wird es durch die Harmonie des innern Lebens, *durch die energische Ruhe*, befriedigt, in dem dritten *wechselt* Streit mit Harmonie, wechselt Ruhe mit Bewegung. Dieser dreifache Empfindungszustand gibt drei verschiedenen Dichtungsarten die Entstehung, denen die gebrauchten Benennungen *Satire*, *Idylle*, *Elegie* vollkommend entsprechend sind, sobald man sich nur an die Stimmung erinnert, in welche die unter diesem Namen vorkommenden Gedichtarten das Gemüt versetzen, und von den Mitteln abstrahiert, wodurch sie dieselbe bewirken.[13]

Sentimental is distinguished from naive poetry, namely, in that it refers actual conditions, at which the latter halts, to ideas, and applies ideas to actuality. Hence it has always . . . to contend simultaneously with two conflicting objects, i.e., with the ideal and experience, between which neither more nor less than just these three following relationships can be conceived of. Either it is the *contradiction* with actual conditions, or it is the *correspondence* with the ideal, which is the preferred attitude of mind, or it is divided between the two. In the first case, it is satisfied by the force of the inner conflict, by *energetic movement*; in the second, it is satisfied by the *harmony* of the inner life, by *dynamic calm*; in the third, conflict *alternates* with harmony, calm alternates with motion. This triadic state of feeling gives rise to three different modes of poetry, to which the customary names, *satire*, *idyll*, *elegy*, correspond exactly, provided only that one recalls the mood into which the poetic species known by these names place the mind, and abstracted from the means by which they achieve it.[14]

Two points in particular should be noted from Schiller's somewhat schematic view of the poetic genres in terms of such modes of feeling. First, the factors determining these modes of feeling are only two: the real and the ideal in their relation to each other. The third, or mixed, genre, which is called "elegy," consists of a variation or modulation between both extremes of conflict and harmony. Second, the legitimacy of such generic modes is determined, so Schiller indicates at the end of the passage, by the effect they have on the reader or listener. The particular kind of poem must convey to the reader that specific mode of feeling, which it represents or manifests, in such a way that the same mood or feeling is evoked in the reader. I would suggest that Hölderlin was struck by the implications of both these points in his response to Schiller's theory, but that he rejected both as too simplistic and reductive, substituting instead a more complex and differentiated tripartite scheme for his own theory.

Here we need to acknowledge a further complication to Hölderlin's thinking about the threefold forms of poetry, its tones and its emotional effects, which also derives in large measure from Schiller. As discussions of "tonal variation" recognize, the poet introduces a developmental model to his distinction of tones, which reflects both the essential experiential stages of the individual self and the larger stages of historical development for humankind in general. The Naive, on the one hand, is projected back upon the innocence and simplicity of childhood; the Ideal, on the other hand, is projected forward toward the goal of all education and culture, where the opposition between thought and experience will be resolved into a higher harmony and unity of the imagination. In the middle, defining the inevitable condition of opposition and conflict, is that state which is designated the Heroic.

Hölderlin's scheme for the tones and the genres thus also implies both a theory of education (*Bildung*) for the self and a model of history for humankind as a whole. The origin of such a tripartite view, of course, is to be found in the tradition of salvation history from the Biblical myth of the fall and redemption, or at least in eighteenth-century secular versions of the same, which were important for the development of Idealism. The immediate source for Hölderlin, however, was no doubt the view of history presented by Schiller in the opening section of his *Briefe über die ästhetische Erziehung des Menschen* (*Letters on the Aesthetic Education of Man*) (letters 1 to 9, published in the opening issue of *Die Horen*, January 1795), as well as the adaptation of that same view to the dialectical development of the individual self presented in the second group of letters (10 to 15, published in the second issue a month later), which owe a singular debt to Fichte's *Wissenschaftslehre*. All this occurred for Hölderlin, as Henrich has demonstrated exhaustively, during the early months of 1795 when the poet was in Jena.[15]

The immediate philosophical outcome of such thinking for Hölderlin, where a whole range of factors come together, most particularly the poet's critical reaction to Fichte—again it is Henrich who has clarified this process—may be found in the brief manuscript fragment labeled "Urtheil und Seyn" ("Judgment and Being"), published by Friedrich Beißner in 1961 (*SA* 4; 1: 216–17) and first discussed by Henrich in his essay on the fragment published in the *Hölderlin-Jahrbuch* for 1965–66.[16] Of interest also is the brief discussion of the same fragment by Ryan in his book, who indicates that Beißner made it available to him prior to its initial publication.[17] Ryan has little to say about the historical context of its origin or even about its significance, as Henrich was later to argue, but he places it side by side, correctly I think, with Hölderlin's later essay fragment "Über den Unterschied der Dichtarten" ("On the Difference of the Poetic Genres") (*SA* 4; 1: 266–72).

A further complication is introduced by Hölderlin in the premise that poetic modes may directly effect mental attitudes and moods. This notion has a long history in Western poetics, extending back at least to Plato, who in his discussion of the education of the guardians in the *Republic*, criticizes poetry and music for the emotional effects they may have upon the mind. Precisely such effects define the central hermeneutical function of art for Romantic theory, as has been generally recognized with regard to the shift from a mimetic to an expressive model of meaning (indicated by the title of the seminal study by M. H. Abrams, *The Mirror and the Lamp*).[18] But Hölderlin recognizes, as Hegel was subsequently to do as well under the heading of "Romantische Kunst" in his *Aesthetics*,[19] that the correlation of the work of art with its emotional effect is equally defined by discontinuity and conflict. This insight is central to Hölderlin's poetics as a radical revision of Schiller, yielding a complex theory of poetic metaphor, which is at least as important as the emergence of a theory of irony in the critical writings of Friedrich Schlegel during those same years just before the turn of the century.[20]

This insight, which I am inclined to regard as originating with Hölderlin—though it may well be that his study of the Platonic dialogues during his months in Waltershausen in 1794 prepared the way—stands behind the distinction in the later theoretical essays between three separate terms: *Grundton* ("ground tone"), *Kunstcharakter* ("artistic character") or *Sprache* ("language"), and *Wirkung* ("effect"). This hierarchy of discontinuity between the constitutive components of the poetic tones yields a principle of figuration, or metaphor, whereby the semiotic and the hermeneutic functions of language in a poem come into conflict with each other. This conflict, furthermore, is the basis for the dynamic development of modulating tones through the sequence of statements which constitute the poem. Hölderlin thus appropriates the basic insight of Schiller's genre theory with regard to the sentimental poem according to *Empfindungsarten* ("modes of feeling"), but he complicates the principle of verbal style in such a way that the developmental basis for distinguishing between the tones is built into the structure of the poem as a constitutive element. What is the result of these complicating devices?

Hölderlin's theory of poetic structure according to a principle of "tonal variation" may be primarily philosophical in its origin, but in its effects it is fundamentally musical. Too little attention has been given by critics of Hölderlin's mature poetry to the central metaphor of tone, or even song (*Gesang)*, as a principle of music in the experience of the poem as event. I submit that this poet approached the challenge of poetic creation, regardless of his philosophical insights, essentially as a composer concerned with questions of tonality, as also with the rhythmic and harmonic complexity of utterance, in ways that we have not yet learned to interpret. The basic difference between a hymn of Hölderlin and, say, a symphony of Beethoven—and this of course is crucial—is found in the challenge of reflective thought which constitutes the language of the "sentimental", or Romantic, poem.

In a musical composition the interrelationship of tones constitutes a dynamic system of sound which is essentially self-referential and autonomous. In the case of a poem by Hölderlin, by contrast, the dynamics of modulating tones serve the cognitive and reflective purposes of the language as figurative, or metaphoric, form, so that the essential referent of the poem, regardless of its particular content or subject matter, is always the movement of thought which constitutes it and which it symbolizes or represents. In this way the poem, conceived by Hölderlin to be a modulation of tones through a dialectical structure of metaphor (*die durchgehende Metapher*, to use the phrase from Hölderlin's marginal note about the structure of "Der Rhein" [*SA* 2; 2: 730]), constitutes a system of verbal signs, which signifies through a complex sequence of tension and conflict, as *progress* and *regress* (again borrowing Hölderlin's own terms), the movement of poetic thought itself. This movement stands behind the language as *Grundton*, much in the same way that the tonic key underlies an entire work of music in classical harmony, and is externalized or expressed as verbal event in what Hölderlin calls the *Kunstcharakter*, so that the entire process may be communicated to, and thus comprehended by, a reader who is competent to respond.

3

This brief survey of the intellectual context within which Hölderlin's theory of tonal modulation emerged will probably seem much too simplistic and schematic to those who are familiar with the essay fragments, and opaque and esoteric to those who are not. Nor can there be any attempt made here to clarify more precisely how the influence of such a mentor in theory as Schiller might have affected Hölderlin's thinking about poetry during the four or five years in question subsequent to his departure from Jena. That alone, if thoroughly researched, might yield another book as long as the work by Henrich. The question is what to make of all this.

My own conviction is that Hölderlin must receive priority as a poet, at least insofar as we wish to read his later poetry as poetry. Whatever role his philosophical insights may have played in his attempt to draft a poetics of composition—and I have no doubt that it was crucial—the poems which result from that effort do not in any way depend upon such a theory or derive from it. The theory in itself remains schematic and abstract. But no poem could ever afford to be so, if it is to succeed as poetry. That Hölderlin seems to have known this despite his preoccupation with philosophy is one measure for the legitimacy of his vocation as a poet. And it is as poetry, not as philosophy, that his poems succeed and must be interpreted. For this reason, however arbitrarily, I turn my attention in the remainder of this essay to a brief consideration of the elegy as poetic genre and to Hölderlin's poem "Elegie" or "Menons Klagen um Diotima" as a representative instance—admittedly not yet quite so complex and difficult as the later poems, above all the Pindaric hymns—in order to demonstrate a few details from the text which benefit from an inquiry into the poetic theory which stands behind it.

A few preliminary remarks, which derive in part from the argument of Ziolkowski's book on the classical German elegy, will serve to introduce the focus of my comments. On one hand, as Ziolkowski demonstrates persuasively, the generic model which Hölderlin may be understood to have had in mind was Schiller's greatest poem, initially entitled (as Hölderlin's poem was initially as well) "Elegie" and published in *Die Horen* in the tenth issue of its first year, that is, in October 1795 (just one issue prior to the first section of *Über Naive und Sentimentalische Dichtung*), and subsequently retitled "Der Spaziergang."[21] The essential experiential model for the genre is there established as the journey of the poetic mind through a landscape and its ascent of a mountain to the summit, where a broad vista opens itself to view, bringing illumination and clarification for the thought which motivates it. In part also, as critics generally acknowledge, Hölderlin's own earliest attempt in the genre of elegy, "Der Wanderer" ("The Traveler"), composed for Schiller and published in *Die*

Horen in 1797, shows the influence of his mentor's generic prototype, if not yet with any success in using the principle of tonal variation.

Another crucial point, which Ziolkowski, so far as I know, was the first to make, concerns an immediate example of elegy, from which Hölderlin's poem derives its sense of occasion. Goethe's greatest contribution to the genre of classical elegy as constituted by Schiller is his poem "Euphrosyne," which appeared in Schiller's *Musen-Almanach für das Jahr 1799* (published in 1798), along with two of Hölderlin's short odes. In that poem, an encounter occurs high up in the landscape of the Alps between the poet and the spirit of a recently deceased young actress from the Weimar Theater, Christiane Neumann Becker, who had been one of Goethe's favorite ingenues. This figure of the female tutelary spirit, as Ziolkowski terms it, may be taken to serve as the model for the poet's encounter with the spirit of the dead Diotima in Hölderlin's elegy.[22]

Yet one further source text for the "Elegie" comes readily to mind, though no scholar of Hölderlin has ever noticed it, as far as I am aware. The concluding moment of the novel *Hyperion*, in the final letter of volume 2, also involves an encounter with the spirit of the dead Diotima, whose voice speaks to Hyperion—or seems to do so ("Mir war, als hört ich Diotimas Stimme")—in response to his longing for her, expressed in the midst of a beautiful spring day in an idyllic landscape (*SA* 3: 158–60). The response of Hyperion to this apparent visitation is a long and complex utterance, addressed in part to nature and in part to Diotima, which concludes the novel. I submit that this incident and the speech of Hyperion constitute the true basis for the subsequent elegy (which might thus be titled "Hyperions Klagen um Diotima"). It may well be that Goethe's "Euphrosyne" was also the immediate stimulus for the conclusion of Hölderlin's novel, which must have been completed for publication by Cotta during the early months of 1799. At any rate, this literary and generic situation, not the memory of Susette Gontard, from whom the poet had separated in Frankfurt some time earlier, should be regarded as the source and motive for the poem. Critics have often been inclined to attribute biographical factors as the motivation for a kind of poetry which, I submit, has nothing whatever to do with *Erlebnisdichtung* ("the poetry of experience").

So far as I can surmise, based on a footnote in the essay on this poem by Karl Viëtor, "Hölderlins Liebeselegie,"[23] it was Wilhelm Böhm, in his monumental study of the poet, who first suggested that "Elegie" / "Menons

Klagen" could be understood as a paradigm for the tragic ode according to the theory of tonal variation, though Böhm does not elaborate in any detail.[24] Ryan concurs with this judgment, outlining in a brief survey of the poem by stanzas (nine altogether in the later published version) precisely where the shifts of tone would fall.[25] Ziolkowski picks up on this view, acknowledging Ryan to be his source, and offers a more differentiated schematic outline for the elegy—which I present here in slightly emended form—distinguishing between *Grundton* and *Kunstcharakter* through the seven stages of modulation:[26]

[tonal sequence]	I	II	III	IV	V	VI	VII
Basic Tone:	ideal	naiv	her	naiv	her	ideal	naiv
Artistic Effect:	Her	Ideal	Naiv	Her	Ideal	Naiv	Her
[structural dynamics]	[opposition]		[catastrophe]			[resolution]	
Stanzas	1&2	3	4	5	6	7&8	9

What can be said about such an elaborate structural scheme? I submit that very little, if anything, can be gained for an understanding of the poem based on such abstraction. Nor do I accept the notion that for Hölderlin in composing the poem this tonal scheme could have been more than a point of departure. In order to discover what function this model may serve for the poem itself, we need to examine several verbal details of the elegy in more detail. What follows will necessarily be selective, and I can justify my choices only on the basis of my own reading of the poem. This reading results from an intimate and intensive engagement with the text, heightened above all by the extensive, often tedious exercise of producing a metrical translation (to the best of my prosaic ability), not from any direct, conscious concern with the theory of tonal modulation. That theory, in my view, is destined to remain either a prior or a posterior schema to be applied before or after the act of reading itself.

One central structural feature of this seven-stage sequence of tonal variation is the threefold repetition of the basic tonal determinant of the verbal style, or *Kunstcharakter* ("artistic effect"), of the poem, namely at the beginning, middle and end, respectively (positions I, IV, and VII). What precisely constitutes that quality of the poem as utterance in these sections in such a way that we may call the tone *Heroic*? The simplest answer is found in the figure of wandering for the isolated self, which seeks along an erring path for an "other" (line 1). This is made apparent in the opening couplet of the poem; again—this slight discrepancy of position should be noted—in the final couplet of stanza 4: "Darum irr' ich umher, und wohl, wie die Schatten, so muß ich / Leben" (lines 55–56); and, third, in stanza 9, though less apparent as an image for the journey of life, the assertion of a path (*Bahn*), which must eventually, so the singer now believes, lead to the gods (lines 117–18). The differences between these instances are at least as important as any continuities.

The opening stanza thematizes the pain and the urgent longing of the poet's solitude, using an elaborate epic simile of the wounded stag (*das getroffene Wild* [line 5]), which is also a traditional trope from the pastoral tradition. That quality of pain and longing for the recovery of what has been lost may define with reference to the poet's state of mind what Hölderlin called the *Grundton* of the "ideal." In stanza 5, following the couplet cited above at the end of 4, the consciousness of self-isolation returns with a ruthless and almost obsessive intensity, as indicated only by the frequent use of the first-person pronoun in the singular (thirteen occurrences in twelve lines). This central section of the poem presents an extreme subjectivity or exclusive self-concern on the part of the speaker, which may reflect that quality of *Grundton* which is designated as "naive." The final stanza, by contrast with both of the above, affirms and celebrates a condition of the self as wanderer that is emphatically plural. The pronominal subject of the extended final sentence is "we" (*wir, uns*). The temporal focus of the conclusion is also in the future, accompanied by invocations of several divine agencies through an appeal that they "abide with us" ("Bleibt so lange mit uns, bis" [line 123]) and followed by a triumphant catalogue of attributes that delineate a realm of bliss, where the poet with his companion hopes to dwell in the presence of the gods. According to the schema for tonal variations, this last stage exhibits as *Grundton* again the "naive"—yet how different that quality in stanza 9 is from the central instance in stanza 5!

Particularly important for the schema of tonal modulation is the central fourth stage, where a categorical reversal, or catastrophe, is argued to occur in the mode of variation (as outlined in the diagram). To follow the designations for *Grundton* and *Kunstcharakter*, *herNaiv* is succeeded by *naivHer*. How and where does this reversal occur? We may locate the change quite precisely, I suggest, in the final four lines of stanza 4, which constitute a transition from the recollection of past bliss, which the poet shared with his lost beloved, and introduce the heightened consciousness that such bliss has been irretrievably lost:

> Aber das Haus ist öde mir nun, und sie haben mein Auge
> Mir genommen, auch mich hab' ich verloren mit ihr.
> Darum irr' ich umher, und wohl, wie die Schatten, so muß ich
> Leben, und sinnlos dünkt lange das Übrige mir.
> (lines 53–56)

This central section of the poem—central in the sequence of modulating tones, though not precisely the central passage by line count—sustains the most powerful moment of self-consciousness for the poet as he speaks within a situation of absolute. This solitude defines his central condition and the overriding motive force for his poem as *lamentation* (*Klagen*). Few passages in all of Hölderlin's work offer such a sustained sense of existential isolation and suffering. His lament indeed achieves a climax of intensity in the conclusion of stanza 5, which is unsurpassed anywhere else in the poem:

> Aber mir in schaudernder Brust die beseelende Sonne,
> Kühl und fruchtlos mir dämmert wie Stralen der Nacht.
> Ach! und nichtig und leer, wie Gefängnißwände der Himmel
> Eine beugende Last über dem Haupte mir hängt!
> (lines 65–68)

Few sections of the elegy offer less apparent difficulty for a reading in accord with the schema of tonal modulation than the third stage, which is introduced at the outset of stanza 4 by the second epic simile of the poem, where the poet compares the condition of love he once enjoyed with the absent Diotima to a pair of swans as they rest upon the surface of a lake and are rocked gently by the waves. As with the wounded stag, there is a sense of traditional iconography to the image, in a way that thematizes the convention of love evoked by the poet as norm for the *Kunstcharakter* of the Naive. But the *Grundton*, according to the schema is "heroic," presumably because of the awareness throughout that this state of love is only being evoked through the power of memory. The function of temporal distance, according to a familiar dialectic in Hölderlin between *einst* ("once") and *jetzt* ("now"), is to heighten the awareness of absolute contrast in the section shortly to follow, which is introduced (as argued already) by the lines which conclude the stanza.

One word more should be said about the dynamics of modulation in the opening sections of the poem before we turn to the more complex and elaborate movement of the latter half. A central rhetorical device throughout the poem is defined by shifts in modes of address. Given the dominant consciousness of isolation and solitude for the poet, it is crucial that he achieve a mode of discourse that projects his voice beyond such solitude into at least a hypothetical mode of dialogue and pattern of communication. In the final couplet of the opening stanza, cited as the motto to this paper, the poet asks with some urgency whether there is anyone who can remove from his brow the burden of his mournful dream. I take this to be, as often in Hölderlin, a reference to his own poem as visionary act, which seeks to convey that vision through language to its readers. For this reason I am inclined to refer the otherwise enigmatic phrase of direct address at the beginning of the couplet—"ihr Lieben!" (line 13)—to the hypothetical audience of the poem, including ourselves as readers, who might in truth provide the relief, which the poet seeks, from his mournful dream.

The second stanza introduces in its opening line a very different focus of direct address: "ihr Todesgötter!" (line 15), subsequently elaborated further (two lines later) as "ihr Bösen." An imagined condition of bondage in the realm of death is thus projected by the poet upon his solitary state. Within this imagined place his discourse then assumes a mode of self-address, where the poet is able to speak in the second person to his own soul—"o meine Seele!" (line 23)—before returning in the final lines to the more subjective and solipsistic focus upon himself as *ich*, using the first-person singular.

All the more surprising, therefore, when the third stanza—signaling the first modulation of tone according to the schema (from *ideal-Heroisch* to *naivIdeal*)—begins with yet another apostrophe of direct address: "Licht der Liebe!" (line 29). This could be taken as a literal reference to the sunlight, especially since an elaborate sequence of invocations to objects and agents in nature follows shortly. Yet the final line of the stanza offers a definitive signal that the entire stanza is to be regarded at least hypothetically as an address to the absent beloved. At that point—it is the only instance in the entire poem (as critics have noted)—the beloved named in the title is addressed directly, "*Diotima!*" (line 42).

Of particular interest, if this link of apostrophic moments is valid, is the series of no less than six discrete invocations in as many lines (31–36) and the question of how to relate these referents of the poet's address to the beloved. The sequence may quickly be surveyed: *liebliche Gärten, ihr abendröthlichen Berge, schweigende Pfade des Hains, hochschauende Sterne, stille Rosen und . . . Lilien* (= *ihr schönen Kinder des Maitags* and, presumably, *ihr Liebenden*). What this sequence establishes implicitly is a naturalistic context, or even a sense of spiritual community within the agencies of nature, where the poet's address to his absent beloved might be possible and valid. This proves to be the case, as the latter sections of the elegy show.

Ultimately, of course, the poem establishes a sense of presence in stanzas 7 and 8, which constitutes the penultimate stage in the tonal modulation, where the dead beloved appears again to the poet in the realm of the living. This has led some readers to argue that the poet has affirmed the power of his own art to resurrect the dead, a claim which in my view is absurd. In fact, Hölderlin's purpose is quite the opposite, in keeping with his central plea that his mournful dream be lifted from his brow. What he affirms through his vision of the beloved as presence is precisely the realization of that vision as verbal artifact *for its reader* (that is, for us), who are obliged to affirm, if anyone can, the validity of that vision as more than a dream in solitude. The final stanza begins with an invocation to the gods as heavenly ones (*ihr Himmlischen* [line 109]) in the form of thanks. As the remainder of the couplet affirms, this form of thanks is in truth what the poem itself, as the poet's prayer, was intended to achieve from the start and now has truly become:

> So will ich, ihr Himmlischen! denn auch danken, und endlich
> Athmet aus leichter Brust wieder des Sängers Gebet.
> (lines 109–10)

The final affirmation of the poem includes the figure of an oracle spoken as if by a god from a temple located upon the silvery summits of Apollo's mountain range (lines 111–14). The exclamation which follows—is it meant to be the actual words of the oracle?—asserts that the dream has been resolved, the wound of the stag in the initial simile has been healed in the bleeding wings (*Fittige*) of the swan from the second simile, so that, in accord with a prophecy of rebirth in a third simile (which we will consider shortly), all the hopes of the poet have been rejuvenated:

> Komm! es war wie ein Traum! die blutenden Fittige sind ja
> Schon genesen, verjüngt leben die Hoffnungen all.
> (lines 115–16)

What has happened in the interval extending from the ruthless negative self-consciousness in stanza 5 to the affirmation of a festival of communion in stanza 9? The key is found in stanza 6, which constitutes the fifth stage in the tonal sequence (designated *heroischIdealisch*). The stanza begins with a question, which is answered by yet a third epic simile, more elaborate and oblique than the two which precede, and concludes with an equally elaborate twofold conditional statement concerning the necessary prerequisite for the affirmation of spiritual presence to follow. A few remarks are needed to clarify what is happening here, in order to relate this stage in the sequence to the design of the poem as a whole.

Once more at issue in the opening question is the imaginary path which the poet's thought pursues, searching for a way back, praying that he may through his poem (as the vehicle of his prayers) retrieve what is lost, here apostrophized as "youth" (*o Jugend* [line 69]). The metaphor of the wandering or erring path, which recurs at intervals throughout the poem, signifies, in ways that can only be intuited from the movement of the poem itself as a dialectic of thought, precisely what the sequence of modulating tones is intended to achieve through its eccentric path from the initial going out until the final prospect of return.

The simile which fills the center of the stanza (lines 71–77) evokes a myth of violence and death at the festive tables of the gods. This may recall the story of Tantalus or perhaps the tale of the centaurs at the wedding feast of the Lapiths. More generally, it may allude to the conflict of divinity with the destructive power of the Titans as analog to the poet's own suffering and loss. The details of this mythical allusion are less important for my purpose than the implication of the simile for the poet's own state of mind. He asks, to paraphrase, whether he also will share the fate of those "godless ones" who once were guilty of excess and violence as guests at the tables of the gods and who were therefore banished into a silence and a sleep below the breezes and the Earth, where they will one day be forced by the power of a miracle to return and to move anew upon the green Earth. The simile signifies a process of death and rebirth that may also be associated figuratively with the cycle of the seasons in nature. But the simile provides the poem with a figure for the transformation and renewal of spirit that the poet himself hopes to achieve in and through the imaginative path of the poem. This transformation, I submit, is precisely what the modulation of tones itself is intended to effect, at least within the verbal domain of the poet's art, as an experience which we as readers may also share as poetic event through the act of reading. In this way, the simile itself answers the question which the poet was asking, almost to the point (indicated by the dash at the end of line 77) where the mythical event referred to by the simile is intuited as an immediate actuality in and for the poem.

An illusion of presence is thus evoked powerfully in the line which begins the final section of the stanza: "Heiliger Othem durchströmt göttlich die lichte Gestalt" (line 78). The "form" of the dead Diotima has been given the power to breathe again with "sacred breath," like a goddess. Such an affirmation of presence would be premature at this juncture, as indicated by the series of qualifying conditions, which define the specific prerequisite circumstances that must prevail before this "luminous form" will once again "flow" with "sacred breath." Festival, floods of love, the rushing of a living stream, resonances of sound, the payment of treasure by night, and the sparkling of buried gold from rivers: these various figures of renewal suggest an entire fabric of metaphoric associations, where the mythical, the natural, the historical (I note an allusion to the legend of Croesus and the gold of the River Pactol in Asia Minor, also cited earlier by Hölderlin in his novel *Hyperion* [*SA* 3: 20–21]), and the spiritual are all united within a process that may best be termed musical or poetic. At some point the attentive reader will perceive the extent to which Hölderlin here pursues a form of self-reflection, where the figures which constitute the language of the poem also signify precisely those mental or experiential processes which they are intended to represent and make manifest. No other purpose is served by the complex sequence of modulated tones that constitutes the poem itself in its full articulation as verbal construct, and it is to that elaborate process of articulation that the sequence of tones refers as the hermeneutical figure of its own performance.

4

Within the confines of this essay, it is impossible to discuss the richness of celebration and affirmation in the sixth stage of the elegy, extending through stanzas 7 and 8, where a sense of revelation is proclaimed, even if it remains only a fiction of the poetic imagination. The sequence of attributes for the spirit of the beloved, arranged throughout these twenty-six lines in a complex pattern of epiphanic apostrophe, indicates the mood or tone (designated as *idealischNaiv*): *Götterkind!*; *Freudiger Schutzgeist!*; *o Heldin!*; *o Gütige; freundlicher Geist!* The entire sequence is also constituted in the form of dialogue as a direct address by the poet as *ich* to the beloved as *du*. Increasingly, as the celebration of presence proceeds, a dramatic sense of immediacy in the tense of the verbs is established, which reaches its extreme in the couplet where the beloved is identified as an Athenian, almost as if she herself were to coalesce in epiphany with the goddess Athena:

> Ja! noch ist sie es ganz! noch schwebt vom Haupte zur Sohle,
> Stillherwandelnd, wie sonst, mir die Athenerin vor.
> (lines 101–2)

We may return in conclusion to the question with which we began, namely the role of Hölderlin's philosophical thought as a contributing factor in the formal and stylistic organization of his later poetry. Insofar as the elegy "Menons Klagen um Diotima" may be interpreted as fulfilling a structural and compositional program defined by the seven-stage sequence of tonal modulation, we may ask whether the philosophical suppositions which lie behind such a poetics directly effect the experience conveyed by the poem to us as readers. I confess that I have no answer to this question. No one is likely to confuse Hölderlin's philosophical thought with his actual practice as a poet. Yet clearly the organizing principle and the structural complexity of such an elegy as this reflect a fundamental hermeneutical concern. Not only is the *Grundton* which underlies the poem as a whole translated through a dialectical process of conflict and reversal into the sequential pattern of modulations that constitutes its *Kunstcharakter*. Such strategies also pertain to the challenge of expression and articulation, whereby the spiritual context of the poem is translated into the fabric of its language in order to reproduce and communicate this experience to its readers as *Wirkung* ("effect"). In this sense, the primary function of the tonal modulation is hermeneutical, imposing upon the reader an obligation to participate in the full cognitive complexity of the poem as performance, like a symphony of reflective thought. The dialectic of tonal variation thus effects a correlative mode of self-consciousness *in us* that will affirm the poet's vision of renewal and hope for revelation, not as mere dream or imagined possibility, but as an immediate poetic event, which takes place for us here and now, in and through the act of reading the poem. We should not forget that in all the diagrams which Hölderlin sketched for the sequence of tonal modulation where the criterion of "effect" (*Wirkung*) was also included, the sequence of these effects is always identical to the sequence of the *Grundton*. This would seem to indicate that a norm of perfect understanding is the goal for the entire process.

With regard to the implicit developmental basis for the fundamental poetic tones, furthermore, we may also define a process of temporal internalization as constitutive of the poet's reflective or meditative form of thinking. The Naive tone, which predominates in stanza 4 of the elegy, is projected back upon a remembered state of bliss in the past of his experience. Yet the act of remembering this state only serves to heighten the consciousness of present solitude, suffering, and loss, which is essentially Heroic. The Ideal tone is likewise projected into the future as a form of expectation or conviction concerning that which is to come. What the elegy seeks to affirm, however, through the dialectical movement of its thought—its conflict, reversal, and resolution—is an affirmation of this future conviction as a present condition *for the poem* as the embodiment of such thought. This event of affirmation, the supreme fiction of the poetic imagination, may be called—as Hölderlin frequently did in his theoretical essays—the metaphor of an Intellectual Intuition (*die Metaphor emir intellektuellen Anschauung*). We may understand this figure in "Menons Klagen um Diotima" to be the affirmation of presence with reference to the beloved as the reality upon which the entire meaning of the poem is based. Our task as readers is to comprehend precisely how such an affirmation may be possible and valid for the poet's thought and experience as articulated in and through the language of the poem.

Finally, the question of the poem as metaphor or figure of thought remains as the central challenge of Hölderlin's poetics. Everything is figure and thus fiction; nothing in all his mature work can be regarded as mimetically true. Even where the external reference for his poetic message seems to be geographically exact and real, as in the evocation of the River Rhine in the hymn of that name or in the recollection of the landscape of Bordeaux in "Andenken" (*SA* 2; 1: 142–48 and 188–89, respectively), the poetic function of his language translates all description and recollection into the metaphoric continuum of the tonal sequence. Who would argue that the landscape through which the poet travels in "Heimkunft," on his return to Swabia from the Alps, or the historical account of the battle of Salamis in "Der Archipelagus" could be understood apart from the mode of poetic representation achieved in Hölderlin's own language and in the formal strategies of his poem (*SA* 2; 1: 96–99 and 103–12, respectively)? If we may thus speak of the tonal modulations which make up his poetry under the heading of metaphor or poetic figuration, we mean by that no more and no less than the process whereby the language of poetry embodies and represents the processes of the mind itself, as figures of a capable imagination, which reflect back upon the very dynamics of thought through which they come to be and which alone they signify and trace.

To some degree this kind of creative transformation may always pertain in poetry. All great poets know that one way or another the medium of their art is only the fictional construct for the cognitive processes that underlie it. It is always the task of their readers to discover precisely how such concerns may contribute to the challenge of reading and interpreting their poems. In Hölderlin's case, however, there was a philosophical foundation for his entire poetic career which was both more original and more profound than for any other comparable writer of his own time or any other. We are still struggling at an early stage to discover what such philosophical concerns may imply for the achievement of his poetry. Yet, as readers, our obligation to the poetry remains the same as it ever was: we must endeavor to read the poems with sufficient awareness and insight to claim even a preliminary legitimacy for our understanding. In the case of Hölderlin, as I expect even the most philosophical of his critics would agree, we still stand near the beginning.

Appendix

Menons Klagen um Diotima

1.

Täglich geh' ich heraus und such' ein Anderes immer,
Habe längst sie befragt alle die Pfade des Lands;
Droben die kühlenden Höhn, die Schatten alle besuch' ich,
Und die Quellen; hinauf irret der Geist und hinab,
Ruh' erbittend; so flieht das getroffene Wild in die Wälder,
Wo es um Mittag sonst sicher im Dunkel geruht;
Aber nimmer erquikt sein grünes Lager das Herz ihm,
Jammernd und schlummerlos treibt es der Stachel umher.
Nicht die Wärme des Lichts, und nicht die Kühle der Nacht hilft,
Und in Woogen des Stroms taucht es die Wunden umsonst.
Und wie ihm vergebens die Erd' ihr fröhliches Heilkraut
Reicht, und das gährende Blut keiner der Zephyre stillt,

So, ihr Lieben! auch mir, so will es scheinen, und niemand
  Kann von der Stirne mir nehmen den traurigen Traum?

2.

Ja! es frommet auch nicht, ihr Todesgötter! wenn einmal
  Ihr ihn haltet, und vest habt den bezwungenen Mann,
Wenn ihr Bösen hinab in die schaurige Nacht ihn genommen,
  Dann zu suchen, zu flehn, oder zu zürnen mit euch,
Oder gedultig auch wohl im furchtsamen Banne zu wohnen,
  Und mit Lächeln von euch hören das nüchterne Lied.
Soll es seyn, so vergiß dein Heil, und schlummere klanglos!
  Aber doch quillt ein Laut hoffend im Busen dir auf,
Immer kannst du noch nicht, o meine Seele! noch kannst du's
  Nicht gewohnen, und träumst mitten im eisernen Schlaf!
Festzeit hab' ich nicht, doch möcht' ich die Loke bekränzen;
  Bin ich allein denn nicht? aber ein Freundliches muß
Fernher nahe mir seyn, und lächeln muß ich und staunen,
  Wie so seelig doch auch mitten im Laide mir ist.

3.

Licht der Liebe! scheinest du denn auch Todten, du goldnes!
  Bilder aus hellerer Zeit leuchtet ihr mir in die Nacht?
Liebliche Gärten seid, ihr abendröthlichen Berge,
  Seid willkommen und ihr, schweigende Pfade des Hains,
Zeugen himmlischen Glüks, und ihr, hochschauende Sterne,
  Die mir damals so oft seegnende Blike gegönnt!
Euch, ihr Liebenden auch, ihr schönen Kinder des Maitags,
  Stille Rosen und euch, Lilien, nenn' ich noch oft!
Wohl gehn Frühlinge fort, ein Jahr verdränget das andre,
  Wechselnd und streitend, so tost droben vorüber die Zeit
Über sterblichem Haupt, doch nicht vor seeligen Augen,
  Und den Liebenden ist anderes Leben geschenkt.
Denn sie alle die Tag' und Jahre der Sterne, sie waren
  Diotima! um uns innig und ewig vereint;

4.

Aber wir, zufrieden gesellt, wie die liebenden Schwäne,
  Wenn sie ruhen am See, oder, auf Wellen gewiegt,
Niedersehn in die Wasser, wo silberne Wolken sich spiegeln,
  Und ätherisches Blau unter den Schiffenden wallt
So auf Erden wandelten wir. Und drohte der Nord auch,
  Er, der Liebenden Feind, klagenbereitend, und fiel
Von den Ästen das Laub, und flog im Winde der Reegen,
  Ruhig lächelten wir, fühlten den eigenen Gott

Unter trautem Gespräch; in Einem Seelengesange,
  Ganz in Frieden mit uns kindlich und freudig allein.
Aber das Haus ist öde mir nun, und sie haben mein Auge
  Mir genommen, auch mich hab' ich verloren mit ihr.
Darum irr' ich umher, und wohl, wie die Schatten, so muß ich
  Leben, und sinnlos dünkt lange das Übrige mir.

5.

Feiern möcht' ich; aber wofür? und singen mit Andern,
  Aber so einsam fehlt jegliches Göttliche mir.
Diß ist's, diß mein Gebrechen, ich weiß, es lähmet ein Fluch mir
  Darum die Sehnen, und wirft, wo ich beginne, mich hin,
Daß ich fühllos size den Tag, und stumm wie die Kinder;
  Nur vom Auge mir kalt öfters die Thräne noch schleicht,
Und die Pflanze des Felds, und der Vögel Singen mich trüb macht,
  Weil mit Freuden auch sie Boten des Himmlischen sind.
Aber mir in schaudernder Brust die beseelende Sonne,
  Kühl und fruchtlos mir dämmert wie Stralen der Nacht.
Ach! und nichtig und leer, wie Gefängnißwände der Himmel
  Eine beugende Last über dem Haupte mir hängt!

6.

Sonst mir anders bekannt! o Jugend, und bringen Gebete
  Dich nicht wieder, dich nie? führet kein Pfad mich zurük?
Soll es werden auch mir, wie den Götterlosen, die vormals
  Glänzenden Auges doch auch saßen am seeligen Tisch',
Aber übersättiget bald, die schwärmenden Gäste
  Nun verstummet, und nun unter der Lüfte Gesang,
Unter blühender Erd' entschlafen sind, bis dereinst sie
  Eines Wunders Gewalt sie, die Versunkenen, zwingt,
Wiederzukehren, und neu auf grünendem Boden zu wandeln.—
  Heiliger Othem durchströmt göttlich die lichte Gestalt,
Wenn das Fest sich beseelt, und Fluthen der Liebe sich regen,
  Und vom Himmel getränkt, rauscht der lebendige Strom,
Wenn es drunten ertönt, und ihre Schäze die Nacht zollt,
  Und aus Bächen herauf glänzt das begrabene Gold.—

7.

Aber o du, die schon am Scheidewege mir damals,
  Da ich versank vor dir, tröstend ein Schöneres wies,
Du, die Großes zu sehn, und froher die Götter zu singen,
  Schweigend, wie sie, mich einst stillebegeisternd gelehrt;
Götterkind! erscheinest du mir, und grüßest, wie einst, mich,
  Redest wieder, wie einst, höhere Dinge mir zu?

Siehe! weinen vor dir, und klagen muß ich, wenn schon noch,
  Denkend edlerer Zeit, dessen die Seele sich schämt.
Denn so lange, so lang auf matten Pfaden der Erde
  Hab' ich, deiner gewohnt, dich in der Irre gesucht,
Freudiger Schuzgeist! aber umsonst, und Jahre zerrannen,
  Seit wir ahnend um uns glänzen die Abende sahn.

8.

Dich nur, dich erhält dein Licht, o Heldin! im Lichte,
  Und dein Dulden erhält liebend, o Gütige, dich;
Und nicht einmal bist du allein; Gespielen genug sind,
  Wo du blühest und ruhst unter den Rosen des Jahrs;
Und der Vater, er selbst, durch sanftumathmende Musen
  Sendet die zärtlichen Wiegengesänge dir zu.
Ja! noch ist sie es ganz! noch schwebt vom Haupte zur Sohle,
  Stillherwandelnd, wie sonst, mir die Athenerin vor.
Und wie! freundlicher Geist! von heitersinnender Stirne
  Seegnend und sicher dein Stral unter die Sterblichen fällt;
So bezeugest du mir's, und sagst mir's, daß ich es andern
  Wiedersage, denn auch Andere glauben es nicht,
Daß unsterblicher doch, denn Sorg' und Zürnen, die Freude
  Und ein goldener Tag täglich am Ende noch ist.

9.

So will ich, ihr Himmlischen! denn auch danken, und endlich
  Athmet aus leichter Brust wieder des Sängers Gebet.
Und wie, wenn ich mit ihr, auf sonniger Höhe mit ihr stand,
  Spricht belebend ein Gott innen vom Tempel mich an.
Leben will ich denn auch! schon grünt's! wie von heiliger Leier
  Ruft es von silbernen Bergen Apollons voran!
Komm! es war wie ein Traum! die blutenden Fittige sind ja
  Schon genesen, verjüngt leben die Hoffnungen all.
Großes zu finden, ist viel, ist viel noch übrig, und wer so
  Liebte, gehet, er muß, gehet zu Göttern die Bahn.
Und geleitet ihr uns, ihr Weihestunden! ihr ernsten,
  Jugendlichen! o bleibt, heilige Ahnungen, ihr
Fromme Bitten! und ihr Begeisterungen und all ihr
  Guten Genien, die gerne bei Liebenden sind;
Bleibt so lange mit uns, bis wir auf gemeinsamem Boden
  Dort, wo die Seeligen all niederzukehren bereit,
Dort, wo die Adler sind, die Gestirne, die Boten des Vaters,
  Dort, wo die Musen, woher Helden und Liebende sind,
Dort uns, oder auch hier, auf thauender Insel begegnen,

Wo die Unsrigen erst, blühend in Gärten gesellt,
Wo die Gesänge wahr, und länger die Frühlinge schön sind,
Und von neuem ein Jahr unserer Seele beginnt.

Meno's Lament for Diotima

1.

Every day I go out and always search for an other, [ideal Heroic]
Long ago have I asked every path of the land;
Cooling heights above and all the shadows I visit,
And the springs; my mind wanders thus upward and down,
Begging for rest; so flees a wounded stag to the forests,
Where he was used to rest safely at noon in the dark;
But his green retreat refreshes his spirit no longer;
Moaning and sleepless, the barb goads him forever about.
Not the warmth of the light and not the coolness of night helps,
Washing in waves of the stream never will succor his wounds.
And as the Earth presents him in vain with her merciful healing
Herbs, and his boiling blood none of the zephyrs can soothe;
So, you loved ones, it seems so also to me, and will no one
Ever take from my forehead such a sorrowful dream?

2.

Yes, it is useless, you gods of death, when you once have
Made him captive and hold fast such a soul in defeat,
When into horrid night, you evil ones, you have dragged him,
Then to search, to implore, or to be angry with you,
Or indeed to dwell under fearsome ban with such patience,
And with a smile from you, study your sobering song.
If it must be, forget your health and soundlessly slumber!
Yet still a sound wells up hopeful within your own breast,
Even now, oh my soul, you cannot yet, still not grow yet
Used to it all, and you dream deep in an iron repose!
Festival have I none, yet I long to be crowned with the laurel;
Am I then not alone? but some kind thing from afar
Must be near me still, and I need to smile and to wonder
How I enjoy such bliss here in the midst of my grief.

3.

Light of love, do you shine for the dead then also, you golden? [naive Ideal]
Visions from brighter times, do you enlighten my night?
Lovely gardens, be welcome, you mountains crimson at sunset,
May you be welcome and you, pathways of quiet retreat,

Witnessing heavenly joy, and watching on high, all you starlights,
 Often you granted me then visages filled with such bliss!
You, so loving as well, you beautiful children of springtime,
 Silent roses and you, lilies, I often proclaim!
Seasons of spring pass by, and one year displaces another,
 Changing and struggling, thus time rages above and beyond,
Over my mortal head, yet not for the eyes of the blessed,
 And all the lovers receive gifts of a different life.
For they all, the days and years of the stars, they were also,
 Diotima! with us, inward, eternally one;

4.

Yet we two, like the loving swans, together and peaceful, [heroic Naive]
 Gently rocked by the waves, when they're at rest on the lake,
Gaze down into the water, where clouds of silver are mirrored,
 And an ethereal blue rises up under the boat,
Thus upon Earth did we wander. Even if winter's blast threatened,
 Hostile to lovers' joy, bearing laments, and from boughs
Fell all the leaves, and down flew the rain on the windstorm,
 Both of us smiled at ease, feeling the god who was ours
In our intimate conversation; in one single soul-song,
 Wholly at peace with ourselves, childlike and joyful alone.
Yet my house is desolate now, my eyes have been taken
 From me, and even myself now has been taken with her.
Therefore I err about, and truly, like spirits, I still must
 Live, and whatever is left over seems senseless to me.

5.

Let me celebrate; yet for what? and sing with the others, [naive Heroic]
 Yet in such loneliness everything godlike is gone.
This it is, my affliction, I know it, that I am so crippled,
 Cursed in my sinews and thrown downward, where'er I begin,
So I sit all the day unfeeling and mute, just like children;
 Only a tear from my eyes often still slips coldly down,
And the flowers of the field and the song of birds make me mournful,
 Since in their joy they too bring us good news from the gods.
But for me in my shuddering heart the enlivening sunlight
 Dawns so fruitless and cold, just like the flashes of night,
Ah! and for naught and blank, the heavens, like walls of a prison,
 Hang down over my head, bending me under the load!

6.

Once it was known to me otherwise; oh youth, and will prayers [heroic Ideal]
 Never bring you again? never a path lead me back?

Will it also happen to me, like the godless, who also
  Sat once with shining eyes round at the heavenly board,
But were surfeited all too soon, those fanatical feasters,
  Now they are mute and now, under the song of the breeze,
Under the blossoming Earth are passed away, until someday
  Forced by a miracle's strength, all those who sank in the earth,
Come again and wander anew upon fresh grassy meadows.—
  Godlike, a sacred breath blows through the luminous form,
When the feast is inspired, surge upward floods of desire, and
  Drenched with heaven's rain, rivers sing triumphs of life,
When it resounds from the deep, and night exhibits her treasures
  And from the riverbeds glitters the long buried gold.—

7.

But you showed me even then, at our crossroads of parting, [ideal Naive]
  When I was sinking away, beauty as solace more fair.
You, who once taught me to see what was great with silent conviction
  And more gladly to sing, silent like them, of the gods;
Do you, child of the gods!, appear as before and thus greet me,
  Speak to me once again, promising higher rewards?
Look! I weep and lament for you, though I still can
  Think about nobler times, I in my soul am ashamed.
For so long, so long, I have searched for you, in my erring,
  Knowing you on the Earth's tedious paths, but in vain,
Joyful guardian spirit, and years have already slipped past us,
  Since we expectantly saw evenings around us aglow.

8.

You, just you, your light sustains, heroic, in light and
  Loving, your patience sustains you, my benevolent friend!
And you're not alone at all; there are playmates in plenty,
  Where you blossom and rest under the roses of spring;
And the father himself, through the muses' delicate breathing,
  Sends you his tender songs, lullabies only for you.
Yes! she still is wholly herself! still quietly moving,
  Head to foot, as before, stands the Athenian here.
Friendly spirit!, and as from your cheerful forehead, reflective,
  Blissful and certain, your rays shine on this mortal domain;
You bear witness for me and you tell it to me, so that I may
  Tell it to others, for they cannot believe it is true,
That, in the end, more immortal than any outrage and worry,
  Joy and a golden day daily at last still can be.

9.

So, you heavenly ones! I wish to give thanks, and at last there [naive Heroic]
Breathes from a lighter heart once more the singer's appeal.
Just as once, when I stood with her, on a luminous hilltop,
Giving me life, a god speaks from his temple within.
For I too want to live! Things now ripen! And as from a sacred
Lyre Apollo calls, echoed from silvery peaks,
Come! it was all like a dream! My bleeding wings are now truly
Healed, and all of my hopes live once again and are new.
Much, so much, is still left, to find what is great, and whoever
Loved so, his path, it must, leads ever on to the gods.
If you accompany us, you sacred seasons, you solemn,
Youthful ones! oh abide, holy surmises, and you
Pious appeals! and you inspirations, and all other kindly
Geniuses, who are glad just to be near those in love,
Stay with us long enough, until upon ground shared in common,
There, where the blessed ones all are prepared to return,
There, where the eagles are and the stars, with news from the father,
There, where the muses and where heroes and lovers reside,
There we shall meet, or here as well, on some temperate island,
Where our own kind are first gathered in gardens abloom,
Where the songs are true, and springtime is beautiful longer,
And it all starts anew, yet one more year of our soul.

"Brod und Wein"

FROM THE "CLASSICAL" FINAL VERSION TO THE LATER REVISION

Bernhard Böschenstein

There are three possible ways to inquire into the radical break Hölderlin made with his classical production of the year 1800 after his return from Bordeaux (1802) and during his second stay in Homburg (1804–6): one can concentrate on the study of the late hymn fragments, on the detailed analysis of the deviations of his translation of *Antigone* from the Greek original, or on a comparative reading of the "classical" and the revised versions of the great elegies. This essay takes the third path with the help of the most productive example, the elegy "Brod und Wein" ("Bread and Wine").

The result is breathtaking. Hölderlin separates himself from sections of the poem that speak of an ancient or modern mediation between gods and men, of a symbiosis of Dionysus and Christ, and of the priesthood of Dionysus that Heinse sets up as a model for Hölderlin. In place of such a dialogue we find images of a destructive invasion of the divine fire and of a ravaged Earth, as well as the attestation of the ascetic funeral art (in the spirit of Hölderlin's reinterpretation of the Danaë stanza of the fourth stasimon in *Antigone*) and of a defense which counters, with the help of laws founded in mathematics, dangers from above and below. Accordingly, the balanced speech of classical distichs breaks apart and turns into brittle, piled-up blocks of language that give rhythmic and musical indications of the break with the past. A parallel to this is the appeal to St. Paul as principal authority, who in Ephesus condemned the cult of ancient statues. In contrast, Dionysus, who in the "classical" version embodied ancient drunkenness, festiveness, and conciliatory community between humans and gods, now retreats, laughing, into a wilderness rich with vegetation.

The radical nature of these innovations has editorial consequences as well, and these should be kept in mind in the context of the following observations.

In retrospect it becomes clear how dangerous Hölderlin's utopian hope for a return of ancient religiousness, established anew and fitting for his time, was for the poet, who now feels safe from elemental excess only within strict borders of cognition. It is toward this perspective that the shattered poem must become readable, and readable with a precision that must be derived by the individual variations in words and images. Thus, comparative examination can sharpen our consciousness of what illusory basis Hölderlin's classicism was erected on.

I do not have the ambition of establishing a new text construction for the late revision of "Bread and Wine," because I am still of the opinion that it is not possible to construct reliably, from the nonrevised and the revised parts of the elegy, a text that would have corresponded to Hölderlin's intention in this heterogeneous form.[1] Rather, I am convinced that the revised parts corresponded to a much more urgent need for change in Hölderlin than the ones that remained unchanged, but that the process of revision was not in any way completed, and that the situation that is now documented by the last handwritten entries captures a phase in a process that was interrupted by external circumstances. Within Hölderlin's interventions I believe I can recognize regular patterns, which have already been pointed out by, especially, Jochen Schmidt, Wolfram Groddeck, and Werner Almhofer.[2] I will, however, adopt their interpretations only in part; I will also change them decisively, and complete them and expand them by considering new aspects. The new perspectives—especially in the last triad—should have priority, although of course I cannot interpret every single late variant.

Why did Hölderlin make so few changes in the first triad of the elegy and so many in the second and third?

The first and second stanza have to do with the night, and there does not seem to be any direct representation there of the contact with the gods. On the contrary, their concealment is the decisive work of the night. Precisely this concealment, however, responds to the new needs that help us understand the late revision: "Aber das Saitenspiel *tönt fern* aus Gärten" ("but the music of strings *sounds from afar* out of gardens"); "der Mond / Kommet *geheim* nun auch" ("the moon / now also comes in secret"); and the night is called "die *Fremdlingin* unter den Menschen" ("the stranger among men").[3] Thus the first stanza establishes three times a distance between an "other" presence and humanity. In the second stanza it is no longer this distance that is presented, but rather the night itself in its effective obscuring of the highest God and his world of day, with a clear emphasis on a mysterious motivation, and with the contrasting determination of whether things are classified under memory or forgetfulness, under the memory of the dead or sleep, under pausing or heightened life, in the medium of the Dionysian, ecstatic wakefulness. Since these passages have to do mainly with the security found "in darkness," the dangerous side of the poem, the side about the contact with the gods, is not yet expressed. Thus the second stanza could, like the first, remain untouched. The third, however, will show us the first two major changes, both of which are of great consequence.

Instead of "Göttliches Feuer auch treibet, bei Tag und bei Nacht, / Aufzubrechen" ("the divine fire also drives, by day and night, to start") (32: 18–20), the poem now reads (I follow now Dietrich Uffhausen, not D. E. Sattler and Wolfram Groddeck) "Zeichen des Himmels auch singen, bei Tag und bei Nacht, / Witterungen" ("weather manifestations also sing signs of the heavens, by day and night").[4] The adjective *göttlich* ("divine") now designates something too dangerous to be kept in the poem. The fire, in the form of lightning from heaven, would also now appear destructive if it had an effect on human beings. Three protective measures become necessary: weather manifestations (as subject), signs (as objects), and song (as predicate) replace the divine fire driving to a start. One can stress the close relationship between "singing" and sign, so that the song can be understood now starting with the sign. The function of thunder as a sign of the supreme father is attested in Hölderlin in multiple ways. But that its appearance is described as a singing, in connection with its quality as a sign, produces an analogical connection to the poetic deciphering of the writing on the heaven (it means the writing of the clouds on the sky) (*SA* 2; 1: 209, lines 3–4) and clarifies the new conception of song, in Hölderlin's later poetry, as a protective sign language.

The second, even weightier change concerns Dionysus himself. He is, at the end of the third stanza, no longer the coming god that points backward; rather, he "laughs, transplanted, the god" (*lachet verpflanzet, der Gott*) (33: 2–3). In this way, not only is the central theme of the colony (which will form the most important statement in the last stanza) mentioned for the first time, but the laughter sets a counterweight to the mortal danger of the Apollonian Sun God. Beside the plantation of the God who migrates from the southeast to the northwest, the plant realm is also emphasized, a realm which belongs to the environment that protects him, benevolently providing a defense against the elemental. The transplant points to an extended Hesperia, not only Germany but also America, as is confirmed by "Andenken" ("Remembrance").[5] The parallel with the Hercules of "Ister," who seeks shade by traveling toward the northwest, is also significant. Dionysus stands for the further course of history; he is, precisely because he is a god that is constantly "transplanted" anew, a "coming god," and his future dimension is now much more concretely actualized than in the "classical" version.

In the second and third triad the changes are much more comprehensive, because here the topic is humanity's encounter with the gods and with the demigod Dionysus. All evocations of their coming closer to each other, especially in the realm of religious and artistic concretization, are replaced with new statements in which the danger of a meeting is confronted. The temples and the vessels filled with nectar, metaphors for song, seen from the perspective of the later versions are quotations from a culture forever past, whose recollection would be without relation to the new horizon opened now to an extended future made relevant to the present. What is significant here is not the closeness to the heavenly beings, who were present in the historically real temples of the Greeks, but rather the *Geseze* ("laws") made by men as earthly beings and that are possible also in a new historical situation, in the sense of what is *gesezt* ("set") (which designates Creon's meaning in *Antigone*)—that is, that which is fixed or regulated by human beings in the relation between them and gods: the laws that limit and soberly define their contact with the gods. Suited to this imposing of discipline are the measured "steps" and "angles" (33: 11–12), which are notations for a mathematically ordered song that can no longer be understood as nectar and from now on cannot borrow from Pindar;[6] otherwise the ancient crisis of Greek religion would become a danger again—now for those who have learned nothing from the past. Even clearer is the replacement of the oracular pronouncements attributed, as lethal sunray-arrows, to the Sun God, Apollo, with "bäurisch sinnige Sprüche" ("boorish practical sayings") (33: 14–15), in which earthly sobriety is once again opposed to the heavenly element, perhaps originating from the folksy spirit of Hesiod. Perhaps it is also legitimate, in this context, to think of Apollo's deadly arrows in *Oedipus the King*. Delphi is now merely *Schaale* ("a shell") (it is not the adjective *schal*—flat, stale—that is meant); the content of the divine pronouncements, which make Apollo's presence clear, is no longer given by the Greek past, but is rather to be derived anew, today, in a differentiating, earthly process of cognition.

Through this new work of thought, something that used to be accessible as epiphany must prove itself to be truth, in the sense of clarified unconcealment:

> Schaal ist Delphi, begreifts besser, erfüllet es sich
> Daß es wahr wird.
> (33: 16–20)

> A shell is Delphi, it understands better, it fulfills itself
> So that it becomes true

Not all the images connected with the now no longer allowed epiphany were replaced; the work on this stanza was certainly not finished. Following the same direction as the changes we have observed, the following line about the arrival of fate could have been deleted as well. The naming of the deadly danger deriving from contact with the fire of the ether is all the more trenchant as a result: "verzehrt und strebt, wie Flammen, zur Erde" ("consumes and strives like flames, toward the Earth") (33: 23). The latter transforms itself, in a fiery reply to the fire from above, into a patch of roses. The flowers here, as metaphor, are at the same time a heightening and a protective circumscription for the presence of the elements that, unfettered, are inappropriate to the connection of human beings and gods and exhaust life on Earth. This awkwardness (*das Ungeschikte*) is here also the expression for the continuing lack of human propriety (*Schiklichkeit*) (see *SA* 5: 270, line 7), that is, the attitude that should actualize through art the differentiating dealings with the gods. The ecstatic community of those imbued with powers of life and spirit, who are united in a jubilant choir, appears—to begin with—to have been deleted. Then, however, we face a great, disconcerting surprise: Hölderlin writes a new line that partly recreates that same ecstatic community of human beings kept united by

God that he had apparently left behind. Proceeding by association from the flaming rose patch, he writes: "Menschen aber, gesellt, theilen das blühende Gut" ("But men, in companionship, share the blooming goods") (33: 34). How should one understand this line that takes up the very same, otherwise rejected, images of the earlier version? The next few words give us the answer: the expression "Das Verzehrende" ("that which consumes") (33: 36), added in apposition, annihilates the apparently harmonious scene of sharing. One should not, then, separate *Blühend* ("blooming") from the flames here: blooming is connoted by the dangers of fire, in accordance with later parallels.[7] It is not possible, then, to argue, following Groddeck, that the grammatically accurate rewriting of the earlier text, "So kommt Himmlisches" ("thus something heavenly comes"), indicates that a "mixture of the earlier text with the new [text] is the author's intention,"[8] since in fact the meter becomes completely jumbled: "Das Verzehrende. So kommt Himmlisches, tiefschütternd gelangt so" ("that which consumes. Thus something heavenly comes, shaking deeply it arrives thus") (33: 36–37). In this attempt to combine the new text with the earlier one I see, rather, a conception that constitutes an exception and is not taken further, but is instead abruptly interrupted. Thus this single attempt to return to the "classical" version to rework it can be neither classified nor evaluated. It is yet another piece of evidence for the provisional state of the revisions.

Who are the *Seher* ("seers") who "interpret here and there and raise their heads" ("aber deuten dort und da und heben die Häupter") (33: 31–32)? Here it is probably the poets, but in the last stanza it is unambiguously Dionysus himself, the mantis of the *Bacchae*.[9] They too are made insecure, and in the seventh stanza there will be talk of their failure again. There is nothing here establishing a continuity with the groundedness of Tiresias's seeing powers, since no comparably clear act of providence reaches human beings anymore.

The excess of the divine arrival is no longer a felicitous event to which one can get accustomed step by step; much rather, it is a blow to the only side of man that can be reached, since he is turned away: his *Rükken* ("back") (33: 43). But here, too, I am not sure whether such a blow can be reconciled with the expression *Glük* ("luck," "happiness") (33: 44), that is left untouched, even though the revision does not put anything else in its place. The same thing occurs here on a smaller scale as occurs on a larger scale in the fourth, sixth, eighth, and ninth stanza: two clearly distinguishable layers remain side by side because the revision could not progress to its final completion. Just as the seers—that is, presumably, the poets—earlier were no longer able to interpret with certainty, now it is the demigod who is no longer able to see with clarity. He is near to these seer-poets in rank. In the earlier version this certainly is not a reference to Dionysus, but in the new version it does mean him, since the lines now say, "und ist" ("and is") like, in the last stanza, "Trüb Feuer, um diesen, und Schlaf" ("cloudy fire around him, and sleep") (33: 48–49). This would mean, then, that in the new version Semele's burning is already hinted, especially since she stood for the poet himself when Hölderlin wrote his first Pindaric hymn, "Wie wenn am Feiertage . . ." The fact that the demigod now does not *see with his eyes*, while in the earlier version he hardly *knows what to say* (33: 45–47), is again related with the image of the demigod Dionysus as seer (*mantis*). Unlike in the "classical" version, here the Earth, as the limited and limiting entity, resists the spreading tendency toward excess; it is the Earth that enables the containment of the heavenly blessing through *Geseze* ("laws") (33: 9) and *Regel* ("rules") (35: 38). Moreover, *Geschik* ("fate") intervenes by opposing to the heavenly fire the counterweight of night, where, as in a grave, overextravagance comes to rest: "Aber zu ruhn, reißt hin ewig in Nacht das Geschik" ("but fate pulls eternally toward night, to rest") (33: 59–60). The gods are no longer patient with excess; their activity now, just like that of the Earth, is also directed toward fettering it (33: 63). Epiphany itself is now presented for the first time—in the medium of the poetry that announces it: "Lang und schwer ist das *Wort* von dieser Ankunft" ("long and heavy is the *word* of its arrival") (34: 7–8, emphasis added). The rhythm of the verses slows down to such an extent that this arrival is no longer able to have anything overpowering, but can only take place in its painstaking preparation by the clarifying, cognitive spirit of the poet, just as in the lines in "Mnemosyne":

> Lang ist
> Die Zeit, es ereignet sich aber
> Das Wahre.[10]
>
> Long is
> Time, the true however
> Occurs.

It is significant in this context that the poets are called "Diener der Himmlischen" ("servants of the gods") (34: 10). As is often the case in Hölderlin, this expression can be made clear by a detour through Greek. *Therapeuein* means both "to serve" and "to honor the gods." It is important that the poets now understand their task from the point of view of the Earth, that is, from the point of view of the *Gesez* ("law"), the *Regel* ("rule"): in the measured movement, compatible with art and knowledge, toward the abyss, not dragged toward it to fight the Titans through their hope in the future—the Titans, who, as the hymn "An die Madonna" makes clearest, are *das Alte* ("that which is old") (see *SA* 2; 1: 213, line 70), that which lives here in *Finstres* ("in the dark"), *in den Tiefen* ("in the depths"), and hate all that is childlike and young. The most important perspective that can be derived from these lines is the work in the service of the future. From this it is possible to build a bridge to the theme of the new "colonies" that are supposed to be founded under the auspices of Dionysus's migration from southeast to northwest.

The sixth stanza was altered in less fundamental ways. It is important that, right at the beginning, the earlier honoring of the gods is now understood as a *Behalten* ("keeping") (34: 20). This expression contains the active hold in the mind, the work of *Andenken* and *Mnemosyne*, the establishment of the *Bleibenden* ("remaining"), the motto "Vieles aber ist / Zu behalten" ("there is, however, much / to keep").

The new seventh stanza, to which the last and newly inserted partial verse of the sixth leads—"In Ephesus aber ist Tempel und Bild" ("In Ephesus however is temple and image") (34: 47)—can be brought together with its "classical" models only in a remote way. Of all the parts that were rewritten, this is the most autonomous and, in parts, the most surprising. The irritation with the temple and the image in Ephesus relates to the goldsmith Demetrius's indignation against St. Paul, who says: "Es sind nicht Götter, welche von Händen gemacht sind" ("gods made by human hands are not gods").[11] Paul's disrespect also toward the temple of Diana in Ephesus is another reason for the mortal danger that he faces. "Auch Geistiges leidet" ("even something spiritual suffers") (34: 50) refers to Paul. The presence of gods, which ignites like fire, and which brings *Versuchung* ("temptation"), *Trunkenheit* ("drunkenness"), *Leidenschaft* ("passion") (35: 1–3), is connected with the destructive power of spiritual fire, which breaks up all the reconciling gestures of antiquity harmonizing the human world and the world of gods, and thus precisely Christ's incarnation in the context of the end of antiquity. Instead of a consoling completion, an incurable conflict breaks out, freeing the pneumatic divine spirit. An equally extreme answer results:

Sich
sein
Eine Kunst, Grab sinnt, doch klug mit den Geistern, der Geist.
(35: 6–8)

For itself
it's
an art the spirit meditates, a grave, yet shrewd with the spirits.

Are the spirits, as is often the case in Hölderlin, the gods? Or, as Schmidt suggests, the "men open to the divine"?[12] Or an intermediate entity, like the Greek *daimones*? The alternative of grave and art as protection from *Trunkenheit* ("drunkenness") is significant. Dionysus is, in this late reworking, no longer the god of the *Heiligtrunkenen* ("holy drunken") (although this word remained in the second stanza, as though the poem did not continue as it does!). The *frohlokkende Wahnsinn* ("rejoicing madness") of the third stanza cannot be reconciled with the dangerous, passionate "drunkenness" of the spiritual fire that is presented here. What breaks up here is the earlier poles of *Natur* ("nature") and *Kunst* ("art"), radicalized into *gesezlos* and *gesezt* ("lawless" and "law-abiding") in the "Anmerkungen zur *Antigone*." So it is possible to connect the spiritual *Trunkenheit* ("drunkenness") and *Leidenschaft* ("passion") with Antigone's *Wahnsinn* ("madness") (*SA* 5: 267, line 17), and *Kunst* ("art") and *Grab* ("grave") with her tomb as well as with the *Todtenkammer* ("death chamber") of Danaë (*SA* 5: 245, line 985), who "zählete dem Vater der Zeit / Die Stundenschläge, die goldnen" ("counted for the father of time / the striking of the hours, the golden ones") (*SA* 5: 245, lines 987–88, and 268, lines 4–5). This "Vater der Zeit oder: Vater der Erde" ("father of time or: father of the Earth") (*SA* 5: 268, lines 8–9) drives the spirit into art and grave. This could mean, concretely, the ruins of ancient Ephesus, city of temples, as protection and sanctuary for the spirit. If "art" also means the *gesezlichen Kalkul* ("lawful calculus") (*SA* 5: 195, line 19) of the work of poetic composition, then the equation "art" = "grave" is highly instructive, not only with regard to the first letter to Böhlendorff, where the flame stands against the container, but most of all with regard to "Mnemosyne," a funeral poem, whose first conception reads "Am Feigenbaum / Ist mir Achilles gestorben" ("My Achilles died by the fig tree" [*SA* 2; 2: 817, lines 15–16]) and which ends with Mnemosyne's death. One should explore, starting from this, the conception of Western funerary art in this late phase of Hölderlin's poetry. The weight given in the *Anmerkungen* to the penultimate chorus song in *Antigone* can be explained in this way:

> Der Leib auch Danaes mußte,
> Statt himmlischen Lichts, in Gedult
> Das eiserne Gitter haben.
> Im Dunkel lag sie
> In der Todtenkammer, in Fesseln;
> Obgleich an Geschlecht edel, o Kind!
> Sie zählete dem Vater der Zeit
> Die Stundenschläge, die goldnen.
> (*SA* 5: 245, lines 981–88)

> Danaë's body too must,
> instead of heavenly light, patiently
> have the iron grate.
> In darkness she lay
> in the death chamber, in chains;
> although noble in lineage, o child!
> she counted for the father of time
> the striking of the hours, the golden ones.[13]

It is well known that Hölderlin derives from this stanza his designation of Zeus as "Vater der Zeit oder: Vater der Erde" ("father of time or: father of the Earth"), "weil sein Karakter ist, der ewigen Tendenz entgegen, *das Streben aus dieser Welt in die andre* zu kehren zu *einem Streben aus einer andern Welt in diese*" ("because his character is, against the eternal tendency, to turn striving from this world to another into the striving to come from another world to this one"; emphasis added) (*SA* 5: 268, lines 8–11).

What is crucial here is that *Zeit, Erde, diese Welt* ("time," "Earth," and "this world") are linked with a *Todtenkammer* ("death chamber"); in contrast, *das goldenströmende Werden* ("the golden-streaming becoming"), interpreted as *Stralen des Lichts* ("light rays"), is transformed into golden *Stundenschläge* ("striking of the hours"). Thus Apollo's fire is transformed into "lawful" time, a time that follows rules, and similarly, in the same chorus, Dionysus punishes Lycurgus's *Wahnsinn* ("madness") and his *blühenden* ("blooming") (i.e., fiery) *Zorn* ("ire")[14] (*SA* 5: 246, lines 997–99), while in Hölderlin's version Lycurgus is *gedeket* ("covered"), as by a grave, "von den stürzenden / Steinhaufen" ("by the falling heaps of rocks"). In each case, *Trunkenheit* ("drunkenness") and *Leidenschaft* ("passion") are answered by the *Grab* ("grave") and by a new form of *Kunst* ("art") suitable to it.

We must keep this in mind if we want to understand the last lines of the new seventh stanza. At the same time, we must still leave open the possibility of understanding the verse—

> Sich
>   sein
> Eine Kunst, Grab sinnt, doch klug mit den Geistern, der Geist.

> For itself
>   it's
> an art the spirit meditates, a grave, yet shrewd with the spirits.

—as referring to Good Friday: then "der Geist" would name God the Father and "die Geister" Christ and his brothers. The following verse also seems to be about God the Father, who is protected by "art," in the shape of the regulated and protective linguistic construct of a "prayer" from immediate influence on the world of men: "denn immer hält den Gott ein Gebet auf" ("because a prayer always holds the god back") (35: 10). The suffering of the spirits caused by the God's contact with the Earth can refer to Christ's and Heracles' death and to Dionysus's capture: all three of them, as earthly sons of God the Father, represent his coming into contact with the Earth. I read the continuation as follows—

>   Schatten
> Nimmer eigenen und die süßen Pfade der Heimath
> (35: 14–15)

>   Shadows
> Never one's own and the sweet paths of the home country

—and not, as Groddeck suggests, mixing the new and the old text, "Nur zu Zeiten erträgt eigenen Schatten der Mensch" ("Only at times man can bear his own shadow") (*FA* 6: 261, line 118).

The balance between sun and shadows and the dimension of one's own earlier lifetime are canceled in the crisis that is being delineated and are replaced by their opposite, barrenness and *Wildniß* ("wilderness") (35: 28). Specifically, it is the artificiality—alienated from its natural origin—in man's dealings with nature, that has led to the crisis: an artificiality that is the correspondence and continuation of the human narrowing of the relationship to the divine visible in the cult of Demeter in Ephesus. The excessively ordered trees, which looked like temple columns and repeated in the realm of "nature" the fault in the realm of "art," are now, just like art, destroyed and replaced by barren heathland that nourishes only the horse and the wolf. In accordance with the loss of the *eigenen Schatten* ("one's own shadow") and of the home country is also the loss of the dimension, not to be delimited by rules, of Christ's secret: the dimension of his birth (*des Ursprungs*, "of the origin"), of his *Wunder* ("miracles"), of the *Palmen* ("palms") as signs of the branch brought to him on the day of his prediction of his own death, and finally of his *Tod* ("death") itself, summarized in the word *Geheimnisse* ("secrets") (35: 27–32). In the wilderness that overgrows all tradition, the *Seher* ("seers") (35: 34), who here probably stand again for the poets, are no longer able to interpret the time of the founding of this Christian tradition, *der Jugend Haus* ("the house of youth"). The same word *fassen* ("to hold"), refers back to the *schwache Gefäß* ("the weak vessel") of the earlier version (35: 11), which at times can no longer hold the gods. *Der Jugend Haus* ("the house of youth") is a protective rewriting for it. There is no more talk of heroes and of the *heilige Priester* ("holy priests") of the *Weingott* ("God of wine") (35: 43); the dialogue with Heinse, which led to the identification of the questioner with the ancient followers of the God, is eradicated without remainder, and with it also every personal reference, especially to a concrete addressee. In their place we find the highest level of abstraction, namely the new definition of the meaning that the night of God's distance now has for the poet: "Die Regel, die Erde. / Eine Klarheit, die Nacht" ("The rules, Earth. / A clarity, night."); the *Ruhige* ("calm"); the length and depth, measured geometrically, of the night's godly character, that is known to the poet, now called *ein Verständiger, ein Fürstlicherer* ("a sensible, a more princely one") (35: 37–45). *Fürstlich* ("princely"), refers back in spite of everything to the heroic dimension proper to Hercules (*SA* 2: 753, line 3), and to the propriety in dealings with the gods. What is significant here is the negative perspective on the application of a rule in the realm of nature, in contrast to the positive perspective on its application in the fundamental realm of Earth and night. Perhaps this is explained by my previously offered interpretation of the reference to Antigone's tomb and Danaë's death chamber, which supports a pure recognition of the mathematically understood divine laws, without a natural or artistic manifestation of the divine: *für sich, allein, in reiner Regel* ("for itself," "alone," "in pure rule") (35: 35 and 38), without any of the mixing that was very much characteristic of the Wine God and his priests. The one who is the *Verständig* ("sensible one") knows what is the *Feindlich* ("inimical") to night (35: 39); it is probably again the old Titanic element, threatening in the abyss, an element that here is also connected to the *Wildniß* ("wilderness"), whose meaning is always double: covering and growing over on the one hand, holding the promise of future life on the other hand. It is so here as well: in the last stanza the differently understood Dionysus is "vergnügt . . . in der Wildniß" ("pleased . . . in the wildness") (36: 17).

None of the stanzas was rewritten as thoroughly as the seventh. It was only approximately possible and permissible to use the "classical" version as the basis for our interpretation. Only in outline can the relation of the chosen men to the gods, constantly explored anew, serve as a point of departure for the new version. We need to disregard completely the possibility that *die Noth und die Nacht* ("necessity and night") strengthen the heroes so much that as a result *Herzen* ("hearts") become "an Kraft, wie sonst, ähnlich den Himmlischen" ("similar to the gods in strength, as they once were") (35: 19–23), as well as the return of the gods as a consequence of this endurance in the Iron Age. The poet of the late revisions has totally lost this hope. An expectation of the divine such as the one pictured in "Der blinde Sänger" has now become inconceivable. The earlier Dionysus similarly seemed to promise to his priests that such a new epiphany would take place if they commemorated his birth, the coincidence of "der Götter und Menschen Werk" ("the work of gods and men") (*SA* 2; 1: 119, lines 8–48).

In the eighth stanza the interventions are again much more limited, so that once again the deletions attract our attention. The lines about the departure of the gods remain untouched, and so also those about Christ's "himmlisch / Tröstenden . . . stillen Genius" ("heavenly / consolatory . . . still genius") (35: 53–54). Why are they untouched, when a similar wording was deleted at the end of the sixth stanza? The difference is significant. The lines in the sixth stanza, "nahm des Menschen Gestalt an / Und vollendet und schloß tröstend das himmlische Fest" ("took on man's shape / and completed and closed, consoling, the heavenly feast") (34: 48), are already meaningfully distinct from the formulation in the old lines of the eighth stanza—"welcher des Tags Ende verkündet und schwand" ("who announced the end of the day and disappeared") (35: 54)—and especially from the establishing of *Zeichen* ("signs") of memory (35: 55), which tie together past and future. The emphasis is no longer on the presence of the divine, as shape and feast, in the earthly, but on its disappearance, and the language of signs corresponds to the stage of the later hymn fragments, as we have already pointed out in the context of the formula "Zeichen des Himmels auch singen . . . Witterungen" ("weather manifestations also sing signs of the heaven"). New verses are introduced, instead, when it comes to human joy over the divine giving of signs, and to the prospect of a generation of mediating heroes—a vision that had already been vehemently rejected in the seventh stanza.

What now appears as a catastrophic disruption of the peace between heavenly gifts and human joy should be seen in connection with Apollo's Sun chariot, which now breaks up. In counterpart to this is the inability of understanding to master this event by interpreting it. Whether the understanding *entschläft* or *sich krümmt* ("falls asleep" or "doubles up") (35: 64–65), we have here a clear opposition to the *reine Regel* ("the pure rule") and the *Klarheit* ("clarity") of the Earth and night, which the *Verständige* ("sensible one") knows. The fiery god of the heavens is now, in contrast, just as incomprehensible as the *Jugend Haus* ("house of youth") was before, the early time of intimate connection between divine and human. The excessive power of the spirit is reminiscent of the lines in the seventh stanza, "Himmlischer Gegenwart zündet wie Feuer, zulezt" ("the heavenly presence, in the end, ignites like fire"), and of the *Trunkenheit* and *Leidenschaft* ("drunkenness" and "passion") in this danger; it is reminiscent, moreover, of the fifth stanza and the *Geschik* ("destiny") pulling *in Nacht* ("into the night"). The understanding, now overburdened, experiences itself as *eine Wüste* ("a wasteland") (35: 65), thus creating a connection to the lines in "Der Enzige" ("The Only One"): "Ohnediß ist / Gewaltig immer und versuchet / Zu sterben eine Wüste voll / Von Gesichten" ("anyway a wasteland full of faces is still violent and attempts to die") (*SA* 2; 2: 745, lines 8–11). Even *sterben* ("to die") can be found here, in correspondence to *entschläft* ("falls asleep") in the other poem. In place of the barren heathland with small flowers we just had, we find here a consuming excess that replaces the verse "Denn zur Freude mit Geist, wurde das Größre zu groß" ("since, for the joy with spirit, the greater thing became too great") (35: 63). In Hölderlin, *Freude* ("joy") has a connection with the fiery spirit from very early on;[15] it is Dionysus's most important characteristic, and the latter is as a result called *Freudengott* ("the god of joy") (*SA* 2; 1: 46, line 1; and *SA* 5: 253, line 1169). Now *Verstand* ("understanding") or *Erkenntnis* ("cognition"), should step in to take the place of the mediating heroes, but understanding is not up to the task. The gift of bread and wine, withdrawing from any closeness with the gods, is the only possibility for a backward- and forward-directed *Andenken* ("memory"), as a way to think of the gods. Every name handed down from antiquity must disappear, both *donnernde Gott* ("the thundering god")[16] and *Weingott*; Dionysus is henceforth called *Weinsgeist* or *Herbstgeist* (spirit of wine or of autumn) (36: 5–7), and in the last stanza simply *Geist* (36: 25 and 29).

In its "classical" version, this last stanza is in part devoted to qualifying Dionysus more precisely. This is even more the case in the revised version. The first lines are not modified, since they do not speak about the mediation between the gods and the godless, but about the unchanging regularity of the course of the Sun, and most of all about the two plants that overcome all the oppositions between seasons: the spruce and the ivy. "Weil er bleibet" ("because it remains") (36: 16): with this motto the God of joy becomes the God of endurance and of the defense against the extremes of up and down, of the fire of heaven and that of the underworld. Both these realms, which are threatening to human beings, are equally deleted in the revision. Thus the last version contains no more "Spur der entflohenen Götter" ("trace of the escape of the gods") (36: 16), and the darkness of godless men is also no longer there. The Earth is *erkrankt* ("fallen ill") (36: 14); the God who radiates health holds it like a *Veste* ("fortress") (36: 15). Where does this idea originate? I trace it back, among other things, to the Dionysus song of the last chorus of *Antigone*, whose final stanza starts: "Jetzt aber, da von gewaltiger / Krankheit die ganze Stadt / Ist befangen, müssen wir / Der Buße Schritte gehen" ("But now that the whole city / is held by a violent / illness, we must make steps toward penance") (*SA* 5: 254, lines 1189–92).[17] Dionysus, who is called on there, is in Hölderlin's version "an Ismenos kaltem Bach" and "an den Zäunen" ("by the cold stream of Ismenos" and "by the fences") (*SA* 5: 253, lines 1171–72)[18] that protect against dragons; Castalia becomes, for him, a wood, and the ivy-covered slopes of Nysa, fountains—all deviations from the original that indicate a cooling down and a setting of limits. The illness that he heals becomes, for the late Hölderlin a universal situation, just as it already is in the second chorus of *Antigone*, where, in contrast to Sophocles, humanity *die Geseze kränkt* ("offends the laws") and with them the Earth and the *Naturgewalt'ge* ("the violent forces of nature"), the gods.[19]

This differently understood Dionysus, whom we already got to know at the end of the third stanza as the laughing, transplanted god, is *verg-*

nügt ("pleased") here as well (36: 17), and the wilderness which blocks access to the *Ursprung* ("the origin") and the house of youth, is for him, who himself embodies this origin and this house of youth, a place of regeneration, where *Schlaf, Garten, Mahl, unsre Blumen, und die Schätten unserer Wälder* ("sleep; garden; meal; our flowers; and the shadows of our woods") (36: 21 and 31–33) all bring the coolness that was lacking in the lightning birth, as well as protection and healing for him who was almost burned by Zeus's lightning. The latter experiences in our country and later also in farther away colonies the recompense which in turn enables him to be for the *erkrankte Erde* ("sick Earth") a god of what remains: more than that, the now necessary *Beseeler* ("giver of life") (36: 35), who also brings reconciliation by showing his now milder fire to advantage in the more Sun-deprived parts of the world. What was vertical movement, a descent of the *Vater Aether* ("Father Ether") (36: 30) to the heartless shadows of the underworld, (it is as these shadows that human beings presented themselves in the "classical" version), becomes now a horizontal movement, distant travels that lead from southwest to northeast, just as the ones endured by Hercules in "Ister" and by the sea traveler in "Andenken." *Frucht von Hesperien* ("fruit of Hesperia") (36: 23)—this is now more than the fulfillment of history in the German Occident: it is, as already mentioned, also America, where the Sun of the Orient sets. Only then is the plural *Kolonien* ("colonies") fully actualized (36: 29). But most of all, this last version leaves behind the image of the Syrian *Fakelschwinger* ("torchbearer") that combined Dionysus and Christ (36: 34). Both this analogy, as the seventh stanza already made quite clear, and also the redeeming gift of fire, are no longer permissible. Heaven and the underworld—burning zones—are replaced by the Earth of flowers and woods. Titans and hellhound, the signs of the underworld's dangers, disappear and give way to the real *Hunde* ("dogs") of an earthly night (36: 44). Correspondingly, the soul is no longer prisoner on Earth and its eye no longer open to the light; instead, the soul manifests itself in the *Augen der Erde* ("eyes of the Earth") (36: 41) that are not indebted either to heaven or to the underworld, but that now participate in God's sleep. The elegy ends, much more clearly than before, with this sleep, and now returns full circle, back to its beginning with the resting city and the people going home to rest. The garden, the bees, the meal, all emphasize that the Dionysian has taken on earthly characteristics. This idyll leaves nothing out, excludes nothing: it gives a new, safe form to the *Freudengott* ("god of joy") whose original *Lust* ("pleasure") (36: 18) is concretized in earthly images reminiscent of Virgil.[20] Here the *Wildniß* completely loses its anarchic, threatening character, and achieves instead the status of protective abundance of vegetation. Enjoyment and pleasure, sleep and meal stand in opposition to the *Regel* ("rule") of the Earth and to the clarity of night, and also to the calm of the sensible one. The Dionysian idyll can also be seen as the late version of the *heilige Fülle* ("holy abundance"), of the "thörig göttlich . . . gesezlose" ("foolish godly . . . lawless") side of the god in "Der Rhein" ("The Rhine"), which rests in the Rousseauistic *Schatten des Walds* ("shadows of the forest") (*SA* 2; 1: 146, lines 144–46; and *SA* 2; 1: 147, line 162). The fundamental images of happiness in companionship that are found at the end of the second stanza of the elegy "Stutgard"—wine, table, bees—are kept, but no human community is mentioned, since Dionysus has now ceased to be *der gemeinsame Gott* ("the common god") (*SA* 2; 1: 87, line 31) who unites human beings into a chorus. Dionysus's political reference to the *Thaten der Welt* ("the acts of the world") (cf. *SA* 2; 1: 119, line 30), is also no longer there; in the "classical" version of the elegy the *Helden* ("heroes") (35: 21) and the *Starken* ("strong ones") (35: 68) indicate this connection, but they are no longer talked about in the new version.

Most conspicuous of all, however, is that an open, visible and loving community between humanity and the heavenly powers given by "Father Ether," a community prepared and directed in part by Dionysus, can no longer be realized; that the exchange and the power of the omnipresent word and the worship of the gods can no longer express themselves; and that, in fact, any expectation of a coming of the gods disappears. Christ has put away his conciliatory traits; he now brings *Ärgernis* ("offense") and *Leiden* ("suffering"). For its part, antiquity has lost its festive, ecstatic character; in place of "Göttern zu Lust der Gesang" ("song for the pleasure of the gods") (33: 13), in place of *Jubel* ("jubilation") (33: 30), lawlike steps of cognition are opposed to the threatening dangers of consuming flames. The new lines deal mainly with these dangers and with the poet's response to them. In parts they show an earthly scene devoid of life, an intervention from above that destroys human culture, even human spirit. But this new experience has not gained control of the entire poem. Some parts were not yet revised, or were revised only in part. The lines that remain often contradict what the later lines say. I balk at any editorial unification of these fragments of stanzas, which are often mutually exclusive. The mixed text that is constructed in this way cannot explain how "das strömende Wort, das, wie die Liebenden, sei" ("the flowing word that should be like those who love"), how "kühneres Leben" ("bolder living") (32: 9–10), how "frohlokkender Wahnsinn" ("rejoicing madness") which "in heiliger Nacht plözlich die Sänger ergreift" ("in a holy night suddenly takes hold of the singers") (32: 27–30), can be brought together with the new Dionysus in wilderness, sleep and forests, or how the invocation "Seeliges Griechenland! du Haus der Himmlischen alle" ("Blessed Greece! you home of all the heavenly ones") (33: 5) can make sense without the temples and the pronouncements of the oracles or, finally, how the presence of the revealed gods can be connected to the defense against dangerous flames from above. The fact that Christ comes once as consoler, but otherwise brings spiritual pain, or that, in an exact reversal, he no longer appears as a torch bearer but is instead replaced by images of beneficial protection against the burning caused by Dionysus, is just as little likely to result from a unified conception. It is much rather the case that all the revisions that we have discussed in more detail come from a perspective that is coherent in itself and which always shows as its basis a moment of extreme danger. All the new lines—whether they are about protective signs, laws, and rules, or whether they present their destructive effects—oppose the realization of a divine presence arising from heavenly fire, whether this realization is direct, or brought about by the harmonically thought mediation of heroes and poets. Thus images of endangerment and destruction stand side by side with images of sensible, regulating reflection. The new relaxed and happy state in this landscape of unhappiness is reserved exclusively for Dionysus, a regressive, pleased god of the northern wilderness, a god apparently without history. However, through the key word *colonies* he represents the development toward the open expanse of new continents and thus paves the way to the images of the Atlantic poet Hölderlin documented by "Andenken" ("Remembrance") and "Kolomb" ("Columbus"). All in all, it seems right to me to start with the premise that the processes found in the "classical" version of the elegy of preparation, concretization, departure and fundamentally new valuation of a forward- and backward-looking connection to the divine presence, developed into an instance of extreme danger, because of Hölderlin's heightened experience of the spiritual presence of God during the weeks he spent in southern France. The defense against and the overcoming of this danger caused enormous strain on the poet now writing in new versions that were never able to come to expression for more than a few lines, or, less frequently, a stanza. Thus, to whatever extent the new text now overlaps—or, in fact, does not overlap—with the old one, the work that came about is not unified and also in no way shaped to the end; it is a work that is so strongly meaningful precisely because it resists any systematization that would level it out. For long stretches it is possible to understand or surmise the distress for which an answer was attempted or given, but the individual moments of this attempt to answer the dangers alluded to do not all simply converge into well-balanced congruence. They are partly indications of the drama, partly consequences of it.

It is as though one were privy to a process of psychic annihilation *during* the catastrophe, *immediately following it* or *after a long interval*, when it was already possible to practice new rules of conduct. The simultaneous presence of these three phases on *one* manuscript page constitutes already in itself an exciting exception. Even more worthy of consideration, however, is the fact that these three phases always play themselves out against the background of a text that had been completed several years before, a text to which these later phases stand in a very tense relation that often documents a radical reversal. These new parts therefore cannot come together with the ones that were not revised in a context governed by intention, as though they belonged together in spite of everything. The processes described here betray a violent break with the earlier state, a break that must be reflected also in the editorial practices. This means that the incomparability of the two main stages must have priority, in every presentation of the text, over possible attempts at producing a combination such as those shown by the text constructed by the *Frankfurter Ausgabe*.[21]

Translated by Georgia Albert

Turns and Echoes

TWO EXAMPLES OF HÖLDERLIN'S POETICS

Arne Melberg

In his epistolary novel *Hyperion*, Hölderlin made some famous philosophical remarks about "beauty," one of which includes an attempt at a demarcation between poetry and philosophy. The passage occurs in a dialogue toward the end of the first book, in which Hyperion discusses the relation between philosophy and poetry in ancient Greece, stating that "without poetry the Greeks would never have become a philosophical people!" Poetry is the "beginning and the end" of philosophy, since harmonious beauty, as realized in poetry, precedes its analysis by philosophical reason:

Das große Wort, das ἕν διαφέρον ἑαυτῷ (das Eine in sich selber unterschiedene) des Heraklit, das konnte nur ein Grieche finden, denn es ist das Wesen der Schönheit, und ehe das gefunden war, gabs keine Philosohie.

Nun konnte man bestimmen, das Ganze war da. Die Blume war gereift; man konnte nun zergliedern. (*SA* 3: 81)

The great saying, the ἕν διαφέρον ἑαυτῷ (the one differentiated in itself) of Heraclitus, could be found only by a Greek, for it is the very being of Beauty, and before that was found there was no philosophy.

Now it was possible to define, the whole was there. The flower was opened; it could be analyzed.

The words of Heraclitus are taken from fragment 51, known in two versions. In none of these, however, does "the great saying of Heraclitus" have the same wording as in *Hyperion*. The version regarded as most reliable, recorded by Hippolytus, reads: "They do not apprehend how being at variance it agrees with itself [literally: how being brought apart it is brought together with itself]: there is a back-stretched connection, as in the bow and the lyre."[1] Using the bow and the lyre as examples, Heraclitus's argument aims at harmonious balance while insisting on harmony as something determined by tension and difference. Hölderlin did not only turn it into a formula, but also adjusted the word for "variance" from a passive construction (διαφερόμενον), instanced in both existing versions, into an active (διαφέροι). It might be argued, that is, that he emphasized the differential tension, inherent in Heraclitus, between "connection" and "variance"—the principle of harmony based on the movement of difference. Such emphasis on tension and division, at any rate, provides a dramatic departure from the considerably more harmonious and harmonizing conception of beauty found in Kant, Schiller, and Schelling. Hölderlin's notion of beauty takes us from transcendental or unitarian utopias into the human world, in which division and difference is all. Such particular division is termed *Ur-theilung*, or "primordial division," in "Seyn Urtheil Möglichkeit," a philosophical fragment from the period of *Hyperion* (*FA* 17: 156). After 1800, Hölderlin's views on beauty as tension within an unnameable as well as unapproachable unity gave way to a poetics of difference and division, in which rhythm, measure, alternation, caesura, inversion, and turning provided guiding concepts. The poetic practice developed under these auspices may be considered an effort to express—in poetic turns—what had been termed, philosophically, "primordial division," and been understood as the tension inherent in beauty. In the following, two examples of this practice will be given.

Turns

Hölderlin composed the famous elegy "Bread and Wine" around 1800, writing it down as one of the first entries in his so-called *Folioheft*. He kept working on the elegy by adding several layers onto the first lines, thereby causing his future editors immense philological difficulty. (A facsimile of the last stanza is given in Illustration 1.) Together with the two ensuing lines, the underlined passage added toward the end of the last stanza was deciphered and published as a separate fragment in the 1930's.

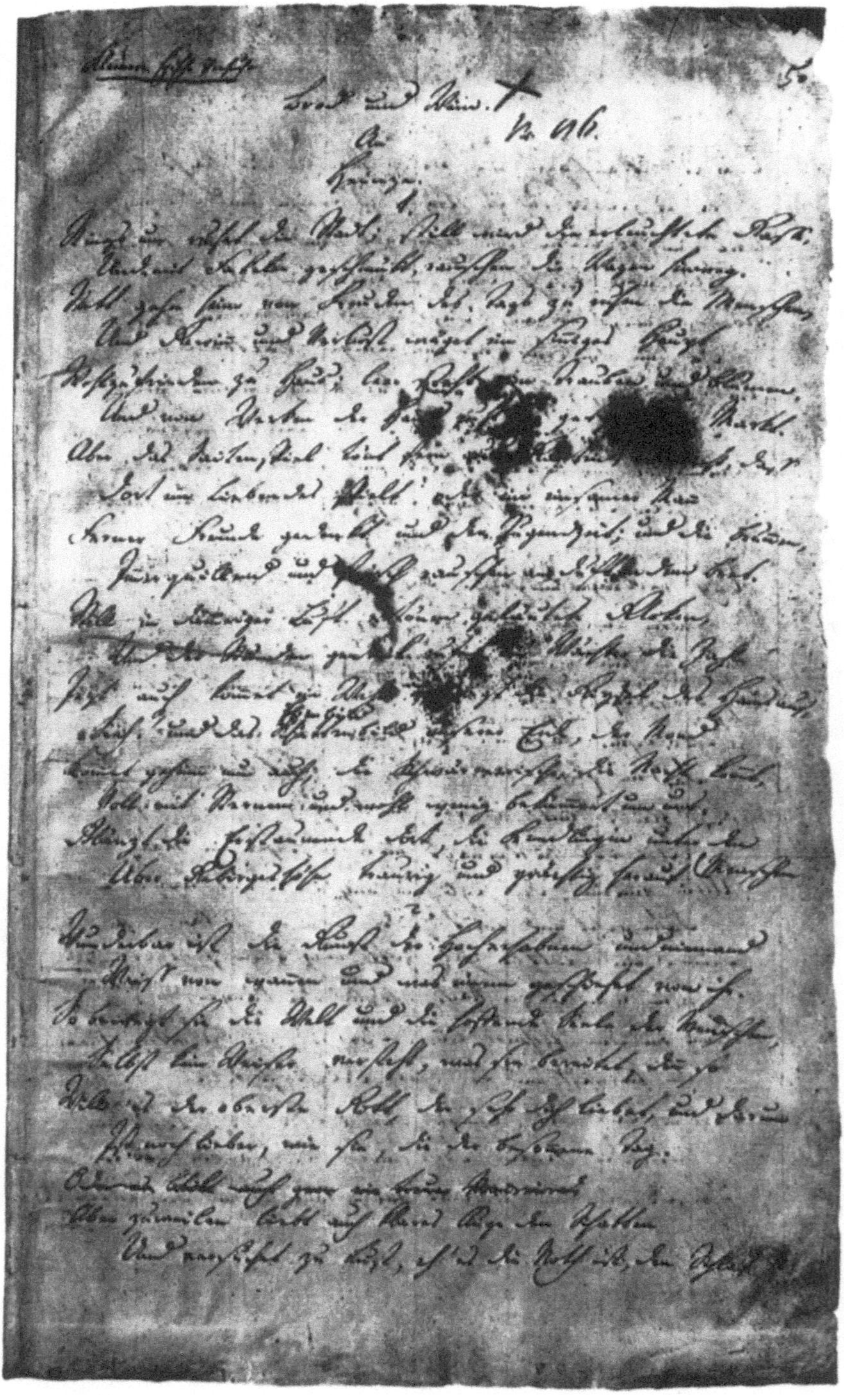

Last stanza of "Brod und Wein" (© Stroemfeld Verlag, Basel and Frankfurt am Main)

In Beißner's Stuttgart edition, they have been placed in the commentary and read as follows:

> . . . nehmlich zu Hauß ist der Geist
> Nicht im Anfang, nicht an der Quell. Ihn zehret die Heimath.
> Kolonien liebt, und tapfer Vergessen der Geist.
> Unsre Blumen erfreuen und die Schatten unserer Wälder
> Den Verschmachteten. Fast wäre der Beseeler verbrandt.
> (*SA* 2; 2: 608)

> . . . for the spirit is at home
> Not in the beginning, not at the source. He is consumed by the
> homeland.
> Colonies loves, and brave forgetting the spirit.
> Our flowers please and the shadows of our woods
> The languished. Almost was the inspirer burnt.

These cryptic lines have been given much attention, in particular by Martin Heidegger, who devoted a major portion of his analysis of another poem ("Andenken") to an interpretation first published in 1943 and now included in his *Erläuterungen zu Hölderlins Dichtung*.[2] This interpretation begins by making the spirit, *der Geist*, into the hero or protagonist of the well-known story of its exile and return to itself. The spirit is not "at home" at the "source." While it is "consumed" by its "homeland," it loves the "colonies" and is bravely "forgetting" it—thus becoming one of "the languished," that is, those who are pleased by the "shadows of our woods." Heidegger terms this spirit a "poetic spirit" (*dichtender Geist*), and asks why it is not "at home" at the "source."[3] According to his understanding of the passage, the "source" spouts out in the "homeland" which, in the second line of the fragment, "consumes" the spirit, thus suggesting that it has in fact been expelled by the *Heimath*. Yet such expulsion does not imply that the relation between "spirit" and "homeland" is broken. On the contrary, for Heidegger, being "unhomely" (*das Unheimischsein*) is to be a part of the homeland, but the very part that the poetic spirit seeks, finds, and refers to as "the colonies." Its love for these "colonies" is, then, a hidden or implicit love for the "homeland," according to Heidegger, just as the word "colony," in his liberal translation, signifies "daughterland" and thus hides and preserves the "mother" of the motherland. In addition, the "brave forgetting" mentioned in the same line refers to the secret love of the spirit, the love of the origin that is the love of the "homeland," "our flowers," and "our woods." Heidegger concludes: "The brave forgetting is the knowing courage to experience the foreign in order to appropriate the own in the future. Meanwhile, the courageous poetic spirit has experienced the long journey to the foreign. It has come home because of its love for the colonies."[4] The turn, the return, the turning toward home, and the coming back provide, then, the guiding principles for Heidegger's understanding of *der dichtende Geist*. And with its coming back, the spirit accomplishes the possibilities which were earlier attributed to the colonies; the spirit is transformed into poetic spirit, returning to its proper self. In fact, it turns creative—creating not only poetry, but also history: "The historicity of history has its essence in the return to the proper, and this return can only take place as a journey to the foreign."[5]

Heidegger's interpretation of the fragment treats it as a narrative of the kind that, in the tradition of Hegel, would be considered a "phenomenology of the spirit." The poetic spirit must experience the power of the negative, by being expelled from origin and homeland. By turning, returning, and coming back, however, it accomplishes itself, thus acquiring what is proper to it by way of what is foreign. Formulated in 1942—a year in which one would expect not only Heidegger to put additional emphasis on terms such as "turning" and "homecoming"—the story receives an odd allegorical twist. In his lectures, Heidegger in fact suggests a Greek term for the fragment's turn from homeland to the colonies and back: καταστροφή. And he does not hesitate to situate the home and destination of the spirit in Germany—more specifically, in the "fire" of the German "hearth."[6]

Particularly indicative of this somewhat capricious and ideological interpretation are two aspects. First, Heidegger uses certain notions as if they belonged to the fragment, though they simply cannot be found there: for example, the fragment speaks of a "homeland" and "beginning," but never of an "origin," let alone what is "proper" or "one's own"; it mentions the "spirit," but nothing like a "poet" or "poetic spirit"; and while it refers to the "colonies," it makes no mention of "the foreign."[7] Second, Heidegger overlooks the fact that the fragment may be no fragment at all. It may of course be possible that Heidegger ignored the five lines in relation to the rest of the stanza to which they belong (and, by extension, to the remaining part of the elegy) simply because they were published as a separate fragment. Yet, regardless of philological acuity, the conditions for an interpretation become quite different when the lines are situated within the elegy. Granted, it is not easy—perhaps not even possible—to determine if they constitute a commentary to the first fair copy or rather amount to the beginning of an entirely new version. (The editors of the Frankfurt edition prefer the latter alternative, providing a new emendated version.) Yet regardless of how the passage is read, it is difficult to avoid reading the "coming god"—the central concept of the elegy—not as "spirit" in general, and certainly not as a Heideggerian "poetic spirit," but as the half-god who, according to the mythology espoused by the poem, would be Christ, Dionysus, or both. (The text seems to keep the alternatives open.) If so, this would imply that the last version—our "fragment"—has decided on Dionysus. It makes sense, at any rate, to read it in light of the mythology associated with the wine god, since it might indeed be claimed of Dionysus that he was born "almost burnt" (by the lightning that killed his mother Semele); that he was, moreover, nursed by the nymphs of the "woods"; that, as a wandering—or, in Hölderlin's terms, "coming"—God, "loved the colonies"; and that he, as god of wine, is associated with "brave forgetting."

Reading the "spirit" mentioned in the passage as Dionysus (or Dionysian) certainly disturbs the particular dialectic that was Heidegger's main concern: the journey to the foreign and the return of what is one's own. It is more reasonable to imagine "our woods" as visited by a Dionysian "spirit" that has its "source" more to the east, perhaps in the "Orient"—"in the beginning" and "at the source" being literal translations of *in oriente*.[8] Heidegger leaves out the Egyptians, as Andrzej Warminski remarks in his reading of Heidegger's argument.[9] The "spirit" of our "fragment," that is, moves from its "source" *in oriente* to the "colonies," which are simply identical with "our woods," and the movement in such direction implies "forgetting." Thus Hölderlin's "spirit" in fact never comes home; it only gets away. Hölderlin's god may "come," but it does not come back.

Elementary philological reasons seem to refute Heidegger's reading of the "fragment." In contrast to his poeticizing German-Greek-German alternative, the Dionysian version, moving from east to west, would have the advantage of having no "home" and no finality, thus offering better resistance to (German) ideology. Heidegger's reading remains Eurocentric, since it allows the Orient to be incorporated only as "colonies" by the intellectual superpowers Greece and Germany. The otherness of these "colonies" is domesticated as "daughterland." His reading is thus dialectical in a harmonizing sense: it leads us to a confrontation with "Oriental" or "colonial" otherness, but quickly returns home and to the homestead.

While this concern with the turning and returning of the "spirit" indicates a sense of *katastrofe* which harmonizes Hölderlin's "primordial division," it may nonetheless—and better, at that, than a "Dionysian" reading—suggest something about Hölderlin's poetic *turns*. In Hölderlin's late writings, *inversion* and *caesura* are used among other terms according to the particular poetics for "hymnic" language that Hölderlin attempted at the time. Inversion is discussed in the so-called "Reflection," whereas the caesura receives clarification primarily in the notes to the *Oedipus* translation (*SA* 4; 1: 233 and *SA* 5: 195, respectively). The latter term is referred to as the "pure word, the counter-rhythmic interruption," of a tragedy, and toward the end of his remarks on Sophocles' play, Hölderlin writes about the "categorical turn" (*kategorische Umkehr*) as the very instant of the confrontation between man and God, the "beginning and end" of tragedy (*SA* 5: 202).

The terms used—inversion, rhythm, caesura, turn—all direct our attention to the poetic line and the breaks (or turns) between lines; that is, to the construction of verse. *Inversion* is used to suggest a *poetic* organization of syntax preceding the *logical*. The *caesura*, employed interchangeably with the "pure word," signifies an exclusively rhythmic aspect of the verse, an invisible gap in one line or between two lines. The *categorical turn* also seems to indicate a rhythmic figuration carrying a heavy metaphysical burden: like the "pure word," it describes a form of caesura. It does not occur in the verse or line, however, but rather in the *turn* of the work from beginning to end. It is in this turn, Hölderlin writes, the that "instant" confronts "time" as "man" is confronting "God" (*SA* 5: 202). Such confrontation, installed by the poetic turn as caesura, involves a break as well as a connection. Perhaps we approach, here, a formula for the strange appeal exerted by Hölderlin's late poetry: the impression that the horizon of meaning widens while the poem itself turns correspondingly cryptic, to the point of being incomprehensible. Hölderlin's *poetic* organization breaks or interrupts communicative meaning, and it may be the very elimination of paraphrasable content by way of such cultivation of *poetic* qualities that, paradoxically, charges it with (an other) meaning. Hölderlin is, in fact, and perhaps first, approximating the poetological persuasions we know from Modernism in its Mallarméan version, where silence and void seem to coincide with the most dense expression of meaning.

Regardless of his possible status as precursor, such density would imply, at any rate, that careful attention must be paid to the rhythmic figuration of his poetry. To take the example of the five lines from the late version of "Bread and Wine": the entire poem is composed according to a strict symmetry, indicating that its categorical reversal of "beginning" and "end" hence might be found in the proper middle of the text. In the middle of the fifth stanza, then, the following turn occurs between lines 81 and 82.

> . . . dann aber in Wahrheit
> Kommen sie selbst . . .
> (*SA* 2; 1: 92)

> . . . but then in truth
> Come they themselves . . .

The ones who are "coming" are, of course, the gods, who in this moment of "truth" meet man in "happiness." Jochen Schmidt, who has offered the most elaborate reading of the poem, suggests that the historical process of civilization articulated in the poem now has reached its peak, and that the "truth" of the poem signifies the Greek ἀλήθεια in the full sense of the word: that is, "truth" as a "coming into day" or "becoming visible." To this I would like to add only a rhythmical observation: the turn from *Wahrheit* to *Kommen*, the break of the verse, a turn which constitutes the very caesura—or categorical turn—of the *whole* poem (as well as the whole of civilization), must be termed a soft break: it is a break, that is, which does not really break, uniting instead of dividing. The lines run smoothly and fluently into each other, without any trace of the *harte Fügungen* typical for Hölderlin's late poetry. The turn is elegiac rather than hymnical; yet it remains "categorical," since it aims for the tense unity of opposites which defines beauty according to *Hyperion*: God confusingly blending with man.

In Hölderlin's late poetry, this kind of categorical confrontation turns less optimistic and more brutal, emphasizing difference and division rather than harmony and reconciliation. His hymnic diction opens for the break *as* break, division, confrontation, and ultimately silence. This transformation may be studied in the late fragmentary lines from the final stanza. A comparable break or turn—possibly a reply or an echo of the earlier elegiac one—here occurs between second and the third lines:

> . . . Ihn zehret die Heimath.
> Kolonien liebt, . . .

> . . . He is consumed by the homeland.
> Colonies loves, . . .

"He," we recall, here refers to the "spirit," which—bravely "forgetting"—was expelled from its homeland, taking up residence in the "daughterland" of colonies, according to Heidegger. In a "Dionysian" reading, by contrast, the "spirit" would be read as coming from an Eastern homeland to the colonies in the West. Yet rather than emphasizing geography, which ought to be subordinated to poetics in this case, the crux of the passage lies in the shift from one line to the next. In contrast to the break occurring in the center of the poem, this particular turn remains rough and austere, illustrating the *armonía austirá* that the first publisher of Hölderlin's late poetry, Hellingrath, used in order to characterize his style.[10]

This rough transition expresses Hölderlin's categorical turn: *Heimath. / Kolonien.* In order to accomplish the break, Hölderlin not only sacrifices the normality of syntax, but also tightens the grammatical inversion into a chiasmus: *Geist/zehret/Heimath* is followed by *Kolonien/liebt/Geist,* thus forming a symmetrical and mirroring juncture of opposites. The figure of a *kategorische Umkehr* is visually emphasized by the full stop employed as a barrier between *Heimath* and *Kolonien,* thus creating the impression not only of *Heimath* being the very opposite of *Kolonien* (*pace* Heidegger), but also of the "spirit" as reversed and torn between opposites ("consumed by . . . / . . . loves"). The full stop functions, in fact, as a caesura by constituting a stark confrontation or an abyss of silence between *Heimath* and *Kolonien.* If not disregarded, such a full stop ought to have prevented Heidegger from reading *Kolonien* as "daughterland."

In both cases of reversal, here as well as earlier, the categorical turn concerns a joining of opposites, be it that of man and God or of self and other. Nonetheless, an almost polemical difference between the two caesuras seems to prevail. The former, early, and elegiac turn of the fifth stanza is "soft" enough to be read as reconciliatory; by contrast, the latter, later, and hymnical, present the "spirit" as split in a chiasmatical turn without any possibility of reconciliation. Approaching collision, the opposites of *Heimath* and *Kolonien* remain categorically separated.

Echoes

Hölderlin's late hymns often seem chaotic, yet the twenty-six texts he did finish, or almost finish, just as often exhibit a perplexing degree of method and symmetry. Most of them are organized in triadic or "dialectical" patterns, indicating his reading of Pindar (strophe–antistrophe–epode). Some of them, however, are based on a different figure, whose mirroring or dialogical pattern may be referred to, simply, as *echo*.

The term occurs in a famous stanza of "Mnemosyne," and merits reflection in this context. The object of intense philological scrutiny in recent years, the hymn has become a virtually new poem: what was once presented in three different versions in the Stuttgart edition, each containing three stanzas, has now become one poem in four stanzas with seventeen lines in each. Even the title of the text has been altered "The Nymph Mnemosyne." The earlier versions suggest a triadic principle of composition, and most of interpretations thus far have followed dialectical or teleological principles, no doubt also influenced by Beißner's analysis of the text as Hölderlin's "last hymn."[11] The "new" version, now established thanks to the efforts of George, Roland-Jensen, Reuß, and Uffhausen,[12] opens for decidedly different interpretations. Although impossible to give credit within the present context, particular attention must be brought to the stanza in which an "echo" occurs, since, *mutatis mutandis*, it may suggest some general conditions of Hölderlin's late poetry.

An artless rendering of the stanza, following the German wording as closely as possible, would read:

Ein Zeichen sind wir, deutungslos
Schmerzlos sind wir und haben fast
Die Sprache in der Fremde verloren.
Wenn nehmlich hoch über Menschen
Ein Streit ist an dem Himmel und gewaltig
Die Monde gehn, so erkranket
Das Meer auch und Ströme müssen
Den Pfad sich suchen. Zweifellos
Ist aber Einer. Der
Kann täglich es ändern. Kaum bedarf er
Gesez. Und es tönet das Blatt. Eichbäume wehn dann neben
Den Firnen. Denn nicht vermögen
Die Himmlischen alles. Nemlich es reichen
Die Sterblichen eh' an den Abgrund. Also wendet es sich, das Echo
Mit diesen. Lang ist
Die Zeit, es ereignet sich aber
Das Wahre.

A sign we are, without interpretation
Without pain we are and have nearly

Lost the language in foreign land.
For when high above the humans
There is a discord in heaven and the moons
Proceed in force, then also the sea
Goes down and rivers must
Find their way. But without doubt
Is One. He
Can change it daily. He hardly needs
Law. And the leaf sounds. Oak trees then sway beside
The glaciers. For not everything
Is in the power of the heavenly. Mortals would sooner
Reach for the abyss. Thus it turns, the echo
With them. Long is
The time, but it occurs
The true.

The first thing to notice is that the second half of the stanza may be read as a reply to the first: that is, as a magnifying though distorting echo. The point of turning between first and second halves occurs in the exact middle, in lines 8 and 9, containing a sentence which points out and (almost) names the "One" who is apparently an "other" or—in Hölderlin's terms—"God." The naming of the "One" breaks the stanza into two parts, the first containing two long statements. The break interrupts and inverts the tempo into a staccato of nine laconic sentences. In addition, as far as the thematic content of the second half is concerned, it replies to the first: demonstrating human, phenomenal, and historical activities as well as occurrences, it contrasts sharply to the regressive forgetfulness of the first sentence and the cosmology suggested by the second. The last sentence of the stanza is a gnomic repetition of the preceding movements.

Spanning three full lines, the first sentence describes how "we" are in a state of passivity. This is the last mention of a "we" in this stanza (it returns later in the poem). Being in a state "without pain," as well as "nearly" without "language," might be understood as a state of inhuman alienation. When "we" have become incomprehensible "signs," and language has been reduced to "nearly" nothing, it would imply that "we" are without memory, mourning, and history—without echo, so to speak.

The second sentence (lines 4–8) appears to offer an explanation to this state: "above the humans" there is a "discord" of cosmic dimensions, with elementary lawlessness as its consequence, indicated by wayward "rivers" and the "sea" going down. (The latter expression may seem odd: to judge from the manuscript, Hölderlin has hesitated as to what verb to employ for the elementary confusion: the manuscript has both "speaks" and "be angry," but *erkranket* is underlined). The reply to the situation is given in the second half of the stanza, in which another natural phenomenon is described: "Oak trees . . . beside / The glaciers" (lines 11–12). This rather odd constellation—oaks do not tend to grow close to glaciers—indicates an utopian order of the very kind that was destroyed in the confusion described in the second sentence. Oak trees / glaciers unite what cannot be united and keep opposites together in tension. In condensed and expressive form, the constellation suggests both the originary division *and* the unity of opposites which provided the definition of beauty in *Hyperion*. Oaks/glaciers is not an observation of how things are in normal physical reality, but rather the posited order of a poetical truth. This truth borders on the unhuman—glaciers are close to heaven and not meant for human survival—but involves nonetheless the natural growth of oak trees. The boldness of the constellation may be attributed to the reference to a physical near-impossibility stated as a poetical possibility, providing a reply to the cosmological confusion of the second sentence. It may, however, also be considered a barely possible constellation (oaks/glaciers) which installs or posits a law. As may be gleaned from the preceding sentences (lines 8–11), the "One" (or "other" or "God") who marks the turn of the stanza "hardly needs Law." This is a thought often recurring in Hölderlin: the "gods" are beyond or above "law," the implication being, of course, that "law" is the task and domain of man. So is language and interpretation, as could be inferred from the first sentence of the stanza; so is mourning and memory, as may be deduced from the rest of the poem; and so is "the true," to judge from the last line of the stanza under consideration.

The naming of the "One" as "He" (or perhaps "It," *Der*) is performed with a maximum of rhythmical emphasis: "He" occurs alone at the end of the line, after a marked caesura, thereby showing his/its being beyond or at the edge of what is comprehensibly human. Reading this turning point of the stanza, we are again confronted with the center of Hölderlin's thought, the "primordial division," expressed as one version of the discordant relation between gods and humans. Later on in the stanza, we read that "the power of the heavenly" (line 13) falls short and that "mortals" at least have the privilege of the "abyss" (line 14). From this statement, it may be inferred that language and law belong to the human domain of the "abyss"—and perhaps what is "true."[13]

Occurring between the "One" and the "mortals" is not only the sentence that posits the perhaps utopian order of oaks/glaciers, but also the elliptically cryptic sentence about a "leaf" that "sounds." Perhaps we are already listening to the "leaf" of the "oak tree," but why then only one *Blatt*? Perhaps because *Blatt* here could also refer to a sheet of paper and "sounds" (*tönet*) also might suggest something like "proclaims." In an earlier version, Hölderlin actually had written that it is the "script" (*Schrift*) that "sounds."[14] This would indicate a thematic inversion of great dramatic effect: whereas in the beginning of the stanza, "we" had "nearly Lost the language", it now "sounds" from the "leaf" (or "sheet"), thus keeping the impossible opposites of oaks and glaciers together. Read in this manner, it would suggest a turn from utter passivity to activity, and hence a reply and a return of posited law as well as human voice and language.

What may have caused this remarkable turn? The stanza does not seem to suggest an answer, but merely a figurative description—or rather: a poetical version of the turn. Nor is an explanation given in the remaining portion of the poem, but merely additional figurations of the same turn. Nor is an explanation to be found in any of Hölderlin's other late poems, but merely the repeatedly expressed wish for a remedy or a means of handling the turns of both life and verse. His usual term for this is "measure" (*Maß*), alluding to the Greek μετρόν and—characteristically—having an equally moral and metrical significance. (A case in point would be the first version of "Der Einzige", in which it is stated: "Nie treffe ich, wie ich wünsche, Das Maß." ["Never do I hit, as I wish, The measure"] [*SA* 2; 1: 155].) Yet no "measure" is called for in the stanza in question. Instead there is only a name for the turns figured by the poem: "echo" (line 14). After the "mortals" (rather than "the heavenly") have reached for "the abyss," the echo "turns" "With them" (line 15). Who or what is here referred to as "them"? Taking into account Hölderlin's parataxis, "them" could refer to any plural noun mentioned, and since the "One" earlier marked the turn of the stanza (line 9), it may even be expected that the echo now turns with "the heavenly." Considering that man seems to reply by echoing the silence of the "heavenly" or the signs of the "One," the "mortals" may well be read as the agents of the echo—that is, as answering the incomprehensible signs that are without "law." As in the myth about the nymph Echo, who replicates "sounds" anew, they even create a meaning which was never there before. And just as the voice of the mythical nymph only repeated fragments of what she heard, they are inevitably forced to express not so much something different, as difference itself. Echo is the name given by the poem for the turns of the poem itself—turns that may be traced from the "heavenly" to the "mortals," as well as the turn from forgetting and passivity into sounding activity. This structure of echoing consists of division and turns, and—in this poem entitled "Mnemosyne"—it is the name for memory and history.

This is summed up and shown again, to great effect, in the last sentence of the stanza, short in words but spanning three lines (lines 15–17). The first words—"Long is The time"—may be considered to express the Other or "heavenly," while the rest of the sentence replies, as echo, by turning the "long" time into an event that "occurs," with the result that "time" is turned into "the true." The mention of "time" as "long" presents time as something ongoing and undifferentiated, like a state prior to Hölderlin's "primordial division," and hence related to the prehumanly Other. (In a different context, but during the same period from which "Mnemosyne" dates, Hölderlin claimed that God "is nothing but time" [*SA* 5: 195].) After the comma dividing the sentence into two parts we meet instead a time which must be termed historical in the sense that something "occurs" or "takes place" (*ereignet sich*).[15] Here, time as a continuum appears broken (turned, echoed, replied) by an occurrence or event; the break installs division, figured by the comma, and it is by way of such dividing break that "the true" may occur.

The relation, or even dialogue, between the two parts of the sentence is rhythmically emphasized: the former half, before the comma, is articulated in a heavy, falling rhythm, whereas the latter half dramatically rises. The passivity of continuous time is broken by the activity of the event. The whole construction aims to direct maximal attention to the last line: "the true." If such reading, led by an echo, has any merit, it would indicate that "the true" is the final answer to the address given by time. *Das Wahre* would be what takes place in history as a result of human activity, being—like language, memory, and mourning—the domain of man.

In Hölderlin, the structure of echo consists of division and turns. Considering the terminology here employed—"echo" including not only division and turns, but also central notions such as inversion, caesura, and measure—it could perhaps be objected that such turn itself is too restricted to poetic construction and technique. Meaning is expected and required. To such objection it may only be replied that, in Hölderlin's own development, poetry and poetics supersede philosophical meaning. A significant part of the reason for this is his belief in poetry as primary and, by extension, that its rhythmical patterns are primary to any paraphrasable content. Which is to say, too, that Hölderlin's poetry includes a philosophy of division and beauty—includes by expressing it, and expresses it by figuration.

Still more remarkable is the fact that Hölderlin seems to have considered human history, too, as included in what he understood by poetry and poetics. That is: it takes place in division and turns. In Hölderlin, humanity creates history—making "the true" occur—by echoing the laws of lawless gods. These laws and echoes find their densest expression in poetic inversions, catastrophes, caesuras, turns, and reversals. A sad contradiction may well be hiding in this rhythmical version of historical time: the human activity making "the true" "occur" and history happen is but an echo of inhuman possibility. Listening attentively enough to this rhythm, we may even hear something of what Benjamin termed *Ursprung*, describing it as the *Rhytmik* of "the becoming and vanishing"[16]—or what Heidegger repeatedly would evoke as the movement of Being as *Lichtung* and *Verbergen*.[17]

Thus, perhaps the most inclusive term to designate Hölderlin's poetics—and thus his language and history—would simply be "rhythm." For Hölderlin, poetic and historical *meaning*, too, are subordinated to rhythm—no doubt a view which helped him to create the almost (but not fully) incomprehensible and highly expressive late poetry. During this period, around 1806, he is reported to have stated, oracularly, that *all is rhythm*.[18] With incomparable rigor, Maurice Blanchot has located such rhythm in Hölderlin to the "first turning—a noun that must be heard as a verb, as the movement of a turning, a vertigo wherein rest the whirlwind, the leap, and the fall." Relating this "first turn, the original structure of turning", to poetry—or, literally, to the turning of lines into verse—he proceeds to quote Hölderlinian lore: "Hölderlin said (according to Sinclair and Bettina): 'All is rhythm; the entire destiny of man is a single heavenly rhythm, just as every work of art is a single rhythm.'"[19] Needless to say, these claims on behalf of poetry as rhythm are either exaggerated or displaced. Philosophy, history, and meaning could do better than that.

But then, the statement was not quoted for philosophical purposes, but in order to summarize Hölderlin's poetics. As such, it seems impossible to read Hölderlin's texts without following their proper rhythm, from the "primordial division" of the early philosophy to the turns and echoes of the late poetry.

Hölderlin's Marginalization of Language

Hans-Jost Frey

Margin and frame belong together.[1] A margin frames what is written. The margin of the page is the frame of the web and bounds the shape of the text; however, the edge of the text does not coincide with that of the page. Between the two is a white strip of no-man's-land, a kind of moat, which fortifies the text and upholds it in its insularity. Its task is to ensure that the unity of the text is not infringed upon.

However, the margin itself is not immune to being touched. By definition, it is not yet inscribed; but this means that it can be written on. The writing in the margin is a special one among all forms of writing because it is the one that most obviously lacks the innocence of beginnings. It is, in every case, an attack on the integrity of the text in whose margin it takes place. Whether it expands, changes, explains, or criticizes the text, the marginal note always has an unavoidable impact on it. By intruding upon the text, it tears it up, opening it toward the outside, and disturbs the truce between the intratextual relations. The marginal text ignores the margin of the text. It begins where the text stops and blurs the text's borders. Where the text's margin is lost because it also becomes text, the finished whole opens up and discovers in itself the insufficiency of the fragment. The illusion of the self-sufficiency of the framed text falls apart when writing encroaches on the margin. Marginalia, as Poe and, in a most extreme form, Coleridge have practiced them, indicate a conception of the text that allows continued writing of it, because no border that the text comes to is the end: rather, it is a border precisely because it is possible to cross it.

Marginalia are the graphic illustration of the impossibility of bringing a text to a close. But they are not only that.

Marginalia are possible because the margin is not an edge. Beyond the edge of the text there is still space: the space between the edge of the text and the edge of the page. Marginalia are always between them, and are therefore related to both at once: on the one hand, inward, to the writing to which they are appended; on the other, outward, to the surroundings of the page to whose edge they extend.

That the marginal text is outside the text appears to indicate that it does not belong there. If it is a reader's statement about the text, it can document his pretense at controlling a text in which he has no part. But what is said in the margin about the text is inextricably intertwined with it. The relation to the marginalia from this point on is part of the text, even when their content is quite foreign to it. Stendhal was in the habit of using whatever book he happened to be reading to make journal entries. The dates of notable events that he recorded on the margin of the pages have no connection with what he was reading except the time of the reading. Nonetheless the marginal notation impresses upon the text a relation to the state of mind of the author of the notation, the text's reader, and induces the later reader to include this relation in his own reading. Thus, marginalia are not simply external to the text, but entangle it in a net out of whose mesh it can no longer be isolated.

But the marginal text is not only intertextually woven into the text on which it borders: it is also in relation to the edge of the page, which is an edge of a different kind. While the border of the text allows for a text beyond it, the possibility of which is vouched for by the white margin, the edge of the page marks the border of the furthest possible extension of the writing. The edge of the page is no longer the passage from one text to another—or at the very least to its possibility—but the passage from language to speechlessness. The edge of language is a more radical border than that of the text. The issue there is no longer how a certain piece of writing relates to another, but the very possibility of writing. This reveals a new dimension of the marginal text. The latter is not just, from the perspective of the text, a pushing back of its border or an addition of whatever kind to the text; rather, by venturing onto the edge of the page, it at the same time leads to the border of speechlessness and thus brings up the question of language.

This interpretation of the edge of the page as the edge of language could be easily refuted by taking another page and writing further; this action, however, would disregard the distinction between the edge of the text and the edge of the page. What happens on a sheet of paper can be given a graphic shape. It is not only the writing that is impacted by this, but also the white surface on which it settles. The parts of the sheet that have no writing on them are lent significance by the marks that appear elsewhere on the page. Language is what is on the paper, but the paper itself is also language because its silence also becomes included in the system of language and starts to speak. However, writing endows silence with speech only up to the edge of the page, since this edge, while it also still means, means only the end of the linguistic order that the page as a whole represents. On the edge, the possibility of linguistic shaping to which the page owes its meaning comes to an end. Beyond this border, the lack of language is no longer signified and also no longer signifies: it just sets in.

The edge of the page cannot be overwritten. This fact manifests itself in the need, when the edge is reached, to start a new line so that language does not simply stop. The written text is kept inside the scope of the page. But the written text also organizes this scope, giving the empty spaces diverse and specific meanings. In the case of a system of writing that goes from left to right, the left margin precedes language and the right margin comes after it. The right margin is the end margin. It signifies the end of language, although it still leaves language room so long as the edge of the text and the edge of the page do not coincide. The marginal text on the right of the page follows the end of language, just as the marginal text on the left precedes its beginning. Both overstep the borders of language. They are a violation: language where language is no more or not yet, language that, because it speaks where it has already ended or not yet begun, becomes for itself the problem of its own possibility.

Hölderlin wrote two texts about language—"Aber die Sprache . . ." ("But language . . .") and "Im Walde" ("In the Forest")—on the right margin of sheets on which he had already written.[2] The fact that a linguistic utterance about language is marginalized in this way can signify something even when other passages about language are not singled out graphically in the same way. To the extent that language becomes a question for itself at the margin that puts it in question, these texts must be read with an eye to the edge of the language they border. As though they were attempting to find a hold, however, they are also interlocked with

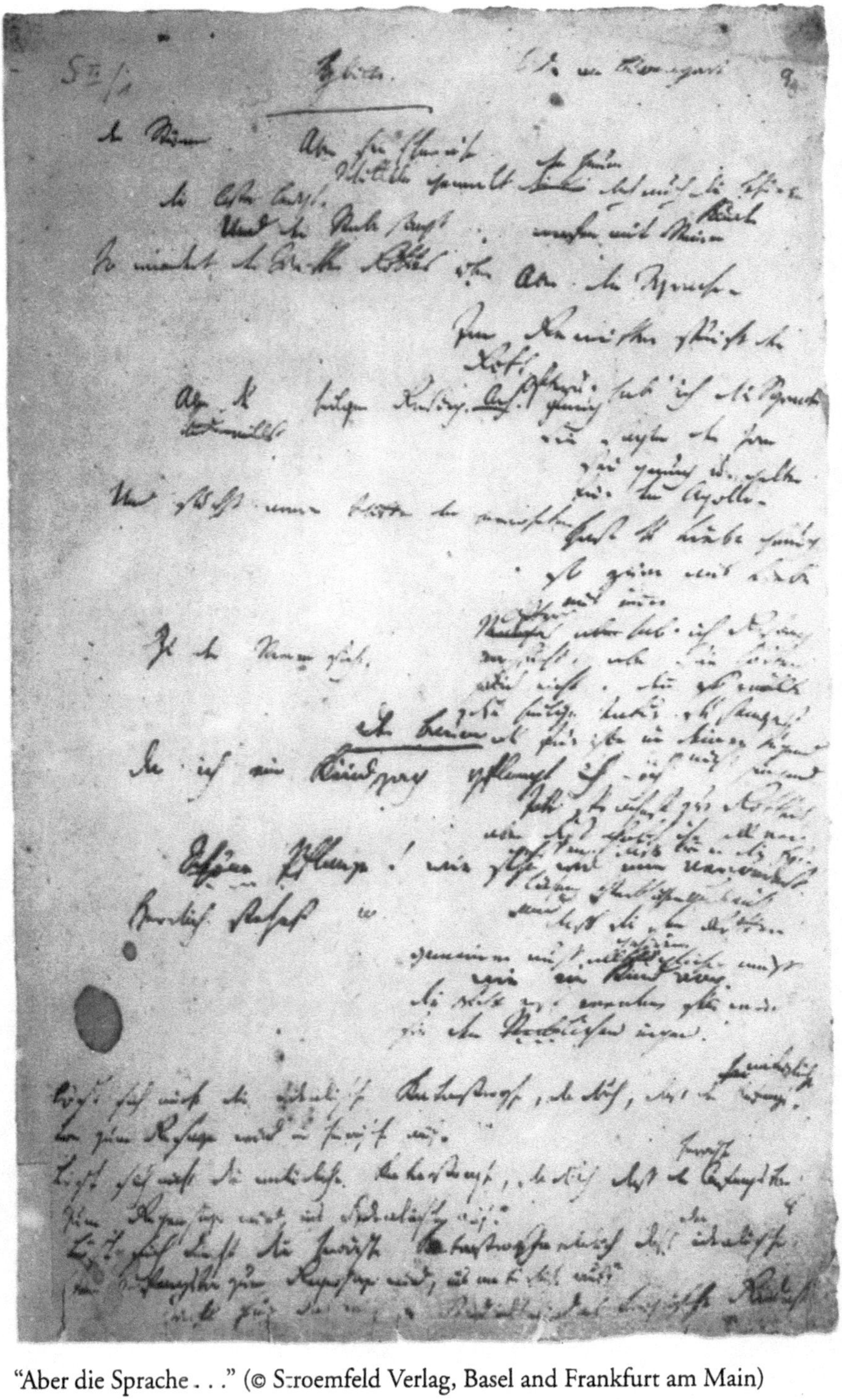

"Aber die Sprache . . ." (© Stroemfeld Verlag, Basel and Frankfurt am Main)

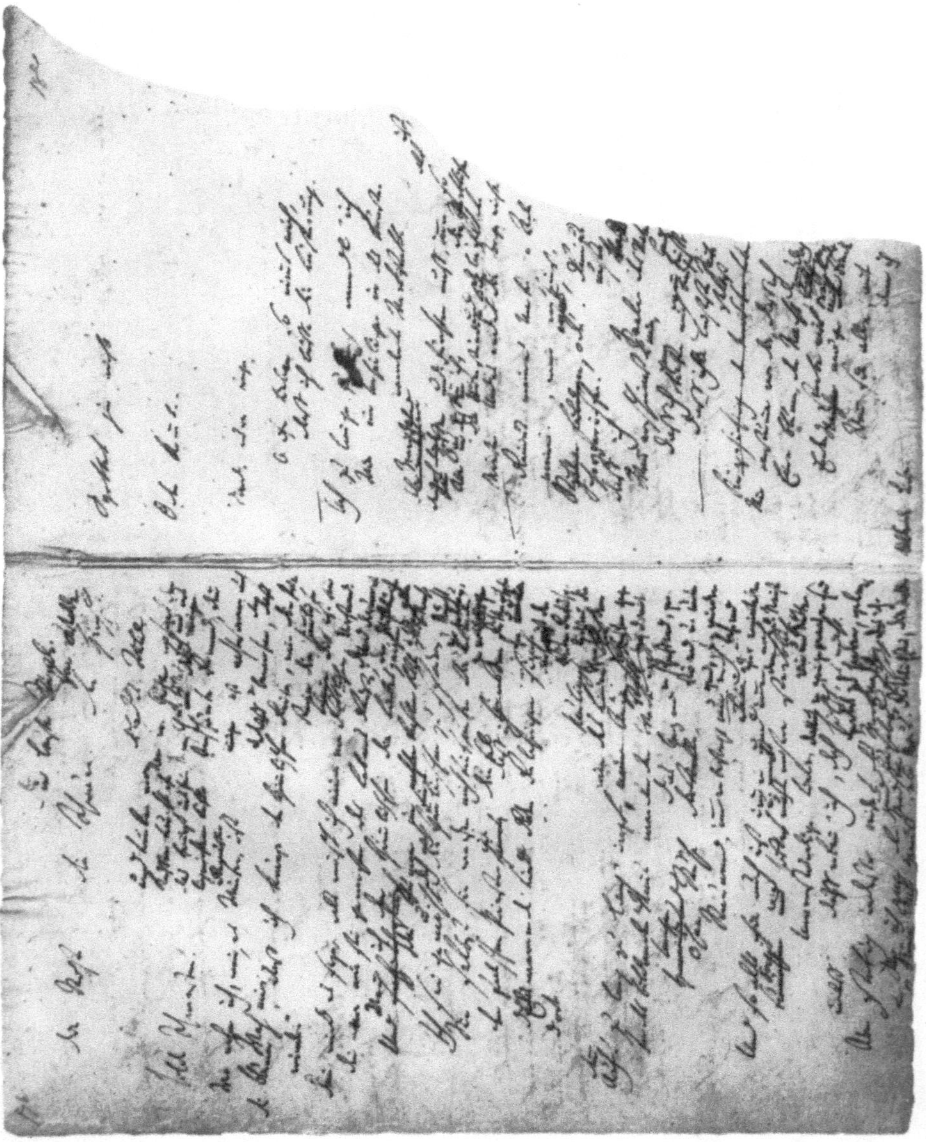

the texts on whose margin they were drafted and out of which they appear to have grown. It should be possible to read them as elucidating the relation of the texts on whose margin they appear to this margin, and as having something to say about language on the margin and about the edge of language.

There appears to be, in Hölderlin, a certain reticence to speak about language. The manuscripts that reconstruct the steps of the genesis of the ode "Ermunterung" indicate that the writing of the poem was interrupted every time at the moment that the word *Sprache* ("language") suggested itself, that the subsequent continuation every time circumvented and postponed the naming of language, and that when the concluding stanza was finally reached, only the privative word *sprachlos* ("speechless") appeared in it.[3] This restraint in speaking about language may have something to do with the fact that naming gives control over the named entity, while language can be mastered only within limits. If language gives control over other things, but it is not itself controllable, it is easy for it to give the impression of being under control when it is its own object. One could counter this by speaking about language, which becomes controllable when it is its own object, in such a way that language appears as something that escapes mastery. Then, on the one hand, language would be brought under control by being spoken about; on the other hand, what is said about it would deny this control.

"Aber die Sprache . . ." says regarding language that it is attempted. Not that something is attempted by way of language as a means of saying something: language is itself the object of the attempt. The one who attempts language does not so much make use of it as feel his way into it. The attempt at language is not one in which one employs language. Much rather, in this attempt language is experienced as a task around which one tries to find one's way, a task which shapes the one to whom it proposes itself as a task, and who proposes it to himself as a task, just as much as he shapes it. This is why in this text no speaker is granted with certainty the ability to use language to his advantage. "Öfters hab' ich die Sprache . . . Öfters hab' ich Gesang versucht, aber sie hörten dich nicht" (From time to time I [attempted] language . . . From time to time I attempted song, but they didn't hear you). In the passage from the first to the second person, the "I" comes face to face with itself, but this self-objectification does

---

Opposite. "Im Walde" (© Stroemfeld Verlag, Basel and Frankfurt am Main)

not result in self-control; rather, it causes the I to become foreign to and slip away from itself. Its speech, even if it is its own, does not belong to it, not only because it is directed to God, to whom it therefore belongs, but also because the language in which this speech takes place is one that is to be attempted and as such is part of a sphere that is not accessible to the I and prevents it from grasping itself as the one speaking.

At the very moment that there is an attempt to get control over language by thematizing it, writing reaches out of the frame it has been given and encroaches upon the margin. This can signify the temptation that is tied to this attempt: to cross over the border of one's own attempt to get control, thus getting this border under control. But the marginalization of the discourse about language can also mean the opposite, namely, that this discourse only occurs on the margin, on the side; therefore, that it precisely opposes the temptation toward presumption, and that it does not happen without reticence. In any case, the fact that language is spoken about on the margin is appropriate, since it is on the margin that the question of how far language can go poses itself. About the fact that language goes beyond what is possible for man the text leaves no doubt: the only promise it makes to man is the possibility that he may get used to something that transcends him.

"Im Walde" clarifies in a different way how, in one's dealings with language, it is necessary to accept that one is just as much vulnerable to it as one makes use of it. Language is *der Güter Gefährlichstes* ("the most dangerous of goods"). It is a constitutive aspect of danger that one does not know whether one will ever be done with it. It is precisely this uncertainty that makes any experience into the experience of a danger. Language is a good that is given to man. It is also, from the beginning, not understood as an instrument that man has procured for his own self-determined use, but rather as a gift that can become dangerous to him if he handles it in the wrong way. "Im Walde" undertakes to determine what is the right employment of language, the only one that is able to avert danger. This, however, does not happen in the form of an isolated reflection, but in close relation to the other texts that are held together on the page. That "Im Walde" originates right here, on the margin of this sheet, is not irrelevant for its understanding; rather, it reveals that unexpected intertextual relations had a hand in the constitution of the marginal text.

The upper half of the page contains a draft of the conclusion of the hymn "Wie wenn am Feiertage. . . ." Connected to it is another draft that came out of it; bringing together the two titles "Die Rose" ("The Rose") and "Die Schwäne" ("The Swans"), it becomes an early stage of "Hälfte des Lebens." Under the third title—"Der Hirsch" ("The Deer")—only the phrase *Du edles Wild* ("you noble wild animal") appears. (Whether this phrase was later integrated in the marginal text "Im Walde" is open to question, but not improbable to the extent that *wild* and *Wald* are etymologically related. The wood, *Wald*, is wilderness in the sense of land with no buildings on it, in contrast to the huts in which man lives.) On the lower half of the page are three stanzas of the ode "An die Deutschen" in a preliminary form. A clear interval separates them from the interrupted draft of the hymn.

The end of the hymn talks about the false priest that the poet who does not fulfill his task is to be regarded as. This task is to mediate for man the language of the god who manifests himself in the storm.

> Doch uns gebührt es, unter Gottes Gewittern,
> Ihr Dichter! mit entblösstem Haupte zu stehen
> Des Vaters Strahl, ihn selbst, mit eigner Hand
> Zu fassen und dem Volk ins Lied
> Gehüllt die himmlische Gabe zu reichen.
> (*SA* 2; 1: 119–20)

> But it befits us, under the storms of the god,
> you poets! to stand with bare heads
> to take the ray of the father, him even,
> with our own hand and give the heavenly gift,
> enveloped in song, to the people.

The false priest is not content with mediating. He attributes to himself the divine quality that should speak through him, and for this the gods punish him just as they do Empedocles.

> . . . Denn es haben
> Die Götter seine Kraft von ihm genommen,
> Seit jenem Tage, da der trunkne Mann
> Vor allem Volk sich einen Gott genannt.
> (*FA* 13: 702)

> For the gods
> have taken away his strength,
> since that day when the drunken man
> called himself a god before all the people.

And:

Es haben ihn die Götter sehr geliebt.
Doch nicht ist er der Erste, den sie drauf
Hinab in sinnenlose Nacht verstossen,
Vom Gipfel ihres gütigen Vertrauns
Weil er des Unterschieds zu sehr vergass
Im übergrossen Glük, und sich allein
Nur fühlte . . .
(*FA* 13: 703)

The gods did love him much.
Yet he is not the first whom they later
pushed down into a night devoid of senses,
for he forgot the difference too much
in his enormous happiness, and now felt
only himself . . .

The one who feels only himself forgets the difference and makes himself into a god. Instead of expressing the divine language in human language and transmitting it, he is a false priest and misuses language just like Empedocles who "called himself a god before all the people." The danger of language consists in bringing the one who is only similar to the gods too close to the divine, so that he attributes divinity to himself instead of making it accessible to human beings. The false priest is the one who succumbs to this danger.

To the false priest of the hymn on the upper half of the page is opposed the "layman" of the ode "An die Deutschen" on the lower half. The layman is the non-priest; in fact, he belongs to the people to whom the poet is supposed to give the divine enveloped in a song. In contrast to the false priest, the layman does not come too close to the divine, but rather remains too far away. In its later form, the stanza that describes the layman is still closely related to the context in which the sketch appears:

Schon zu lange, zu lang irr ich, dem Laien gleich,
In des bildenden Geists werdender Werkstatt hier,
Nur was blühet, erkenn ich,
Was er sinnet, erkenn ich nicht.
(*FA* 5: 533)

Too long, too long already have I wandered, like a layman,
In the emergent workshop of the creative spirit,

I recognize only what blooms,
what the spirit ponders I do not recognize.

The layman recognizes only what has already originated (that which blooms), but he has no access to the process of origination (the workshop of the creative spirit). This withdrawal of the spirit is similar to the winter of which the draft of "Hälfte des Lebens" speaks:

Wo nehm ich, wenn es Winter, ist
Die Blumen, dass ich Kränze den Himmlischen winde?
Dann wird es seyn, als wüsst ich nimmer von Göttlichen,
Denn von mir sei gewichen des Lebens Geist.

Where do I get flowers when it is winter,
so that I may bind a garland to the heavenly ones?
Then it will be as though I no longer knew about the gods
because the spirit of life has gone away from me.

Looked at from the standpoint of the opposition between the false priest and the layman, the sheet begins to reveal a rough structure. Between the excessive closeness to God of the false priest and the excessive distance from God of the layman is the place of the priestess of "Im Walde," who has to preserve the heavenly flame and to whom man, who has to preserve the spirit, is compared. It appears as though the marginal text were attempting to find the right way to deal with language between the extreme positions of the false priest and the layman. This entails a peculiar reversal of the graphic structure of the sheet on the level of content, since in terms of their content the texts in the middle of the page describe marginal positions and the marginal text a middle position. But the marginal text, if we look at it in a corresponding fashion, has a mediating function even in its graphic appearance. It runs continuously in an up-down direction and covers also the otherwise blank space between the hymn and the ode. In this way it becomes a bracket that connects the two poems to each other. The real connection there, that which bridges over the gap between the texts by spanning the linguistic arch from the margin of the one to the margin of the other on the margin of the sheet, is "the most dangerous of goods, language." Language appears in the empty space between the poems as the mediation between the false priest, who does too much, and the layman, who does too little; yet it is at the same time, as this middle, on the margin. The reason for language being situated at the same time in a central and an eccentric position can be understood from the way the texts are intertwined. "Im Walde" is a reflection on the way to deal with language, which preserves and mediates the divine spirit. The danger associated with this, namely, that the mediator may confuse himself with what he mediates, is presented in the hymn through the false priest who succumbs to it; in the marginal text, however, it is recognized and expressed when language is characterized as "the most dangerous of goods." This recognition of the danger, however, in no way eliminates it. Instead, it is all the more exposed to the danger precisely because any discussion of language that says what language is carries within itself the unavoidable ambition to get it under control. That the statement takes the form of a marginal text allows it to retain the temporary nature of what has not been mastered and fosters a humble attitude toward language. The false priest would not express himself in marginalia.

However, the marginal text can also be read as the attempt to get a handle on language *in extremis* and from a position almost already beyond language. The ambivalence of the margin is irreducible and should not be forgotten even in favor of a preferred reading.

Both the marginal texts speak about language as that which enables and produces the relation between man and the divine. That the first fruits of the year belong to the gods points in the same direction as the fact that language is given to man to attest to the higher entity from which he has inherited and learned. Language appears here not on the horizontal plane of communication between human beings, but as the vertical link to the divine. This vertical direction, important to Hölderlin's understanding of language, is graphically realized in the marginal texts on the basis of the formal conditions that the marginal strip makes available. In the case of "Aber die Sprache . . . ," the fact that writing is applied perpendicularly is brought together especially clearly with the orientation, on the level of content, of the texts gathered on the page. On top of prose foundations consisting of the beginning of the reflections on the modulation of tones, one finds two poetic drafts, "Sybille" and "Der Baum." Through the pronouncements of the oracle, the sibyl mediates between the gods and men: she is, therefore, quite bound to verticality, which also appears in the phrase *Zu den Sternen siehe* ("look to the stars"), a phrase that obviously refers to the sailor who hopes to acquire some clarity about his horizontal path by looking upward. The *Wetter Gottes* ("God's weather"), which reappears, clarified, in the marginal text as *Im Gewitter spricht der Gott* ("in the storm speaks the god"), introduces verticality. Moreover, the tree that has its own draft further down—a draft that, with the phrase *Herrlich stehest* ("you stand magnificently"), also stresses the perpendicular—appears at this point already. The general graphic look of the page can be read without difficulty as a tree whose branches, the lines of the poetic drafts separated by large gaps, grow out of the vertical trunk of "Aber die Sprache . . ." Here too, then, in a certain way the margin becomes the middle (the trunk) and the middle the margin (the branches). This is justified on the level of content to the extent that the marginal text names language as that which connects vertically and which is the premise for the texts that are planted in the middle of the page. The importance of the verticality of the linguistic relation is reaffirmed in "Aber die Sprach . . ." also by the total exclusion of the horizontal direction. *Du sprachest zur Gottheit* ("you spoke to the divinity") is the successful vertical relation, while the horizontal relation to man fails: "Öfters hab ich Gesang versucht, aber sie hörten dich nicht" ("from time to time I attempted song, but they didn't hear you").

The context of "Im Walde" represents not so much the vertical relation as its disturbance. Neither the false priest nor the layman fulfills the task, formulated in the marginal text, of attesting to having inherited and having learned. The false priest fails because of his arrogance, which makes him forget the difference, yet he still must sing a song of warning to those who are quick to learn, a song that, although it keeps present the relation to the divine, does so not by giving access to the divine, but by showing the danger of coming into contact with it. Quickness to learn, to which the song is directed, and which appears again in the learning men of "Im Walde," seems to be lacking in the three ode stanzas on the lower half of the sheet. The layman wanders around the workshop without completing the work and hesitates, while the master, who is sure of his ground, gets it done. The second part of "Im Walde," which is next to these stanzas, can be read as an answer to them. While in the ode the student is unable to learn from the master, in the marginal text the mediation of the "Meisterin und Mutter" ("mistress and mother") is successful. From her is learned "ihr Göttlichstes, die allerhaltende Liebe" ("her most divine attribute, all-sustaining love"), which for its part is in clear opposition to the "sterblichen unverständigen Liebe" ("mortal, uncomprehending love") of the poem. The uncomprehending love does not achieve that which the marginal text calls man's "Verstand" ("understanding"): "dass er bewahre den Geist, wie die Priesterin die himmlische Flamme, dies ist sein Verstand" ("that he should preserve the spirit as the priestess does the heavenly flame: this is his understanding").

A passage at the beginning of "Hyperions Jugend" reveals that it is love—though not the uncomprehending kind—that preserves the spirit (*FA* 10: 210). The pure spirit cannot preserve itself, because it does not possess itself: "er ist alles, und darum ist er nichts für sich" ("it is everything, and therefore it is nothing for itself"). It is, because it cannot feel itself, dependent on man, who, to the extent that he is lacking something, feels a drive toward completion and knows it. "Die Armut der Endlichkeit ist unzertrennlich in uns vereinigt mit dem Überflusse der Göttlichkeit" ("the poverty of finiteness is inseparably united in us with the excess of divinity"). The "daughter of poverty and excess," however, is love. "Als unser Geist . . . sich aus dem freien Fluge der Himmlischen verlor und sich erdwärts neigte vom Aether, als der Überfluss mit der Armut sich gattete, da ward die Liebe" ("when our spirit . . . lost its way from the free flight of the gods and turned toward Earth from the ether, when poverty coupled with excess, love was born"). Love, born and understood this way, is the all-sustaining kind that "Im Walde" speaks of: only in this love does mortal man feel the divine in himself, and only in this love can the divine feel itself. "Sich aber nicht zu fühlen, ist der Tod" ("but not to feel oneself is death"). It is for this reason that love is sustaining. This love that man feels toward that within himself that transcends him has its source in this higher element, and in order to attest to love's origin on a higher plane man is given language. This task also implies the danger to which Empedocles and the false priest are exposed. Since the god cannot feel himself, "Muss wohl, wenn solches zu sagen / Erlaubt ist, in der Götter Namen / Teilnehmend fühlen ein Andrer" ("then another, if saying such a thing is allowed, / must feel, sympathetically, in the name of the gods" [*SA* 2; 1: 145])—because not to feel oneself is death. The danger consists in the possibility that the other may believe himself to be the person on whose behalf he feels: that he may forget the difference and not want to "tolerate any inequality" ("und nicht Ungleiches dulden") (*SA* 2; 1: 145).

The all-sustaining love that is the topic of the marginal text is, since it is attested to linguistically, the connection of the upper and the lower spheres. It functions on an essentially vertical axis. The relation to the divine that it puts in place makes possible two things: in it the spirit is preserved, which otherwise, since it cannot not feel itself, would be dead; but moreover, man himself also comes to his senses for the first time in it by experiencing what is beyond him. It is therefore possible to read the conclusion of "Im Walde" in two ways. Language is given man "damit er zeuge, was er sei / geerbet zu haben" ("so that he may attest that he has inherited what he is"). This means that man is not what he is by his own agency, and that he can only attest to the higher entity through which he is what he is. It also means, however, that man, by attesting the presence in him of what is beyond him, thereby preserving the spirit, begets himself as that which he is: "damit er zeuge, was er sei" ("so that he may beget what he is").

The attempt to lend meaning to the verticality of the marginal text, as well as the reflections on the border of the text and the margin of the page, is based on an allegorization of the graphic appearance of the text. This is suspicious like every allegorization, or even untenable when one demands proof, and meaningful when one allows for what is just possible. It belongs to the general traits of this allegorization that in it the edge of the page can be interpreted as the border onto speechlessness. Beside the danger of language is the danger of loss of language. Of what does it consist? This can be derived from the only two passages in Hölderlin's poems where the word *sprachlos* ("speechless") occurs. The first passage is the last stanza of the ode "Ermunterung":

Und er, der sprachlos waltet und unbekannt
   Zukünftiges bereitet, der Gott, der Geist
      Im Menschenwort, am schönen Tage
         Kommenden Jahren, wie einst, sich ausspricht.

(*FA* 5: 769)

And he who speechless prevails and, unknown,
   prepares the future, the god, the spirit
      in the word of men, on the fine day
         of coming years, as he once did, expresses himself.

Here the speechlessness is that of the divine spirit who waits to express himself in the word of man, and who, as we have shown, depends on this possibility to feel himself. The second passage is the published version of "Hälfte des Lebens":

Weh mir, wo nehm' ich, wenn
Es Winter ist, die Blumen, und wo
Den Sonnenschein
Und Schatten der Erde?

Die Mauern stehn
Sprachlos und kalt, im Winde
Klirren die Fahnen.
(*SA* 2; 1: 117)

Woe is me, where will I get, when
it is winter, flowers, and where
the sunshine
and the shadows of the Earth?
The walls stand
speechless and cold, in the wind
crackle the flags.

Here speechlessness is man's winter, about which the preliminary version says: "Dann wird es sein, als wüsst ich nimmer von Göttlichen, / Denn von mir sey gewichen des Lebens Geist" ("then it will be as though I no longer knew about the gods / because the spirit of life has gone away from me"). In the first passage, then, the speechlessness is that of the god without man, in the second that of man without the god. Speechlessness is the condition in which the relation to the divine is broken off. If the god without man and man without the god are speechless, there is language only in their relation to each other. Language is this relation.

Taken together with our remarks about the margin, this means that the edge of the page, to the extent that it signifies the border onto speechlessness, also marks the failure of the relation to the divine. Now, all the poems that are in the process of originating on the sheet under discussion are in one way or another about the endangering of this relation if not about the actual loss of it, a loss which, according to this understanding of speechlessness, should lead to the silence which the hymn "Wie wenn am Feiertage . . ." comes closest to in its definitive interruption. Since the relation between the divine and the human is the theme of all the poems united on this sheet, the question arises whether and to what extent the marginal text differentiates itself from them. So long as the endangering of the relation is discussed, language, which is what makes this discussion possible, is not in danger. It is, however, in danger where it comes to its own edge: namely, on the edge of the page. Where language ceases, the connection to the divine breaks off.

In the marginal text, the margin is made language. This is, already almost beyond language, the affirmation of language as that which connects. This affirmation occurs here, no longer just because language takes place in the form of a discussion of the relation that it founds: here, where language threatens to cease, it becomes its own object. On the margin of language, in order to postpone speechlessness, language speaks about itself.

In the absence of language, the relation to the divine falls apart. The edge of the page stands for this break. Language is given to man in order for him to attest to the divine origin of the all-sustaining love. This proof of a love that brings together finitude's poverty with the god's excess produces the connection that is threatened by the edge of the page seen as the edge of language. That is why the impossible happens here: the all-sustaining love oversteps the edge of the page. At this point, and only here, the writing goes over to the neighboring sheet. It can do this because the edge of the page here is only linguistically represented by the fold that divides the sheet in the middle. By testifying to the "aller // haltende Liebe" ("all // sustaining love"), language bursts open the frame assigned to it and allows what is unavailable to it to express itself in it by speaking out into it.

Whether the attempt to read the graphic appearance of manuscripts and to include it in the exegesis of the text is a workable one is not a priori certain, and requires a reflection on the status of the results obtained in this way. Texts like "Aber die Sprache . . ." and "Im Walde" have been edited without regard to the shape they had in the manuscript and can be read in this form. This, however, does not exclude that what surrounds a text on the sheet of paper on which it is created makes accessible layers in it that remain buried in a conventional printed version. In the case of the marginal text, its publication, if it does not designate it as such, neglects at least two kinds of possible relations in which the text takes part. First of all, the marginal text always has a context, because a margin on which something can be written comes into existence by the very fact that the sheet has been written on. Beside this intertextual relation, however, one also needs to ask about the relation of the marginal text to the margin as such. The issue is not only its relation to the text on whose margin it is, but also the fact that it is on the margin, and the question whether and how its content is related to its marginal position.

That the possibility of such contextual relations is often not considered important enough to be respected in printed editions is due to the fact that it cannot often be anchored with certainty to authorial intention. It would not occur to anyone to print Mallarmé's "Un coup de dés" as continuous prose, since here there is no doubt about the intention. In contrast, the concomitant presence of different drafts on one of Hölderlin's manuscript pages appears, and perhaps even is, accidental. But leaving aside the fact that this does not exclude meaningful connections between the texts, the opposition of intentional and accidental does not hold unless one makes room between them for a broad area of indeterminability, in which the categorization as one or the other is not possible. There is no need to impoverish the text by limiting it to what is certainly intentional and can be reduced to the author, unless, that is, one is less interested in the text than in the author. But even when one does not make one's decision about whether to admit an insight dependent on whether an authorial intention can take responsibility for it, one must ask oneself in the case of every textual relation how it is produced and what sanctions it.

Several apparently independent texts that are next to each other on a sheet of paper are in a relation of simultaneity for the reader. Their very proximity to each other can make some parts in them attract attention and form constellations. The false priest, the layman and the priestess are characters in three different texts, but because the texts are next to each other, these characters enter into a contextual relation that can be verified even though it is difficult to pinpoint what kind of relation it is. It remains unknown whether or not the author intended or even noticed it.

Nonetheless the question of how such a contextual relation occurs is not irrelevant. It can be the reader who creates it by reading a kind of over-text out of the individual texts, a procedure that also plays a role in the attempt to think about the entire work of an author in its interrelations. A genetic relation between the texts, however, is also conceivable. The texts did not originate simultaneously but one after the other, so that one should consider whether and how the earlier texts may have contributed to the way in which the later ones took shape. The proof must be produced for each individual text and cannot always be found. Thus it is easy to assume that the word *layman* in "An die Deutschen" was inspired by the spatially close word *priest* in "Wie wenn am Feiertage . . ." Since *Laie* ("layman") does not appear in any of the earlier drafts of the ode that are written down elsewhere, this assumption cannot be refuted. It can just as little be proven. It is supported by the chronology of the texts, by their appearance on the page and by the meaning that can be extracted from the relation, but there is no criterion that could ascertain that Hölderlin understood the ode stanzas drafted here as counterpart to the hymn that he had interrupted earlier. In other cases, the encroaching of the texts onto each other can be proven. A good example is the exclamation *Weh mir!* ("Woe is me!"), which crosses over from the draft of "Wie wenn am Feiertage . . . ," to which it first belongs, to the poem "Hälfte des Lebens," which was drafted later; in this case the graphic appearance of the texts shows that it is their proximity that triggered the takeover. The poem "Hälfte des Lebens" would not have become what it is if it had not been created on this sheet.

Even where it is not possible to prove such a genetic connection, the contextual relation of neighboring texts remains. The impossibility of making a sharp distinction between what is intended and what is accidental is no reason to reject what is possible. The possible relations between texts are the ones that produce meaning. The question of whether they are intended is subordinate to the question of meaning. Moreover, limiting the author to his intention is arbitrary. This act is governed by the understanding of the text as something that is made and can be mastered, and for which someone is responsible down to every single ramification. But the author is exposed to language and to the circumstances of its production. The meaning of his text does not remain within the borders of what is intended: it emerges through the accidents of unpredictable contextual relations. If someone has a feel for it, he may want the accident and intend what cannot be planned. To what extent this was the case with Hölderlin is again impossible to decide; however, it cannot be disputed that certain manuscripts favor cross-links about which it is seldom possible to indicate with any certainty to what degree they are controlled.

These links include not only those between individual texts, but also those of the texts to the paper on which they are written. The allegorization of the graphic appearance of the text can neither be attributed to the author nor can it manage without him. The possibility of interpreting the edge of the page as the edge of language is not tied to a specific author, text, or sheet of paper: it is implicit in the conditions of writing on paper. One can attempt to put this interpretation to use for the understanding of any marginal text; yet in most cases this tactic will not yield much. However, when two texts that are on the margin speak of language, the place where they are written can produce meaning. It would be rash to want to legitimate the allegorizaton of the margin by attributing it to Hölderlin's intention, but it would be just as arbitrary to ignore that he was the one who spoke about language on the margin. It is not possible to know with certainty what he meant by it. However, that meaning was produced from this, and that these texts mean differently when they are read as marginal texts than when one just reads what they say, has to do, in a manner that is difficult to determine, with the understanding of those texts that are tied to the name "Hölderlin."[4]

Translated by Georgia Albert

The Highest

Thomas Schestag

In the summer of 1805, in Homburg vor der Höhe, Hölderlin writes the later so-called Pindar commentaries. They first translate, and then interpret, nine fragments—or fragments of fragments—by Pindar. To be precise, however, the translations that precede the commentaries are part of the interpretation. They already interpret. The nine Pindar commentaries expose, quietly, the belief in the possibility of positing and translating two languages, Greek and German, by substituting phrases and words of the one with the other. By interpreting the interpretation offered by the other text, the commentaries split [*teilen*] the appearance that both languages are simply "given."

The fifth and middle one of the nine commentaries, titled "Das Höchste" ("The Highest"), starts out from the interpretation, in the Greek text—

> *νόμος ὁ πάντων βασιλεὺς*
> *θανατῶν τε καὶ ἀθανάτων*
> *ἄγει δικαιῶν τὸ βιαιότατον*
> *ὑπερτάτα χειπί . . .*[1]

—of the *nomos*:

> Das Höchste.
> Das Gesez,
> Von allen der König, Sterblichen und
> Unsterblichen; das führt eben

Darum gewaltig
Das gerechteste Recht mit allerhöchster Hand.

Das Unmittelbare, streng genommen, ist für die Sterblichen unmöglich, wie für die Unsterblichen; der Gott muß verschiedene Welten unterscheiden, seiner Natur gemäß, weil himmlische Güte, ihret selber wegen, heilig seyn muß, unvermischet. Der Mensch, als Erkennendes, muß auch verschiedene Welten unterscheiden, weil Erkenntniß nur durch Entgegensezung möglich ist. Deswegen ist das Unmittelbare, streng genommen, für die Sterblichen unmöglich, wie für die Unsterblichen.

Die strenge Mittelbarkeit aber ist das Gesez.

Deswegen aber führt es gewaltig das gerechteste Recht mit allerhöchster Hand.

Die Zucht, so fern sie die Gestalt ist, worinn der Mensch sich und der Gott begegnet, der Kirche und des Staats Gesez und anererbte Sazungen (die Heiligkeit des Gottes, und für den Menschen die Möglichkeit einer Erkenntniß, einer Erklärung), diese führen gewaltig das gerechteste Recht mit allerhöchster Hand, sie halten strenger, als die Kunst, die lebendigen Verhältnisse fest, in denen, mit der Zeit, ein Volk sich begegnet hat und begegnet.

"König" bedeutet hier den Superlativ, der nur das Zeichen ist für den höchsten Erkenntnißgrund, nicht für die höchste Macht. (*FA* 15: 355)

The highest.
The law,
King of all, mortals and
Immortals; just for this reason it wields
Powerfully
The most rightful law with the very highest hand.

The immediate, strictly speaking, is impossible for the mortals as it is for the immortals; the god must differentiate between different worlds, according to his nature, because heavenly goodness, on its own account, must be holy, unmixed. Man, as knowing, must also differentiate between different worlds, because knowledge is only possible through opposition. For this reason the immediate, strictly speaking, is impossible for the mortals as it is for the immortals.

Strict mediacy, however, is the law.

But this is why the law wields powerfully the most rightful law with the very highest hand.

Discipline, to the extent that this is the form in which man encounters both himself and the god, the law of church and state and inherited statutes (the holiness of the god, and for man the possibility of a moment of knowledge, of explanation), these wield powerfully the most rightful law with the very highest hand; they hold even more strictly than art does the living relationships in which, with time, a people has encountered and encounters itself.

"King" means here the superlative, which is the sign only for the highest ground for knowledge, not for the highest power.

The lines following what, at first glance, appears to be, under the title "Das Höchste," the German translation of a Greek fragment, are about a third element, or something *like* a third element that is impossible—impossible for what purpose, that remains open—for both of the sides that are related to each other in a certain relation of exchange, mortals and immortals, gods and human beings: the immediate [*vom Unmittelbaren*]. The immediate is impossible for both because both, to be possible *as* such and to exist, to become real, to come to themselves, and to remain by themselves, different from others, must differentiate. In the case of man as knower, this differentiating—a condition of his possibility—is accomplished by opposition. Knowledge of different things, and thus also knowledge of himself as one of other different things identical to themselves, of other Ones identical to themselves, is only possible for man to the extent that what is different—different from itself—does not just come up to him as an other but as *his* other: that it does not just *encounter* him but is his *counterpart.*

Another knowledge, of what is different, a knowledge other than that obtained by opposition, must be ruled out for both, immortals and mortals, in order not to endanger the ability of both to be. The thrice repeated *muß*—"muß . . . unterscheiden / heilig seyn muß / muß auch . . . unterscheiden" ("must . . . differentiate / must be holy / must also differentiate")—suggests, however, that neither the mortals nor the immortals, neither man nor the god, *are*, to be precise; rather, they are *on their way* to encounter each other, to oppose each other "on their own account." They are on a journey that is not a teleological progress—namely a return—toward the joining of one with itself—its other—into the self in and for itself, a joining in itself already completed, but rather—without ruling out the interpretation of this journey as a return—less comprehensibly, errancy. The positing of man, which *is* not, but is called for—*muß auch . . . unterscheiden* ("also must differentiate")—depends on the opposition of a god who also must differentiate. The possibility of the one, the possibility of being posited undifferentiated from oneself, depends on the difference, as opposition, of one—of another—whose positing depends on the difference, as opposition, of that one. The positing, to be knowable *as* such—as differentiated from other positings—must be an opposition. Position . . . opposition. The positing of the one comes about only through the opposition of the other, the positing of the other only through the opposition—of the one: the other of that other. Both sides come about—oddly undifferentiated—only through opposition. The opposition, however, is not identical to itself, is not, undifferentiated from itself, difference itself, but rather, as a modification of differentiating in general, different from—but not in opposition to—itself.

This differentiating from oneself, an undetermined same [*Selbes*], in which the possibility of opposition appears but remains impossible, a same [*Selbes*] which neither *is* nor is *opposed*, does not make unavailable simply *itself* to the knowledge that proceeds through opposition and ascertains, but rather, more initially, disfigures both knowledge and the status of the knower. The possibility of knowing is tied to differentiating as the condition for its impossibility, "weil Erkenntniß nur durch Entgegensezung möglich ist" ("because knowledge is only possible through opposition"). The possibility of differentiating exceeds and splits [*teilt*] the capacity to know through opposition. What comes into view and seems known *as* something—opposed—even a sheet of paper, even this sheet, "Das Höchste", before our eyes, remains different [*unterschieden*]: unforeseeably divisible [*teilbar*]. The ends related through opposition, mortals and immortals, trace too close to each *other*, in a manner too different [*unterscheidbar*] from themselves, to be, opposed, *as* such—undifferentiated from themselves. They cannot rule out the condition for their possible impossibility, namely differentiating, without remaining marked by it. Both do not, really, come about, because the middle [*Mitte*], which produces the possibility for both to be mediatable [*vermittelbar*] and communicable [*mitteilbar*], remains too immediately [*unmittelbar*] differentiated to be knowable *as* such, through opposition. It is the becoming aware, in the knowers' turning to themselves, of the fact that they can still be differentiated from themselves—from the explanation that defines them as knowers—without this awareness leading to knowable information [*Aufschluß*] about themselves who do not get resolved in being interpreted as knowers, it is this becoming aware that frees in them the possible inability to be identified with knowers.

In the experience of the impossibility of knowing differentiating in general *as* such through opposition, in the experience of the possible impossibility of their being posited and opposed as knowers, the sides of the opposition break up toward what the first sentence calls the *immediate*. The more impossible the knowledge about the knower, as well as about knowledge, becomes in the knower's turning toward himself—a turning that drives him apart—the more possible (not in the sense of a possible *knowledge*, but in a more differentiated way) the immediate becomes: indeterminate middle that dissolves, yet holds up, the possibility of communicable extremes [*Enden*]. The impossibility of a moment of knowledge, an impossibility that can be experienced by the knower in his turning toward himself, in a failure of knowing, leaves open the possibility, however, of making knowledge be defeated on *something*, to put this something into words, to assert that it is a figure and an entity; and leaves open the impossibility of knowing it—the ground or lack of ground of knowledge—to formulate it as knowledge about it. Thus, in the first sentence the immediate, a substantivized adjective, appears as a placeholder for the impossibility, which can be experienced in turning toward oneself, of knowing knowing, and thus of knowing. Where the immediate is taken *as* something, however, it becomes possible. Strictly speaking, therefore, it must be contrasted, by way of the attribute "impossible," with the possibility that man and god are and know. However, the possibility for both mortals and immortals to be there and to know depends, when the experience of the failure of knowledge remains alive, on that immediate that is called impossible. This is why this immediate is elevated to that which, *as* unknowable ground of knowledge, produces the possibility of mediation and of that which is mediated: this is why it is elevated to law. "Die strenge Mittelbarkeit ist aber das Gesez" ("Strict mediacy, however, is the law"). The immediate, *strictly speaking* [*streng genommen*, "taken strictly"], is *strict* mediacy: this sentence describes, precisely, the ambiguous sketch [*Aufriß*] of the law. The experience—in the turn toward himself—of the knower, the experience of unregulated capacity for differentiation, in which appears the impossibility not only of knowing but also of the *decision* which is based on faith in indivisible, given magnitudes, and also of *undecidability*, a divisible apotropaic element against differentiation, is *taken strictly* [*streng genommen*]; hypostatized into law.

It is, however, the last sentence that suggests what *streng genommen* ("taken strictly, strictly speaking")—a taking that necessitates the sketching of the law—means in these lines: the sentence that refers to the law, personified as king, as something extreme: " 'König' bedeutet hier den Superlativ, der nur das Zeichen ist für den höchsten Erkenntnißgrund, nicht für die höchste Macht" (" 'King' means here the superlative, which is the sign only for the highest ground for knowledge, not for the highest power"). Law, *von allen der König* ("of all the king"), traces in this sentence the contours of a rhetorical figure, the oxymoron. The phrase *höchster . . . grund* ("highest . . . ground") describes the oxymoronic outline of the law. The word *Erkenntniß* ("knowledge"), which is to be inserted in the phrase, underlines, in the compound *Erkenntnißgrund* ("ground of knowledge"), the ambiguity of the law: the phrase brings into view at least two mutually incompatible possibilities for its interpretation, which, measured up against the will to decide between them, since an indeterminate same—the highest ground of knowledge—produces them, remain undecidable. *Der höchste Erkenntnißgrund* can mean both the highest, most certain knowledge about the ground of all knowledge and the highest, most unknowable ground that is subject to all knowledge—in which knowledge, since the phrase *highest ground* is impossible for it, is suspended. The differentiation experienced in the failure of knowledge, a differentiation that is not resolved in oppositions [*Entgegensetzungen*], but rather splits the appearance of -position [*-setzung*], and does not develop into any position or figure that would not for its part remain divisible, should be banned in and as the law and made into a decidedly undecidable figure of the experience of unregulated divisibility [*Teilbarkeit*]. In the sketch of the law, there is a correspondence to the experience of the failure of knowledge such that the ends of the most extreme opposition, mediate/immediate, are brought together into a figure: into *strict* mediation.

The sketch of the law narrows, *strictly speaking*, the two related ends of the opposition into an oxymoron. The contracted ends, bound together into law, should be *bindingly separated* in it. Namely in such a way that the immediate, brought in to serve, underlines the unmediated and unmediatable undifferentiation of mediacy from itself, the autothesis of the law. The *strictness* of mediacy has its ground in the fact that the impossibility of positing the law [*Gesetz*] as something posited [*Gesetztes*], implicit in the assertion of the law's immediacy (for the possibility of positing emerges only in the interpretation of what is different as something opposite, only in one's original nonidentity *as* something—law or "I"—that attempts to posit itself), must be faded out in order to declare [*erklären*] and recognize it *as* law: as the condition of *possibility* of all mediation, through opposition. In reverse, and too much in the same way [*zu gleich*], however, the law, when the assertion of its immediacy fades out, becomes for its part something only possible among other things, and it is cut off—differentiated—from the condition it was supposed to be. The plausibility of a declaration of the law [*de Erklärung des Gesetzes*] as condition of *possibility* depends on the fact that the *condition* of possibility is unconditional: it depends on the immediate independence of the law. In other words: the binding force of the declaration is proportional to how unbinding it is. This declaration would be at its most stringent and strictest, and the independence of the law at its most clearly explained and clearest, in a case in which it was not there: where the law needed no declaration. The more unalterable the declaration [*Erklärung*]—of the independence of the law—the more dependent the declared independence is from the declaration, and the more impossible it becomes to unify the act of explaining with that which is explained: the easier it is to differentiate, the more difficult to unify into an declaration, the declaration from itself.

The declaration of *independence* depends on the *declaration* of independence; the law [*Gesetz*]—on its opposite [*Entgegengesetz*]. Declaration and knowledge, however, just like opposition, depend on differentiation, which does not get resolved in the knowledge about it, because what is known remains possible to differentiate. As something immediate, strictly speaking, the law, *condition* of the possibility of god and man, should remain impossible and inaccessible; as strict mediacy, however, it should be the condition of the *possibility* of both, god and man, to be *there* and to know. The more immediate the law, the more inaccessible it becomes to the possibility whose condition it is supposed to be, and the more inseparable possible and impossible become from each *other*: from differentiating. What is tearing [*das Rissige*]—the irrepressibility of differentiation—in the sketch [*Aufriß*] of the law, its possible impossibility, just as the impossibility of the possibilities for which the law is a condition, man and god, is brought into view by the last sentence, which adds to it the synonyms *Zucht* ("discipline"), *Gestalt* ("shape"), and *Sazungen* ("statutes"):

> Die Zucht, so fern sie die Gestalt ist, worinn der Mensch sich und der Gott begegnet, der Kirche und des Staats Gesez und anererbte Sazungen (die Heiligkeit des Gottes, und für den Menschen die Möglichkeit einer Erkenntniß, einer Erklärung), diese führen gewaltig das gerechteste Recht mit allerhöchster Hand, sie halten strenger, als die Kunst, die lebendigen Verhältnissen fest, in denen, mit der Zeit, ein Volk sich begegnet hat und begegnet.

> Discipline, to the extent that this is the form in which man encounters both himself and the god, the law [*Gesez*] of church and state and inherited statutes (the holiness of the god, and for man the possibility of a moment of knowledge, of explanation), these wield powerfully the most rightful law [*Recht*] with the very

highest hand; they hold even more strictly than art does the living relationships in which, with time, a people has encountered and encounters itself.

The sentence gives the law the meaning of a shape [*Gestalt*]. It is for the first time in it, as in a *frame*, that man and god are there [*da*]—or gone [*fort*]—*as* such, able to be known and to know. The law, an actual, proper shape, which something differentiated and differentiating understood as though mediated [*vermittelt*] and mediate [*mittelbar*], man or god, tries to attain or—for the sake of its distinction and formation into something shaped and an *image* [*Bild*], for the sake of its education [*Bildung*]—should be obliged to do so, can be inferred only from the encounter, as opposition, of something differentiated to another shape.

This inference or conclusion, from another, encountered shape to the one belonging to whomever attempts to see, does not make certain, however, as retrospective inference [*Rückschluß*], something already given, already possessed: rather, it remains open. In this conclusion/inference the "backwards" [*Zurück*] is drafted in a nonorientable way. The supposed one, one's own, does not get resolved in its interpretation as the other of the other, as *its* other, since the other results *otherwise*: from the attempted conception of what is different as what is opposed, without the shape that rises out of that remaining undifferentiable. That which is encountered comes to the law, conceived as a form that is not just to be one shape, one among others, but its own shape, the one proper to it, and thus conceived also as letters and words—which, just like shapes, appear opposed and posited—that which is encountered comes to the law only *strictly* speaking. The phrase *streng genommen* ("strictly speaking, taken strictly"), however, brings into view the fact that this strict taking [*das strenge Nehmen*], as a way of taking among others, does not capture the others in itself, but rather, measured up to the will to shape, strives to subordinate or exclude them as inauthentic [*uneigentlich*]. Thus also that taking that is called here *art*: "sie halten strenger, als die Kunst, die lebendigen Verhältnisse fest" ("they hold even more strictly than art does the living relationships"). The "living relationships," however, they and the discussion of "encountering" [*Begegnen*], draw our attention, in this second-to-last sentence, to the possible impossibility of the law. Encountering does not get resolved in going toward [*Entgegengehn*] and opposing [*Entgegensetzen*]; instead it attracts, in the *Be* of *Begegnen*, the *Bei* ("by") of an unpredictable closeness of things to each *other* that are different and unforeseeably possible to differentiate.

A closeness that is intended to be conceived [*gefasst*], through opposition, as a (bridgeable) cleft, as a neutral between, as a predictable middle of the mediation of shapes, but remains without frame [*fassungslos*], not only abysmal but also undermining, because the look and the hand, which on the detour through the attempted conception of something differentiated as a shape, where the shape remains possible to differentiate, also remain possible to differentiate: do not see in the look one's own look, the one longed for; do not, in grasping, capture one's own hand.

It becomes impossible to determine what the *look* and the *hand* mean where they seem to be, *as* words, conceived as shapes, that is, seen and touchable. Seeing [*erblicken*] remains impossible to see completely, grasping [*Begreifen*] ungraspable [*unbegreiflich*]. Just the same for the *living relationships.* They do not result, as dialogic-dialectic mediation of positions and persons, from the presupposition [*Vorausgesetztheit*] of laws [*Gesetzen*]; rather, they suggest a closeness of different things to each *other*—indicated by the word *living*—not unambiguously to be interpreted as shape. In it, in this closeness, the hand, which takes and holds the living relationships even more strictly than art, because it, the hand, where it grasps, does not get a grasp on itself or have the upper hand on it, the hand becomes inseparable [*unscheidbar*] from the living relationships, from differentiation beyond measure. *Inseparable* does not mean "identical with the living relationships," but rather—in a more reserved way—nothing other than the living relationships: *possible to differentiate* [*unterscheidbar*]. A letter to Casimir Ulrich von Böhlendorff of December 4, 1801, four years before the writing of this Pindar commentary, is also concerned with the living relationship: "Ich habe lange daran laborirt und weiß nun daß außer dem, was bei den Griechen und uns das höchste seyn muß, nemlich dem lebendigen Verhältniß und Geschik, wir nicht wohl etwas *gleich* mit ihnen haben dürfen." ("I have labored long over this and know by now that, with the exception of what must be the highest for the Greeks and for us—namely, the living relationship and destiny—we must not have anything *in common* with them").[2]

The highest, circumscribed as "living relationship" and *Geschik* in this letter, is not, as may appear obvious when we first look at "The Highest," to be equated to the law, which holds, "with the very highest hand," the living relations from which it results [*entspringt*], without being able to hold and offset the fissure [*Sprung*] that pervades the shape. Closer to the highest than the will to shape, the very highest *hand*, which the law, personified as king, needs to be able to use, a strong hand. The hands: an end of the shape. The hand, which, among other things, writes "The Highest"—means, without getting resolved on this meaning, the possibility, yet impossibility, of the law: a *Geschik* that remains impossible to interpret unambiguously as *its* destiny, that of the law. In the strict taking hold [*Zugriff*] that attempts to straiten [*engführen*] relationships [*Ver*hält*nisse*] to standpoints and clues [*An*halts*punkte*], to straiten relations to positions, in order to have grasped them and to hold them *as* such—straitened, expressed in an image, in order to find in them the guarantee for the well-defined There of something that grasps and acts [*Handelnden*], the accent, which in *Geschik*, more differentiated, gets out of the way toward propriety [*Schicklichkeit*] and "dexterity" [*Geschicklichkeit*]: the hand—escapes the hand, which conceives *Geschik* as "fate" [*Schicksal*] as the unavoidably linear complex of inherited fitting [*schickend*] and sent [*beschickt*] entities. Strictly speaking, the conception of *Geschik* as the unavoidable—law and fate—cannot be contrasted enough *against* the possibility, decipherable in *Geschik*, to avoid this interpretation, to proceed otherwise. This conception cannot avoid avoidance of the interpretation of destiny as the unavoidable. The conception remains differentiated: open for other readings; indivisible from dexterity and propriety, which accompany it. The impossibility—*a limine*—not to avoid (another word for differentiate), concerns, in the Pindar commentary, also the phrase under which the law, strict mediacy, is brought into words in the most general and most determinate, the most encompassing [*umfassend*] tendency: *mit der Zeit* ("with time"). And it also concerns the sketch of that shape of shapes, bound to time and to *inherited* statutes, the shape that "mit der Zeit . . . sich begegnet hat und begegnet" ("with time . . . has encountered and encounters itself"): a people. In other words: a language.[3]

For, as little as the highest can be equated with the law, just as little—on this "The Highest" leaves no doubt—can the living relationships be equated with time: "die lebendigen Verhältnisse . . . *in denen*, mit der Zeit, ein Volk sich begegnet hat und begegnet" ("the living relationships . . . *in which* with time a people has encountered and encounters itself"). Time, the will to the most encompassed inside, the most set off outside, the will to the dichotomy of inside and outside, remains outside itself. It encompasses the living relationships out of which it springs, not without a fissure, but rather it deviates—from time. It is not otherwise with the people (a language), which, in each of its members (sign, letter, word), where these members encounter and, strictly speaking, are opposed—known, deciphered—appear, remains possible to differentiate. Irreconcilable with its interpretation as people, as language. It encounters wherever it has encountered and encounters *itself*, in avoidance, an *other*. In the look, which meets its mark, what is met, just as what meets, is split. In the instant when the encounter, a meeting, seems to be ordained into opposition, taxed and syntaxed, the ends of the encounter, which meet each other, break up. This breaking up, on the spot, that which seems to be there and given, posited, opposed, known, and translatable, is what "The Highest" writes and describes.

"The Highest" does not so much belong to the approximate genre "translation with commentary," as it discusses the appearance of the possibility of positing, substituting, opposing, translating (*der Setz- —Er-, Entgegen-, Übersetz- —barkeit*), in a singular way. For the same reason, and just like every encountering entity interpreted as text, it cannot finally be attributed to an author (the reachable or missing guarantor of the intention kept safe in a text), whether it be a man or a god, an immortal or a mortal. Just as little does an encountering entity coagulate/develop into a reading entity—one shaping and tying together meaning—into the given, arising, disappearing or failing shape of the reader. "The Highest" exposes toward the middle, toward an indeterminate and unpredictable third, the possibility, but impossibility, of dichotomizing. And it does that from the (eccentric) middle of the fragment: out of the phrase *streng genommen*. It cuts, underhandedly, the supposed unquestioned translation (as substitution of a word with another) of *nomos* with *Gesez* ("law"). For *nomos*—noun of action for the verb *nemein*: *Nehmen* ("to take")—states more precisely that the words synonymous with the phrase *streng genommen* ("taken strictly, strictly speaking"), with the accent on *streng*, the words around *Setzen—Entgegensetung* ("opposition"), *Sazungen* ("statutes")—and thus also the translation [*Über*setzung] of *nomos* with *Gesez*, putting the accent on . . . *genommen*, more untranslatable, because too close to each other—states more precisely that these words are a modification of *Nehmen* ("taking"). It becomes clearer from this angle that *Gesez* is not a translation among others, but rather it attempts to straiten [*engführen*] and hold taking—*nemein*—to taking *strictly*: to positing; it attempts to straiten and to hold the taking [*Nahme*], *nomos*, to the law, attempts to make it proper [*vereigentlichen*], in order to have outlined and conceived in the *nomos*, as law, the immediate requirement for the possibility—strictly mediate—of translatability and substitutability in general, and also of knowing, of the differentiated, through opposition. The expansion, however, of the participle *genommen* ("taken") by the adjective *streng* ("strict, strictly"), takes away the limits of the tendency to conceive taking as positing, *nomos* as law, not so much strictly as precisely, as one taking among others. In this middle of the taking—dexterous with something: different than only taking strictly—under the phrase *streng genommen*, discussing the leaning toward law [*Gesez*], statute [*Sazung*], and opposition [*Entgegensezung*], "The Highest" breaks up.

Repeatedly taken up and interrupted, attempts to read—and more exactly, to take—"The Highest" run through Martin Heidegger's "Erläuterungen zu Hölderlins Dichtung." Thus, the middle of the interpretation "'Wie wenn am Feiertage . . . ,'" published for the first time in 1941, comes around, after a long citation, to speaking about "The Highest":

> That which is present [Gegenwärtige] in everything beforehand gathers all that is isolated in one single presence [Anwesenheit] and mediates [vermittelt] to allow each thing to appear. Immediate omnipresence is the mediator for all that is conveyed [vermittelt] and that means for that which is mediate. The immediate is itself never something mediate; in contrast, the immediate, strictly speaking, is mediation [die Vermittlung], i.e., the mediacy of the mediate, because it makes the mediate possible in the latter's essence. "Nature" is the mediacy that conveys everything, it is the "law."[4]

In a manner fitting to the emphasis given to its meaning, the phrase *streng genommen* is taken over from the Pindar commentary, and implanted in the middle of the interpretation in such a way that it is this phrase that causes the ambiguous outline—to immediate mediacy of the law, an oxymoron: The immediate, *strictly speaking*, is *strict* mediacy. Heidegger: "The immediate, strictly speaking, is mediation, i.e., the mediacy of the mediate."

In an essay published in 1955, "Heidegger's Exegeses of Hölderlin," Paul de Man, after quoting these lines (and translating *streng genommen*, avoiding *genommen*, with "dans un sens rigoureux" ["in the strict sense"]), objects to this interpretation that repeats the sketch of the law which "The Highest" traces:

> This passage, which is the turning point of the demonstration, is contradictory. It states that mediation is possible thanks to the very immediate that is its agent;

indeed, in this movement, it is the immediate that appears as the positive and moving element. It does not follow, though, that the immediate, as the sole agent of action, must be identifiable to mediation itself, which defines the multiple structure of the action. If the word *mediation* has any meaning, it is that what mediation ends up with is never identical to one of the two elements in presence to the exclusion of the other; it is a third entity that contains them both. To say that the immediate contains the possibility of the mediation of the mediate because it permits it in its being, is correct, but it is not correct to go on to the conclusion that the immediate *is*, therefore, itself the mediating intercession.[5]

The objection seeks to avoid the oxymoronic sketch of the law in order to be able to muster, within the frame of opposition ("contradictory") and equation ("identifiable," "identical"), *against* the attempt ascribed to Heidegger to equate the immediate and mediacy (entrusted to the guardianship of the immediate, which de Man explains as synonymous to being), the (so it is claimed) proper content (ascribed to Hölderlin), namely mediacy, differentiated from the immediate, equated with the law —in de Man's French text translated and replaced by *loi*. The argument leads to a sentence that summarizes the interpretations that have been opposed to one another: "When he states the Law [la loi], the poet does not say Being, but, rather, the impossibility of naming anything but an order that, in its essence, is distinct from immediate Being."[6] Without mentioning the difference between mediacy and *strict* mediacy—with the eclipse of the word *strictly*—mediacy, equated with the law, is elevated to the communication [*Mitteilung*] of the poet (Hölderlin) and as essentially, *in its essence*, different from the immediate, contrasted *against* the equation, explained as Heidegger's communication—with the eclipse of the phrase *strictly speaking*—of mediacy and the immediate. A few pages earlier, the essay thematizes the scheme of this contrasting of interpretations:

Hölderlin says exactly the opposite of what Heidegger makes him say. Such an assertion is paradoxical only in appearance. At this level of thought, it is difficult to distinguish between a proposition and that which constitutes its opposite. In fact, to state the opposite is still to talk of the same thing though in an opposite sense. . . . It can indeed be said that Heidegger and Hölderlin speak of the same thing.[7]

What the italic sentence attributes to Hölderlin describes, to speak more precisely, the position of the argument signed "de Man" and claimed not to be different from "Hölderlin." Corresponding more to the scheme of the argument, repeated barely otherwise, this position means: *de Man says exactly the opposite* (about "The Highest") *of what he makes Heidegger say* (about "The Highest"). Both the opposed interpretations, called "Hölderlin" and "Heidegger" in the essay, appear, in the outline of the law that "The Highest" sketches and the middle of the interpretation "Wie wenn am Feiertage . . ." repeats, compressed (undifferentiated but not identical, differentiable but not opposed: indeterminably divisible) into an oxymoron—immediate mediacy. Both the essay from 1955, in which it is missing, and the interpretation of 1939, which takes it over in the middle, and as the middle, of the interpretation, remain blind to the phrase *strictly speaking* [*streng genommen*]. For when it is taken into account, it unsettles both arguments because it defines the *strict* taking, less as the actual or proper taking—this is de Man's and Heidegger's emphasis—than as one of other acts of taking, different only in degree. It unsettles both arguments: both the pure mediacy of the law, not different from itself, in de Man, and the essential indivisibility of mediacy from the immediate in Heidegger (this is why, in Heidegger's interpretation, the phrase *strictly speaking* functions only as a stock phrase: brought to a standstill as a stylistic figure in order to stress that the asserted lack of difference is decisive and beyond question). The former puts his emphasis on essential (not strict) mediacy, the latter on the essential (not taken strictly) precedence of the immediate; both, however, attempt to fade out the unregulated differentiability of each of the explained essential elements—mediacy / the immediate—sketched in "The Highest" in the formulation *strenge Mittelbarkeit* ("strict mediacy") and in the phrase *streng genommen*.

Not differently, but otherwise than de Man (who lends substance to *strict* mediacy, making it mediacy *in itself*), Heidegger, by taking the phrase *streng genommen* in no other way than as a cliché, is forced to consolidate the highest with the law. In 1943, four years after the interpretation "Wie wenn am Feiertage . . . ," he writes in the essay "Andenken":

> The highest being is closest to that which is *the highest thing*. This, however is, *the holy*, the law [Gesetz] that posits [setzt] in a different way than human law. Seen from the point of view of the latter, the way in which the holy sets *statutes* [Sazungen setzt] in front of everyone, because it is the sending [Schicken] of destiny [Schicksal], can *hardly* be called a law.[8]

The implicit declaration as beyond question of the translation of *nomos* as "law" [*Gesetz*], and the asserted undifferentiation of the highest from the law, make it necessary to hold to the positing [*Setzen*]. So that two laws (more precisely: two tendencies in the positing, in the taking strictly) are set [*gesetzt*] against each other—opposite to each other: a holy, positing law [*setzendes Gesetz*], which, because the emphasis is only on the positing, defined as "sending" [*Schicken*]—of destiny [*Schicksal*]—but somehow, no matter how indeterminately, should be posited and given *as* such, "can *hardly* be called a law"; and a human law, which emphasizes the "inherited statutes," the *posited character* of the law. The interpretation explains the oxymoronic outline of the law as a *double bind*, and spreads it out as a narrative of historical-fateful relation between the laws; it has to assert, but to reject, positing as the positing entity, and to keep open for positing, yet declare blind, both man (as the site where the essential priority of positing over that which is posited is reversed into the inessential priority of that which is posited over positing, and where positing itself—or Being—reset as a neutral middle and as means, falls victim to forgetting, yet the memory of forgetting, and therefore the forgetting of forgetting, the reversal of reversal, remains possible) and human statutes, in order not to take both, human and holy law, in a way that would too little differentiate between them, not to de-pose [*entsetzen*] them into differentiability without measure, in the interpreting look to which it, the interpretation, attempts to remain blind, in order to long for it.

This interpretation clarifies its relationship to *destiny* and to the *living relationship* that paraphrase "the highest" in the letter to Böhlendorff by calling it a finding of bearings which, between positing and posited, between *taken strictly* and *taking strictly*, without resolving strictness as a separable modification of taking, attempts to orient destiny and living relationship. Thus the end of the essay "Andenken," following a quotation about the "living relationships" in "The Highest," argues, "If, however, the *living relationships* have never before come to life and originated, i.e., if they are not founded solidly in an origin and are in their essence of poetic origin, then all the discipline in the world, no matter how strict, has nothing solid, nothing that could *hold* it."[9] The living relationships (this is where the interpretation puts its emphasis) must be fitted [*verfügt*] in, and *as*, a space in between—originating, but precisely for that reason, originally, solidly founded—while "The Highest," *almost* in reverse, stresses the inability to be fitted [*Unverfügbarkeit*] and the lack of origin, the nonoriginated, nonorientable aspect of the living relationships, *in which* even time, the will to encompassing form and to the dichotomy of inside and outside, originates, without being able to grasp and hold as such either itself as originated or the presumed origin. This interpretation—placing the living relationships in the vicinity of an origin to which, since they originate from it, they stay close—is supported by the presupposition (which dims the possibility that the appearance of givenness and solidity of that which is grabbed, grabbing, originates for the first time in the action [*Zugriff*]) that the essentiality [*Eigentlichkeit*] of the upholding relation to the living relationships is signified in discipline and in the stricter holding, to the extent that they, "solidly founded in the origin," help to hold on to the origin through essentiality [*Eigentlichkeit*]. This interpretation construes the living relationships as an in-between, between positing and posited, where "The Highest," more difficult to place and closer to the lack of comprehension [*Fassungslosigkeit*] even of the strictest grasping—because it does not understand itself, as self—sketches positing and posited *in them.*

"Destiny" [*Geschik*] is also situated as in an in-between in the relationship of positing and posited. In the two passages (within the published writings) where Heidegger speaks about *nemein* and *nomos* and attempts to translate them, destiny [*das Geschik*], from the angle of the synonymy of *Setzen* ("to posit") and *Schicken* ("to send"), is bound in the context of the tradition of the (positing) law and of its acts of providence [*Schickungen*]. In its inner seam, the "Brief über den 'Humanismus'" explains:

> Only insofar as man, ek-sisting into the truth of Being, belongs to Being can there come from Being itself the assignment of those directions that must become law and rule for man. In Greek "to assign" is *nemein. Nomos* is not only law but more originally the attribution contained in the dispensation of Being. Only this attribution is capable of dispatching man into Being.[10]

And thirteen years later, in 1959, the lecture "Hölderlins Erde und Himmel":

> Earth "follows great laws." The "laws" named here are the *nomoi* in the sense of the directions [Weisungen] of the great destiny [Geschik] that directs [weist] and sends [schickt] wherever each thing is needed according to its essence. Unwritten because impossible to write, they determine the infinite hanging together of the whole relation.[11]

For the same reason for which—avoiding *Nehmen*, to take, and *Nahme*, taking—*nemein*, as allocating, and *nomos*, as allocation, are conceived as the determinations of mediation in a relationship of acts of providence [*Schickungen*], the splitting semantic field of the word *Geschik*, narrowed—*strictly* speaking—to *Schicksal* ("fate") and *Schickung* ("act of providence"), in order to be able to hold to the authenticity of the context of the tradition, must be faded out along with *Geschicklichkeit* ("dexterity"), which causes the taking [*Nehmen*] of *Geschik*, its meeting and splitting, its interpretation as fate.

Heidegger's interpretation of *Geschik* remains tied to what Hölderlin in 1804 calls in the "Remarks on *Antigone*" the "Greek representations," "insofar as their main tendency is to be able to comprehend [*fassen*] themselves, because that was their weakness, while on the contrary the main tendency in the modes of representation of our time is, in order to 'hit' [*treffen*] something, to possess a skill [*Geschik*], since the lack of destiny, the *dysmoron*, is our weakness" (*FA* 16: 418).[12] In the quoted interpretation (Heidegger's) the skill of the Greeks "to be able to comprehend themselves" becomes "understood" [*gefaßt*] as destiny in the "unendlichen Zusammenhang des ganzen Verhältnisses," while the "Remarks," more fittingly [*treffender*], emphasize in this quality the dexterity [*Geschicklichkeit*] that reaches virtuosity in their dealings with forms, and expand this dexterity, in the phrase about the "main tendency in the modes of representation of our time," by a modification of that "skill"—to be able to "hit" something. But they do not make a whole out of it. That which seems to have been brought together [*gefaßt*] into a letter or a word, and especially where a thing believes to have comprehended [*gefaßt*] *itself*, into a shape, is "hit" in the most pregnant sense of this word: shaken up. More markedly or more quietly. This hitting mitigates even the appearance that *time* is given and understood [*gefaßt*], making it differentiability, without measure, of every single form. The skill "to be able to 'hit' something," cuts the *strictness* of the phrase *streng genommen* into the lack of composure [*Fassungslosigkeit*] of "taking" [*Nehmen*]. It cuts it up. It splits the skill of the hands "to be able to grasp/comprehend [*fassen*] themselves," around that which is more—lacking composure, clumsy, groping in it.

The quietly splitting composition [*Fassung*] of the *Geschik* ("fate/skill"), of the word *Geschik*, its not only semantic ambiguity and difficulty in orientation, concerns in the quoted lines of the "Remarks on *Antigone*" not only one among other, more composed [*gefaßter*] words, but rather, in a more underlying fashion, it concerns, in this word, the inconspicuous middle—the separable splintering origin—of a context of tradition: the innermost outermost middle of belief, or the tendency or the will to tradition *as such*, that is to the immemorial *there is* [*Es gibt*] of a historical-fateful happening (the, not only semantic, understanding [*Fassung*] of history [*Geschichte*], of this word, is affected [*getroffen*] by the splintering of *Geschik*, which is too close to it).[13] More precisely: the *discussion*—in the literal, most cutting sense of this word—of *Geschik* ("fate, skill"), in the quoted lines, affects [*trifft*] the usual understanding of language of words and signs as a means, whether of communicating extralinguistic contents or matters (dates, fictions, facts, etc.), or—more imperatively—of commandments, demands, wishes, and so on, whose fulfillment lies in acts that correspond, wordlessly, to what has been professed linguistically. The splintering *Geschik* affects [*trifft*] that understanding of language that conveys—*should* convey—the will to, the belief in, *history* [*Geschichte*]—especially in the sense of *res gestae* and *historia*—within the framework of a continuum of tradition that looks like *Geschik*, namely, like destiny. It affects a language (the language in which, at first glance, the "Remarks" also seem to be written), which should convey what "later in human time," as the "Remarks" say, "is considered a firm notion engendered by divine destiny" (*FA* 16: 419; *Essays*, 114). The "Remarks on *Antigone*," which are not only about the skill [*Geschik*] "to be able to comprehend oneself," about the skill "to 'hit' [treffen] something," but—almost in reverse—peculiarly *comprehend*, and more precisely, "*hit*," or take, *Geschik*, barely a word, expand this, this "hitting," by the discussion of the serializable *Geschik* (. . . Geschik ["fate"], das Schiksaallose ["being without destiny"], the *dysmoron*, *Geschik* ["skill"], *Schiklichkeit* ["dexterity"] . . .), as held around their eccentric middle, in the framework of a—splintering—construction of *history* [*Geschichts*konstruktion], between the "Greek" and "our" (Western, Hesperian) conception of it, or our take [*Nahme*] on it, bound by language. This construction, therefore, of the framework is to be called splintering because history [*Geschichte*], too close to the *Geschik*, less a presupposition that serves as a basis than—more precisely—also affected by the discussion of *Geschik*, is what is taken [*Gegenstand der Nahme*]:

The tragic representation has as its premise that the immediate god is all at one with man (for the god of an apostle is more mediate, is highest intellect in highest spirit), that the *infinite* enthusiasm conceives of itself [sich fasst] *infinitely*, that is, in oppositions, in consciousness which cancels consciousness, separating itself in sacred manner, and that the god is present in the figure of death.

Hence . . . the dialogic form and the chorus in contrast to it, hence the dangerous form in the scenes which, following the more Greek manner, concludes in a necessarily factual manner in the sense that the word becomes *more mediately factual* by taking hold of the more sensuous body; following our time and mode of representation, more immediate by taking hold of the more spiritual body. *The Greek-tragic word is deadly-factual*, for the body which it seizes truly kills. For us . . . the Greek representations change insofar as their main tendency is to be able to comprehend [fassen] themselves, because that was their weakness, while on the contrary the main tendency in the modes of representation of our time is, in order to "hit" [treffen] something, to possess a skill [Geschik], since the lack of destiny, the *dysmoron*, is our weakness. Hence, the Greek also possesses more skill [Geschik] and athletic virtue and needs to have these as genuine qualities and serious virtues, however paradoxical the heroes of the *Iliad* may appear to us. For us, these are more subordinated to propriety [Schiklichkeit]. And thus the Greek modes of representation and poetic forms are also more subordinated to the patriotic ones.

And hence *that which is deadly factual, the actual murder with words has to be considered more a specifically Greek art form, subordinate to a more patriotic one*. As can well be demonstrated, a patriotic one may be more murderous-factual than deadly-factual; it ends not in murder or death, for the tragic must be comprehended [gefasst] herein, but more in the manner of *Oedipus at Colonus*, such that the word coming from the enthusiastic mouth is terrible and kills, not concretely [fasslich] Greek, in the athletic and plastic spirit where the word seizes the body so that the latter kills. (*FA* 16: 417–18; *Essays*, 113–14 [trans. modified])

These lines mark not only the eccentric middle of the "Remarks on *Antigone*," and not only that of Hölderlin's writings; they mark (in a repeatedly, otherwise readable way) the moment of turning toward itself—a turning which drives apart (in a nonorientable manner)—of the context of tradition which circulates, in common language, in the form of the phrases "Western history" and "Western fate [*Geschik*]." This context's caesura, which however cannot be considered *its own*, *proper* to it. With another expression from the "Remarks": "counter-rhythmic interruption" (*FA* 16: 250).

In the seam of this moment—a caesura—in order to learn how to comprehend [*fassen*] it more exactly, to "hit" [*treffen*] it, the following lines also arise. They come from the middle of a letter written to Böhlendorff three years earlier, on December 4, 1801, and are, as it were, the first version of this moment:

We learn nothing with more difficulty than the free use of the national. And I believe that it is precisely the clarity of the presentation that is so natural to us from the beginning as is for the Greeks the fire from heaven. For exactly that reason they will have to be *surpassed* [übertreffen] in beautiful passion—which you have also preserved for yourself—rather than in that Homeric presence of mind and talent for presentation.

> It sounds paradoxical. Yet I argue it once again and leave it for your examination and use: in the progress of education the truly national will become the ever less attractive. Hence the Greeks are less masters of the sacred pathos, because to them it was inborn, whereas they excel in their talent for presentation, beginning with Homer, because this exceptional man was sufficiently sensitive to conquer the Western *Junonian sobriety* for his Apollonian empire and thus to veritably appropriate what is foreign.
>
> With us it is the reverse. This is also why it is also so dangerous to deduce the rules of art for oneself exclusively from Greek excellence. I have labored long over this and know by now that, with the exception of what must be the highest for the Greeks and for us—namely, the living relationship and destiny [Geschik]—we must not share anything *identical* with them.
>
> Yet what is familiar must be learned as well as what is alien. This is why the Greeks are indispensable for us. It is only that we will not follow them in what is own and national to us since, as I said, the *free* use of *what is one's own* is the most difficult thing. (*FA* 16: 425–26; *Essays*, 149–50 [trans. modified])

Close to the Pindar commentary "The Highest," the entrance to the quoted middle of the "Remarks on *Antigone*," devoted to "tragic representation," depicts a peculiar relationship between mediacy and immediacy. The "tragic representation" rests on transgression, or hybris, of that which "The Highest" calls *impossible*: "For this reason the immediate, strictly speaking, is impossible for the mortals as it is for the immortals." Tragic representation lies in an immediate interpenetration, no matter whether feared or longed for—"all one"—of immortal and mortal, man and god. It also lies in this, that the tendency or drive to an immediate joining,[14] the "*infinite enthusiasm*," reaches knowledge of this striving or longing, god and man part—moment of purification, *katharsis*, from overabundance—and the One comprehends [*faßt*] itself in this parting from the other, as *its* other, absolutely, that is, *infinitely*, namely through opposition, because, as "The Highest" formulates it, "knowledge is only possible through opposition." This solution, or parting, in man and god, which grants man life—as destiny or lot, *moira*, and calls God, through opposition, present "in the figure of death"; this solution, through *infinite* comprehension [*Fassung*], (determination of each one through opposition with every other—one—as *his* other), in other words through knowing consciousness,[15] remains, however, problematic. *Hybris* is inherent in *katharsis*: those plastically composed [*gefaßt*] as shapes through opposition, man and god, remain too *identical* to each other, because both, *a limine* impossible to separate from each other, have to be considered as an opposition. The (*dialectical*) solution, knowledge through opposition, remains what triggers the (*tragic*) transgression. The dialectical scheme / —a tragic scheme.[16] Resolution—entanglement. Separation—"entirely one." Man—god. And vice versa. The solution nourishes the drive, which, to speak precisely, actually originates from it, and makes it into the infinite unification of the one, or other, with *its* other: unification into the one *self.*

The parenthetical remark bound into the representation of "tragic representation," ("for the god of an apostle is more mediate, is highest intellect in highest spirit"), interprets the "immediate god" of the tragic representation as a more sensuous, a more touchingly comprehensible shape, and suggests in the more mediate god of the apostle, in Christ, a relationship that, because it is not graspable through opposition, is detrimental to the will or drive to knowledge. The lines that follow this parenthesis describe more precisely this contamination of the dialectic and tragic schemes, and describe also the implanted symptom/sign of another relationship—which cannot be comprehended through opposition or as the scheme of opposition, but more strikingly [*treffender*] unfolds in it (splitting the appearance of composure [*Gefaßtheit*] and of givenness of the scheme): they describe them as a *linguistic* relationship.

The historical framework, whose halves are apparently constituted by the "Greeks" and "us," and in which this turn to the linguistic ground of the relationship between immediate and mediate, Greek-tragic and Christian god, seems bound, is freed of boundaries by the comparatives—"according to the more Greek manner," a "more patriotic art form"—which make it a more finite, separable, and differentiable relationship, in which the mention of a *more* mediate god of the apostle concerns also the opposition mediate/immediate. This freeing of boundaries makes the relationship into that "living relationship" which, beside the "Geschik," as the letter to Böhlendorff says, "what must be the highest for the Greeks and for us," but outside of which "we must not have anything that is *the same* as they." In other words: we must not be too *opposed* to them. This is unfolded by the turn to the linguistic ground of this peculiar closeness.

First of all, this turn relates the more immediate (more Greek) and more mediate (more Western) god, or man, chiasmatically to each other: the Greek, tragic-dialectical drive to immediacy and to the opposition of mediacy and immediacy, is supported by a word that "becomes *more mediately factual* by taking hold of the more sensuous body," the unsettling of this tendency—or the same tendency, to speak more precisely—more reservedly, in the more mediate god, or man, is supported by a word that becomes "more immediately" factual "by taking hold of the more spiritual body." Both times, crossing over toward the middle (or from it), in a chiasmus of a more immediate and a more mediate relationship, an (indeterminate) same [*Selbes*] takes hold: the word.

That which these lines, at first glance, comprehended [*gefaßt*] in the framework of a span of history and fate between "Greece" and "the West," seem to signify as a variable expression of the word or of the word-language, actually describes, to speak with more precision and less history, crosswise, the pull toward and away from each other of the moment or the word in which something is "seen" [*gesehn*], in the strict sense—outlined into a shape—*as* something perceivable, at the same time however, in a *more seeding* [*säender*] way, remains otherwise, and splits, unforeseeably, the contour of a moment or a word. In a "more Greek" fashion, these lines name the tendency toward seeing [*Sehen*] and longing [*Sehnen*], the tendency toward tendency, *intentio*, and transcendence as such: the tendency toward taking the word *as* word, toward comprehending it as the signifying carrier of a signified shape, as shape, and moreover to interpret it as a means or mediator of communication (hence the talk of the *more mediate* factual word). Or, and more precisely, the means or mediator of a "claim" [*Anspruch*], to which should correspond an act that has the more sensuous body as the agent of its fulfillment. This understanding of the word—worn up to the belief in its givenness since time immemorial—as a means for communicating extralinguistic matters, as the carrier of a dictum that knows the sensuous body to be its real addressee and seeks it, is the basis for the tragic, or dialogic-dialectic scheme. It says, in the words of the "Remarks," referring to the "tragic representation" in the strict sense: "The *Greek-tragic* word *is deadly-factual*, for the body which it seizes truly kills."

The "more immediately" factual word—and the series of comparatives emphasizes that it is not opposed to the "more *mediate* factual" word, not identical to it—takes hold less *as* word understood as mediator than as more immediate. The same (indeterminate, neither Greek nor our) word. It unsettles—less composed [*ungefaßter*], less outlined, less interpretable—the understanding of the word as the typified sensuous body. Or, in reverse, as spirit. This irritation, of the tendency toward the Greek-tragic word, is triggered by the turn toward the "more spiritual body." The mention of the "more spiritual body" relates, hardly otherwise but more visibly than the mention of the "more sensuous body," to each other that which both supports and yet checks, unsettles (the drive toward opposition, body/spirit) the will to knowledge. It splits the will to communicate, to know by way of the word as a vehicle of communication, to let know, to oppose, to unify, to establish. The phrase "more spiritual body" does not get resolved, not really, in its interpretation as a word for something signified by the word. It is still possible to feel for it, but it is ungraspable. For it, to speak precisely, does not ascribe the body spirituality or the spirit corporeality as an (absolute) attribute (it splits even the appearance of the construction of "spiritual body" as the rhetorical figure of the oxymoron), but rather it is detrimental, in a more irritating, less graspable way, by speaking of the *more spiritual* body, the appearance of givenness and opposition, and also of the identical combination of the two (under cover of one or the other opposition), of body and spirit, because no measure corresponds to the comparatives *more* spiritual / *more* sensuous.[17]

Referring to the "art form," and pushed close to the "deadly-factual," of the actual murder with words, to the "tragic representation," but (to repeat) declared neither opposite nor identical, the "Remarks" call the more immediately factual word, here too comparatively, "*more* murderous-factual": "It ends not in murder or death, for the tragic must be comprehended [gefasst] herein, but more in the manner of *Oedipus at Colonus*, such that the enthusiastic word is terrible and kills, not concretely [fasslich] Greek, in the athletic and plastic spirit where the word seizes the body so that the latter kills." These lines define the word of the "more spiritual body" more precisely as "*murderous* factual": for the tendency to comprehend in it, as through a signifying word (or body) the signified content (or spirit), grasps (almost) nothing. Rather, the turn of phrase that attempts to comprehend this tendency and to establish it, through comprehension, as opposition, "hits" it in the most disruptive sense of this word. What is murderous, however, in the more immediate word, does not kill, and takes hold of no body "so that the latter kills," but rather mitigates, in a *more* murderous fashion, the appearance of the immediately given mediacy of the word, the appearance of the easy There of every reality which depends on the belief in the authenticity of the mediate word. The passage indicates this by talking about the *more immediate* factuality, or reality, of the deadly-factual word—: the murderous-factual word concerns [*trifft*]—quietly, splintering, anticipating every mediate reality—the tendency toward the real murder with words, which, in order to commit the murder, must try to be blind to every (hardly) other, close, more immediate take [*Nahme*] of the word, that commands murder, for the more immediate reality of the word, which can be taken otherwise than *as* word, otherwise than *as* means, communication, and command. It "hits" in a *more murderous* fashion, that is: it interrupts the usual understanding of killing, which is considered an (immediate) factual process, which corresponds to a communication through a (mediate) word ("Kill!"), and is perpetrated against the more sensuous body. The more murderous-factual word kills, if you will, the understanding of killing that is tied to the word as means of communication. To repeat this otherwise, more paradoxically: in the enfolding of the *more murderous, more* murderous-*factual* word, the tendency or drive toward "real murder with words" dies out, because the belief in the mediacy and in the mediating function of the word splinters. The more murderous word de-interprets [*entdeutet*] the dichotomy of living and dead toward that "living relationship" about which it is said in the letter to Böhlendorff that it, together with "Geschik," "must be the highest for the Greeks and for us," and which—since the tendency to dichotomy is bound to the interpretation of the word as mediate and mediator—does not get resolved in the dichotomy of living and dead. In it, the more murderous word, the knowledge about what dead *means*, what is dead and what is living, stops.

The more murderous-factual "hits" the more deadly-factual word. The word—the word. *Irreversibly*. "Irreversibly" means: not to be unambiguously interpreted as a (symmetrical) relation of exchange, opposition, or dichotomy (all tendencies that are tied to the word *as* word, to the more deadly-factual word). In the more murderous-factual word, in the word more immediate, the interpretation of the word as a means stops, but also starts. This interpretation *dawns* [*bricht . . . an*] in the more murderous-factual word: the tendency or the will toward the more mediately factual, to the word as means of communication, remains dependent on the more immediately factual, on the (indeterminately) *same* [*selbe*] word, as on the condition of its possibility, yet impossibly—because it cannot cancel out the more immediate, which is incalculable, of the word, in the interpretation of the word as something more mediate. The more deadly-factual remains too close to the more murderous-factual word (with a word of the more deadly-factual language: *to itself*). The more deadly-factual precisely describes, *a limine* impossible to separate from, but not identical with, the more mediate, the outline of the more mediate word as an (asymmetrical) coming apart, in which, where this coming apart is unfolded, the "*Greek-tragic*" tendency to the condition not so much of its possibility as its impossibility—which cannot be considered either *its* own or foreign—stops.

Starting out from the peculiar relation between Hesperia and Greece, own and foreign, the Greeks and "us," which appears at first glance to be opposition, the middle of the mentioned letter to Böhlendorff alludes to this irreversibility, or asymmetry. The lines are carried by the assertion that the *free* use of the national, inborn or *own*, is the most difficult, and differentiate what is "own" to the Greeks and "us" in the words: "And I believe that it is precisely the clarity of the presentation that is so natural to us from the beginning as is for the Greeks the fire from heaven." Because, however, the free use of what is one's own is the most difficult, "the Greeks are less masters of the sacred pathos [of the fire from heaven] . . . , whereas they excel in their talent for presentation." They stand out in that which is foreign and therefore more easily accessible to them—or, in the words of the "Remarks," *more mediately factual*—than that which was their own, or the more immediately factual. They practice virtuosity, the mastery of the more mediate reality, the forgetfulness of the more immediate, to such an extent that what is foreign becomes what is own to them. Thus Homer, who succeeded in "veritably appropriating the foreign." What is "foreign" means, in this context, less what is own in the others, that which is "our" own, than the more mediate, the more distant in general: that which appears *as* something, acquires a shape, fails to appear, or disappears. Eyes, trained to see and understand something *as* something. This appropriation of what is foreign, the advancement in rank of the more deadly-factual word into the word in general (the "Remarks" call it the "Greek-tragic" word), leads, as Hölderlin writes elsewhere, to the denial of what is own and to the perishing of Greece. Thus, a letter to the publisher Wilmans, written in 1803, states that "Greek art, which is foreign to us . . . has denied the Oriental [or what was own to it]" (*SA* 6; 1: 434). And a sheet in the Homburger Folioheft from the same year repeats—interrupting the syntactical chain:

> . . . meinest du
> Es solle gehen,
> Wie damals? Nemlich sie wollten stiften
> Ein Reich der Kunst. Dabei ward aber
> Das Vaterländische von ihnen
> Versäumet und erbärmlich gieng

Das Griechenland, das schönste, zu Grunde.
Wohl hat es andere
Bewandtniß jezt.
(*HF* 110)

. . . do you think
the same thing should happen
as did then? Namely, they wanted to found
a kingdom of art. But they failed
in this attempt at the national
and pitifully Greece,
the most beautiful, perished.
There is indeed another reason
for it today.

"With us it is the reverse," the letter to Böhlendorff says, pithily. This reversal, however, is irreversible. It does not signify a symmetrical relationship, as though "we," where the Greeks became masters of what was foreign to them—"clarity of presentation"—were challenged to become masters of what is foreign to us—"the fire from heaven"—since the Greeks denied, by denying what was their own, what is own (the more immediate factuality of the word, namely) in general. They "perished pitifully" because they missed, over the appropriation of the foreign, of the more mediate reality—of the word—the *free* use of what was their own, the more immediate reality, of the word. That which encounters "us" *as* Greek art, that means first of all and most importantly a *language*, the mastery of the "*more mediately factual*," "deadly-factual" word—bound on the syntactical chain, in the catalogues of lexis, of grammatical-rhetorical tropes, on the types [*Typos*], the "Greek letters," which Hölderlin writes about, and attempts to set free from, in a letter to Schiller from 1801 (*SA* 6; 1: 422)—is not that which is own to us, but that which was appropriated by the Greeks, that which was foreign to them. This is why it is "dangerous" to imitate "Greek excellence," because "we" would imitate in it only the imitation of what was originally own to us, and would perish, having missed our own, the *free* use of what is our own, the most difficult.

The "Remarks" clarify and enfold this unforeseeable "reversal"—unforeseeable, because it reverses the figure of inversion, of reversal[18]—in their middle, around the *Geschik*. The reversal is most of all not a turning away from that which is before the eyes *as* Greek art—a language. For a turn back would remain merely declared, and would be performed only in another language, here the one that is before the eyes *as* German, but which—to the extent that it is on the syntactical chain, that it follows the canon of rules for grammatical-rhetorical tropes, *or in reverse*, strives only to free itself, to put up resistance, to refuse a following, in other words it attempts to be under the spell of the faith in the unlimited validity of the more mediately factual sign[19]—remains "Greek-tragic," undifferentiated from the language, or art—*techne*—that is there *as* Greek. It attempts to have a history, a destiny, or (simple opposition) to escape both. The break with the drive and will to imitate happens; this is what is *unheard of* in the "Remarks" and in the text they accompany, which, understood as *translation* in the usual sense (for which—since it is prisoner of the more immediate factual conception of the word—translation is the transport of meaning from one to another particular language, that is, to speak precisely, a transport inside the language which, *as* Greek, determines the outline of both: Greek, German) does not get resolved[20]—the farewell to the drive or will to imitate occurs in that which is present, appears to be present, *as* something, to take tradition (hardly) otherwise, "more immediately factually," and to split—in an unregulated, precise way—the appearance of the context of tradition, a chain of history and fate that depends on the belief in the validity and givenness of the more mediate factual word. The farewell is not the contrast of a time that soars with a time that is dying, a new from an old time, a pre- from a posthistory, or in reverse, of one destiny from another, but rather—with an expression from the Homburger Folioheft—*Der Abschied der Zeit* ("time's farewell") (*HF* 98). Or form in general. Of itself. "Der Abschied der Zeit" names the moment, impossible to date, of the turn to oneself, which—in an unorientable way—drives one apart. A hitting [*Treffen*] that defines time, precisely, as *dawning*. Here—in the middle of the "Remarks on *Antigone*"—in the turn toward that which marks, as *Geschik*, the innermost outermost middle of the comprehension of the deadly, the more mediately factual sign, which is considered to be the Greek language, and moreover, almost inseparably, to be the German language:

For us . . . the Greek representations change insofar as their main tendency is to be able to comprehend [fassen] themselves, because that was their weakness, while on the contrary the main tendency in the modes of representation of our time is, in order to "hit" [treffen] something, to possess a skill [Geschik], since the lack of destiny, the *dysmoron*, is our weakness. Hence, the Greek also possesses more skill [Geschik] and athletic virtue and needs to have these as genuine qualities and serious virtues, however paradoxical the heroes of the *Iliad* may appear to us. For us, these are more subordinated to propriety [Schiklichkeit].

The main tendency of the Greek representations, to be able to comprehend themselves, in other words, to comprehend their *Geschik* and their "athletic virtue," the plastic ability, names one more time the tendency toward the "Greek-tragic word," which is "deadly factual," as a more mediate factual sign, but for the same reason, where tendency is elevated to *main* tendency, and strives towards exclusivity, the missing of one's own, which—unsettling the more mediately factual conception of what is own and of property—consists in not being able to comprehend (oneself): the missing of the more immediately factual word. Its *free* use. While on the contrary" [*Da hingegen*]: the peculiar expression, which names the moment of reversal, causes the conflicting tendencies *hin* ("toward") and *gegen* ("against"), without leading to a decision about whether the expression is heading for something or shrinking back from something. For this reason it *precisely* "hits" the moment: "the main tendency in the modes of representation of our time . . . in order to "hit" [treffen] something, to possess a skill [Geschik], since the lack of destiny, the *dysmoron*, is our weakness."

The Greek main tendency, toward the more mediately factual word, consists in having "more *Geschik*," the main tendency "in the "modes of representation" of our time consists—"while on the contrary"—in "having skill [*Geschik*]." How the word *Geschik* should be understood [*zu fassen sei*] in these lines remains open. They "hit" *Geschik* in such a way that the appearance of semantic and, more exactly, plastic construction of the word, in the look, is unsettled. Quietly trembles. The explanation of *Geschik* as a (more mediately factual) word breaks, because it is not possible to divide strictly between *Geschik* (in the sense of "destiny") and *Geschik* (in the sense of "dexterity"). At the point where, and because, *Geschik* is "hit" *as* a word, that is, it is comprehended, it is "hit": it breaks. For, although the text refers to that which is own to us as the lack of destiny (a paradoxically Greek word, *dysmoron*, which opens up not only what is own *to us*, but what is own in general), and hence destiny appears to be declared to be the main tendency; that is, the main danger, in the modes of representation of our time, in contrast to dexterity, seemingly the main tendency of the Greeks, both destiny and dexterity are run through with a more immediate and a more mediate factuality: the more mediate factuality of dexterity consists in the ability to comprehend oneself, the more immediate in the ability to "hit." Through this more immediately factual skill/fate [*Geschik*] to "hit" something, which is declared to be the main tendency in the modes of representation of our time, what is more saving in the main danger in the modes of representation of our time, the tendency toward destiny, is introduced in such a way that what is tended toward, or intention in general, and more exactly, the tendency toward tendency, which draws the profile of the word *Schiksaal* ("destiny") no less than that of *Geschichte* ("history"), splits up quietly, and the more immediately factual, the breaking apart in the outline of the word, dawns. Where, however, does the more mediate and more immediate reality of destiny lie? It is the one, hardly Greek word *dysmoron* that carries—or holds in itself—the (imponderable) weight of the *turn back* [*Umkehr*] of the Greek-tragic tendency toward reversal [*Umkehrung*]. This word, *dysmoron*, breaks open (the mention of *Oedipus at Colonus*, a few lines further down, quietly suggests this) becoming the more immediately factual, the missed own, of Greece, that is, of the usual, also what is usual *to us*, understanding of language oriented by the priority of the more communicative—more mediately factual word. In the middle of the "Remarks on *Antigone*" this word is transplanted from *Oedipus at Colonus*, where, coming from Oedipus's mouth, it hits the chorus, "is terrible and kills":

'Οιδίπους: *θέος ἴσχετε μηδὲν ὅσ' α'υδῶ.*
Χορός: *ἰὼ ὢ ὢ.*
'Οιδίπους: *δύσμορος.*
Χορός: *ὢ ὢ.*[21]

Dysmoros is in these lines a "more murderous-factual than deadly-factual" word. It interrupts the main tendency of the Greeks "to have more *Geschik* and athletic virtue," "to be able to comprehend themselves." It does not so much let itself be comprehended and grasped, as it contradicts, shatters, the attempt to find a word that would correspond to it. The response which it triggers in the chorus that it hits does not make it to a meaningful word: ὢὢ. Since, however, *dysmoron* does not get resolved in its explanation as a deadly-factual word, in it dawns, in however invisible and impossible to remember a manner, the (missed) own of Greece, that is, inseparably, *our* own. More precisely, the most difficult: the *free* use of what is one's own. Where, in *Oedipus the King* and in *Antigone*, Tiresias's speech always constitute the "caesura" or "the pure word, the counter-rhythmic interruption," as the "Remarks on *Oedipus*" argue, *dysmoron*, hardly a (Greek) word, constitutes the caesura of the "Greek-tragic," or, in another word, of the language of words and signs in general. Where it is hit, and where this hit is unfolded as the condition of possibility *but* impossibility of language, it stops.

That which precedes the word *dysmoron* in the "Remarks on *Antigone*," *das Schiksaallose* ("without destiny") (a word that, taken in isolation, on second look declines *destiny* in the suffix (*-lose* ["-less"], if it is understood as a noun—*Los, Lose* ["lot, lots"]) and is detrimental in another way to the presumed unity of the one word); *das Schiksaallose* is not the only attempt of Hölderlin, not so much to translate *dysmoron*, but to expose as the more immediately factual, unpositable, and untranslatable ground of the word, but exposes, however underhandedly, in that indeterminate middle which unfolds another, a more exact equivalence to the *dysmoron*: to the middle of the relationship between mortals and gods. As that which is without destiny *dysmoron* is entangled in the discussion of the structure of, or of the will to, knowledge through opposition: to posit something *as* something, that is, to oppose, to have determined and comprehended, grasped it through its contrary (and moreover through a part [*Teil*], that should be essentially *Gegen*teil ["contrary"] and, *as* part, *whole* part, should be indivisible). *Schiksaallos* is what the alcaic ode "Palinodie," drafted in autumn 1799, calls the gods *Ihr schiksaallosen Götter* ("you fateless gods"), "in blessed youth over the mortal man," who appear to be related to the mortal man through opposition (*FA* 5: 654–55). The draft takes up the beginning of the ode "Mein Eigentum" ("My Possessions"), written shortly before, whose last stanza, in an almost palinodic manner, ends with the words *nicht ende* ("not end"), and which also names destiny in its last line[22]—here too related to the middle of the opposition of gods and mortals. Here however, under the name *Parze* ("Parca") (from the Latin noun *pars*, "part"), it defines destiny more precisely as part and lot:

Ihr seegnet gütig über den Sterblichen
  Ihr Himmelskräfte! jedem sein Eigentum,
    O seegnet meines auch und daß zu
      Frühe die Parze den Traum nicht ende.

(*FA* 5: 619)

You bless kindly above the mortals
  You heavenly powers! to each his possessions,
    O bless mine also and make it that too
      Soon the Parcae may not end my dream.[23]

In the middle of the relationship between mortals and immortals, and more generally, of knowledge through opposition (triggered and carried by the apophantic . . . semontological "Greek-tragic" outline of the word- and sign-language), own neither to the mortals nor to the immortals, neither to the one nor to *its* other, but actually not foreign [*fremd*], but disconcertingly [*befremdlich*]—more immediately factually—close, destiny and the *Parca* are related to each other in both odes. With it, however, more exactly, the question of the divisibility [*Teilbarkeit*] and inseparability of the part [*Teil*], which nests, almost forgotten, in the Greek-Latin nouns that are commonly translated as destiny and lot. The same is true of *dysmoron.*

Just like the noun *moira, moros* means the part and lot, also that allotted to one, and in particular the parts of the sacrificial victim that have been cut off for the immortals in the sacrifice ceremony, and the meat allotted to each in the communal meal, that of the mortals, where it appears repeatedly—for instance in the *Odyssey* (8.470)—in connection with the verb *nemein*: distribute [*verteilen*] and allocate [*zuteilen*], but also cut apart [*zer*teilen].[24] Correspondingly, the affiliated verb *meiromai* means, among other things, "to take as one's share" [*zum Anteil nehmen*], where *nehmen* (the precarious indeterminate middle, and condition of possibility of all procedures to determine part and lot) for the most part is faded out as receipt of something already *given*, in order to dispense with the disconcerting divisibility of part and lot. About the textual sketch of the part—*moros, moira*—and to the disruptive, more ripping-apart parts (in the ground of the noun *Parca*), which both opens the possibility of the web in the narrower and broader sense, but also, more indispensably, holds up the closing of the text, a *part*iciple, into *text*—a *past* participle, and scatters it; about *moira* and *Parca*, a page of Johann Arnold Kanne's "Ersten Urkunden der Geschichte oder allgemeine Mythologie" (First documents of history, or general mythology), 1808, explains:

Spinning, the Parca *distributed* [vertheilte] destiny to the world in a thread made of linen, wool, or hair. This is why the divinity was called Αισα, although originally Αζα, destiny and lot, and the latter was also used to mean language. Μειρεν, from which derives μοιρα, means *to distribute*, μερος and μοιραι is the lot, but in the spun thread; from there comes also *mermis*, rope. With Oriental duplicity, the Orphics translate the name of the Oriental goddess of destiny Νεμεσις, for this word means, as Herder already saw, the one who distributes (from νέμω). . . . The name of the Latin *Parca* is a form of *partiri, pars,* for *parcere,* whose exact meaning, actually, is to budget, to save, to spare, could also be said *parsere*, as proven by *parsimonia*. The root was *par*, from which, with the *s* in front, comes *sparen*. It meant to divide, to split apart, to destroy, and from it came σπαιρω, σπαρω, and so it has remained in Hebrew, in the form פרם and פרש, *paras*, *partiri*, *dividere*, פרד, *parad*, split apart, *dividere*, *disjungere*, *separare*, Pyh. *dissipare*, σπαιρω, פרם, *param*, to cut through, to cut up, פרץ, *paras*, to shatter, פְּרַט, *perat*, Rabbinical, *particulare* (*partior*), פרךְ, *parach*, to shatter (*parca parco*), פרע, *parag*, of the vengeful Moira *ulscisi*, especially in Chaldaic, and of the hair-spinning Moira פְּרַע, *perag*, *coma*, Lat. *pargo*, *spargo*, פרק, *parak*, to shatter, to rip apart from one another.[25]

The fact that, in the "Remarks on *Antigone*," "being without destiny" is attributed no longer to the gods, as it is still in the ode draft "Palinodie," but to *us* as our own, is not a mark for Hölderlin's allegedly "palinodic thought,"[26] but the indication of an experience: namely, that knowledge through opposition, or palinode (within the brackets of the will to retraction), and also the opposition of mortals and gods, those bound to destiny and those without destiny, remains a proof of the will to the more mediately or deadly-factual, to the "Greek-tragic" word: proof of the will which the "Remarks on *Antigone*," under the angle of the *dysmoron*—of the words around *moros*, *moira* and *Parca*, but also, in a more underlying way, around *nemein*, *Nemesis*, and *Nomos*—define more precisely as the will to the indivisibility of the part, whether that be a sign *as* sign, a word, man, or god. *We*, however, who (as every sign and word) do not get resolved in our interpretation as sign and word, in our interpretation as men or, in reverse, as gods, are not summoned *against* [*gegen*] the will to co-part [*Mit-teil*] and to communication [*Mitteilung*], but rather are related to it as *da hingegen*. What does this mean? *We* are, or is, according to tradition, to the will to tradition, a sign among signs. The *free* use of what is one's own, or of the "fateless," the awareness of the impossibility of a unified interpretation of that part as part, or, repeated otherwise, the experience of unregulated divisibility [*Teilbarkeit*] of the sign or the part, does not lie—and especially not—in the turning of a *we* that is somehow given, signified by words, and endowed with perception, into a part that would be before our eyes, to take it otherwise than *as* part, but rather to speak more exactly, to the extent that *we* also, and especially, within the framework of the "Greek-tragic" will to *Geschik*—understood in a more mediately factual way—and to history, within the framework of the belief in the uncomplicated givenness of the communicating, deadly-factual word, are, or is, considered *as* sign, in the turning toward oneself of that which was considered *as* sign, *as* part, and indivisible. In the more gentle, or more cutting, partaking of the framework. In the more splitting-cutting up-taking [*teilend-zerteilenderen Nehmen*] also of the *we*. This, the *free* use of what is one's own, that does not name any positivity expressed in a word, nothing to be had, but rather, in the word *das Schicksaal*lose ("the fate*less*"), rather an inability or something not granted, which quietly does detriment to the will to have and have in one's hand, to the understanding of what is one's own as a *gift*, something pretended or given up [*Vor- oder Aufgegebenen*], but also to its opposition, not wanting anything, because it speaks of the free *use* of what is one's own, this is what the letter to Böhlendorff in 1801 names as the most difficult.

Another, more literal correspondence to the *dysmoron*, in a different passage, comes close to this most difficult thing, the *free* use of what is one's own. A line that was presumably inserted retrospectively around 1805 under the title "Die Titanen" ("The Titans") in the Homburger Folioheft, "hits" *dysmoron* more exactly to relate *Theil* ("part") and *Nehmen* ("to take"): "Göttliches trift untheilnehmende nicht" ("Those who have no sympathy are not concerned with the divine") (*HF* 54). The line (taking the Greek noun *nomos* more exactly) defines more precisely that which is own *to us*, or that which is own in general, *dysmoron*, as *Untheilnahme* ("lack of participation, lack of sympathy"), and the *free* use of what is one's own, in an unexpressed and irritating way, as *Theilnahme* ("participation, sympathy"). Irritating, for the tendency toward the "Greek-tragic" word, because participation turns the more mediate and the more immediate factualities of destiny into one another to such an extent that both realities, too close to each other, do not become opposites, but can be differentiated: *participation*, that is, the will to *take part* in something, to have a part, in fact to be a part, and as a part, whole, indivisible—*individuum*—is more mediately factual. "On the contrary" [*da hingegen*], *participation* is more immediately factual. It consists, repeated otherwise—in a more dissecting, cutting way—in *taking part*: to *take* otherwise what seems to be there *as* part, what seems to be given. In the *taking* of the part, an expression that "hits" the *free* use of what is one's own, the lack of participation, most exactly, the line is exposed to the long-lost ground of the German *Nehmen* and the Greek *nemein*, which, in the vocabulary of the etymologists, come from the same root. In chapter 21, "Trees and Animals," of the *Deutschen Mythologie* (1835), Jacob Grimm cuts this root:

> *Abnemen* means, in the thirteen century, *mactare*, "to slaughter" (used about livestock) . . . just as we say today *abthun, abschneide*. . . . Schmid's Swabian dictionary, 405, *abnehmen*, "to butcher poultry." This meaning does not lie in the particle, but in the word itself: *niman, neman,* would then mean "to cut, to slaughter, to divide up."[27]

The *free* use of what is one's own, the *free* use of the lack of participation, dawns in the *splitting* [*teilend*], the cutting up of that which is considered to be, and appears given, *as* part—destiny. Two lines of the "Friedensfeier" ("Celebration of Peace")—which, written between 1801 and 1804, accompany the translations, *Antigonä* and *König Oedipus*, as well as the "Remarks"—enfold this *freer* use. Open here as well to the unlocatable middle of mortals and gods, they relate *Theilen* ("to divide") and *Schiksaal* ("destiny") to each other in such a way that the appearance of semantic unity of the word *Theilen*—oscillating between more immediate and more mediate factuality, between *Theilnahme* (in the sense of being part and having a part) and *Theilnahme* (taking the part, cutting up)—no less than the chance of *destiny*, is splintered:

> Einmal mag aber ein Gott auch Tagewerk erwählen
> Gleich Sterblichen und theilen alles Schiksaal.
>
> For once, however, may a god also choose day's work
> like mortals and share all destiny.[28]

These lines, by splitting (up) *Teilen*, take *Nehmen* ("taking"), whose semantic content seemed to be given, comprehended, otherwise, and release mortals and gods (the opposition par excellence of the "Greek-tragic," more communicating word, marked by the will to opposition) from the brackets of opposition and into the same indeterminate middle, to *participation*, out of which the possibility of part [*Teil*] and opposite [*Gegenteil*], share and having nothing granted also, of course, originates. Everything that is comprehended *as* part, however—too differentiable from itself—remains open to *taking* [*Nahme*]. Which can be ascribed neither to a mortal nor to a god, as *given* support or subject of taking, for it is, to begin with, in taking that the possibility of both, their being there and missing, of mortals and gods, appears.

To what extent, more exactly, in *participation*, which turns the more mediate word and the more immediate factual word asymmetrically—in a way open to the unregulated, exact divisibility of the part—into each other, and the free use of what is one's own as something linguistic is touched upon [*angeschnitten*], is suggested by the middle of the hymn "Der Rhein," which can be dated back to the same period—between 1801 and 1804:

> Es haben aber an eigner
> Unsterblichkeit die Götter genug, und bedürfen
> Die Himmlischen eines Dings,
> So sinds Heroën und Menschen
> Und Sterbliche sonst. Denn weil
> Die Seeligsten nichts fühlen von selbst,
> Muß wohl, wenn solches zu sagen
> Erlaubt ist, in der Götter *Nahmen*
> *Theilnehmend* fühlen ein Andrer,
> Den brauchen sie . . .
> (*SA* 2; 1: 145)

> But their own immortality
> Suffices the gods, and if
> The Heavenly have need of one thing
> It is of heroes and human beings
> And other mortals. For since
> The most Blessed feel nothing by themselves
> Another, if to say such a thing is
> Permitted, must, I suppose,
> Participating feel in the name of the gods,
> And him they need . . .[29]

These lines clarify, in the turn toward the *Nahme* ("name"), supplied with an *h* in accordance with Hölderlin's orthography, which follows older linguistic usage, the Teil*nahme*—which here too names the middle of those related to each other, mortals and gods—as a doing, or an occurrence, that irreversibly splits the outline of the word-language: they clarify the appearance of unity of the—more mediately factual—*word* in the *taking* [*Nehmen*] as unforeseeably divisible, driven apart as *Nahme* ("name"). Only when the word-language [*Wortsprache*] is de-interpreted into a language of taking or of names [*Nahmensprache*], where that which appeared part and *as* part is *taken* differently, where the memory—however impossible to remember—is renewed in the coimplication of *Nahmen* ("names") and taking [*Nehmen*], of *Nahme* and *Nahme*, only there does the more immediately factual word dawn, that is, no other word than the more mediately factual, but the same one, *taken* differently, in a more concerned [*getroffen*] fashion. Only the more splitting, more interrupting participation in the handling [*Handhaben*] and use of the more mediate or deadly-factual, "Greek-tragic" word splits: *cesurates* the appearance of the unquestionable, unbroken givenness of the word- and sign-language, just as the belief, tied to it, in the—premised or striven for—indivisibility of part and lot, a belief entertained by the contrasting conceptions of history, destiny, and fate. Only, once more, the more mediate factual interpenetration of participation and *Nahmen*, the awareness of their more immediate factualness, of the word- and sign-language in general, splits also the appearance of translatability of *nomos* (noun of action from *nemein*, "Nehmen") with *Gesez* ("law") and allows to decipher, in the phrase *streng genommen* ("strictly speaking"), another, more exact word for the "Greek-tragic" or deadly-factual, because it defines more precisely the "Greek-tragic" word and in it the skill [*Geschik*] "to comprehend oneself," as a modification of the more mediately factual, more splitting-taking, which is "hit" [*getroffen*] in the skill "to be able to 'hit' something."

That it is in the turn toward *Nahmen* for the first time that the usual, "Greek-tragic" conception of translation as a symmetrical correspondence, word for word, breaks, a conception which takes its bearings from the validity of the more mediate factual word, and that only in this way is "the highest" both reached and disfigured, is enfolded in another passage of the "Remarks on *Antigone*." This passage explains a deviation from the substitution, word for word, through the insertion of the *Nahmen* in a place where it is not there in the Greek text:

> CREON. Wenn meinem Uranfang' ich treu beistehe, lüg ich?
> HÄMON. Das bist du nicht, hällst du nicht heilig *Gottes Nahmen*.

statt: trittst du der Götter Ehre. Es war wohl nöthig, hier den heiligen Ausdruk zu ändern, da er in der Mitte bedeutend ist, als Ernst und selbständiges Wort, an dem sich alles übrige objectiviret und verklärt.

Wohl die Art, wie in der Mitte sich die Zeit wendet, ist nicht wohl veränderlich, so auch nicht wohl, wie ein Karakter der kategorischen Zeit kategorisch folget, und wie es vom griechischen zum hesperischen gehet, hingegen der heilige Nahmen, unter welchem das Höchste gefühlt wird oder geschiehet.

> CREON. If I stay faithful to my first origins, do I lie?
> HAEMON. That you are not, if you do not honor *god's name*.

rather than: "if you trod on the god's honor." It was necessary, it seems, to change the sacred expression here, since it is significant in the middle as seriousness and as autonomous word wherein everything else finds and transfigures itself.

Neither the way in which time reverses in the middle nor how a character categorically follows the categorical time nor how it proceeds from the Greek to the Hesperian are changeable, but the sacred name under which the highest is felt or occurs. (*FA* 16: 413–14; *Essays*, 110–11)

Creon advocates the law [*Recht*]—*dike*—in this passage and the *nomos*, conceived as more mediately factual and translated with *Gesez* ("law"), in another word, the *Nahme* ("taking"), conceived as more mediately factual. Haemon, "on the contrary" [*da hingegen*], refers to the more immediately factual origin of the possibility to hold on to the *law* [*Gesez*] and, more exactly, to the will to the positable and translatable, "deadly-factual" word as to its "first origins" by interrupting, but first of all in the Greek text that is *taken* otherwise, the "Greek-tragic" tendency, the origin of the "Greek-tragic" tendency, toward the more communicating, more meaningful word, to seek the gods in the meaningful word and in the honor (*timas*) signified by the word; he also finds the origin of that possibility in the *Nahmen* ("name"), which, precisely because it is not there in the Greek text, or perhaps got *missed*, does not get resolved in its interpretation as word, -for word, but, more immediately factual, points to the unregulated exact divisibility of the—mediately factual—word.

Seen from the point of view of the "Remarks," the translation does not get resolved in its interpretation as translation, because it exposes to the more immediately factual ground of the more communicating, "Greek-tragic" tendency (from which also translation in general depends, as will to substitution, a word for a word) of the Greek text. It is not so much a translation, as, closer to the "Remarks," where it is said of the name that under it "the highest is felt or occurs," an *occurrence* [*Geschehen*]. Not lastly because in the occurrence the—more interrupting—ground of the words around *Geschichte* ("history"), *Schiksaal* ("destiny"), and *Geschik* ("fate, skill") appears. For no less than the *Nehmen*, more living, exposes to a more cutting-splitting, in *Geschehen*, where it is paraphrased in the *Deutsche Wörterbuch*, a "main meaning" dawns, which irreversibly splits the skill "to comprehend oneself": "to turn suddenly."[30]

Translated by Georgia Albert

REFERENCE MATTER

Notes

FIORETOS *Introduction*

1. Paul Celan, *Last Poems*, trans. Katharine Washburn and Margret Guillemin (San Francisco: North Point, 1986), 189.

2. Paul Celan, "Tübingen, Jänner," in *Werke*, Tübinger Ausgabe, ed. Jürgen Wertheimer (Frankfurt am Main: Suhrkamp), *Die Niemandsrose*, ed. Heino Schmull (1996), 37. "Tübingen, Jänner," in Paul Celan, *Poems*, trans. Michael Hamburger (Manchester, England: Carcanet, 1986), 177.

3. As one typical example, see Charitas Jenny-Ebeling, "Von Deut zu Deut: Pindar—Hölderlin—Celan," *Neue Zürcher Zeitung*, no. 83 (April 10–11, 1993): 61–62.

4. Paul Celan, "Ansprache anläßlich der Entgegennahme des Literaturpreises der Freien Hansestadt Bremen," in *Gesammelte Werke*, ed. Beda Allemann et al. (Frankfurt am Main: Suhrkamp, 1983), 3: 186. "Speech on the Occasion of Receiving the Literature Prize of the Free Hanseatic City of Bremen," in Paul Celan, *Collected Prose*, trans. Rosmarie Waldrop (Manchester, England: Carcanet, 1986), 34. For a brief overview of this period, including references to Hölderlin, see Jost Hermand, *Geschichte der Germanistik* (Hamburg: Rowohlt, 1994), esp. 98–113.

5. Hermand offers a diminutive list of plausible reasons for such avoidance. See *Geschichte der Germanistik*, 116.

6. At present, there are roughly twenty thousand entries of secondary literature. In *"Ein Zeichen sind wir, deutungslos": Die Rezeption Hölderlins von ihren Anfängen bis zu Stefan George* (Stuttgart: Metzler, 1992), Henning Bothe offers a survey of the *Diskursmilieu* of the Hölderlin reception from his own time, over Nietzsche, Haym, Zinkernagel, Lange-Eichbaum, and Dilthey, in particular, through the George circle.

7. See Walter Benjamin, "Zwei Gedichte von Friedrich Hölderlin," in *Gesammelte Schriften*, ed. Rolf Tiedemann et al. (Frankfurt am Main: Suhrkamp, 1977), 2; 1: 105–26.

8. www.uni-tuebingen.de/deutsches-seminar/hoelder/.

9. For references, see the Bibliography.

10. Emery E. George, ed., *Friedrich Hölderlin: An Early Modern* (Ann Arbor: University of Michigan Press, 1972).

11. For these, see the Bibliography.

12. Martin Heidegger, *Hölderlins Hymne "Der Ister"*, vol. 53 of *Gesamtausgabe* (Frankfurt am Main: Klostermann, 1984). Martin Heidegger, *Hölderlin's Hymn "The Ister"*, trans. William McNeill and Julia Davis (Bloomington: Indiana University Press, 1996).

13. See *FA* 13 and 14, as well as "The Deaths of Empedocles," in Andrzej Warminski, *Readings in Interpretation: Hölderlin, Hegel, Heidegger* (Minneapolis: University of Minnesota Press, 1987), 11–17.

14. The claim was made by Bettina von Arnim. See *Die Günderode*, ed. Elisabeth Bronfen (Munich: Matthes & Seitz, 1982), 225.

15. Celan has the unusual verb *zackern* from a letter reproduced by von Pigenot and Seebaß in their edition. On July 11, 1805, Gerning van Knebel, Court Counselor in Homburg, writes: "Hölderlin, who is always half-mad, also toils [zackert] with Pindar." See *Sämtliche Werke* (Berlin: Propyläen, 1923), 6: 373. Cf. Bernhard Böschenstein, "Hölderlin und Celan," *Hölderlin-Jahrbuch* 23 (1982–83): 147.

16. Paul Celan, *Gesammelte Werke* 3: 108.

17. Cf. Jenny-Ebeling, "Von Deut zu Deut," referred to in note 3 above. Although Jenny-Ebeling's interpretation remains vague on certain key issues, such as the relationship between the poem's formal structure and its anticipation of the act through which it is made hermeneutically available (crucially concentrated in the monosyllabic *Deut*), it is nonetheless the first serious attempt to come to terms with the tension between formal and historical structures of signification in Celan's poem. In turns illuminating and suggestive, Böschenstein's and Manger's essays—hitherto the only two extended readings of "Ich trink Wein"—remain resolutely thematic in their orientation. For Böschenstein, see his "Hölderlin und Celan," referred to in note 15; for Klaus Manger, see "Die Königszäsur: Zu Hölderlins Gegenwart in Celans Gedicht," *Hölderlin-Jahrbuch* 23 (1982–93): 156–65.

18. The English translation is taken from Friedrich Hölderlin, *Essays and Letters on Theory*, intro. and trans. Thomas Pfau (Albany: State University of New York Press, 1988), 102. Further references to this edition will appear in the main body of the text as *Essays*, after references to the German original.

19. Paul Celan, "Der Meridian: Rede anläßlich der Verleihung des Georg-Büchner-Preises Darmstadt, am 22. Oktober 1960," *Gesammelte Werke*, 3: 189. "The Meridian: Speech On the Occasion of Receiving the Georg Büchner Prize, Darmstadt, October 22, 1960," *Collected Prose*, 40.

20. Entirely precise, Hölderlin's use of the disjunctive connective "or" indicates the condition of spatio-temporal conditions: without it, neither the possibility of time nor that of space would be given. As a differing and deferring *oder*, the caesura is the setting-free of their possibility. For an extended meditation concerning such an ultratranscendental condition, see Werner Hamacher, "Ou,

séance, touch de Nancy, ici," *Paragraph* 16, no. 2 (1993): 216–31 (part 1); and 17, no. 2 (1994): 103–19 (part 2).

21. Cf. Warminski, "'Patmos': The Senses of Interpretation," in *Readings in Interpretation*, esp. 74–76. For the etymology, see Grimm, *Wörterbuch* (Leipzig: Hirzel, 1854), 2: 1038 (also quoted by Warminski). There is, of course, a vast scholarly literature devoted to Hölderlin's "Patmos" hymn, not to mention the question of the status of interpretation in it, as well as in his oeuvre in general. To give just one example, explicitly devoted to the biblical allusion, contained in Hölderlin's poem, to 2 Corinthians 3: 4–8: Alice A. Kuzniar, "Hölderlin," in *Delayed Endings: Nonclosure in Novalis and Hölderlin* (Athens: University of Georgia Press, 1987), 133–90.

FENVES *Measure for Measure: Hölderlin and the Place of Philosophy*

1. For a thorough discussion of the controversy surrounding the so-called "oldest system program," see Frank-Peter Hansen, *"Das älteste Systemprogram des deutschen Idealismus": Rezeptionsgeschichte und Interpretation* (Berlin: de Gruyter, 1989). In an as yet unpublished essay, Eckart Förster has conclusively shown that Hölderlin must have had a decisive hand in certain aspects of the "program," especially those concerned with "giving wings" to physics.

2. See the letter of July 10, 1794. *SA* 6; 1: 128.

3. This is Andrzej Warminski's exquisite translation of Diotima's complaint, "Aber ihr schweift mir aus." In *Readings in Interpretation: Hölderlin, Hegel, Heidegger* (Minneapolis: University of Minnesota Press, 1987), 51.

4. Theodor W. Adorno, *Noten zur Literatur* (Frankfurt am Main: Surhkamp, 1981), 3: 452. Cf. "Parataxis: On Hölderlin's Late Poetry," in Theodor W. Adorno, *Notes on Literature*, trans. Shierry Weber Nicholsen (New York: Columbia University Press, 1992), 2: 109–49. The text of "Friedensfeier" could be translated as follows: "For sparingly, at all times expert in measure, / A god touches only for a moment the dwellings of men, / And, unforeseen, no one knows it, when? / The impudent may then go over as well / And to the sacred place must come the savage / From ends remote, and crudely handling, exercise his delusion, / And strikes then a fate, but thanks,/ This never follows at once on the god-given gift."

5. Adorno, "Parataxis," 481. For another presentation of Hölderlin's relation to the "dialectic of enlightenment," see Rainer Nägele, "Der Diskurs des Andern," *Le pauvre Holterling* 5 (1980): 61–76. "The Discourse of the Other: Hölderlin's Voice and the Voice of the People," in Rainer Nägele, *Reading After Freud: Essays on Goethe, Hölderlin, Habermas, Nietzsche, Brecht, Celan, and Freud* (New York: Columbia University Press, 1987), 47–65. A lengthy discussion of "Parataxis" in relation to Adorno's interpretative protocols can be found in Gerhard van den Berg, *Adornos philosophisches Deuten von Dichtung. Ästhetische Theorie und Praxis der Interpretation: Der Hölderlin-Essay als Modell* (Bonn: Bouvier Verlag, 1989).

6. Adorno, *Noten zur Literatur*, 490.

7. See Michael Konrad, *Hölderlins Philosophie im Grundriß* (Bonn: Bouvier, 1967), 32–33.

8. See the opening paragraph of Hölderlin's "Anmerkungen zum 'Ödipus'" (*SA* 5; 1: 195).

9. On Hölderlin's "Entgrenzung der lyrischen Sprache," see Gerhard Kurz, *Hölderlin-Jahrbuch* 23 (1982–83): 34–53. Near the conclusion of his dissertation on romantic art criticism, Walter Benjamin notes that Hölderlin's late lyrics parallel those of the early romantics to the extent that they find their fulfillment in a certain prosaic "sobriety." See Walter Benjamin, *Der Begriff der Kunstkritik in der deutschen Romantik, Gesammelte Schriften*, ed. Rolf Tiedemann and Hermann Schweppenhäuser (Frankfurt am Main: Suhrkamp, 1977–85), 1; 1: 104–6.

10. The central lines of this stanza are the "opening" lines for the entire poem, the space of its orientation: "So komm! daß wir das Offene schauen, / Daß ein Eigenes wir suchen, so weit es auch ist. / Fest bleibt Eins; es sei um Mittag oder es gehe / Bis in die Mitternacht, immer bestehet ein Maß, / Allen gemein, doch jeglichem auch is eignes beschieden, / Dahin gehet und kommt jeder, wohin er es kann." (*SA* 2; 1: 91). ("So come! That we see the opening, / That we seek something proper, however far it is. / One thing holds strong; be it midday or let it go into midnight, there always remains a measure / Common to all, yet to each one is its own allotted also, / Thence comes and goes each one toward which it can."

11. Giorgio Agamben has devoted a book to what Hölderlin calls "only each one." See his presentation of *qualunque* in *The Coming Community*, trans. Michael Hardt (Minneapolis: University of Minnesota Press, 1993).

12. See, for example, Werner Marx, *Gibt es auf Erden ein Maß?* (Hamburg: Meiner, 1983).

13. As is his custom, Heidegger first cites Hellingrath's edition of Hölderlin, which reconstructs "In lieblicher Bläue . . ." (hereafter without the ellipses) as a poem. See *Sämtliche Werke*, ed. Norbert von Hellingrath, Friedrich Seebaß, and Ludwig Pigenot (Berlin: Propyläen, 1923); text: 6: 24–27; notes: 6: 490–92. Quoted by Martin Heidegger in *Vorträge und Aufsätze* (Pfullingen: Neske, 1985), 181–98. Heidegger never alludes to the justification for this procedure, that is, Waiblinger's remark that the version of the text he published in his novel *Phaëton* were written "according to the Pindaric manner [nach Pindarischer Weise]." From Waiblinger's perspective this could only mean one thing: he did not understand its metrical rules, since a reliable explication of these rules was not published until later in the century. Precisely because the text lacks a "lawful calculus" Beißner places it among the "doubtful" (2; 2: 991–92). Sattler, by contrast, considers it the luckily preserved remnants of a poem Hölderlin wrote after his "crisis." See *KT* 9: 25. The most incisive recent consideration of the authorship of "In lieblicher Bläue" is, to my mind, that of Sabine Doering, who focuses on the status of its questions. See *Aber was ist diß? Formen und Funktionen der Frage in Hölderlins dichterischem Werk* (Göttingen: Vandenhoeck & Ruprecht, 1992), 163–76.

In a short text written some fifteen years later than ". . . dichterisch wohnet der Mensch . . . ," entitled "Das Wohnen des Menschen," Heidegger confronts the question of Hölderlin's role in the writing of "In lieblicher Bläue." This question is posed with respect to the word *dichterisch*—whether and where Hölderlin wrote down this word. Heidegger twists the question to a point where it gives way. On the one hand, he dismisses the philological procedures by which "Hölderlin-Forschung" secures its texts as representative of the very "undichterish" character of contemporary "Wohnen," but, on the other, he uses precisely these procedures, or a modification of them, to support his contention that Hölderlin—it has to be him, not Waiblinger, for example—wrote the text in question. And this text is quite clearly recognized as prose: "Do the handwritten places of this completed poem ['Der Archipelagus,' an earlier version of which uses the word *undichertisch*, albeit in connection with the 'heavenly ones,' not mortals] diminish or even do away with the doubts about the authenticity of the prose text 'In lieblicher Bläue blühet . . . ,' from which the lines 'Voll Verdienst, doch dicterisch, wohnet der Mensch auf dieser Erde' are taken? If this were so, then the already mentioned difference [between calling mortals and the heavenly ones *dichterisch*] nevertheless remains." Martin Heidegger, "Das Wohnen des Menschen," in *Denkerfahrungen*, ed. Hermann Heidegger (Frankfurt am Main: Klostermann, 1983), 155. Heidegger confronts the question of "authenticity" (*Echtheit*), not of prosody: what it means is that the text is "in" prose. And although he praises the heroic Hellingrath over other Hölderlin editors, he has nothing to say of his reconstruction of a poetic version of "In lieblicher Bläue." The construction of a poem from prose remains outside the problematic of "dwelling, construction, thinking." But in another late text Heidegger does approach these matters. Under the title "Rimbaud vivant" he considers the problem of rhythm and, in connection with a letter by Rimbaud, the difference between ancient and modern rhythm. See *Denkerfahrungen*, 165–67. A reading of this short text—and of the late Heidegger's peculiar relation to Rimbaud—cannot, however, be undertaken in this context.

14. See Martin Heidegger, *Erläuterungen zu Hölderlins Dichtung* (Frankfurt am Main: Klostermann, 1981), 51: "Der hier zugrunde gelegte Text beruht, nach den urschriftlichen Entwürfen erneut geprüft, auf dem folgenden Versuch einer Auslegung." The strangeness of this remark occasioned a letter by Detlev Lüders in 1951. For an analysis of what is at stake in Heidegger's statement and in his response to Lüders, see Samuel Weber, "Reading and Writing *chez* Derrida," in *Institutions and Interpretations* (Minneapolis: University of Minnesota Press, 1987), 100–101.

15. Heidegger, *Vorträge und Aufsätze*, 190.

16. Ibid., 197.

17. See Martin Heidegger, *Kant und das Problem der Metaphysik* (Frankfurt am Main: Klostermann, 1973).

18. See Ibid., 155–65.

19. Heidegger mocks the identification of poetry with mere fantasy (*Phantasie*); but the imagination (*Ein-bildung*) that lets images appear is another matter. See *Vorträge und Aufsätze*, 194–95. On the image of *das Bild* that haunts "In lieblicher Bläue," see Jean-Luc Marion, *L'idole et la distance* (Paris: Grasset, 1978).

20. Heidegger, *Vorträge und Aufsätze*, 197.

21. Poetry is a measure-taking without a given—not with giving. This difference does not, of course, escape Heidegger, and it could be said to be the difference from which he rethinks ontological difference in his last writing, especially, "Zeit und Sein." See *Zur Sache des Denkens* (Tübingen: Niemeyer, 1976), 1–26. Nothing is more important for Heidegger's last writings than the thought of giving (*es gibt*, *Geschenk*, etc.). The context of this essay does not allow for a consideration of the insistence of the gift in the late Heidegger, but there is nevertheless one point to be made: an *Es* ("it") gives, and only insofar as this "Es" gives, does the gift (or the "sending") recede from the given. But, as the essay seeks to show, the *Es* is itself staked out in a certain metrical taking. On the gift and giving in Heidegger and elsewhere, see the remarkable work of Jacques Derrida, *Donner le temps: I. La fausse monnaie* (Paris: Galilée, 1991). *Given Time: I. Counterfeit Money*, trans. Peggy Kamuf (Chicago: University of Chicago Press, 1992).

22. Heidegger, *Vorträge und Aufsätze*, 197–98. The lines in German could be translated: ". . . as long as friendliness, the pure one, still lasts near the heart, man does not unhappily measure himself with the divinity . . ."

23. An important presentation of the close relation between Adorno's critique to Heidegger's thought can be found in Alexander García Düttmann, *Gedächtnis des Denkens: Versuch über Heidegger und Adorno* (Frankfurt am Main: Suhrkamp, 1991).

24. This "taking" comes up again in Heidegger's "Aus einem Gespräch von der Sprache" when it is a question of another translation—the translation of the Japanese word for *Sprache*. See Heidegger, *Unterwegs zur Sprache* (Pfullingen: Neske, 1986), 143. Adorno also insists that one must take words literally (*wörtlich*) if one is to escape the traps through which a text guards its "truth-content" from exposure. See the first chapter of his first "Adornian" critique, *Kierkegaard: Konstruktion des Ästhetischen* (Frankfurt am Main: Suhrkamp, 1974), 19–23. *Kierkegaard: Construction of the Aesthetic*, trans. Robert Hullot-Kentor (Minneapolis: University of Minnesota Press, 1989), 11–13. For a discussion of what happens when Adorno takes Kierkegaard in a "wordly" manner, see my "Image and Chatter: Adorno's Construction of Kierkegaard," *Diacritics* 22 (1992): 100–114.

25. The significance of "taking" in Hölderlin has been emphasized by Thomas Schestag, who, in response to Carl Schmitt, relates *Nehmen* (taking) to *nomos* (law). See the section on Hölderlin's "Das Höchste" in *Parerga* (Munich: Boer, 1991), 15–50, translated in this volume.

26. Heidegger, *Vorträge und Aufsätze*, 191.

27. See Philippe Lacoue-Labarthe, *La poésie comme expérience* (Paris: Bour-

gois, 1986), 156–66. See also Jean-Luc Nancy, "Divine Places," trans. Michael Holland, in *The Inoperative Community* (Minneapolis: University of Minnesota Press, 1991), 122–24.

28. Hölderlin rarely included diacritical mark in *ist's*; particularly in his late works, he simply wrote *ists*. See, for example, the two instances of this contraction in "Der Rhein," lines 83 and 166 (*SA* 2; 1: 144 and 147). Whether the inclusion of a diacritical mark has something to say about the authorship of "In lieblicher Bläue" is an open question.

29. See Jacques Derrida's critique of Heidegger's reliance on the word in *De la grammatologie* (Paris: Minuit, 1967), 31–39. *Of Grammatology*, trans. Gayatri Chakravorty Spivak (Baltimore: Johns Hopkins University Press, 1976), 18–26.

NANCY *The Calculation of the Poet*

1. This is to say that here I will deliberately avoid the Heideggerian interpretation of Hölderlin, remarking only that this interpretation, considered independently of its themes, always leaves Hölderlin's poetics to the side, even in the rare instances when it alludes to it, for example, in the *Erläuterungen zu Hölderlins Dichtung*, *Gesamtausgabe*, vol. 4 (Frankfurt am Main: Klostermann, 1982), 183, and even when it takes precautions concerning the poetic nature of the texts it interprets philosophically, for example, in the Introduction to the course on the hymn "Germania": *Hölderlins Hymnen "Germanien" und "Der Rhein," Gesamtausgabe*, vol. 39 (Frankfurt am Main: Klostermann, 1989), 3–8. I will return elsewhere to Heidegger's relation or nonrelation to the *ars poetica* as such. Moreover, I risk this essay independently of (that is, ignoring) the properly technical studies of Hölderlinian poetics, such as Lawrence Ryan, *Hölderlins Lehre vom Wechsel der Töne* (Stuttgart: Metzler, 1960), pointed out to me by Alexander García-Düttmann. I must acknowledge my debt to the essays by Eliane Escoubas, François Fédier, and Rainer Nägele collected in the *Cahier de l'Herne Hölderlin*, ed. Jean François Courtine (Paris: L'Herne, 1989). More generally, however, I am not claiming to propose a true "interpretation of Hölderlin" and cheerfully leave too many aspects of his thought to one side. Here, Hölderlin is halfway between theme and pretext.

2. Nancy refers to the Pléiade edition of Hölderlin's *Oeuvres*, ed. Philippe Jaccottet (Paris: Gallimard, 1967), "occasionally modifying when it is absolutely necessary." English citations are based on the German original, occasionally modified to reflect the French versions Nancy is using. Passages that appear in the edition of the essays translated by Thomas Pfau—Hölderlin, *Essays and Letters on Theory* (Albany: SUNY Press, 1988)—are given in the text as *Essays*.—Trans.

3. Here, Nancy follows François Fédier in translating Hölderlin's *lebendiger Sinn* as *sens vivant*; cf. "Remarques sur Oedipe" in Hölderlin, *Oeuvres*, ed. Jaccottet, 951–52. Unless otherwise indicated, *sens*—as sense, meaning, direction—has been translated throughout as "sense," "meaning" being reserved for Hölderlin's

Bedeutung, usually translated, once again by Jaccottet and others, into French as *signification.* —Trans.

4. Here, this analysis intersects with Philippe Lacoue-Labarthe's account of the Hölderlinian theory of the tragic in "La césure du spéculatif," in *L'imitation des modernes: Typographies II* (Paris: Galilée, 1986), 39–69; trans. Robert Eisenhauer as "The Caesura of the Speculative," in *Typography: Mimesis, Philosophy, Politics*, ed. Christopher Fynsk (Stanford: Stanford University Press, 1998; orig. pub. Harvard University Press, 1989), 208–35. One might also consult the continuation of this analysis by Arnauld Villani in "Figures of Duality," and by Lacoue-Labarthe in "Poetry's Courage," both in this text, as well as Jean-François Courtine's study of Hölderlin's break with speculative Idealism in "Hölderlin au seuil de l'idealisme allemand," in *Extase de la raison: Essai sur Schelling* (Paris: Galilée, 1990), 15–43.

5. As the Hegelian analysis of the epic, to which I will return elsewhere, shows. This is so de jure in Hegel, because poetry de facto resists its speculative sublation. Paradoxically, however, this "fact" belongs to the principle of dialectical reason. On this point, there is contact, albeit without dialogue, between Hegel and Hölderlin.

6. Hölderlin, *Oeuvres*, ed. Jaccottet, 815.

LACOUE-LABARTHE *Poetry's Courage*

1. Martin Heidegger, *Hölderlins Hymnen "Germanien" und "Der Rhein," Gesamtausgabe*, vol. 39 (Frankfurt am Main: Klostermann, 1980), 213–14. Henceforth cited in the text as *GA* 39. All translations in this essay have been prepared for this volume, and all references and notes have been added by the translator.

2. The reference is to the French translation of "Hölderlin und das Wesen der Dichtung," "Hölderlin et l'essence de la poésie," in Martin Heidegger, *Qu'est-ce que la métaphysique* (Paris: Gallimard, 1937). As Michel Deguy and François Fédier noted upon its re-edition, this translation "magisterially inaugurated the epoch of Heidegger translation in France, and truly initiated a first public to his readings." See their comments in the French translation of *Erläuterungen zur Hölderlins Dichtung, Approche de Hölderlin* (Paris: Gallimard, 1973), 39.

3. Martin Heidegger, *Hölderlins Hymnen "Germanien" un "Der Rhein," Gesamtausgabe*, vol. 39 (Frankfurt am Main: Klostermann, 1980), 1.

4. Martin Heidegger, " 'Nur noch ein Gott kann uns retten.' *Spiegel*-Gespräch mit Martin Heidegger am 23. September 1966," in *Der Spiegel*, May 31, 1976, 207.

5. Martin Heidegger, "Brief über den 'Humanismus' " in *Wegmarken, Gesamtausgabe*, vol. 9 (Frankfurt am Main: Klostermann, 1996), 338–39.

6. See Martin Heidegger, *Einführung in die Metaphysik* (Tübingen: Niemeyer, 1987), 131.

7. See the letter to Casimir Ulrich Böhlendorff of December 4, 1801, in *SA*, 4; 1: 425.

8. Heidegger, "Brief über den 'Humanismus,' " 9: 338.

9. Heidegger, *Hölderlins Hymnen*, 39: 290, for example.

10. See, for example, Lacoue-Labarthe, *Le mythe nazi* (La Tour d'Aigues: Editions de l'Aube, 1991), esp. 32–50; trans. Brian Holmes as "The Nazi Myth" in *Critical Inquiry* 16: 2 (1990): esp. 296–303; "The Essence of National Socialism and Its Destiny," trans. Simon Sparks, in Philippe Lacoue-Labarthe and Jean-Luc Nancy, *Retreating the Political*, ed. Simon Sparks (London: Routledge, 1996), 148–56.

11. See Heidegger, "Der Weg zur Sprache," in *Unterwegs zur Sprache, Gesamtausgabe*, vol. 12 (Frankfurt am Main: Klostermann, 1985), 234–38.

12. Pierre Klossowski, "Entre Marx et Fourier," *Le Monde*, May 31, 1969, supplement: special page devoted to Walter Benjamin.

13. Martin Heidegger, *Erläuterungen zur Hölderlins Dichtung, Gesamtausgabe*, vol. 4 (Frankfurt am Main: Klostermann, 1988), 36–37.

14. Ibid., 4: 37.

15. Ibid.

16. Walter Benjamin, "Zwei Gedichte von Friedrich Hölderlin," in *Gesammelte Schriften*, ed. Rolf Tiedemann and Hermann Schweppenhauser (Frankfurt am Main: Suhrkamp, 1977), 2; 1: 105–26. All references to Benjamin are to this edition, henceforth cited by volume and page number.

17. "'*Timidité ou 'gaucherie*.'" Lacoue-Labarthe's point of reference here is the French translation of the essay by Maurice de Gondillac, who chooses *timidité*. See "Deux poémes de Friedrich Hölderlin," in *Mythe et violence* (Paris: Denoël, 1971), 51–78.

18. Benjamin, 2; 1: 105.

19. Since, for reasons he outlines, Lacoue-Labarthe maintains de Gondillac's translation of *das Gedichtete* by *le dictamen*, a word as "foreign" to French ears as to English, the term "dictamen," rather than "poematized," has been retained throughout.

20. Benjamin, 2; 1: 105–6.

21. Ibid., 1; 1: 125.

22. Ibid., 2; 1: 105.

23. Ibid., 2; 1: 105–6.

24. Immanuel Kant, *Kritik der reinen Vernunft*, ed. Raymund Schmidt (Hamburg: Felix Meiner, 1956), A 141; B 181.

25. Benjamin, 2; 1: 106–7.

26. See Martin Heidegger, "'. . . dichterisch wohnet der Menschen . . . ,'" in *Vorträge und Aufsätze* (Pfullingen: Neske, 1954), 181–98.

27. Benjamin, 2; 1: 107.

28. Thomas Mann, "Freud and the Future," in *Essays of Three Decades* (New York: Philosophy Library, 1947), 425.

29. Benjamin, 2; 1: 107.

30. Ibid., 2; 1: 108.

31. Ibid., 2; 1:109–10.
32. Ibid, 2; 1: 114.
33. See Theodor W. Adorno, "Parataxis," in *Noten zur Literatur* III, *Gesammelte Schriften* (Frankfurt am Main: Suhrkamp, 1971), 9: 447–91.
34. Benjamin, 2; 1:123.
35. Ibid., 2; 1: 124.
36. Ibid.
37. Ibid., 2; 1: 125.
38. Ibid., 2; 1: 125–26.
39. Ibid., 2; 1: 126.
40. Ibid., 1; 1: 100–101.
41. See Jean-Christophe Bailly, *Le fin de l'hymne* (Paris: Bourgeois, 1991).
42. Jacques Lacan, *L'ethique de la psychanalyse: Le séminaire de Jacques Lacan* VII (Paris: Seuil, 1986).

DE VRIES *'Winke': Divine Topoi in Hölderlin, Heidegger, Nancy*

1. See Jean-François Courtine, "Les traces et le passage du Dieu dans les *Beiträge zur Philosophie* de Martin Heidegger," *Archivio di Filosofia* 1–3 (1994), 519–38, esp. 530; as well as Manfred Frank, *Der kommende Gott: Vorlesungen über die neue Mythologie* (Frankfurt am Main: Suhrkamp, 1982), 1: 245–342.
2. Jean-Luc Nancy, *Qu'est-ce que Dieu? Philosophie/Théologie: Hommage à l'abbé Daniel Coppieters de Gibson (1929–1983)* (Brussels: Publications des Facultés Universitaires Saint-Louis, 1985). Republished in a separate edition as *Des lieux divins* (Mauvezin: Éditions Trans-Europe-Repress, 1987), the text ("Of Divine Places") has been translated by Michael Holland in Jean-Luc Nancy, *The Inoperative Community*, ed. Peter Connor (Minneapolis: University of Minnesota Press, 1991), 110–50. In the following, page references are to English translation followed by that of the French re-publication.
3. Compare Jean-Luc Nancy, *Une pensée finie* (Paris: Galilée, 1990), 353–64.
4. Nancy, "Of Divine Places," 112; 3–4 (trans. modified). Nancy shies away from Jean-Luc Marion's suggestion that it is precisely in this discovery of the structural insufficiency of the principle of reason—in the uncovering, that is, of a "principle of *in*sufficient reason," culminating in the death of the ontotheological and moral conception of God—that theology, as the privileged discourse of difference or, rather, "distance," can take its chances and finally conquer its proper place. An elaboration of this view is to be found in Marion's reading of Hölderlin. See *L'idole et la distance* (Paris: Bernard Grasset, 1977), 115–79.
5. Nancy, "Of Divine Places," 113; 4.
6. Ibid., 113; 5.
7. Ibid.
8. Ibid., 142; 40.
9. Ibid., 114; 5–6.

10. Ibid., 114; 6. For an exemplary analysis of the relation between religion and place, compare the study of Jonathan Z. Smith, *To Take Place: Toward Theory in Ritual* (Chicago: University of Chicago Press, 1987).
11. Nancy, "Of Divine Places," 114; 6.
12. Ibid., 133; 29.
13. See his contribution to this volume.
14. Nancy, "Of Divine Places," 133; 29.
15. Ibid., 117–18; 10 (emphasis added).
16. Ibid., 118; 11 (trans. modified).
17. Martin Heidegger, *Denkerfahrungen 1910–1976* (Frankfurt am Main: Klostermann, 1983), 176.
18. Ibid., 178.
19. Ibid.
20. Ibid.
21. Ibid.: "In Wahrheit is das Wort [*Seinsvergessenheit*] der Name des Geschicks der Lichtung des Seins, insofern dies als Anwesen nur offenkundig werden und alles Seiende bestimmen kann, wenn die Lichtung des Seins, die ἀλήθεια, an sich hält, sich dem Denken vorenthält, was im Anfang des abendländischen Denkens und als dessen Anfang geschah und seitdem die Epochen der Seinsgeschichte bis in das heutige technologische Weltalter kennzeichnet."
22. Ibid., 179.
23. Ibid.: "Solange uns der Wegblick dafür versagt ist, dass und wie auch im Entzug und im Vorenthalt eine eigene Weise des Anwesens waltet, solange bleiben wir blind und unbetroffen vom bedrängenden Anwesen, das dem Fehl eignet, der den Namen des Heiligen und mit ihm dieses selbst in sich birgt, und jedoch verbirgt. Nur ein Aufenthalt in der offenen Gegend, aus der her der Fehl anwest, gewährt die Möglichkeit eines Einblickes, in das, was heute *ist*, indem es fehlt."
24. On these matters, see also Jean Greisch, "Hölderlin et le chemin vers le sacré," in Michel Haar, ed., *Cahier de l'Herne: Heidegger* (Paris: Herne, 1983), 543–67, and Jean-Louis Chrétien, "La réserve de l'être," in the same volume, 233–60.
25. The background of any such reading should be the writings of Peter Szondi, notably his *Poetik und Geschichtsphilosophie I* in the *Studienausgabe der Vorlesungen* (Suhrkamp: Frankfurt am Main, 1975), 2: 184–214. With respect to Hölderlin's so-called later poetry, in particular the free-verse hymns, see Szondi's "Interpretationsprobleme," in *Einführung in die literarische Hermeneutik* (Suhrkamp: Frankfurt am Main, 1975), 5:193–402, and *Hölderlin Studien*, notably the essays "Über philologische Erkenntnis," "Der andere Pfeil: Zur Entstehungsgeschichte des hymnischen Spätstils," "Er selbst, der Fürst des Fests: Die Hymne *Friedensfeier*," and "Überwindung des Klassizismus: Der Brief an Böhlendorff vom 4. Dezember 1801," in *Schriften* (Suhrkamp: Frankfurt am Main, 1978), 1: 263–86; 289–314; 315–42; and 345–4[illegible]2, respectively.
26. Richard Sieburth, introduction to Friedrich Hölderlin, *Hymns and Frag-*

ments, trans. Richard Sieburth (Princeton, N.J.: Princeton University Press, 1984), 21–22.

27. Ibid., 20–21.

28. Ibid., 15.

29. Compare Ibid., 39–40:

> In an early essay "On Religion," Hölderlin defined the "mythic" as a dialectical synthesis of the local and the universal, of the particular and the general, of the concrete and the abstract. The geography of his late hymns is "mythic" in a similar sense. . . . For Hölderlin, writing was a kind of mapping—legend has it that the walls of his room were covered with maps of the four quarters of the earth. . . . Hölderlin's landscapes are most frequently imagined as a pattern of vectors or moving energies. . . . All trace a process of becoming, a motion at once toward and away from the source.

30. Ibid., 21.

31. Ibid., 20–21, note 4. Sieburth notes that although Hölderlin "distinguishes between 'God' (Father Aether, Zeus the Thunderer, the Almighty or Most High, a remote and often wrathful sky-god) and 'the gods' (*die Himmlischen*, the celestial or heavenly ones), the distance between singular and plural is far from absolute, indicating an open sweep of theological possibilities that includes both polytheism and monotheism, both paganism and Christianity, as well as their eventual reconciliation."

32. Ibid., 38.

33. See Françoise Dastur, *Hölderlin: Tragédie et modernité* (Paris: Encre Marine, 1992), 25.

34. Nancy, "Of Divine Places," 140; 37.

35. Martin Heidegger, ". . . Poetically Man Dwells . . . ," in *Poetry, Language, Thought*, trans., Albert Hofstadter (New York: Harper & Row, 1971), 225. ". . . dichterisch wohnet der Mensch . . . ," in *Vorträge und Aufsätze* (Pfullingen: Neske, 1985), 193.

36. Sieburth, *Hymns and Fragments*, 42.

37. Paul de Man, "Hölderlin's Riddle," in *Critical Writings, 1953–1978*, ed. Lindsay Waters (Minneapolis: University of Minnesota Press, 1989), 202. See also his essays "The Image of Rousseau in the Poetry of Hölderlin" and "Wordsworth and Hölderlin," both in *The Rhetoric of Romanticism* (New York: Columbia University Press, 1984), 19–45 and 47–65, respectively. Some relevant remarks on Hölderlin may also be found in this book's first essay, "Intentional Structure of the Romantic Image" (1–17), as well as in the essays "Patterns of Temporality in Hölderlin's 'Wie wenn am Feiertage . . .'," "Hölderlin and the Romantic Tradition," and "Heaven and Earth in Wordsworth and Hölderlin," all of which have been published in *Romanticism and Contemporary Criticism: The Gauss Seminar and Other Papers*, ed. E. S. Burt, Kevin Newmark, and Andrzej Warminski (Balti-

more: The Johns Hopkins University Press, 1993), 50–73; 123–36; and 137–46, respectively.

38. Ibid., 60, note 15.

39. Ibid., 211.

40. One example of Heidegger's diagnosis must suffice here.

> Nature, separated from being by science, what happens to it through technology? The growing or rather the simple destruction of "Nature" that rolls towards its end. What was nature before? The momentary place for the arrival and stay of the gods, when it [nature] was still phusis and when it rested in the essencing of *Seyn* itself. Since then it rapidly became the counterpart to "grace" and after this degradation it was fully exposed in the determining force of calculable exploitation and economy. And what remained, finally, was "countryside" and "an opportunity to relax." [Die Natur, herausgesondert aus dem Seienden durch die Naturwissenschaft, was geschieht ihr durch die Technik? Die wachsende oder besser einfach zu ihrem Ende abrollende Zerstörung der "Natur." Was war sie einst? Die Stätte des Augenblicks der Ankunft und des Aufenthalts der Götter, als sie, noch φύσις, in der Wesung des Seyns selbst ruhte. Seitdem wurde sie alsbald ein *Seiendes* und dann gar das Gegenspiel zur "Gnade" und nach dieser Absetzung vollends herausgesetzt in die Verzwingung der berechenbaren Machenschaft und Wirtschaft. Und schliesslich blieb noch "Landschaft" und Erhohlungsgelegenheit.]

From *Beiträge zur Philosophie: Vom Ereignis*, Gesamtausgabe (Frankfurt am Main: Klostermann, 1989), 65: 277 (trans. mine).

41. Paul de Man, "Heidegger's Exegeses of Hölderlin," in *Blindness and Insight: Essays in the Rhetoric of Contemporary Criticism* (Minneapolis: University of Minnesota Press, 1983), 256.

42. Ibid.

43. Ibid.

44. Ibid.

45. Ibid.

46. Ibid., 257.

47. Ibid., 255.

48. Ibid., 257.

49. Ibid., 258.

50. Ibid., 258–59.

51. Ibid., 250 (emphasis added).

52. Ibid., 252 and 254. Yet, de Man *does* seem surprised: "It is in the commentary to 'Andenken' at most that one may find, and not without applying some force, the traces of a dialogue. In all of the other commentaries, dialogue is out of the question, and Heidegger's attitude differs fundamentally from the one he adopts toward all metaphysicians, the pre-Socratics included." Ibid., 254, note 3.

53. Jacques Derrida, "Mnemosyne," trans. Cecile Lindsey, in *Memoires for Paul de Man* (New York: Columbia University Press, 1986), 7.

54. Paul de Man, "Heidegger's Exegeses of Hölderlin," 261, cited in Derrida, *Mémoires*, 7–8.

55. Nancy, "Of Divine Places," 123; 16.

56. Ibid.

57. Ibid., 123; 17. The lines in question figure prominently in a host of critical writings on Hölderlin, Heidegger's *Erläuterungen*, and so on. For example, we find them in Maurice Blanchot, *L'espace littéraire* (Paris: Gallimard, 1955), 374; Marion, *L'idole et la distance*, 123; Dastur, *Hölderlin: Tragédie et modernité*, 91; and also Beda Allemann, *Hölderlin und Heidegger* (Zürich: Atlantis, 1954), 169.

58. Nancy, "Of Divine Places," 124; 17.

59. Jean-Luc Nancy, *L'expérience de la liberté* (Paris: Galilée, 1988), 131 and 131, note 1. *The Experience of Freedom*, trans. Bridget McDonald (Stanford, Calif.: Stanford University Press, 1995), 102 and 195–96, note 6.

60. Ibid., 131, note 1; 195–96, note 6.

61. Ibid., 200; 156–57.

62. Ibid., 157.

63. Nancy, "Of Divine Places," 126; 19.

64. Ibid., 124; 17.

65. Ibid., 124; 17–18.

66. See Andrzej Warminski, *Readings in Interpretation: Hölderlin, Hegel, Heidegger* (Minneapolis: University of Minnesota Press, 1987), 47 and 64.

67. ". . . indem Hölderlin das Wesen der Dichtung neu stiftet, bestimmt er . . . eine neue Zeit. Es ist die Zeit der entflohenen Götter *und* des kommenden Gottes. Das ist die *dürftige* Zeit, weil sie in einem gedoppelten Mangel und Nicht steht: im Nichtmehr der entflohenen Götter und im Nochnicht des Kommenden." Martin Heidegger, *Erläuterungen zu Hölderlins Dichtung* (Frankfurt am Main: Klostermann, 1971), 47. Or, in "Building Dwelling Thinking": "Mortals dwell in that they await the divinities as divinities. They wait for intimations of their coming and do not mistake the signs of their absence. They do not make their gods for themselves and do not worship idols. In the very depth of misfortune they wait for the weal that has been withdrawn." Martin Heidegger, *Poetry, Language, Thought*, trans. Albert Hofstadter (New York: Harper & Row, 1971), 150. ("Die Sterblichen wohnen, insofern sie die Göttlichen als die Göttlichen erwarten. Hoffend halten sie ihnen das Unverhoffte entgegen. Sie warten der Winke ihrer Ankunft und verkennen nicht die Zeichen ihres Fehls. Sie machen sich nicht Götter und betreiben nicht den Dienst an Götter. Im Unheil noch warten sie des entzogenen Heils." *Vorträge und Aufsätze*, 145.)

68. Heidegger, *Vorträge und Aufsätze*, 144.

69. For instance the one proposed by Christopher Fynsk, who, in the context of a discussion of the same poetic fragments that concern us here, "What is God?" and "In lovely blueness," and with reference to Heidegger's ". . . Poetically Man

Dwells . . . ," writes that here "the holy is defined more precisely as the condition for the reception of divine presence (the godhead), which, in turn, names the self-disclosure of the deity (*der Gott*)." *Heidegger: Thought and Historicity* (Ithaca, N.Y.: Cornell University Press, 1986), 222, note 33. Fynsk goes on to cite David White's *Heidegger and the Language of Poetry*, which, he claims, "offers what is probably the most exact and helpful analysis of these terms," by defining the holy as "*the dispositional capacity in all that is other than the deity to receive the appearances of the divine presence.*" For White, see *Heidegger and the Language of Poetry*, 127 (cited in Fynsk, *Thought and Historicity*, 222, note 33) (emphasis added). With this definition of the holy as a possible in terms of a "dispositional capacity," it is suggested here, one would also be able to capture the relation Heidegger describes in the "Letter on Humanism," and according to which "the holy, which alone is the essential sphere of the godhead, which in turn alone affords a dimension for the gods and for God, comes to radiate only when Being itself is beforehand and after extensive preparation has been illuminated and is experienced in its truth." "Brief über den Humanismus," *Wegmarken*, 338–39. By the same token, Fynsk claims, White's definition seems to adequately rephrase what is at stake in "What are Poets for?," where Heidegger states:

> Poets are the mortals who, singing . . . sense the trace of the fugitive gods, stay on the god's tracks, and so trace for their kindred mortals the way toward the turning. The ether, however, in which alone the gods are gods, is their godhead [*Gottheit*]. The element of this ether, that within which even the godhead itself is still present, is the holy. The element of the ether for the coming of the fugitive gods, the holy, is the track of the fugitive gods. But who has the power to sense, to trace such a track? Traces are often inconspicuous, and are always the legacy of a directive that is barely divined. To be a poet in a destitute time means: to attend, singing, to the trace of the fugitive gods. This is why the poet in the time of the world's night utters the holy. (*Poetry, Language, Thought*, 94; Martin Heidegger, *Holzwege* [Frankfurt am Main: V. Klostermann, 1977], 25)

70. Heidegger, *Beiträge zur Philosophie*, 24, cf., 416:

> Originating in a disposition towards being that has been determined by "metaphysics," it is only with difficulty and slowly that we will come to know *the Other* that appears neither in the "personal" nor in the "collective" "experience" of the god, but only in the abysmal "space" of *Seyn* itself. [Herkünftig aus einer durch die "Metaphysik" bestimmten Stellung zum Seienden werden wir nur schwer und langsam *das Andere* wissen können, dass weder im "persönlichen" noch im "masseinweise" "Erlebnis" der Gott noch erscheint, sondern einzig in dem abgründigen "Raum" des Seyns selbst.]

For Courtine, see "Les traces et le passage du Dieu," 529–30.

71. Nancy,"Of Divine Places," 125; 19.

72. Cf. Martin Heidegger, *Hölderlins Hymnen "Germanien" und "Der Rhein"* (Frankfurt am Main: Klostermann, 1980), 54 and 110.

73. Nancy, "Of Divine Places," 115; 7.

74. See Courtine, "Les traces et le passage du Dieu," 533.

75. Ibid., 524, note 20.

76. In the *Beiträge zur Philosophie*, Heidegger writes: "If we do not even understand 'death' in its extremity, how then can mature enough for the rare nod of the last god?" (Wenn wir schon den "Tod" in seinem Äussersten so wenig begreifen, wie wollen wir dann schon dem seltenen Wink des letzten Gottes gewachsen sein?), 405. However, it would seem that the motif of radical finitude is at the same time inspired by the experience of a certain indelible elusiveness, a relation without relation—*religion*—or, at least, by a finitude that is experienced and thought as absolute, in the etymological sense of the word, that is, as that which *infinitizes* itself in the very measure with which we approach it.

77. Heidegger, *Erläuterungen zu Hölderlins Dichtung*, 46:

> Das Sagen des Dichters ist das Auffangen dieser Winke, um sie weiter zu winken in sein Volk. Dieses Auffangen der Winke ist ein Empfangen und doch zugleich ein neues Geben; denn der Dichter erblickt im "ersten Zeichen" auch schon das Vollendete und stellt dieses Erschaute kühn in sein Wort um das noch-nicht-Erfüllte vorauszusagen. So ". . . fliegt, der kühnen Geist, wie Adler den Gewittern, weissagend seinen Kommenden Göttern voraus."

See also the poem, the *Dichtung* or, more precisely, the *Winke*, entitled *Winke*, in *Denkerfahrungen*, 162–63:

> Je aufdringlicher die Rechner,
> je massloser die Gesellschaft.
> Je seltsamer Denkende,
> je einsamer Dichtende.
> Je notvoller Ahnende,
> ahnend die Ferne
> rettender Winke.
> Die das Selbe denken
> im Reichtum seiner Selbigkeit,
> gehen die mühsam langen Wege
> in das immer Einfachere, Einfältige
> seiner im Unzugangbaren
> sich versagenden Ortschaft.

78. Nancy, "Of Divine Places," 115; 7. See also Jean Greisch, *La parole heureuse: Martin Heidegger entre les choses et les mots* (Paris: Beauchesne, 1987), 108, on a "weisendes Offenbarmachen," and 109: "Every donation of a sign refers in the final

instance to an arche-sign, which Heidegger calls *Wink*." Greisch gives yet another citation from Heidegger that suggests an intimate link between poetry, the response—indeed, the deference and also deferral—with respect to godly hints, and the founding acts in which Being (or *Seyn*) manifests itself: "Poetry—to endure the nods of the gods—foundation of Being" (Dichtung—Aushalten der Winke der Götter—Stiftung des Seyns). 109.

79. Nancy, "Of Divine Places," 119; 12.

80. Ibid.

81. Ibid.

82. For example, one that reads:

> Mahnt je das Seyn
> den werklosen Gott
> in die Ankunft entfernten Bleibens? . . .
> Anfänglich naht verhüllt die Zeit
> Der Inständigen des letzten Gottes.
> (Heidegger, *Denkerfahrungen*, 32)

83. Ibid., 33:

> Die "Winke" sind keine Dichtungen. Sie sind auch nicht eine in Verse und Reime gebrachte "Philosophie." Die "Winke" sind Worte eines Denkens, das zu einem Teil dieses Aussagen braucht, aber in ihm sich nicht erfüllt. Dieses Denken hat im Seienden keinen Anhalt, denn es denkt das Seyn. Dieses Denken findet im Gedanken keinen Vorbild, denn das Gedachte denkt das Seiende. Das Sagen des Denkens ist im Unterschied zum Wort der Dichtung bildlos. Und wo ein Bild zu sein scheint, ist es weder das Gedichtete einer Dichtung noch das Anschauliche eines "Sinnes", sondern nur der Notanker der gewagten, aber nicht geglückten Bildlosigkeit.

84. Compare the "Enleitung" to "Was ist Metaphysik?": "Aletheia könnte das Wort sein, dass einen noch nicht erfahrenen Wink in das ungedachte Wesen des *esse* gibt." Ibid., 11.

85. Nancy, "Of Divine Places," 130; 25.

86. Ibid.

87. Cf. Sieburth, *Hymns and Fragments*, 30–31:

> In an influential essay on Hölderlin's late poetry, Theodor Adorno has defined its most outstanding stylistic feature as *parataxis*, that is, the juxtaposition, without explanatory connectives, of various syntactical and grammatical elements (as opposed to hypotaxis, the subordination or coordination of phrase or clause). . . . The style of parataxis, as Auerbach's *Mimesis* reminds us, is as ancient as it is modern: Hölderlin is only so *new* because he has travelled so far back to poetry's most archaic roots. Norberth von Hellingrath was perhaps the first to demonstrate that Hölderlin's late hymns, far from constituting an aberration, rejoin the tradition of "austere harmony"

as defined by Dionysius of Holicarnassus in reference to Homer, Pindar, Alcaeus, and Sophocles.

88. Nancy, "Of Divine Places," 130; 25.

COURTINE *Hölderlin's Christ*

1. "Der Einzige" ("The Only One"), third version, *SA* 2; 1: 164: ". . . Beautiful / And good is it to compare."

2. The publishing information for these books by Tilliette is, respectively: Paris: Desclée, 1986; Paris: Cerf, 1990; and Paris: Desclée, 1992.

3. Cf., in particular, *Der Streit um den Frieden: Beiträge zur Auseinandersetzung um Hölderlins "Friedensfeier"*, ed. Eduard Lachmann (Nürnberg: Glock und Lutz, 1957), with contributions by Eduard Lachmann, Erich Przywara, Walter Bröcker, Heinrich Buhr, Walter Hof, Ludwig von Pigenot, and Alois Winklhofer. Cf. also Heinrich Buhr and Walter Bröcker, *Zur Theologie des Geistes* (Pfullingen: Neske, 1960).

4. See Wilhelm Michel, "Hölderlin's abendländlische Wendung," in *Hölderlin's abendländlische Wendung* (Weimar: Feuer, 1922), reprinted in *Hölderlin's Widerkunft* (Wien: Gallus, 1943), 57–109.

5. Martin Heidegger, *Hölderlins Hymnen "Germanien" und "Der Rhein"* (Frankfurt am Main: Klostermann, 1980).

6. See Hölderlin, "Versöhnender der du nimmergeglaubt . . .": "Sei gegenwärtig Jüngling, jezt erst, denn noch ehe du ausgeredet / Rief es herab . . ." ("Be present, youth, now for the first time, for before you spoke / A call came down") (*SA* 2; 1: 131). Cf. also "Friedensfeier," *SA* 3: 534.

7. See Hölderlin, "Patmos":

. . . sie
Zusammensaßen, zu der Stunde des Gastemals
Und in der großen Seele, ruhigahnend den Tod
Aussprach der Herr und die lezte Liebe, denn nie genug
Hatt' er, von Güte, zu sagen
Der Worte, damals, und zu schweigen, da
Ers sahe, das Zürnen der Welt.
Denn alles ist gut. Drauf starb er. Vieles wäre liebes
Zu sagen . . .

. . . they sat together, at the banquet hour,
And in his great soul, calmly foreboding
The Lord spoke death and the final love, for he never
Had enough words to speak of goodness
At that time, and to remain silent, when
He saw it, the wrath of the world.

For all things are good. Thereupon he died. There would be much
good to say . . .
(*SA* 2; 1: 175).

Rather than seeing, along with Wolfgang Binder, the clear witnessing of a "Christ without Passion and without reconciliation" in this extraordinarily reserved account, we can also read there the expression of that "modesty" [*pudeur*], the affective disposition that for Hölderlin responds strictly to the necessary divine "infidelity." *Hölderlin-Aufsätze* (Frankfurt am Main: Insel, 1970), 363–64.

8. See "Griechenland": "Alltag aber wunderbar zu lieb den Menschen / Gott an hat ein Gewand. / Und Erkentnissen verberget sich sein Angesicht . . ." ("But everyday, wonderful for the sake of men / God wears a garment. / And his face is hidden from knowing . . ."). (*SA* 2; 1: 257).

9. I will thus keep myself from simply opposing an Empedoclean principle characterized by the aspiration for unity and for the immediate return to Nature in its divine overpowerfulness to a Hesperidian principle that is attentive to mediation and to the "statute." In fact, Hölderlin's work on "modern" tragedy, that is *Empedocles*, leads him to superimpose with increasing precision the "christological" traits on the royal figure of the sovereign of Agrigente. In order to move quickly, allow me to refer to the indications I have given in my study "Qui est l'Empédocle de Hölderlin?," in *Nietzsche, Hölderlin, et la Grèce* (Paris: Belles Lettres, 1988), 19–32.

10. The English translation is from Friedrich Hölderlin, *Essays and Letters on Theory*, trans. and ed. Thomas Pfau (Albany: State University of New York Press, 1988), 102 (trans. modified). All further lengthy translations from Hölderlin's prose will be from this edition, are noted parenthetically after the German citation as *Essays* and are sometimes modified.—Trans.

11. *Mittelbarkeit* is also, as we know, the key term in the commentary given by Hölderlin to Pindar's fragment 7 (Hellingrath). See *SA* 5: 265. We find in Gerhard Kurz's work *Mittelbarkeit und Vereinigung: Zum Verhältnis von Poesie, Reflexion un Revolution bei Hölderlin* (Stuttgart: Metzler, 1975) a good study of this theme, in which, however, the "theological" dimension is curiously lacking.

12. The "Occidental and Junoian sobriety" that Hölderlin evokes in the first letter to Böhlendorff. See *SA* 6: 425

13. Beda Allemann, *Hölderlin und Heidegger*, rev. ed. (Zürich: Atlantis, 1954), 50.

14. Ibid., 51.

15. "Der Einzige" (*SA* 2;1: 153–56, 157–60, and 161–64); "Versöhnender der nu nimmergeglaubt . . ." (*SA* 2;1: 130–37); and "Patmos" (*SA* 2;1: 165–72). For "Friedensfeier," see Allemann: "The poem 'Patmos' cannot be named, properly speaking, a 'Hymn to Christ.' On the contrary, this hymn sings of John the Evangelist, the hero who protects from uprooting" (*Hölderlin und Heidegger*, 52). For a study of all the christological hymns in the context of German Idealism,

see Jochen Schmidt's very rich book, *Hölderlins Geschichtsphilosophische Hymnen: "Friedensfeier," "Der Einzige," "Patmos"* (Darmstadt: Wissenschaftliche Buchgesellschaft, 1990).

16. Hölderlin, "An die Madonna" ("To Madonna") (*SA* 2; 1: 211–16); "Was ist Gott? . . ." ("What is God? . . .") (*SA* 2; 1: 210); and ". . . Der Vatikan . . ." ("The Vatican") (*SA* 2; 1: 252–53).

17. See Hölderlin, "Der Vatikan": "Gott rein und mit Unterscheidung / Bewahren, das ist uns vertrauet" ("To keep God pure and with distinction / Is entrusted to us") (*SA* 2; 1: 252).

18. See Hölderlin, "Brod und Wein": "Oder er kam auch selbst und nahm des Menschen Gestalt an / Und vollendet' und schloß tröstend das himmlische Fest" ("Or he came himself and assumed human figure / And consolingly completed and closed the divine feast") (*SA* 2; 1: 93).

19. Luke 24: 29: "Stay with us, for it is toward evening and the day is now far spent."

20. Cf. also "Patmos":

Doch trauerten sie, da nun
Es Abend worden, erstaunt,
Denn Großentschiedenes hatten in der Seele
Die Männer, aber sie liebten unter der Sonne
Das Leben und lassen wollten sie nicht
Vom Angesichte des Herrn
Und der Heimath.

Yet they were sad, now that
It had become evening, astonished,
For something greatly determined had the men in their souls
But under the Sun they loved
Life and did not want to leave
The Lord's face
And their homeland.
(*SA* 2; 1: 167–68)

21. See the beginning of the ninth strophe of "Brod und Wein":

Ja! sie sagen mit Recht, er söhne den Tag mit der Nacht aus,
  Führe des Himmels Gestirn ewig hinunter, hinauf,
Allzeit froh, wie das Laub der immergrünenden Fichte,
  Das er liebt, und der Kranz, den er von Epheu gewählt,
Weil er bleibet und selbst die Spur der entflohenen Götter
  Götterlosen hinab unter das Finstere bringt.

Yes, they rightly say that he reconciles day with night
  Leads the star of heaven up and down without end

> Always glad, like the leaves of the evergreen pine-tree
> That he loves, and the wreath he picked from Epheus
> Because it endures and even brings the trace of the fled gods
> Down to the godless below amidst the darkness.
>
> (*SA* 2; 1: 94).

22. Cf. Bernhard Böschenstein, *"Frucht des Gewitters": Zu Hölderlins Dionysos als Gott der Revolution* (Frankfurt am Main: Insel, 1989), 88–90.

23. Cf. also Jochen Schmidt, *Hölderlins Elegie "Brot und Wein": Die Entwicklung der hymnischen Stils in der elegischen Dichtung* (Berlin: de Gruyter, 1968), 160–72.

24. On the one hand, "sichtbare, bildhaft erscheinende Religion" ("manifest religion appearing in images"), on the other, Christianity, spiritual or pneumatic religion, which is lacking *weltliche Glanz* (worldly glitter). Cf. Schmidt, *Hölderlins geschichtsphilosophische Hymnen*, 197.

25. Cf. Hölderlin, "Der Einzige": "Viel hab' ich schönes gesehn, / Und gesungen Gottes Bild, / Hab' ich, das lebet unter / Den Menschen . . ." ("I have seen much that is beautiful / And sung God's image / have I, which lives among / Men") (*SA* 2; 1: 153–54).

26. Cf. "Patmos": "Drum, da gehäuft sind rings / Die Gipfel der Zeit, und die Liebsten / Nah wohnen, ermattend auf / Getrenntesten Bergen . . ." ("Therefore, since the summits of Time / Are heaped around, and the most beloved / Live near, exhausted upon / The most separate mountains . . .") (*SA* 2; 1: 165).

27. Cf. Hölderlin, "Der Einzige": ". . . Und freilich weiß / Ich der dich zeugte, dein Vater ist / Derselbe . . ." ("And indeed do I know / He who begot you, your father is / The same who . . .") (*SA* 2; 1: 163).

28. Ibid.: "Nie treff ich, wie ich wünsche, / Das Maas . . ." ("Never do I strike, as I would like to, / The measure . . .") (*SA* 2; 1: 155).

29. Ibid.: "Es hindert aber eine Schaam / Mich dir zu vergleichen / Die weltlichen Männer" ("But a shame stops / Me from comparing you / To the men of this world") (*SA* 2; 1: 155).

30. Ibid.

> Und kühn bekenn' ich, du
> Bist Bruder auch des Eviers, der
> An den Wagen spannte
> Die Tyger und hinab
> Bis an den Indus
> Gebietend freudigen Dienst
> Den Weinberg stiftet und
> Den Grimm bezähmte der Völker.
>
> And I boldly admit you
> Are also the brother of Evius, who

To his chariot harnessed
The tigers and right down
To the Indus
Offering joyful service
Planted the vineyard and
Tamed the wrath of the peoples.
(*SA* 2; 1: 154).

31. "Ich weiß es aber, eigene Schuld / Ists! . . ." ("I know, however, my own fault / Is it!") (*SA* 2; 1: 154).

32. See "Brod und Wein":

Unempfunden kommen sie erst, es streben entgegen
  Ihnen die Kinder, zu hell kommet, zu blendend das Glük,
Und es scheut sie der Mensch, kaum weiß zu sagen ein Halbgott,
  Wer mit Nahmen sie sind, die mit den Gaaben ihm nahn.
  (*SA* 2; 1: 92)

Unperceived at first they come, and toward them
  Struggle the children, too bright apprears, too brilliant, joy,
And man is full of fear, a demi-god could hardly
  Name those who approach him with gifts.

On this Hölderlinian "modesty," Tilliette notes very correctly: "To be sure, the presence of Christ in Hölderlinian poetry is a veiled presence, against the light, not immediately recognizable. . . . He hides in the company of the gods, his half brothers . . . , but this is neither a blending nor even ambiguity" (*La semaine sainte des philosophes*, 29).

33. "Gott rein und mit Unterscheidung / Bewahren, das ist uns vertraut" ("To keep God purely and with differentiation / Is confided in us"), Hölderlin will write in the outline of ". . . Der Vatikan."

34. It seems to me that Allemann goes off track here when he interprets the mistake as the forgetting of the difference between mortals and immortals: "This is not the shame of failing Christ with a profane comparison. It is the unique and supreme shame that falls to poets—the shame that keeps one from forgetting the difference between men and the gods" (*Hölderlin und Heidegger*, 54). In truth, Allemann's entire interpretation of the figure of Christ is at stake here: "Christ aspires to return to the Empire of the gods by the shortest path. That is why the poet feels the shame that is connected to the purity of the differentiation at the moment in which he is going to compare the royal demigods, Heracles and Dionysus, to Christ, an Empedoclean demigod. The worry of this shame concerns the men of this world, not Christ" (Ibid.). Cf. also: "The singularity of Christ is connected to his Empedoclean right to meet death demonstratively . . . and to return without delay to the community of the gods. . . . [Christ] is the demigod

who escapes this earth elated" (Ibid.). The systematic opposition of the Royal and the Empedoclean here leads, it seems to me, entirely to error.

35. See also "Friedensfeier":

> Wo aber wirkt der Geist, sind wir auch mit, und streiten
> Was wohl das Beste sei. So dünkt mir jetzt das Beste,
> Wenn . . . nur der Liebe Gesez,
> Das schönausgleichende gilt von hier an bis zum Himmel.
> (*SA* 3: 535–36)
>
> But where spirit works, we are also, and strive
> For what is best. It seems to me the best
> When/If . . . only love's law,
> The equally beautiful, reigns from here to heaven.

36. Hölderlin, "Versöhnender der du nimmergeglaubt . . . "(*SA* 2; 1: 131). Cf. also the ending of *Hyperion*: "Versöhnung ist mitten im Streit und alles Getrennte findet sich wieder" ("Reconciliation is in the midst of conflict and everything that is separated joins up again") (*SA* 3: 160).

37. Cf. Hölderlin, "Was ist Gott? . . .": "Was ist Gott? unbekannt, dennoch / Voll Eigenschaften ist das Angesicht / Des Himmels von ihm" ("What is God? Unknown, and yet / The face of the heavens are / Full of his features") (*SA* 2; 1: 210).

38. Hölderlin, "Der Einzige," *SA* 2; 1: 751. Cf. Böschenstein, *"Frucht des Gewitters"*, 17.

39. Cf. Hölderlin, "Stimme des Volkes" ("Voice of the People"): "Ins All zurük die kürzeste Bahn" ("Back into the All by the shortest path"); "Das wunderbare Sehnen dem Abgrund zu" ("The miraculous longing for the abyss") (*SA* 2; 1: 51).

40. Cf. the evocation of the temptation in the desert (Matthew 4: 1–4), where there is still a trace of a word (*Spur des Wortes*) that is called to the rescue: ". . . Es bleibet aber eine Spur / Doch eines Wortes; die ein Mann erhaschet. Der Ort war aber // Die Wüste" (". . . But a trace remains, / Yes, a trace of a word, which a man seizes. The place, however, was // The desert") (*SA* 2; 1: 163). On this difficult passage from the third version of "Der Einzige," refer to Beda Allemann's study, " 'Der Ort war aber die Wüste,' " in *Martin Heidegger zum siebzigsten Geburtstag* (Pfulling: Neske, 1959), 204–16.

41. Wolfgang Binder, "Hölderlin's Patmos-Hymne," *Hölderlin-Aufsätze* (Frankfurt am Main: Insel, 1970), 362–402, esp. 363.

42. Cf. Hölderlin, "Friedenfeier," *SA* 3: 534: "Denn schonend rührt des Maases allzeit kundig / Nur einen Augenblik die Wohnungen der Menschen / Ein Gott an, unversehn, und keiner weiß es, wenn?" ("For sparingly, always knowing the measure, / Only for a moment will a god, unforeseen, touch men's dwelling / And no one knows when").

43. The disciples believed they found in the immediacy of the divine in Christ

the object of their desire, their homeland. But therein lies the danger. By way of his kenosis, his departure, he saves: "Thereupon he died. / Yet nonetheless they saw the figure inclined before God / Of he who renounces himself."

44. Detlev Lüders, *Die Welt im verringerten Maß. Hölderlin-Studien* (Tübingen: Niemeyer, 1968), 49.

45. Allemann, *Hölderlin und Heidegger*, 55.

46. Lüders, *Die Welt im verringerten Maß*, 49. Cf. the difficult late variant of "Patmos" (*SA* 2; 1: 176):

> . . . lassen wollten sie nicht
> Vom Angesichte des Herrn
> Und der Heimath. Eingeboren war,
> Wie Feuer im Eisen, das, und ihnen gieng
> Zur Seite der Schatte des Lieben.
> Darum auch sandt' er ihnen
> Den Geist, und freilich bebte
> Das Haus und die Wetter Gottes rollten
> Ferndonnernd über
> Die ahnenden Häupter.

> . . . they did not want to leave
> The face of the Lord
> And their homeland. Inborn
> Like fire in iron, was this, and beside them
> Walked the shadows of the beloved. Therefore he sent them
> Spirit and mightily shook
> The house, and God's weather rolled
> Thundering from afar over
> The foreknowing heads.

47. Cf. Ruth-Eva Schulz, "Herakles, Dionysos, Christus: Interpretationen zu Hölderlins Hymne 'Der Einzige,'" in *Die Gegewart der Griechen im neueren Denken*, Festschrift für Hans-Georg Gadamer zum 60. Geburtstag, ed. Dieter Henrich, Walter Schulz, and Karl-Heinz Vokmann-Schluck (Tübingen: Niemeyer, 1960), 233–60, esp. 241.

PANKOW *Epistolary Writing, Fate, Language: Hölderlin's 'Hyperion'*

1. Cf. *SA* 3: 296–335.

2. See the letter to Christian Ludwig Neuffer from July 1793, in *SA* 6; 1: 87, and Friedrich Hölderlin, *Essays and Letters on Theory*, ed., intro. and trans. Thomas Pfau (Albany: State University of New York Press, 1988), 122. Further references to this translation will appear as *Essays* after a semicolon.

3. The German title reads "Hyperion oder der Eremit in Griechenland." Friedrich Hölderlin, "Hyperion; or The Hermit in Greece," in *"Hyperion" and Selected*

Poems, ed. Eric L. Santner, trans. Willard R. Trask, adapted by David Schwarz (New York: Continuum, 1990), 1–133. All further references will be given to the Beißner edition, abbreviated as *Hyp*; page references are first to the German original, then to the English translation.

4. Lawrence Ryan, *Hölderlin's "Hyperion": Exzentrische Bahn und Dichterberuf* (Stuttgart: Metzler, 1965).

5. There are only a few remarks on the issue by Lawrence Ryan, mainly in the introduction of his study. He interprets the epistolary form of Hölderlin's novel as a means which "forces" Hyperion to write in a "more 'objective' mode." Ibid., 5. His account of the "inherent necessity" of epistolary writing may be called vague at best, framed by quotation marks at moments when essential—or "objective"—claims are made. Even more impatient with Hölderlin's epistolary writing is Walter Silz, who believes that the choice of form was unreasonable: "There is something self-contradictory about letters as used here, retrospectively. Letters are a vehicle of direct, immediate communication. . . . There is no reason for cutting off a *past* story into letters, instead of telling it as a continuous narrative in the first person." *Hölderlin's "Hyperion": A Critical Reading* (Philadelphia: University of Pennsylvania Press, 1969), 18. Friedrich Aspetsberger's voluminous study, *Welteinheit und epische Gestaltung: Studien zur Ichform von Hölderlins "Hyperion"* (Munich: Fink, 1971), touches only marginally on the issue of epistolary writing. When it does, it argues along the line of Ryan's "objective" proposition: the epistolary form is understood as a means to detach Hyperion from the confines of "inwardness" (*Innerlichkeit*) (274).

An interesting anomaly in the literature on *Hyperion* offers Jean-Luc Nancy's essay "La joie d'Hypérion," in *Hölderlin*, ed. Jean-François Courtine (Paris: L'Herne, 1989), 200–216. Nancy turns the widespread disregard for the epistolary character of Hölderlin's novel around and presents his meditation in the form of an epistolary correspondence among several participants, including one named "Bellarmin." The essay is an innovative gesture in the right direction. However, Nancy leaves his epistolary mimesis without a discussion of the epistolary form as such. Thinking and epistolary writing still entertain an arbitrary relationship in his essay, as is the case—albeit for different reasons—in the work of Ryan, Silz, and Aspetsberger.

6. Charles Baudelaire, "Le voyage," in *Oeuvres complètes*, ed. Claude Pichois (Paris: Gallimard [Pléiade], 1975), 1: 130.

7. Beißner suggests *The Statesman* (268e–274e) as Hölderlin's probable point of reference. See *SA* 3: 450.

8. Cf. Peter Szondi, "Hölderlins Überwindung des Klassizismus: Der Brief an Böhlendorff vom 4. Dezember 1801," in *Schriften*, ed. Jean Bollack et al. (Frankfurt am Main: Suhrkamp, 1978), 1: 345–66. Peter Szondi, "Hölderlin's Overcoming of Classicism," trans. Timothy Bahti, *Comparative Criticism* 5 (1983): 251–70. In later references to this work, the translation will be cited after a semicolon.

9. See Gerhart Mayer, who considers Hyperion to move along the "path of personal self-fulfillment." "Hölderlins *Hyperion*: Ein frühromantischer Bildungsroman," *Hölderlin-Jahrbuch* 19–20 (1975–77): 249. A different view of *Hyperion* as a *Bildungsroman* is offered by Ulrich Gaier, who treats the novel as a pedagogical text seeking to reform the "life of the reader." "Hölderlins *Hyperion*: Compendium, Roman, Rede," *Hölderlin-Jahrbuch* 21 (1978–79): 90.

10. Lawrence Ryan, "Hölderlins *Hyperion*: Ein 'romantischer' Roman?," in *Über Hölderlin*, ed. Jochen Schmidt (Frankfurt am Main: Insel, 1970), 177.

11. See Wilhelm Dilthey, *Das Erlebnis und die Dichtung. Lessing, Goethe, Novalis, Hölderlin* (Göttingen: Vandenhoeck & Ruprecht, 1970), 271–86.

12. Cf. *SA* 3: 463.

13. See Ryan, *Hölderlin's "Hyperion"*, 135–56; Aspetsberger, *Welteinheit und epische Gestaltung*, 42–81; and Nancy, "La joie d'Hypérion," 202–12.

14. In his "Anmerkungen zum *Ödipus*," from 1803, Hölderlin speaks even more radically about the "insane questioning for a consciousness." *SA* 5: 200; "Remarks on *Oedipus*," *Essays*, 105. Hyperion, too, compares his experience with that of Oedipus. See, for example, 153; 128–29.

15. My reading differs here from the one proposed by Dieter Henrich in his voluminous study *Der Grund im Bewußtsein: Untersuchungen zu Hölderlins Denken (1794–1795)* (Stuttgart: Klett-Cotta, 1992). Henrich's outline of the "philosophical conception" of Hölderlin's novel—which "explicitly leaves the interpretation of the text . . . to the work of others" (266)—seeks, in an almost classical way, to establish philosophical hegemony over a literary text. The notion that any successful reading of the novel "presupposes" (266) the construction of such a "philosophical conception" in fact takes up Hyperion's position before the fall, and is clearly refuted by Hölderlin's text itself.

16. Wolfgang Binder, "Hölderlins Namenssymbolik," *Hölderlin-Jahrbuch* 12 (1961–62): 162; trans. mine.

17. Immediately after the evocation of some past *jouissances*, Rousseau writes: "Indeed, true happiness is quite indescribable; it can only be felt, and the stronger the feeling the less it can be described, because it is not the result of a collection of facts but a permanent state." *Les confessions*, ed. Bernhard Gagnebin et al., *Oeuvres complètes*, ed. Bernhard Gagnebin and Marcel Raymond (Paris: Gallimard [Pléiade], 1959), 1: 236. Jean-Jacques Rousseau, *The Confessions*, trans. J. M. Cohen (Harmondsworth, England: Penguin, 1953), 224–25. See also the beginning of the sixth book of the *Confessions*, where Rousseau evokes another instance of the indescribable.

18. In this respect Hölderlin's hermit proves again very close to the occupant of the *Ermitage*. Thus Saint-Preux writes to Julie: "Besides, it is not the use of rhetorical figures, but the choice of them, which I blame in Marini. If a man has the least warmth of imagination, he must necessarily use metaphors and figurative expressions to make himself understood. Even your own letters are full of them, without

your knowing it; and I will maintain it, that none but a geometrician or a blockhead can talk without a metaphor." *Julie, ou La novelle Héloïse*, ed. Henri Coulet, annotated by Bernard Guyon, *Oeuvres complètes*, 2: 241. Jean-Jacques Rousseau, *Eloisa; or A Series of Original Letters*, trans. William Kenrick (Oxford: Woodstock, 1989), 2: 61. The coincidence between *Hyperion* and certain traits of *Julie* may not be accidental. In a letter to Neuffer of June 4, 1799, the year in which the second volume of *Hyperion* is published, Hölderlin considers writing an essay on Rousseau's epistolary novel for his projected monthly magazine. See *SA* 6; 1: 323.

19. Rousseau, "Discours sur les sciences et les arts," in ed. François Bouchardy, *Oeuvres complètes*, 3: 26.

20. Karlheinz Stierle, "Dichtung und Auftrag: Hölderlins Patmos-Hymne," *Hölderlin-Jahrbuch* 22 (1980–81): 50.

21. Ibid.

22. In his otherwise groundbreaking book from 1920, Georg Lukács thus starts out with assumptions similar to the characteristics of Greek art developed by Winckelmann one and a half centuries earlier. Once again, Greek antiquity is made to represent a time when "the soul does not yet know any abyss within itself." *The Theory of the Novel*, trans. Anna Bostock (Cambridge: MIT Press, 1971), 30. Hölderlin effectively suspends this highly romantic image when he claims "Junonian sobriety" as essential for the development of Homer's poetry.

23. A closer look at Szondi's essay may be instructive, since it allows me to outline some differences with respect to an interpretation that has influenced the study of Hölderlin for many years after its initial publication in 1964. Szondi writes: "Hölderlin perceives the origin of Greek art in the 'fire from heaven,' in 'holy pathos,' that of Hesperian art in the 'sobriety,' in 'clarity of representation.' Hölderlin also calls this natural origin the 'national,' that which is 'one's own' [*Eigene*]. . . . Greek art, for which Homer is seen as exemplary, is not nature, does not originate in itself. It rather has its origin in 'holy pathos.' "Hölderlins Überwindung des Klassizismus," 348; 255 (trans. modified). According to Szondi, Hölderlin recognizes in his first Böhlendorff letter two distinctively different origins for Greek and modern—or more precisely, Hesperian—art. This claim will play a central role in the later polemic against interpretations which construe an essential continuity between Hölderlin's view of the Greeks and his own contemporary efforts. Hölderlin, writes Szondi, radically breaks with the belief in such a continuity and identifies the "fire from heaven" and "holy pathos" as the origin of Greek art, and "soberness" and "clarity of representation" as the origin of Hesperian art.

An astonishing conclusion indeed, since there is, in fact, nothing in Hölderlin's letter to support such a claim. At the point where Szondi has Hölderlin write about the origin of art, the letter to Böhlendorff is concerned with something very different, namely with the elaboration of an *analogy* between what Hölderlin perceives to be originally natural for the Greeks and "us": "And I believe it is precisely the clarity of representation which is originally as natural for us as the fire from heaven for the Greeks." *SA* 6; 1: 425–26; *Essays*, 149 (trans. modified). Nothing in this sentence allows us to believe that Hölderlin regarded "clarity of representation" and "fire from heaven" as sufficient to serve as origin of art—Greek or non-Greek. "Clarity of representation" and "fire from heaven" figure as *gifts of nature*; a notion of Hölderlin's, to be sure, which stretches more familiar intuitions of what comes naturally. Clearly, Szondi was not prepared to follow Hölderlin's outline of Greek and Hesperian nature. This becomes evident once again when his reading of the Böhlendorff letter establishes an opposition between Greek nature and "holy pathos" in the course of an argument intended to state the origin of Greek art: "Greek art, for which Homer is seen as exemplary, is not nature, does not originate in itself. Rather, it has its origin in *holy pathos*." "Hölderlins Überwindung des Klazissismus," 348; 255. Szondi assumes here that "holy pathos" functions as the non-natural origin of Greek art. Again, the letter to Böhlendorff speaks an entirely different language. More than once, Hölderlin makes it a point to stress the originally natural character of the "holy pathos" for the Greeks; a few lines after his initial contention, he even insists that "holy pathos" must be regarded as something "inborn" (*angeboren*). For Hölderlin, then, "holy pathos" alone does not make Greek art.

While Szondi confuses the phenomenal description of nature and culture in Hölderlin's letter, he is more successful in his investigation the general structure of Hölderlin's argument. This is especially true for Hölderlin's careful qualification of linguistic and artistic nationalisms. Indeed, for Hölderlin, "Greek art . . . does not originate in itself." Szondi's philological and political insight—almost unparalleled in the Hölderlin scholarship of the 1950's and early 1960's—is clearly in evidence when he demonstrates that Hölderlin's break with Winckelmann's classicism was not conceived as a return to the homegrown virtues of native German art. It should be noted, in addition, that Szondi's was published in the same year, 1964, as Theodor W. Adorno's Hölderlin essay. See "Parataxis. Zur späten Lyrik Hölderlins," in *Gesammelte Schriften*, ed. Rolf Tiedemann (Frankfurt am Main: Suhrkamp, 1974) 11: 447–91. Szondi's recently published correspondence, in fact, indicates a close collaboration with Adorno. See, in particular, the long letter to Adorno, dated December 5, 1963, in *Briefe*, ed. Christoph König and Thomas Sparr (Frankfurt am Main: Suhrkamp, 1993), 135–42. Szondi's essay makes unmistakably clear that Hölderlin's dictum about that which is one's own—"But that which is one's own has to be learned as well as that which is foreign" (426; 150 [trans. modified])—effectively precludes any romantic idea of a return to some lost substantiality. (Parenthetically, it may be noted that Hölderlin capitalizes the word "foreign," *das Fremde*, while "that which is our own," *das eigene*, makes a more modest appearance in lower case—as if to underline the general precedence of the other.) As is true for *Hyperion*, too, the letter to Böhlendorff suggests that with respect not only to the other, but also to that which Hölderlin calls one's own, translation ("learning") is a necessary condition of understanding—a condition, moreover, which in some dimly understood way has always already taken place.

24. See Andrzej Warminski, "Hölderlin in France," in *Readings in Interpretation: Hölderlin, Hegel, Heidegger* (Minneapolis: University of Minnesota Press, 1987), 23–44. While Warminski does not notice some of Szondi's obvious misreadings, he focuses on what he perceives to be an "attempt at a dialectical mediation of Greeks and Hesperians, that which is proper and that which is foreign, *das Eigene* and *das Fremde*" (30). According to Warminski, Szondi assimilates Hölderlin to a "philosophy of history in dialectical terms: that is, in terms of an opposition between *das Eigene* and *das Fremde* mediated by a history of self-consciousness that contains the aesthetic, self-representation in art, as an essential (and sublatable) moment" (32). No doubt, the critical intervention would be crucial if substantiated. Warminski's "reading" of Szondi's "interpretation," however, fails on several accounts and in fact often proceeds by mere interpolation. He writes:

> Szondi explicitly formulates the scheme [of a Hegelian philosophy of history] in terms of Hegelian self and other later in the essay when he characterizes those things that can have a destructive effect on the poet—the given, positive, inherited, art of the Greeks—but which he should nevertheless not avoid, for "they are, in Hegel's terminology, his other, without which he is also not himself (*Sie sind, in Hegels Terminologie, sein anderes, ohne das er auch nicht er selber ist*)." (32)

Warminski's claim is misleading since it insinuates that Szondi's reference to Hegel takes place in the context of a discussion of Hölderlin's relation to "the given, positive, inherited, art of the Greeks." Nothing could be further from the truth. In the passage of the essay quoted by Warminski, Szondi recapitulates part of Hölderlin's letter to Neuffer from November 12, 1798. Neither in this letter, nor in Szondi's account of it is there any mention of the "art of the Greeks." Cf. *SA* 6; 1: 288–91 and "Hölderlin's Überwindung des Klazissismus," 363–65; 265–66, respectively. Szondi takes up the letter to Neuffer to underscore Hölderlin's conviction that, in order to achieve true liveliness (*Lebendigkeit*), the poet should seek to take advantage of those things that have had a destructive effect on his life. Szondi names these elements, experienced as destructive by Hölderlin, with two quotations from the letter to Neuffer: timidity in the face of the " 'common and ordinary in real life' " and, more generally, the " 'ice-cold history of the day.' " "Hölderlin's Überwindung des Klazissismus," 364; 265. These are topics very different from "the given, positive, inherited, art of the Greeks," that Warminski imagines Szondi to be writing about. It is, however, with reference to these topics alone that Szondi writes: "They are, in Hegel's terminology, his other, without which he is not even himself."

A compilation of various source materials is also at work when Warminski attempts to sustain the charge against Szondi a few lines later.

> Indeed, Szondi speaks of a "mirror-symmetry" (*Spiegelsymmetrie*) between Greek and Hesperian nature and art. On the basis of this "mirror-symmetry" he can interpret that which is our own (*das Eigene*) as Hegelian *das Positive* (given, immediate, abstract, etc.), the "freedom" of "the *free* use of *that which is our own*" (*der* freie *Gebrauch des* Eigenen) as meaning "with consciousness" (i.e., "with self-consciousness") and the task of the poet as "mediating the opposites" (*Vermittlung der Gegensätze*). ("Hölderlin in France," 33)

Warminski here mixes quotations from texts by Szondi, Hölderlin, and Hegel without proper indication of the source, an arrangement used to supply the evidence for Szondi's alleged Hegelianism. While Warminski has Szondi interpret the "task of the poet" with the help of a quote from Hegel as " 'mediating the opposites,' " Szondi himself defines the relation of antiquity and modernity in Hölderlin's Böhlendorff letter in terms of an "aporia"—a structural positioning of elements beyond the possibility of mediation. "Hölderlins Überwindung des Klassizismus," 346; 254. In fact, Szondi never uses the expression "mediation of the opposites" in the entire essay. He only speaks of *Rettung*—"rescue," "recovery," "salvation," "redemption"—to characterize Hölderlin's interpretation of the Greeks, a term borrowed from Walter Benjamin. Cf. Szondi's essay "Hoffnung im Vergangenen: Über Walter Benjamin," in *Schriften* 2: 293–94, where he refers to Benjamin's collection of letters, *Deutsche Menschen. Eine Folge von Briefen. Gesammelte Schriften*, ed. Rolf Tiedemann and Hermann Schweppenhäuser (Frankfurt am Main: Suhrkamp, 1972), 4; 1: 149–233. *Deutsche Menschen* also includes Hölderlin's second letter to Böhlendorff, with a short commentary by Benjamin (cf. 171–73). Benjamin published his collection in 1936, under a *nom de guerre*, Detlef Holz. Holz, it will be remembered, is another name for Hölderlin; Rudolf Magenau makes use of it in a letter to Neuffer in which he mentions Hyperion's love for freedom. Cf. *SA* 3: 296. Among other things, then, both philological and political, Szondi's notion of *Rettung* indicates a form of criticism whose dialectic remains negative and unsublatable—not unlike what Adorno termed "negative dialectic." Its bearing belongs so little in the tradition of a philosophy of self-consciousness that it could be taken, however falsely, to be its mere opposite. Cf. Jürgen Habermas "Bewußtmachende oder rettende Kritik: Die Aktualität Walter Benjamins," in *Zur Aktualität Walter Benjamins*, ed. Siegfried Unseld (Frankfurt am Main: Suhrkamp, 1972), 173–223. What Warminski's commentary does not take fully into account is precisely what Szondi's essay and some of Hölderlin's theoretical writings have in common: a guarded use of the terminology of German Idealism that nevertheless succeeds in suspending the systematic conclusions of its original philosophical formalization.

25. Szondi, "Hölderlins Überwindung des Klassizismus," 366; 266.
26. Ibid.
27. Warminski, "Hölderlin in France," 34.

28. The fact that Hölderlin later speaks about *Geschik*, not *Schiksaal*, does not necessarily indicate a meaning different from fate. Hölderlin often uses *Geschik* synonymously for *Schicksaal*. See, for example, the letter to Friedrich Wilmans from December 8, 1803: "Your kind attention has pleased me very much. I value it as a true and happy fate [*Geschik*] to have established a connection with you." *SA* 6; 1: 435.

29. The choice of a *Greek* term, including a latinized Greek spelling, in order to describe Hölderlin's view of the relation between Greek and Hesperian poetry is not unproblematic. It suggests a privileged position of Greek language, if not Greek thinking, in Szondi's and Warminski's essays at precisely the point when Hölderlin's letter to Böhlendorff puts such a privilege into question.

30. While it is true that Hölderlin praises Böhlendorff's play for its technical innovation—Böhlendorff treats the drama "in a more epic manner" (*epischer*)—it is also true that this hybrid of epic and dramatic elements is not something Böhlendorff could possibly have "in common" with the Greeks. In a study on the notion of narrative in Hölderlin's work, Eric L. Santer observes: "'Epic' in this context cannot have the same meaning as it did for Homer, whose point of departure was not insulation from *das Feuer vom Himmel*, but rather the exposure to it." *Friedrich Hölderlin: Narrative Vigilance and Poetical Imagination* (New Brunswick: Rutgers University Press, 1986), 74.

31. See, for example, Hölderlin's letter to Christian Gottfried Schütz from the winter of 1799–1800, where he takes issue with a conception that perceives Greek art "much more like a well calculated pleasure than a holy adherence to destiny [eine heilige Schiklichkeit], with which they [the Greeks] had to proceed in divine matters." *SA* 6; 1: 381.

32. Szondi, "Hölderlins Überwindung des Klassizismus," 365; 266 (trans. modified).

33. Warminski, "Hölderlin in France," 34.

34. Cf., e.g., Ibid., 29. The atemporal character of a reading that bases the transformation of nature and culture on the figure of a chiasmus becomes especially obvious when Warminski turns the chiasmus one more time around in order to extend its form to the constellation of nature and culture in Egypt. "In short, to put it as brusquely as possible, *we are the Egyptians*: that is, in terms of the first letter to Böhlendorff, the relation between nature and culture for us and the relation between nature and culture for the Egyptian is the same," "Endpapers: Hölderlin's Textual History," in *Readings in Interpretation*, 19. Less brusque, but no less reductive, is Philippe Lacoue-Labarthe proceeding when he writes about "the chiasmic structure that shapes history." "Hölderlin and the Greeks," trans. Judi Olson, in *Typography: Mimesis, Philosophy, Politics*, ed. Christopher Fynsk, intro. Jacques Derrida (Cambridge: Harvard University Press, 1989), 247. Both Lacoue-Labarthe and Warminski ignore that Hölderlin's reading of the chiasmus in the light of what he calls *Schiksaal* relates the formal structure of the reversal

toward its own decline. Both critics place the chiasmus "outside of fate"—a position Hyperion's *Schiksaalslied* reserves for the gods.

35. It is in this sense that Diotima may be called an "image." For a slightly different reading, see Marlies Janz, "Hölderlins Flamme: Zur Bildwerdung der Frau im 'Hyperion,'" *Hölderlin-Jahrbuch* 22 (1980–81): 122–42.

36. Silz, *Hölderlin's "Hyperion"*, 22.

37. "I have known what the Greeks did not: uncertainty." Jorge Luis Borges, "La lotería en Babilonia," in *Ficciones* (Buenos Aires: Emecé, 1956), 67. "The Babylon Lottery," in trans. Anthony Kerrigan, Jorge Luis Borges, *Ficciones*, ed. and intro. Anthony Kerrigan (New York: Grove Press, 1962), 65.

38. Silz, *Hölderlin's "Hyperion"*, 221.

39. Henrich, *Der Grund im Bewußtsein*, 215; trans. mine.

40. Ibid.

VILLANI *Figures of Duality: Hölderlin and Greek Tragedy*

1. Friedrich Nietzsche, *The Birth of Tragedy and The Case of Wagner*, trans. Walter Kaufman (New York: Vintage, 1967), 33.

2. The Nietzschean couple of Dionysus and Apollo is confirmed in recent studies by Marcel Detienne: a murderous Apollo, thirsty for sacrificial blood, patron of butchers, black and impure, doubles the Apollo of the rites of purification, who is luminously rational. In this way, the interchange of the two divinities is easy. See "Apollon de noir vêtu: Le bel homicide de Delphes" (unpublished manuscript).

3. This idea was emphasized by Schelling:

> A mortal destined by fatality to become a criminal, even struggling *against* fatality, and yet punished terribly for a crime that was the work of destiny! . . . [In this] struggle of human freedom with the power of the objective world, the mortal, when this power is overpowering, must necessarily succumb, and yet, since he does not succumb without a struggle, must be punished even for succumbing.

From *Philosophische Briefe über Dogmatismus und Kriticismus*, in *Ausgewählte Werke*, (*Schriften von 1794–1798*), 6: 216. See Philippe Lacoue-Labarthe, La césure du spéculatif," in *L'imitation des modernes*, Typographies II (Paris: Galilée, 1986), 39–69. "The Caesura of the Speculative," trans. Robert Eisenhauer, in Philippe Lacoue-Labarthe, *Typographies: Mimesis, Philosophy, Politics*, ed. Christopher Fynsk, intro. Jacques Derrida (Cambridge: Harvard University Press, 1989), 208–35.

4. Angelus Silesius, *Selections from the Cherubinic Wanderer*, trans. and intro. J. E. Crawford Flitch (London: Allen & Unwin, 1932), 1: 7. "I know that God cannot live one instant without Me: / If I should come to nought, needs must He cease to be" (1: 8); "I am as great as God, He is as small as I" (1: 9); "As much as you cede to God, / This much can He be for you" 1: 22).

5. There is no use returning to the famous chorus in *Antigone*, interpreted, in particular, by Martin Heidegger in *An Introduction to Metaphysics*, trans. Ralph Manheim (New York: Doubleday, 1961), and whose form is already the whole meaning, the "user's instructions."

6. "The Caesura of the Speculative" is, in this sense, a rich text. The author relies on Derrida's programmatic study *Glas* (Paris: Galilée, 1974); Jacques Derrida, *Glas*, trans. John P. Leavey, Jr., and Richard Rand (Lincoln: University of Nebraska Press, 1986). Page references are to the French edition first, then the English edition. "The logic of the *Aufhebung* (re)turns itself at each instant into its absolute other. Absolute appropriation is absolute expropriation. Onto-logic can always be reread or rewritten as the logic of loss or of spending without reserve" (188; 167). There is a hiatus between Hegel and Bataille, between spending and "sublation." Rather than supposing a "speculative dialectic" that Hölderlin could only have repeated in parodying, in the course of a long "sapping work" that would offer "no occasion to install any difference whatsoever" (Philippe Lacoue-Labarthe, *La césure du spéculatif* [Paris: Bourgois, 1978], 191), why not take off at once from the "suspension" in which Lacoue-Labarthe very correctly sees the spirit of Hölderlin's thought: "The closer it is the further away it is . . . the maximum appropriation . . . is the maximum depropriation" (*Hölderlin: L' "Antigone" de Sophocle*, followed by *La césure du spéculatif*, 216–17). By way of the radical economy of the dialectical process would emerge an original space of thought in which hesitation, suspense, balance, and "pure loss" would become effective—the return of Heraclitus.

7. G. W. F. Hegel, *Wißenschaft der Logik*, vol. 1 of *Encyklopädie der philosophischen Wißenschaften im Grundrisse*, 1817 ed. (Heidelberg: A. Osswald), paragraphs 192 and 193; 1830 ed. (Heidelberg: A. Osswald), paragraph 243. English trans.: *Hegel's Science of Logic*, trans. A. V. Miller (London: Allen and Unwin, 1969).

8. Cf. Jacqueline de Romilly, *La tragédie grecque* (Paris, PUF, 1970).

9. Cf. Louis Marin, *Utopiques: Jeux d'espace* (Paris: Minuit, 1973).

10. The French word *répons* refers to a song, drawn from scripture, in which the choir repeats part or all of a song sung by a soloist.—Trans.

11. A similar problem is posed with the origin of information and the debate between order and disorder, entropy, and negentropy. See Raymond Ruyer, *La cybernétique et l'origine de l'information* (Paris: Flammarion, 1954).

12. Edmund Husserl, *Cartesian Meditations*, trans. Dorion Cains (The Hague: Nijhoff, 1960), 23–24 (paragraph 10). It should be quite clear here that the crucial moment of the birth of the "positivities" in Greece is in no way in question. But these positivities are not born without a struggle: secret/promulgation, hierarchy/equality, family/city. . . . As Derrida says, "this war is not one war among others; it is *the* war." *Glas* 166; 146. This war opposes (Hegel) the law of singularity, which is feminine, divine, familial, natural, and nocturnal, and the law of the universal, which is virile, human, political, spiritual, and diurnal.

13. Franz Kafka, "Description of a Struggle," trans. Tanio and James Stern, in *The Complete Stories*, ed. Nahum N. Glatzer (New York: Schocken, 1971), 9–51.

14. Parmenides was inspired by a just *aidôs* when he made a caesura not between thought and being, but between "thought/Being" and "non-Being."

15. The English translations of all the plays cited have, for the most part, been retranslated from the French, which often differs significantly from existing English versions. However, I have consulted the translation of Aeschylus by Frederic Raphael and Kenneth McLeish found in Aeschylus, *Plays One* (London: Methuen, 1991), 337–41.—Trans.

16. The thematics of the bridge is recurrent, profound, and secret, and there are nice examples in the enigmatic oeuvre of Raymond Roussel, which takes care to join the poles of the duality. See my "De la métagrammatologie: Roussel et la philosophie contemporaine," in *Mélusine* (Lausanne: L'Âge d'homme, 1984), 221–43.

17. See Martin Heidegger, "Building Dwelling Thinking," in *Poetry, Language, Thought*, trans. Albert Hofstadter (New York: Harper and Row, 1971), 143–61, and the interview between Heidegger and Jean Beaufret in *Les lettres nouvelles* (Paris: Denoël, 1974), 11–39.

18. This is a figure of dichotomy and its normal outcome: A excludes B in order to respect the tautology A = A or to incorporate it. In the most pernicious case, A, under the cover of a pseudorivalry with B, established even more its tyranny of appearing double. Baudrillard showed this mechanism in *L'échange symbolique et la mort* (Paris: Gallimard, 1976).

19. See Pierre Fontanier, *Les figures de discours* (Paris: Flammarion, 1968): "This return of a name upon itself in these expression in a sacred style is a Hebraism: Slave of slaves, Canticle of canticles, Vanity of vanities. . . . And what energy, since thereby the name finds itself at the highest degree of signification [and expresses] the most exaggerated and most emphatic things we can imagine" (291). That is a textual mimesis of acts, a *μίμησις πράξεως* (Aristotle, *Ars poetica*, beginning of chapter 6). Aristotle was familiar with this procedure: *Poetics*, chapter 14: ἀδελφὸς ἀδελφόν. See also, *Rhetoric* 3.9: "The briefer the turn of phrase, the more antithetical, the more we savour it." An example: *τὸν ξένον ξένου.*

20. Through a merely acted suppression of differences—Yahweh passes along (nonetheless) to Moses, for whose stopped mouth Aaron fills in (for better and for worse), the people having been assimilated to the chosen people by their decimation (*grosso modo*)—*Exodus* aims at a paradoxical union of Yahweh and the Jewish people.

21. Parmenides, Heraclitus, and Aeschylus give analogous descriptions of men. In Heraclitus, they "fail to be aware of what they do after they wake up" and "know nothing of the real nature of gods and heroes." In Parmenides, *δίκανοι*, they wander in a total *ἀμηχαινή*.

22. *Glas*, 166; 147.

23. Beda Allemann, *Hölderlin und Heidegger*, 2d ed. (Freiburg: Atlantis, 1954).

24. The braid is a *symbolon* because it intersects two contrary movements. On the importance of braiding (στρέφειν) in the Greek imagination, see Marcel Detienne and Jean-Pierre Vernant, *Les ruses de l'intelligence* (Paris: Flammarion, 1974), 49–50.

25. These three terms seem to be thought in a unity by the Greeks. There are numerous exchanges between the two radicals that phonetics postulates to explain *time* (+ κωεῖ) and *poine* (+ κωεῖ).

26. Bifurcation is a triplicity in the Greeks: "O ye three roads, and thou secret glen,—thou coppice, and narrow way where three paths met," *Oedipus the King* 1198–99, 716, τριπλαῖς. In the "second family" of manuscripts, a diagram in the margins represents a *y*, the image of the dyad. We would need to look for the origin of the symbolic in this adherence of the one to the two. The cutting in two is also in *Oedipus at Colonus* 900–901: δίστομος. The same adjective qualifies Philoctetes' grotto with two openings (*Philoctetes* 18). See also *Ajax* 268 and 277 and *Libation-Bearers* 1415 for examples of duality.

27. The oracles, at this time, play cunningly with language in order to remain truthful: they are true only if we assume that there are two meanings at once. The relation between Truth and Error, however, is not as distinct in Sophocles as it will be in Plato: illusion is *Apate* as a positive ruse, that of Orestes and Ajax, just as much as it is negative, that of Hyllus and Ulysses.

28. *Oedipus the* King 353. Retranslated from the French.—Trans. There is a trace of a magical-religious *Aletheia* here. See Marcel Detienne, *Les maîtres de vérité dans la Grèce archaïque* (Paris: Maspero, 1979), 51–80.

29. *Trachiniae* 1136 (trans. modified).—Trans.

30. Bifurcation is the point of hesitation, possibility, and inversion. On the importance of inversion, see the end of *Philoctetes*; line 505 of the same play; the final lines of *Oedipus the King*; *Trachiniae* 109; *Oedipus at Colonus* 609; and *Ajax* 131. See also the Greek Sages: Hermann Diels and Walter Kranz, *Die Fragmente der Vorsokratiker* (Berlin: Weidmann, 1960), 1: 10 (73) a.

31. On *philos*, see Emile Benveniste, *Le vocabulaire des institutions indo-européennes* vol. 1 (Paris: Minuit, 1969). The *philia* of Ajax and Hector unites through the enemies' pact. See *Iliad* 6.230 and 7.303. On the poisoned gift, see Louis Gernet, *Droit et institutions en Grèce antique* (Paris: Flammarion, 1982), 17 and 35; and Gernet, *Anthropologie de la Grèce ancienne* (Paris: Flammarion, 1982), 36. See also Euripides' *Medea*, and the *pharmakon* of Medea's gift: a poison that mimics coupling by making bodies melt.

32. Concerning Derrida's well-known study of the *pharmakon*, one will notice that writing is at once a *doron* and a *pharmakon*.

33. In Fontanier's *Les figures de discours*, this is a "reversion" (381, note).

34. See Derrida: "The *Mehr* in question is neither equal nor unequal to any object; it is *as* nothing." *Glas*, 79; 67. Against this, see Plato *Philebus* 24c.

35. For Hegel, see "The Spirit of Christianity and its Fate," in *On Christianity: Early Theological Writings by Friedrich Hegel*, trans. T. M. Knox (New York: Harper, 1961), 182–301: "Objectively considered, then, the bread is just bread, the wine just wine; yet both are something more" (249). For Baudrillard, see *L'echange symbolique et la mort* (Paris: Grasset, 1983), 170–74, and *Les stratégies fatales* (Paris: Gallimard, 1976), 77.

36. *Trachiniae* 734 and 817. *Electra* 261: "I hate the mother who bore me more than anything else in the world."

37. See Heraclitus, *Fragmente der Vorsokratiker*, 2: 162, 164, and 169 (fragments 54, 60, and 80).

38. Cf. Fontanier, *Les figures de discours*, 352.

39. Heraclitus, *Fragmente der Vorsokratiker*, 2: 162 and 164 (fragments 54 and 62).

40. See Hölderlin, "In lieblicher Bläue," *SA* 2: 372–74.

41. See the discussion by Moses I. Finley and the current specialists of the "city" in antiquity. *Le monde d'Ulysse* (Paris: La Découverte, 1983), 235.

42. The two columns of *Glas* are consecrated, one to the family, the other to the *gl* of the blow. See Derrida, *Glas*, 164–96 (right column); 147–74.

43. See Heraclitus's fragment 103. *Fragmente der Vorsokratiker*, 2: 174.

44. Karl Reinhardt, *Sophokles* (1933; reprint, Frankfurt am Main: Klostermann, 1947), 11.

45. G. W. F. Hegel, *Aesthetics*, trans. T. M. Knox (Oxford: Oxford University Press, 1975). Hereafter cited as *Aesthetics*.

46. I do not have the space here to detail the incredible complexity of the symbolic machine of *Antigone*. Think, for example, that it is Antigone who makes of Creon a tragic hero.

47. We see this in *Oedipus at Colonus* 1414.

48. As Paul Mazon notes in his translation. Sophocles, *Antigone* (Paris: Belles Lettres, 1967), 65.

49. Nietzsche expresses this with the couple of Dionysus and Apollo, Parmenides with fragment 5 of the *Poem*.

50. *Heraclidae*, nonetheless starting like *Suppliants*, does not give an essential place to the dilemma. The inversions that end several of Euripides' plays are stereotyped. However, the couple of the mixed-up bodies of the bride and of the father in *Medea* and the tearing apart of Pentheus in *Bacchantes* recall the eminent place of duality in Greek thought (1127 and 1136).

51. *Aesthetics*, 2: 1037–38.

52. Ibid., 2: 960.

53. Ibid., 2: 1039.

54. Ibid., 2: 1214–15.

55. Cf. Ibid., 2: 1217–18.

56. Hölderlin, "Reflection," *SA* 4; 1: 233. *Friedrich Hölderlin: Essays and Letters on Theory*, trans. and ed. Thomas Pfau (Albany: State University of New York

Press, 1988), 45. All citations of Hölderlin's prose are from these editions. References to the English translations will appear after a semicolon as *Essays*. The translations are sometimes modified—Trans.

57. See Nietzsche, *Thus Spoke Zarathustra*, trans. R. J. Hollingdale (Harmondsworth, England: Penguin, 1961), 184.

58. Heraclitus, *Fragmente der Vorsokratiker*, 2: 152 (fragment 8).

59. The translation has been modified extensively to approach more closely the author's French translation, which brings out the doubleness (*zweifach*) more explicitly.—Trans.

60. The translations of *deinon* were, for Hölderlin, *Gewaltige*, then *Ungeheuer*, and, for Heidegger, *Unheimliche*.

61. The Aristotelian reading of *purifying* is not up to the exceptional level of Hölderlin's thought. *Furor* is not a *passion*.

62. An excessive but positive *aidôs* is recurved into infidelity and makes room for an infinite respect (Empedocles). An *aidôs* negative as fear devalorizes man and annuls the god; it is the fact of the "deteriorating priest" ("Death of Empedocles").

63. See the double interiorizing movement of the "negative communication" in Kierkegaard, *Concluding Unscientific Postscript to the Philosophical Fragments*, ed. and trans. Howard V. Hong and Edna N. Hong (Princeton, N.J.: Princeton University Press, 1992).

64. Cf. the letter to Neuffer on November 12, 1798 (*SA* 6; 1: 288–91).

65. Cf. Hölderlin's letter to his brother, January 1, 1799 (*SA* 6; 1: 302–7, esp. 306; *Essays*, 136–40; 139).

66. On this problem, see Friedrich Beißner, *Hölderlins Übersetzungen aus dem Griechischen* (Stuttgart: Kohlhammer, 1961), and the notes to Hölderlin's translation of *Antigone* by Lacoue-Labarthe, 155–81.

67. Jean Beaufret, "Hölderlin et Sophocle," introduction to *Remarques sur Oedipe et sur Antigone*, trans. François Fédier (Paris: Bourgois, 1965), 1: 30–31. Lacoue-Labarthe, 109, 173 (note on lines 987–88).

68. See Hölderlin's poem "Hälfte des Lebens" (*SA* 2; 1), and Fédier's note in Hölderlin, *Oeuvres* (Paris: Pléiade, 1967), 1233.

69. Savageness is an addition. Sophocles says "the graves of the dead" and Hölderlin "Wildniss des Gestorbnen." See Lacoue-Labarthe, 105, 172 (note on lines 956–7).

70. Hölderlin's letter to his brother (*SA* 6; 1: 302–4; *Essays*, 136–37).

71. Where the French translation reads *contour*, the English reads "character." I have therefore retranslated the final sentence from the French and rendered *contour* as "outline or contour" in order to make the author's argument clearer.—Trans.

72. Cf. Sophocles, *Antigone*, 744–45; Hölderlin, *Antigone*, *SA* 5: 203–62 and "Remarks on *Antigone*" (*SA* 5: 266–67; *Essays*, 111–12). Lacoue-Labarthe, 170 (note on lines 773–74).

73. *Antigone* 460 (retranslated from the French). Lacoue-Labarthe, 53 (note on line 467). Reinhardt and Schadewalt add further to what Hölderlin finds in these lines.

74. Empedocles, the unlimited being, entered the desert the day he pronounced himself equal to the gods. His *deinon* is precisely an interval in latency: "A terrible being inhabits him that metamorphoses everything" (*SA* 4; 1: 3), and "I flowered solitary from the time that the world still slept" (*SA* 4; 1: 5).

75. See Hölderlin's letter of December 4, 1801, and the letter to Böhlendorff (*SA* 6; 1: 425–8; *Essays*, 149–51).

76. Hölderlin, "In lieblicher Bläue" ("In Lovely Blueness") (*SA* 2; 1: 372). The author's French translation is a free rendering of the German, "dichterisch, wohnet der Mensch auf dieser Erde."—Trans.

77. Cf. Hölderlin's letters to Wilmans on September 28, 1803, and April 2, 1804 (*SA* 6; 1: 434; 438–39).

WARMINSKI *Monstrous History: Heidegger Reading Hölderlin*

1. Martin Heidegger, *An Introduction to Metaphysics* (New York: Doubleday, 1961), 123.

2. Martin Heidegger, *Hölderlins Hymne "Der Ister,"* Gesamtausgabe (Frankfurt am Main: Klostermann, 1984), 53: 70.

3. Ibid., 75.

4. Ibid., 79–80. And to talk "American" is according to these lectures of 1942 the worst thing one can do. What the lectures call *Amerikanismus* is bereft of history (*geschichtslos*) and unhistorical (*ungeschichtlich*), and even "Bolshevism" is only a degenerate form (*Abart*) of *Amerikanismus.*

5. Ibid., 80.

6. Ibid., 155.

7. Philippe Lacoue-Labarthe, "La césure du spéculatif," *L'imitation des modernes*, Typographies II (Paris: Galilée, 1986), e.g., 64. "The Caesura of the Speculative," trans. Robert Eisenhauer, in Philippe Lacoue-Labarthe, *Typographies: Mimesis, Philosophy, Politics*, ed. Christopher Fynsk, intro. Jacques Derrida (Cambridge: Harvard University Press, 1989), e.g., 231. See also "Hölderlin et les Grecs," *L'imitation des modernes*, 71–84. "Hölderlin and the Greeks," trans. Judi Olson, in *Typographies: Mimesis, Philosophy, Politics*, 237–47.

8. Heidegger, *Hölderlins Hymne "Der Ister,"* 100.

9. Ibid., 61.

10. Ibid., 60.

11. Ibid., 113–14.

12. We pluralize the translation of *Seiendes* (as "beings") in order to distinguish it (aurally) from Being.

13. Heidegger, *Hölderlins Hymne "Der Ister"*, 86.

14. Ibid., 87.

15. Ibid., 182.

16. Ibid., 179.

17. As quoted by Lacoue-Labarthe (and translated by Robert Eisenhauer) in "La césure du spéculatif," 64–65; 231–32. Cf. *FA* 16: 257.

18. Cf. Carol Jacobs, "The Monstrosity of Translation: Walter Benjamin's 'The Task of the Translator,' " in *Telling Time: Lévi-Strauss, Ford, Lessing, Benjamin, de Man, Wordsworth, Rilke* (Baltimore: Johns Hopkins University Press, 1993), 128–41.

19. Heidegger, *Hölderlins Hymne "Der Ister"*, 179.

20. Peter Szondi, "Hölderlins Überwindung des Klassizismus: Der Brief an Böhlendorff vom 4. Dezember 1801," in *Schriften*, ed. Jean Bollack et al. (Frankfurt am Main: Suhrkamp, 1978), 1: 345–66. Peter Szondi, "Hölderlin's Overcoming of Classicism," trans. Timothy Bathi, *Comparative Criticism* 5 (1983): 251–70. Cf. our reading of Szondi in "Hölderlin in France," in *Readings in Interpretation: Hölderlin, Hegel, Heidegger* (Minneapolis: University of Minnesota Press, 1987), 23–44.

21. I have done so in three previous essays on Hölderlin. See "Endpapers: Hölderlin's Textual History," "Hölderlin in France," and "Heidegger Reading Hölderlin," contained in *Readings in Interpretation*, 3–22, 23–44, and 45–71, respectively.

22. See, for example, the itinerary of the eagle—the eagle of history—in "Germanien": from the Indus, over the Parnassus, high above Italy, and then on to the Alps. *SA* 2; 1: 149–52.

23. See our "Endpapers: Hölderlin's Textual History," referred to in footnote 21, for more on this impossible "identification" as well as the scheme below.

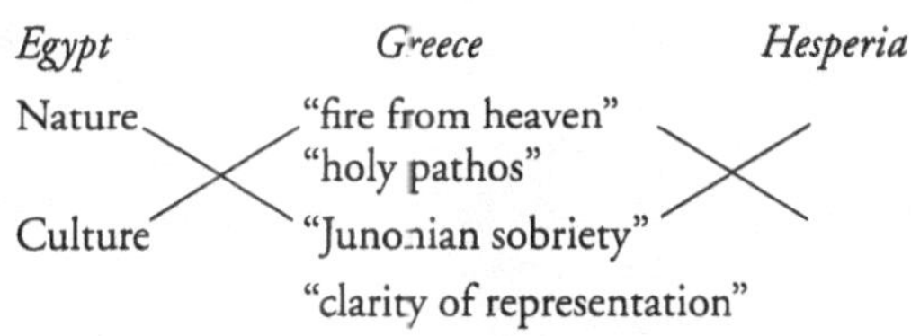

24. Maurice Blanchot, *De Kafka à Kafka* (Paris: Gallimard, 1981), 94–131.

25. "Monstrous" how? Most abruptly: as the allegory of a disjunction of self and other that cannot be mediated by *any* history of sense or sense-making.

26. Paul de Man, "The Riddle of Hölderlin," in *Critical Writings, 1953–1978*, ed. Lindsay Waters (Minneapolis: University of Minnesota Press, 1988), 211.

27. Ibid., 212.

28. Perhaps this is the direction in which Heidegger thinks in a remark of the "Letter on Humanism" from 1946: "We have still hardly begun to think the mysterious relations to the East, that were put into words in Hölderlin's poetry." *Platons Lehre von der Wahrheit: Mit einem Brief über den "Humanismus"* (Bern: Francke, 1947), 85. Something like an "Oriental," "Egyptian," reading of Heidegger—these words being read in a Hölderlinian sense—is Jacques Derrida on Heidegger on Trakl in his Paris seminar of spring 1985, as well as in the essay "La

main de Heidegger (*Geschlecht* II)," in *Psyché: Inventions de l'autre* (Paris: Galilée, 1987), 415–51. Jacques Derrida, "*Geschlecht* 2: Heidegger's Hand," trans. John P. Leavey Jr., in *Deconstruction and Philosophy: The Texts of Jacques Derrida*, ed. John Sallis (Chicago: University of Chicago Press, 1987), 161–96.

CORNGOLD *Disowning Contingencies in Hölderlin's 'Empedocles'*

1. Franz Kafka, *Beim Bau der chinesischen Mauer und andere Schriften aus dem Nachlaß, Gesammelte Werke in zwölf Bänden*, ed. Hans-Gerd Koch (Frankfurt: Fischer, 1992), 6: 237. *Dearest Father*, trans. Ernest Kaiser and Eithne Wilkins (New York: Schocken, 1954), 40 (trans. modified).

2. Friedrich Hölderlin, letter to Christian Ludwig Neuffer, July 3, 1799. *FA* 12: 18.

3. See the section "The Deaths of Empedocles," in Andrzej Warminski, *Readings in Interpretation: Hölderlin, Hegel, Heidegger* (Minneapolis: University of Minnesota Press, 1987), 11–17.

4. Jürgen Söring, "Hölderlin's Empedokles: Die Tragödie eines Trauerspiels," in *Bad Homburger Hölderlin-Vorträge 1988–89* (Bad Homburg: Stadt Bad Homburg, 1989), 78.

5. Thomas Mann, *Doctor Faustus* (New York: Vintage, 1948), 502.

6. The word *verläugnen* stands in a critical relation to Hegel's concept of *aufheben* ("to sublate"): Hölderlin's *verläugnen* does foreshadow the antithetical doubleness of this intellectual activity that negates while chiefly preserving and transcending its object—indeed, that negates in order to preserve and transcend its object. But Hölderlin's *Verläugnung* has the more markedly negative connotation, evoking the compulsive, unwilling negative doubleness with which we are better acquainted in Freud's *Verneinung*, "denegation."

7. Indeed in the third version we learn of a wisdom called "Egyptian" stating "that we mortals, just as we strike the eye, are only signs and images" (*FA* 13: 940). If Hölderlin, through Empedocles, does not appear to treat this wisdom with the same earnestness he recommends to his disciple, nonetheless in the "General Basis" he wants to stress that we are at least images. Hence, any attribution of solidity, validity, to what is visibly given is as fraught with risk as it is inescapable.

8. Such is the mood, certainly, of the first version, where Empedocles' adversaries Critias and Hermocrates function as stereotypes without dialectical power. The mood of the third version, which proposes a genuine antagonist, is incipiently different, but the antagonist never actually appears.

9. Glenn Most interestingly connects the "hyperpolitical and calculating attitude" of the Agrigentians with the Roman character, especially as it is condensed in the figure of Jupiter in Hölderlin's ode "Nature and Art or Saturn and Jupiter." The contrast of the Greek Empedocles with the Roman-like Agrigentians dramatizes the distinction between essential and contingent: only the Greek enjoys a fellowship with essence. "The Bait of Falsehood: Studies in the Rhetorical Strategy

of Poetic Truth in the Romantic Period" (Department of Comparative Literature, Yale University, 1980), 259.

10. One could see in this image the dances of Corybantes, a sort of intermediate image between the most contingent of human affairs and contact with the god himself.

11. Or if it is to be claimed that it is not so, then there must be an argument that shows why the opposition between essential and accidental is not itself essential. Certainly a tragic *play* must not fail to empower the image of interpersonal conflict, but what Hölderlin is inclined to call "accidentals" is precisely what is visible as action and hence playable on stage. As we well know, Hölderlin was bent on representation.

12. Pierre Bertaux, *Friedrich Hölderlin* (Frankfurt am Main: Suhrkamp, 1981), 334.

13. It would, moreover, involve a sort of immediacy quite at odds with the semiotics of the *Grund*.

14. Peter Szondi, *Satz und Gegensatz* (Frankfurt: Insel, 1964), 9–10.

15. Both the field that situates the correspondence and the way in which this field is realized by the subject change.

16. This reading opens up the passage for a creative use of the freed subject position. An alternative reading—for many native readers having a more "natural" feel—permits an archaic substantivizing of the adjective *genug*, "enough," on the model of *Genug der vielen Worte* ("enough palaver")—hence, as meaning "a sufficiency"—and understands the closing two lines as follows: "Yet taking revenge, of empty echoing / A sufficiency then aped from the dead breast the foolish one." Yet Grimm gives nothing like it. See "*Genug* as *Genüge*," in *Deutsches Wörterbuch* (Leipzig: Hirzel, 1885), 4; 1–2: 3503. Even this more "natural" reading is a *hapax legomenon* and by any account tortured. My own preference is for the first reading: it cannot be ruled out grammatically, and it produces the more interesting result.

17. Theodor W. Adorno, "Parataxis: Zur späten Lyrik Hölderlins," in Jochen Schmidt, ed., *Über Hölderlin* (Frankfurt am Main: Insel, 1970), 358.

18. Jürgen Söring, *Die Dialektik der Rechtfertigung: Überlegungen zu Hölderlins Empedokles-Projekt* (Frankfurt am Main: Äthenäum, 1973), 204.

19. Warminski, *Readings in Interpretation*, 13.

20. This undecided state of affairs is one which Warminski has himself amply shown to be the major thrust of "Patmos." See the chapter "'Patmos': The Sense of Interpretation," in *Readings in Interpretation*, 72–92.

21. Walter Benjamin, "Zwei Gedichte von Friedrich Hölderlin: 'Dichtermut'—'Blödigkeit,'" in *Gesammelte Schriften*, ed. Rolf Tiedemann and Hermann Schweppenhäuser (Frankfurt am Main: Suhrkamp, 1977), 2; 1: 115.

22. This *Bildung* recurs finally in the medium of the word. The "word" is represented as in itself an unproblematic whole; but as readers of Hölderlin it

is hard to suppress our knowledge of his semiotic knowledge. The passage suppresses the materiality of the signifier (*Zeichen*)—a suppression that constitutes an inexplicit division. Now, in a way something like the earlier passage describing the mocking self-doubling of Empedocles, these lines become a small cosmos of fusions and bare distinctions that does not come to rest.

23. Söring, *Die Dialektik*, 208.

24. Eric Santner ably discusses this development. See *Friedrich Hölderlin: Narrative Vigilance and the Poetic Imagination* (New Brunswick, N.J.: Rutgers University Press, 1986).

25. Benjamin, "Zwei Gedichte," 121.

26. Gerhard Kurz, "Poetik und Geschichtsphilosophie der Tragödie bei Hölderlin," *Text und Kontext* 5, no. 2 (1977): 15–36.

FYNSK *Reading the 'Poetics' After the "Remarks"*

1. The German, even more complex by reason of the syntactical play of the reflexive *sich begreifen* (which prompted me to speak of "absolute reflection"), reads as follows: "Die Darstellung des Tragischen beruht vorzüglich darauf, daß das Ungeheure, wie der Gott und Mensch sich paart, und grenzenlos die Naturmacht und des Menschen Innerstes im Zorn Eins wird, dadurch sich begreift, daß das grenzenlose Eineswerden durch grenzenloses Scheiden sich reiniget. *Tes physeos grammateus en ton kalamon apobrechon eunoun.*" "Anmerkungen zum *Ödipus*," *SA* 5: 201. The Greek phrase is from a Byzantine encyclopedia of the tenth century falsely attributed to Suidas of Athens. As Thomas Pfau notes in his edition of the "Remarks," Hölderlin has altered the phrase by changing the Greek *eis noun* to *eunoun*, giving "Nature's scribe, dipping the well-meaning reed" instead of "dipping the reed into meaning." Friedrich Hölderlin, *Essays and Letters on Theory* (Albany: State University of New York Press, 1988), 174. (Pfau translates "quill" rather than "reed," but *kalamon* clearly indicates the latter, as does evidence regarding Greek writing practices in the fifth century). *Eunoun* is perhaps most difficult to translate here; "benevolent" is appropriate, but there is also a suggestion of sound understanding.

2. Beyond the very concern with tragic *Darstellung*, I am thinking principally of Hölderlin's reworking of the tragic *metabole* and the consequent disruption of Aristotle's famous "organicist" description of the relation between beginning, middle and end (see *Poetics* 1450b.26). At the moment of the categorical turn, Hölderlin writes, "Man forgets himself because he is entirely within the moment, the god because he is nothing but time; and on both sides there is infidelity, time, because in such moments it veers categorically and beginning and end can no longer rhyme with one another at all; man, because in this moment he must follow the categorical reversal and hereby cannot at all resemble the beginning in what follows." "Anmerkungen zum *Ödipus*," 202.

3. Peter Szondi, *Versuch über das Tragische* (Frankfurt am Main: Insel, 1961).

"The Notion of the Tragic in Schelling, Hölderlin, and Hegel," in Peter Szondi, *On Textual Understanding and Other Essays*, ed. Michael Hays, trans. Harvey Mendelsohn (Minneapolis: University of Minnesota Press, 1986), 43–55.

4. Important groundwork for this question is provided by Philippe Lacoue-Labarthe in his two essays "La césure du spéculatif" and "Hölderlin et les Grecs," *L'imitation des modernes*, Typographies II (Paris: Galilée, 1986), 39–69 and 71–84. "The Caesura of the Speculative," trans. Robert Eisenhauer, and "Hölderlin and the Greeks," trans. Judi Olson, in Philippe Lacoue-Labarthe, *Typographies: Mimesis, Philosophy, Politics*, ed. Christopher Fynsk, intro. Jacques Derrida (Cambridge: Harvard University Press, 1989), 208–35 and 237–47.

5. I am referring, of course, to the opening pages of the "Remarks on *Oedipus*" (as well as to the corresponding opening of the "Remarks on *Antigone*") where Hölderlin describes a kind of calculus of human finitude (". . . how a system of sensibility, the entire human being develops as when under the influence of the element") that is characterized in tragedy by an "equilibrium" and represented (in the sense of *Darstellung*) as follows ("Anmerkungen zum *Ödipus*," 196):

> For the tragic *transport* is properly empty and the most unbound.
>
> Thus, in the rhythmic succession of representations wherein the *transport* presents itself, there becomes necessary what in meter is called the caesura, the pure word, the counter-rhythmic interruption, in order to meet the racing exchange of representations at its summit, in such a way that it is then no longer the exchange of representations that appears, but representation itself.
>
> Thereby, the succession of the calculus and the rhythm are divided, and relate to one another in such a way that their two halves, as of equal weight, appear.

6. I will cite the translation of the *Poetics* by Ingram Bywater collected in *The Basic Works of Aristotle*, ed Richard Mckeon (New York: Random House, 1941). Citations of the *Politics* will be drawn from the same volume.

7. "May [music] not have also some influence over the character and the soul? It must have such an influence if characters are affected by it. And that they are so affected is proved in many ways, and not least by the power which the songs of Olympus exercise; for beyond question they inspire enthusiasm, and enthusiasm is an emotion of the ethical part of the soul. Besides, when men hear imitations, even apart from the rhythms and tunes themselves, their feelings move in sympathy. . . . Rhythm and melody supply imitations of anger and gentleness, and also of courage and temperance, and of all the qualities contrary to these, and of the other qualities of character, which hardly fall short of the actual affections, as we know from our own experience, for in listening to such strains our souls undergo a change." *Politics*, 1340a.7–20.

8. *Politics*, 1341a.21: "Besides, the flute is not an instrument which is expressive of moral character; it is too exciting. The proper time for using it is when

the performance aims not at instruction, but at the relief of the passions." *Politics* 1342a: "In education the most ethical modes are to be preferred, but in listening to the performances of others we may admit the modes of action and passion also. For feelings such as pity and fear, or, again, enthusiasm, exist very strongly in some souls, and have more or less influence over all. Some persons fall into a religious frenzy, whom we see as a result of the sacred melodies—when they have used the melodies that excite the soul to mystic frenzy—restored as though they had found healing and purgation. Those who are influenced by pity or fear, and every emotional nature, must have a like experience, and others in so far as each is susceptible to such emotions, and all are in a manner purged and their souls lightened and delighted. The purgative melodies likewise give an innocent pleasure to mankind."

9. Summaries of this research abound. See, e.g., Stephen Halliwell's *Aristotle's Poetics* (London: Duckworth, 1986), 350–56. For Gerald Else, see *Aristotle's "Poetics": The Argument* (Cambridge: Harvard University Press, 1967).

10. Roselyne DuPont-Roc and Jean Lallot, eds., *La poétique* (Paris: Seuil, 1980).

11. DuPont-Roc and Lallot speak of an "expérience émotive épurée." *La poétique*, 193.

12. I use the term *aesthetic* here in view of the modern appropriation of Kant's reflections on the apprehension of form. DuPont-Roc and Lallot refer in this respect to a "contemplation of represented form" and describe catharsis as being of an "aesthetic nature" (*La poétique*, 193).

13. Dupont-Roc and Lallot are inclined to attribute musical catharsis to a mere neutralization by aesthetic pleasure of the pain music produces homeopathically as a mimetic form. A "more immediate" and "less elaborate" form of mimesis than theater, its cathartic effect they acknowledge, may be conceived by "analogy" with that of the latter, where catharsis results from a "contemplation of represented forms." *La poétique*, 193. But the pleasure afforded by musical catharsis is attributed principally to music's "hedonic" effects. This argument is difficult to square with the lines I have cited above (note 8). But more significant, I believe, is the fact that the explanation offered by Dupont-Roc and Lallot fails to account for the important assertion that feelings "move in sympathy" with music. Music, as such, may well play a supplementary role in tragedy, most importantly where it reflects character and action (character itself playing a secondary role in the effect "proper" to tragedy), but also where it contributes to the delight in the representation as a kind of "seasoning." Yet the question I want to raise is whether the mainspring of catharsis is not a kind of rhythmical movement engaged by both theatrical representation and music—that is, whether catharsis is not fundamentally a matter of sympathetic movement and some form of rhythmic ordering of feelings. Such a notion would provide the link that Else finds lacking when he writes, "I can see no bridge leading from the structural-objective concept of the

Poetics to the therapeutic subjective concept of the *Politics*." *Aristotle's "Poetics": The Argument*, 441. Of course, this explanation would not exclude the dimension of pleasure defined by Dupont-Roc and Lallot.

14. Dupont-Roc and Lallot, *La poétique*, 283.

15. Ibid., 284.

16. Emile Benveniste, *Problèmes de la linguistique générale* (Paris: Gallimard, 1986), 329. Emile Benveniste, *Problems in General Linguistics*, trans. Mary Elizabeth Meek (Coral Gables, Fla.: University of Miami Press, 1971), 17.

17. "It [time] is always more [calculable] when it is counted in suffering." Hölderlin, "Anmerkungen zur *Antigonä*," 268.

NÄGELE *Ancient Sports and Modern Transports: Hölderlin's Tragic Bodies*

1. "The Earth resounds with metaphors." "Нашедший подкову," Osip Mandelstam, *Стихотворения* (Moscow, 1928).

2. The translations are generally my own. In the case of the theoretical writings, I have consulted Thomas Pfau's translation, but modified it wherever necessary. See Friedrich Hölderlin, *Essays and Letters on Theory*, ed. and trans. Thomas Pfau (Albany: State University of New York Press, 1988). The comparison between tragedy and athletic games is taken up by Nietzsche again, who posits an analogy between the "creator of the ancient musical drama" (Nietzsche's concept of Greek tragedy) and the *Pentathlos*, the competitor in five events of the contest: wrestling, long jump, running, throwing the discus, and throwing the javelin. See "Das griechische Musikdrama," *Sämtliche Werke*, Kritische Studienausgabe (Munich and Berlin: DTV / de Gruyter, 1980), 1: 531.

3. In the case of Hölderlin, this has been done most systematically and with great seriousness by Rolf Zuberbühler. See *Hölderlins Erneuerung der Sprache aus ihren etymologischen Ursprüngen* (Gießen: Schmidt, 1969).

4. In lines 65 and 69 of the elegy "Brod und Wein."

5. Lacan points at the linguistic doubling of the "thing" in the constitution of the object in the French *chose* (from Latin *causa*) and the *rien* (from Latin *rem*). See "L'instance de la lettre dans l'inconscient," in *Ecrits* (Paris: Seuil, 1966), 498.

6. Friedrich Nietzsche, "Über Wahrheit und Lüge im außermoralischen Sinne," in *Sämtliche Werke*, 1: 873–90.

7. French critics played a major role in the rereading of this essay and in a general semiological and rhetorical turn. The issue of *Poétique* 5 (1971), with essays by Derrida, Lacoue-Labarthe, and Kofman, was as seminal as the ensuing Nietzsche colloquium of Cérisy-la-Salle in 1972.

8. Nietzsche, "Über Wahrheit und Lüge," 880–81.

9. Jürgen Habermas, *Der philosophische Diskurs der Moderne* (Frankfurt am Main: Suhrkamp, 1985).

10. Not even the pages of the *New York Review of Books* recoil in shame when Geoffery Hartman is condescendingly patted on the shoulder and assured that

his long-time occupation with the Holocaust saves him, despite some taintedness with deconstruction, from the idolatry of language and words. Perhaps it has never occurred to these literary rhetoricians and their moral pathos that the Holocaust was an attempt at an annihilation of a people and culture where the letter stood and stands in highest regard, and that it was not least this reverence for the letter that fed the rhetoric of Christian anti-Semitism since St. Paul and St. Augustine.

11. The word *tragedy* already inscribes the sexual body in language as *tragodia*, or "goat song," as the song of the phallic body of the satyr-goat. In the "swell-foot" Oedipus, the phallic destiny coins the form of the paradigmatic tragedy.

12. Another poetological essay on the change of tones in the three genres of the lyrical, epic, and tragic is also written from back to front.

13. Peter Szondi, "Versuch über das Tragische," in *Schriften*, ed. Jean Bollack et al. (Frankfurt am Main: Suhrkamp, 1978), 1: 151. See "The Notion of the Tragic in Schelling, Hölderlin, and Hegel," in Peter Szondi, *On Textual Understanding and Other Essays*, ed. Michael Hays, trans. Harvey Mendelsohn (Minneapolis: University of Minnesota Press, 1986), 43–55.

14. See, for example, "Das lyrische dem Schein nach idealische Gedicht . . . ," *FA* 14: 369–72. Emil Staiger made this linguistic shift the principle of his *Poetik*, defining the conventional three genres of *Lyrik*, *Epik*, and *Drama* as *das Lyrische*, *Epische*, and *Dramatische* in terms of a fundamental difference in the temporality of being.

15. It is this quality that forces Habermas into his construction of a philosophical discourse of modernity that must exile all the major philosophical texts of Modernism into the construct of postmodernism.

16. For a more extended discussion of this transformation see the chapter on "From Aesthetics to Poetics: Brecht, Benjamin and the Poetics of the Caesura," in Rainer Nägele, *Theater, Theory, Speculation: Walter Benjamin and the Scenes of Modernity* (Baltimore: Johns Hopkins University Press, 1991), 135–66.

17. Grammarians differentiate the dative as the indirect object case, referring to the "partner in a verbal act," from the accusative as the direct, "transitive" object case. The name itself is based on the Latin paradigm of *dare* and thus places the object in the symbolic exchange of the gift. Johannes Erben, *Deutsche Grammatik* (Frankfurt am Main: Fischer, 1968), 112.

18. *Fühlbar*—literally "feelable," also "tangible"—is a word recurring in Hölderlin's poetological texts, especially in the essay "Wenn der Dichter einmal des Geistes mächtig . . ." and the text on difference of poetic genres (see *FA* 14: 371). It refers to an effect of the poetic operations that has little to do with so-called spontaneous feelings (which, according to Rilke's Malte, are the property of any dog, but do not make poetry) and more with the complex texture of experience and (re)presentation (*Darstellung*) that constitutes Hölderlin's poetic work at the threshold of Modernism.

19. Jürgen Söring, *Die Dialektik der Rechtfertigung: Überlegungen zu Hölderlins Empedokles-Projekt* (Frankfurt am Main: Athenäum, 1973),1.

20. In paragraph 166 of his *Enzyklopädie der philosophischen Wissenschaften*, Hegel makes a similar etymological argument: "Die *etymologische* Bedeutung des *Urteils* in unserer Sprache ist tiefer und drückt die Einheit des Begriffs als das Erste, und dessen Unterscheidung als die *ursprüngliche* Teilung aus, was das Urteil in Wahrheit ist." *Werke in zwanzig Bänden*, ed. Eva Moldenhauer and Karl Markus Michel (Frankfurt am Main: Suhrkamp, 1970), 8: 316.

21. As if in response to Hölderlin's entrance into the tragic through the arch-separation of the *Urtheil*, Benjamin's book on the *Trauerspiel* ends in the last chapter as a *ponderacion misteriosa* on the constellation of allegory, *Urteil*, and signification. See *Gesammelte Schriften*, ed. Rolf Tiedemann and Hermann Schweppenhäuser (Frankfurt am Main: Suhrkamp, 1977), 1; 1: 406–9.

22. The first outline of the *Empedocles* project, the so-called *Frankfurter Plan*, bases the radical disaccord between Empedocles and his world on the law of succession: "because he . . . is bound to the law of succession" (*FA* 13: 544).

23. This is a slightly simplified rendering of the notations as presented in *FA* 4: 134–35: "Everything is intimately connected—This separates/divides—Thus the poet hides/veils/protects—Frivolously bold one, if you would want to see from face to face the soul—you would go under in flames."

24. Entitled "Das lyrische dem Schein nach idealische Gedicht . . ." in Sattler's edition. See *FA* 14: 343–72.

25. For a systematic exploration of Hölderlin's doctrine of the alternation of tones, see Lawrence Ryan, *Hölderlins Lehre vom Wechsel der Töne* (Stuttgart: Kohlhammer, 1960).

26. This is one of the forms in which Hölderlin notes the permutations of the tones: with capital letters for the manifest tone and small letters for the "ground."

27. These permutations are sketched both in quasi-mathematical tabulations (e.g., *FA* 14: 336–37) and descriptively (e.g., in the first section of "The Ground for Empedocles," *FA* 13: 868, concerning the tragic ode).

28. Albrecht Seifert makes a similar observation on Hölderlin's use of *metaphor*: "Metaphor is used by Hölderlin in an extremely wide sense close to the Greek meaning: transfer, transposition, transport into a beyond, into something opposite." *Untersuchungen zu Hölderlins Pindar-Rezeption* (Munich: Fink, 1982), 162.

29. Hölderlin's tabulation of the sequence of tones indicates precisely such an inversion when, after the third step, the manifest tone of the previous step enters as the ground of the following one.

30. *Innigkeit* is a difficult term to translate. It corresponds to the Latin *intimus*, the superlative of interior, "innermost," and as such it has been reduced more and more to its metaphorical connotation of an intimate, ardent feeling. As the superlative of the spatial position of interiority it designates the thresh-

old and limit where spatiality is condensed to its point of nonextension. At this limit, it designates the point of the pure differential between literal spatiality and metaphor. While much of Romantic poetry tends to erase the literalness of spatiality, Hölderlin emphasizes precisely the spatial condensation in the contraction of total unity which produces at the same time the necessity for spatial expansion in *Darstellung*. It is a linguistic and poetic big bang theory, where an original infinite condensation and concentration explodes into the galaxies and constellations of texts.

31. German differentiates between *der Gehalt* ("substantial," "significant content") and *das Gehalt* ("salary," "wages") where the content is again marked by the signifier of money exchange.

32. Proverbs like *Wer nicht hören will, muß fühlen* leave no doubt about the painful tactile consequences of disobedience to the word.

33. In his second letter to Böhlendorff, Hölderlin refers to himself as one who has been hit and beaten by Apollo: "daß mich Apollo geschlagen" (*SA* 6: 432).

34. See also the first chorus part of *Oedipus* (line 167): *phoibon hekabolon*, which Hölderlin translates as "den Phöbus fernhin treffend" (*FA* 16: 97).

35. For the Pindaric connotations and their traces in Hölderlin's poetry, see Seifert, *Pindar-Rezeption*, 210–11, note 27.

36. *Darstellung* in the sense of "sacrifice" is still common usage at Hölderlin's time. Goethe uses it twice in this sense in *Iphigenie auf Tauris* (lines 843 and 1221).

37. Oedipus, at one point, refers to the murder of Laius as his becoming invisible, *aphantos errei*, translated by Hölderlin as *Unsichtbar ward er* (line 567).

38. This delayed appearance of the dramatic hero on the stage is of course a not uncommon mode of exposition, particularly popular in German drama of the later eighteenth century, where Goethe's *Götz von Berlichingen* provided a famous example.

39. The *selber* is grammatically ambiguous; it could refer either to "you" or to "him," thus blurring the demarcations of identity—indeed an effect of Empedocles on those close to him.

40. Line 25. In line 682, Hölderlin translates the word more literally as *welkend* ("wilting," "fading").

41. G. W. F. Hegel, *Werke in zwanzig Bänden*, 3: 37.

42. The metaphorical difference of fire and water in Hölderlin's and Hegel's presentation of the prelinguistic substratum and linguistic medium deserves a separate analysis.

43. René Girard, *Violence and the Sacred* (Baltimore: Johns Hopkins University Press, 1977), 49.

44. Nietzsche, "Zur Genealogie der Moral," in *Sämtliche Werke* 5: 264–65.

45. The notion of "transport," as Hölderlin uses it in the notes to Sophocles, may well resonate with its French connotations. According to the *Dictionnaire Furetière*, it indicates not only an "[a]ction par laquelle on fait changer de lieu

à quelque chose," but also a medical condition of delirium: "Quand le fiévre est violente, on apprehende le *transport* au cerveau qui cause le delire. Quand le *transport* des humeurs se fait sur une partie, elle l'affige, et souvent la rend impotente." Above all, it designates an excess in the economy of feeling and passion: "Transport, se dit aussi figurément en choses morales, du trouble ou de l'agitation de l'ame par la violence des passions. . . . On dit aussi, un *transport*, une fureur poëtique." Finally, it is the metaphor of metaphor itself, the work of figuration, that is, the transport from literal to figurative meaning: "La metaphore est un *transport* d'un mot propre à un sens figuré." Antoine Furetière, *Le dictionnaire universel* (Paris: S.N.L.-Le Robert, 1978).

FIORETOS *Color Read: Hölderlin and Translation*

1. Edward Young, *Night Thoughts, The Complete Works*, ed. James Nichols (1854; reprint, Hildesheim: Olms, 1968), 6.

2. This is, obviously, a paraphrase of Paul de Man's account of Benjamin's argument in the preface to his translations of the *Tableaux parisiens*. See "Conclusions: Walter Benjamin's 'The Task of the Translator,'" in *The Resistance to Theory*, intro. Wlad Godzich (Minneapolis: University of Minnesota Press, 1986), 82.

3. Goethe, *Maximen und Reflexionen, Werke*, Hamburger Ausgabe, ed. Erich Trunz (Munich: Beck, 1973), 12: 499.

4. Goethe, "Noten und Abhandlungen zu bessern Verständnis des west-östlichen Divans," *Werke*, 5: 255. There are passages in Goethe's work, for example in his preparatory notes to *Kunst und Altertum*, dealing with "European, i.e., world literature," where he suggests that German is, in a sense, the final language or final destination of things Greek. See also Eckermann's account of the conversation on January 10, 1825. For a treatment of these, and related, issues, see, for instance, Franz Strich, *Goethe und die Weltliteratur*, 2d rev. ed. (1946; Bern: Francke, 1957). Fritz Strich, *Goethe and World Literature*, trans. C. A. M. Sym (New York: Hafner, 1949).

5. Ibid.

6. Ibid., 256.

7. Ibid.

8. The Grecisms contained in Voß's rendition of the *Iliad*, however, approaching Goethe's third form of translation, indicated a practice significantly bolder than that of both Goethe and Schiller—accordingly, it was chastised in a review by August Wilhelm Schlegel for having forged an amalgamation of Greek and German much too alien and uncomfortable to accept. One risked, Schlegel maintained, to no longer speak a valid language, recognized as such, but a sort of gibberish invented by oneself. See "Homers Werke von Voß," in August Wilhelm Schlegel and Friedrich Schlegel, *Charakteristiken und Kritiken* (Königsberg: Nicolovius, 1801), 2: 96–197.

9. Wilhelm von Humboldt, "Einleitung zum *Agamemnon*," in *Werke in fünf*

Bänden, ed. Andreas Flitner and Klaus Giel (Stuttgart: Cotta, 1981), 5: 140. For an interpretation of Humboldt's theory of translation, including this passage, see Antoine Berman's chapter on Schleiermacher and Humboldt in *L'epreuve de l'étranger: Culture et traduction dans l'Allemagne romantique* (Paris: Gallimard, 1984), esp. 246–48. *The Experience of the Foreign: Culture and Translation in Romantic Germany*, trans. S. Heyvaert (Albany: State University of New York Press, 1992), esp. 154–55.

10. This assessment, not unaware of the tropological stakes involved in the practice of translation, occurs when the narrator in Hoffmann's "The Sandman" describes the breakdown of an attempt to render the wonders of interiority in the words of exteriority. "Der Sandmann," in *Werke*, rev. ed. Herbert Kraft and Manfred Wacker (Frankfurt am Main: Insel, 1967), 2: 19. "The Sandman," in E. T. A. Hoffmann, *Selected Writings*, ed. and trans. Leonard J. Kent and Elizabeth C. Knight (Chicago: University of Chicago Press, 1969), 1: 148 (trans. modified). Neil Hertz offers a reading of the relationship between coloring and Freud's concept of "deferred action"—thus, by extension and implication, also of translation. See "Freud and the Sandman," *The End of the Line: Essays on Psychoanalysis and the Sublime* (New York: Columbia University Press, 1985), 97–121.

11. Humboldt, "Einleitung zum *Agamemnon*," 140.

12. Friedrich Nietzsche, "Brief an meinen Freund, in dem ich ihm meinen Lieblingsdichter zum Lesen empfehle," *Werke*, ed. Karl Schlechta (Frankfurt am Main: Ullstein, 1981), 3: 805.

13. Hölderlin translates parenthetically: "das Eine in sich selber unter schiedne." *FA* 11: 681. Among the pertinent readings of this notion of "the-one-in-itself-differentiated" are two texts by Philippe Lacoue-Labarthe, "La césure du spéculatif" and "Hölderlin et les Grecs," in *L'imitation des modernes*, Typographies II (Paris: Galilée, 1986), 39–69 and 71–84, respectively. "The Caesura of the Speculative," trans. Robert Eisenhauer, and "Hölderlin and the Greeks," trans. Judi Olson, in Philippe Lacoue-Labarthe, *Typographies: Mimesis, Philosophy, Politics*, ed. Christopher Fynsk, intro. Jacques Derrida (Cambridge: Harvard University Press, 1989), 208–35 and 237–47, respectively. Andrzej Warminski offers further specifications of the stakes involved in his chapter on "Hölderlin in France." See *Readings in Interpretation: Hölderlin, Hegel, Heidegger* (Minneapolis: University of Minnesota Press, 1987), esp. 37–42.

14. Hölderlin, "Anmerkungen zur *Antigonä*," *FA* 16: 414. "Remarks on *Antigone*," in Friedrich Hölderlin, *Essays and Letters on Theory*, ed. and trans. Thomas Pfau (New York: State University of New York Press, 1988), 111. Further references will appear in the main body of the text as *Essays*. In case there is an additional page reference after a semicolon, it will be to Pfau's translation (which I have modified on occasion).

15. "There are two principles concerning translation," Goethe writes about Wieland, "the one demands that the author of a foreign nation should be brought

over [herüber gebracht] to us in such a way that we may consider him one of our own; the other, however, demands from us that we move over [hinüber begeben] and adhere to his conditions, manner of speech, and particular characteristics. The advantages of both methods are sufficiently known, through exemplary cases, to all learned men. Our friend, who also here sought the middle road, endeavored to bring the two together, but as a man of sense and taste, he preferred the former principle in ambiguous cases." "Zu brüderlichem Andenken Wielands," *Werke*, 12: 705.

16. Nietzsche, "Brief an meinen Freund," 804.

17. Quoted in Dietrich E. Sattler, "Zeittafel," *FA* 16:28 and 29.

18. Ibid., 27. The critic was J. G. Gurlitt, writing in the *Neue allgemeine deutsche Bibliothek* 93 (1804).

19. Ibid., 20.

20. Ibid., 23.

21. Ibid., 21.

22. Ibid., 20. The friend was Bernhard Rudolf Abeken.

23. Cf. Albin Lesky, "Zwei Sophokles-Interpretationen," in *Gesammelte Schriften* (Bern: Francke, 1966), 179–89; Karl Reinhardt, "Hölderlin und Sophokles," in *Tradition und Geist* (Göttingen: Vandenhoeck & Ruprecht, 1960), 381–97; and Wolfgang Schadewaldt, "Hölderlins Übersetzung des Sophokles," in *Hellas und Hesperien* (Zürich: Artemis, 1960), 767–824.

24. What is *fremd* to the Greeks may be familiar to the Germans and vice versa, but the *Fremdheit* peculiar to the Greek cannot be that particular to German. Andrzej Warminski traces the logic in detail. See, for example, "Monstrous History: Heidegger Reading Hölderlin," in Part II of the present volume.

25. As Walter Benjamin puts it in a late fragment, the translation must attempt to "render the foreign tongue *in* its own," rather than *as* its own. "La traduction—pour et contre," in *Gesammelte Schriften*, ed. Rolf Tiedemann and Hermann Schweppenhäuser (Frankfurt am Main: Suhrkamp, 1974–89), 6: 160 (emphasis added).

26. Tom McCall has taken this as the point of departure for an analysis of "the theme of burial [in *Antigone*] as exemplification or literalization of translation." "The Case of the Missing Body," *Le pauvre Holterling* 8 (1988): 59. The reading of the play offered here, though indebted to McCall's observations, is less concerned with Hölderlin's allegorization of the conditions of translation, than with his text's relation to the argument outlined in the adjacent remarks.

27. Whereas Hölderlin mentions "red" or "red-like" (*rötlichen*) clouds, Goethe has "rolling" or "thundering" (*rollenden*) ones: "Wenn der uralte / Heilige Vater / Mit gelassener Hand / Aus rollenden Wolken / Segnende Blitze / Über die Erde sät" (*Werke*, 1: 146–47).

28. The letter is dated November 12, 1798 and addressed to Christian Ludwig Neuffer.

29. Elsewhere, I have discussed these efforts at definition in relation to Benjamin's theorization of the legibility of history. See "Contraction (Benjamin, Reading, History)," *Modern Language Notes* 110, no. 3 (1995): 540–64.

30. Sophocles, *Oedipus the King, Oedipus at Colonus, Antigone*, trans. F. Storr (Cambridge: Harvard University Press, 1981), 309–419. For Hölderlin's translation, see *FA* 16: 259–407. Further references will be given to line and in the main body of the text; first to the German version, then to the Greek.

31. I am paraphrasing a suggestion made by Lacoue-Labarthe in a different context: "I would like to ask," he writes, "if the 'modern' was not for Hölderlin something like the après coup, in the strict sense, of Greek art: that is to say, the repetition of what occurred there without ever taking place." "La césure du spéculatif," 53; 221.

32. Inscribed in the Greek καλχαίνω is the name of Calchas, the prophet who foresaw the Greeks' victory at Troy. Cf. George Steiner, *Antigones* (Oxford: Oxford University Press, 1984), 87. An analysis of Hölderlin's effacement of his nominal presence in the word, neither undertaken nor suggested by Steiner, would have to take into account, among other things, the critique of prophetic discourse in Hölderlin's thinking, as well as its bearing on the political. For a brief treatment of Hölderlin's literalizing translation, see Haroldo de Campos, "Hölderlin's Red Word," trans. Albert G. Bork, *20th Century Studies* 11, no. 6 (1974): 4–10.

33. Steiner, *Antigones*, 69.

34. For the distinction between *futurus* and *adventus*, see Alice A. Kuzniar, *Delayed Endings: Nonclosure in Novalis and Hölderlin* (Athens: University of Georgia Press, 1987), 29–39. Kuzniar borrows the terms from Jürgen Moltmann.

35. Cf. also the remark that "[i]n the utmost form of suffering, . . . there exists nothing but the conditions of time or space." "Anmerkungen zum *Ödipus*," 258; *Essays*, 108 (trans. modified). In the drafts of the last version of "Patmos"—one of the poems which most explicitly struggles with a teleological thematics the tropes of which it nonetheless resists as a text—it is said that suffering imparts color to purity. See *SA* 2; 1: 186, lines 164–66. If purity is the only attribution of the caesura conceivable, it seems entirely judicious that Hölderlin, in *Ödipus*, chooses to translate the designation of Tiresias, the incarnation of such "counter-rhythmic rupture," as ἄτεγκτος ("pitiless" or "hard") with *farblos* ("colorless") (line 340).

36. Walter Benjamin, "Die Aufgabe des Übersetzers," in *Gesammelte Schriften*, 4; 1: 13. "The Task of the Translator," trans. Harry Zohn, in Walter Benjamin, *Illuminations*, ed. and intro. Hannah Arendt (New York: Schocken, 1969), 73.

HAMLIN *The Philosophy of Poetic Form: Hölderlin's Theory of Poetry and the Classical German Elegy*

1. "Menons Klagen um Diotima," *KT* 6:74–78; here 74 (lines 13–14). The German original is reproduced in an appendix to the argument, accompanied by my own attempt to achieve a reasonably accurate version in English using, as Höl-

derlin also did, the metrical form of the Classical elegiac dystichon (a dactylic hexameter followed by a pentameter line). I do not delude myself with the claim that my translation can stand on its own as a poem, but the effort to achieve such a version, fraught with endless compromise on details and continuous frustration—as only a fellow translator could possibly know—, provided a splendid diversion during an extended cruise under sail along the coast of Maine during the summer which preceded the writing of the essay. The essay was originally written by invitation for a conference on Hölderlin held at Stanford University in April 1993, organized by Eckart Förster, to commemorate the 150th anniversary of the poet's death.

2. Ernst Cassirer, "Hölderlin und der deutsche Idealismus," in *Idee und Gestalt*, 2d ed. (Berlin: Cassirer, 1924).

3. Theodore Ziolkowski, *The Classical German Elegy 1795–1950* (Princeton, N.J.: Princeton University Press, 1980).

4. Lawrence Ryan, *Hölderlins Lehre vom Wechsel der Töne* (Stuttgart: Kohlhammer, 1960)

5. Karl Viëtor, "Hölderlins Liebeselegie," in *Geist und Form: Aufsätze zur deutschen Literaturgeschichte* (Bern: Francke, 1952), 267–91 and 357–65.

6. I have discussed this episode at fuller length in the expanded, revised version of "The Poetics of Self-consciousness in European Romanticism," contained in my essay collection *Hermeneutics of Form: Romantic Poetics in Theory and Practice* (New Haven, Conn.: Henry R. Schwat, 1998).

7. Dieter Henrich, *Der Grund im Bewußtsein: Untersuchungen zu Hölderlins Denken (1794–1795)* (Stuttgart: Klett, 1992).

8. Friedrich Hölderlin, *Sämtliche Werke*, ed. Norbert von Hellingrath (Berlin: Propyläen, 1913), 4: xi.

9. In particular, see Michael Konrad, *Hölderlins Philosophie im Grundriß: Analytisch-kritischer Kommentar zu Hölderlins Aufsatzfragment "Über die Verfahrungsweise des poetischen Geistes"* (Bonn: Bouvier, 1967), and Gerhard Kurz, *Mittelbarkeit und Vereinigung: Zum Verhältnis von Poesie, Reflexion und Revolution bei Hölderlin* (Stuttgart: Metzler, 1975).

10. See Ryan, *Hölderlins Lehre vom Wechsel der Töne*, 239–42.

11. Let me also acknowledge a specific debt with regard to my reading of this poem to Ziolkowski's remarkable study, particularly in regard to Hölderlin's place in the history of Idealist philosophy. Ziolkowski discusses Hölderlin's "Elegie" briefly in relation to the work of Schiller and Goethe in this genre, and he also offers at least some passing thoughts on the question of "tonal variation." I have learned much from Ziolkowski's book with regard to the history of this specific poetic genre in Germany at the end of the eighteenth century, and I shall use several suggestions from his argument in order to go a few steps further toward a reading of Hölderlin's poem. See *The Classical German Elegy (1795–1950)*, 110–24; esp. 120–21.

12. Ralph R. Read III, trans., *German Romantic Criticism*, ed. A. Leslie Wilson, vol. 21 of the German Library (New York: Continuum, 1982), 221.

13. Friedrich von Schiller, *Sämtliche Werke* (Munich: Hanser, 1959), 5: 745. Schiller, "*Naive and Sentimental Poetry*" *and* "*On the Sublime*," trans. Julius A. Elias (New York: Ungar, 1966), 146.

14. Schiller, "*Naive and Sentimental Poetry*," 146.

15. See Henrich, *Der Grund im Bewußtsein*, 300–329.

16. See Dieter Henrich, "Hölderlin über Urteil und Sein: Eine Studie zur Entwicklungsgeschichte des deutschen Idealismus," *Hölderlin-Jahrbuch* 14 (1964–65): 73–96.

17. See Ryan, *Hölderlins Lehre vom Wechsel der Töne*, 22–23.

18. M. H. Abrams, *The Mirror and the Lamp: Romantic Theory and the Critical Tradition* (New York: Oxford University Press, 1953).

19. G. W. F. Hegel, *Werke in zwanzig Bänden: Theorie Werkausgabe*, ed. Eva Moldenhauer and Karl Markus Michel (Frankfurt am Main: Suhrkamp, 1986), 14: 127–41.

20. Hölderlin's concept of metaphor is further discussed in my essay "The Temporality of Selfhood," also included in my collection *Hermeneutics of Form*, 261–87, esp. 282–86.

21. Cf. Ziolkowski, *The Classical German Elegy (1795–1950)*, 3–26.

22. Ibid., 93–99, 111–14.

23. Viëtor, "Hölderlins Liebeselegie," 357.

24. Wilhelm Böhm, *Hölderlin* (Halle, Germany: Niemeyer, 1930), 2: 133–34, 284, and 329.

25. Ryan, *Hölderlins Lehre vom Wechsel der Töne*, 239–42.

26. See Ziolkowski, *The Classical German Elegy (1795–1950)*, 121. As indicated, these moments of tonal modulation are argued to occur, respectively, at the beginning of stanza 3 (line 29), stanza 4 (line 43), stanza 5 (line 57), stanza 6 (line 69), stanza 7 (line 83), and stanza 9 (line 109). Stanzas 1 and 2 and stanzas 7 and 8, by this view, thus belong together as single units in the tonal sequence. Also apparent is a lack of symmetry in the length of the units, which contain the following number: 28, 13, 14, 12, 14, 26, and 22. (One might note that the earlier version of the elegy, which also exhibits the same tonal variation, is divided by spacing in the manuscript into six unnumbered stanzas with even greater asymmetry of units: stanzas 1 and 2, 22 lines; stanza 3a, 18 lines; stanzas 3b and 4a, 14 lines; stanza 4b, 8 lines; stanza c, 10 lines; stanza 5, 24 lines; stanza 6, 20 lines.)

BÖSCHENSTEIN *"Brod und Wein": From the "Classical" Final Version to the Later Revision*

1. Cf. my earlier statement in "Hölderlin—'work in progress'? Die neue Frankfurter Ausgabe," *Neue Zürcher Zeitung* 296, no. 17 (December 1977): 56, and Wolfram Groddeck's response to this position in "Die Nacht: Überlegungen zur

Lektüre der späten Gestalt von 'Brod und Wein,'" in *Hölderlin-Jahrbuch* 21 (1978–79): 206–24, esp. 207–10.

2. Jochen Schmidt, *Hölderlins Elegie "Brod und Wein"* (Berlin: de Gruyter, 1968), 179–208; Werner Almhofer, "'Wildniß' und Vergnügen: Hölderlins mythologische Bildersprache in den späten Korrekturen von 'Brod und Wein,'" *Hölderlin-Jahrbuch* 26 (1988–89): 162–74.

3. Quotations from "Brod und Wein" are from *HF* and are identified in the text by page and line number. In this case: *HF* 31: 11, 19–20, and 22–23 (emphasis mine). The text is almost identical to *FA* 6: 222–33.

4. Friedrich Hölderlin, *"Bevestigter Gesang": Die neu zu entdeckende hymnische Spätdichtung bis 1806*, ed. and critically annotated by Dietrich Uffhausen (Stuttgart: Metzler, 1989), 78, 5–6. Cf. *FA* 6: 259, lines 40–41.

5. On this thematic, see Bernhard Böschenstein, *"Frucht des Gewitters": Hölderlins Dionysos als Gott der Revolution* (Frankfurt am Main: Insel, 1989), 52 and 145–52.

6. See *SA* 5: 268, lines 29–30; and Beißner's gloss in *SA* 2; 2: 613, lines 23–28.

7. Cf., for example, *SA* 2; 1: 166, lines 37–38, and 2; 1: 231, lines 10–11.

8. Groddeck, "Die Nacht," 213, note 28.

9. Cf. *HF* 36: 36; Euripides *The Bacchae* 298; Cf. Schmidt, *Hölderlins Elegie "Brod und Wein"*, 205–6.

10. See *SA* 2; 1: 193, lines 16–18, and 195, lines 15–17. On this theme see Böschenstein, *"Frucht des Gewitters"*, 187–204.

11. Acts 19: 26.

12. Schmidt, *Hölderlins Elegie "Brod und Wein"*, 191.

13. Robert Fagles's translation of the same stanza runs as follows: "Danaë, Danaë— / even she endured a fate like yours, / in all her lovely strength she traded / the light of day for the bolted brazen vault— / buried within her tomb, her bridal-chamber, / wed to the yoke and broken. / But she was of glorious birth / my child, my child / and treasured the seed of Zeus within her womb, / the cloudburst streaming of gold!" Sophocles, *The Three Theban Plays: Antigone, Oedipus the King, Oedipus at Colonus* (1982; New York: Penguin, 1984), 108.—Trans.

14. On this, see Böschenstein, *"Frucht des Gewitters"*, 66–67.

15. On this, see Böschenstein, *Hölderlins Rheinhymne* (Zürich: Atlantis, 1968), 102–4.

16. The underlining of *donnernden* indicates the poet's intention to replace this word.

17. Fagles' translation runs as follows: "Come, Dionysus! now your people lie / in the iron grip of plague, / come in your racing, healing stride . . ."—Sophocles, *Three Theban Plays*, 119.—Trans.

18. See Böschenstein, *"Frucht des Gewitters"*, 21–24.

19. See ibid., 47–48.

20. Schmidt provides rich and productive material on this in his *Hölderlins*

Elegie "Brod und Wein", 196–99, partly following Momme Mommsen, "Dionysos in der Dichtung Hölderlins," *Germanisch-Romanische Monatsschrift* 13, no. 4 (1963): 345–79.

21. Groddeck's defense of this constituted text was the object of a work group in the context of the Hölderlin conference in May 1978; the results of the work done in that context are written up in the essay cited in the first note.

MELBERG *Turns and Echoes: Two Examples of Hölderlin's Poetics*

1. *The Presocratic Philosophers*, ed. G. S. Kirk, J. E. Raven, and M. Schofield (1957; Cambridge: Cambridge University Press, 1983), 192.

2. Martin Heidegger, *Erläuterungen zu Hölderlins Dichtung* (Frankfurt am Main: Klostermann 1981). A more extensive version of essentially the same interpretation may be found in Heidegger's lectures from 1942 to 1943, published in 1984. See *Hölderlins Hymne "Der Ister"*, Gesamtausgabe (Frankfurt am Main: Klostermann, 1984), 53.

3. Heidegger, *Erläuterungen zu Hölderlins Dichtung*, 90–91.

4. Ibid., 94.

5. Ibid., 95.

6. Heidegger, *Hölderlins Hymne "Der Ister"*, 130, 156, and 170.

7. The terms for what is "proper" or "one's own" and "foreign" are in fact taken from Hölderlin's first letter to Böhlendorff, dated December 4, 1801, which Heidegger simply reads into the fragment.

8. As Jochen Schmidt, among others, has pointed out. See *Hölderlins Elegie "Brot und Wein": Die Entwicklung der hymnischen Stils in der elegischen Dichtung* (Berlin: de Gruyter 1968), 203.

9. See Andrzej Warminski, "Heidegger Reading Hölderlin," in *Readings in Interpretation: Hölderlin, Hegel, Heidegger* (Minneapolis: University of Minnesota Press, 1987), esp. 47–57.

10. Hellingrath borrowed the phrase from Dionysios from Halicarnassus, who applied it to Pindar. Cf. Schmidt, *Hölderlins Elegie "Brod und Wein"*, 28–29.

11. See Beißner's remarks at *SA* 2; 2: 829. For later interpretations, predicated on dialectical principles, see, for example, Anselm Haverkamp, "Error in Mourning—A Crux in Hölderlin: 'dem gleich fehlet die Trauer' ('Mnemosyne')," *Yale French Studies* 69 (1985): 238–53; and Véronique M. Fóti, "Mnemosyne's Death and the Failure of Mourning," in *Heidegger and the Poets: Poiésis, Sophia, Techné* (Atlantic Highlands, N.J.: Humanities, 1992), 60–77.

12. See Emery E. George, *Hölderlins "Ars Poetica": A Part-Rigorous Analysis of the Information Structure in the Late Hymns* (The Hague: Mouton, 1973); Flemming Roland-Jensen, *Hölderlins Muse: Edition und Interpretation der Hymne "Die Nymphe Mnemosyne"* (Würzburg: Königshausen & Neumann, 1989); Roland Reuß, *". . . / Die eigene Rede des andern": Hölderlins "Andenken" und "Mnemosyne"*, (Frankfurt am Main: Stromfeld, 1990); and Dietrich Uffhausen, *"Bevestigter Ge-*

sang": Die neu zu entdeckende hymnischen Spätdichtung bis 1806 (Tübingen: Niemeyer, 1989). These commentators differ in many respects (including the order of the stanzas), but are at least unanimous in crediting the poem with four stanzas. Roland-Jensen has accomplished the most thorough work as to the establishment of the text. In the following, however, I will employ Uffhausen's version, according to which the *echo*-stanza constitues the second stanza of the poem.

13. Heidegger's interpretation suggests a humanizing and "dialectical" understanding, much like his analysis of the "re-turning spirit" discussed earlier, when it relates the "abyss" to the "essence" (*Wesen*) of "mortals." "Wozu Dichter?," in *Holzwege* (Frankfurt am Main: Klostermann 1950), 250.

14. As may be seen in the synoptical version established by Uffhausen, *"Bevestigter Gesang,"* 161.

15. Paul de Man has suggested translating the lines as "What is true is what is bound to take place," thus doing more justice to the fatalistic than the activist version of Hölderlinian rhythm. Carol Jacobs, foreword to *The Dissimulating Harmony: The Image of Interpretation in Nietzsche, Rilke, Artaud, and Benjamin* (Baltimore: Johns Hopkins University Press, 1978), xi.

16. Walter Benjamin, "Ursprung des deutschen Trauerspiels," in *Gesammelte Schriften*, ed. Rolf Tiedemann and Hermann Schweppenhäuser (Frankfurt am Main: Suhrkamp, 1974), 1; 1: 226.

17. See Heidegger, "Der Ursprung des Kunstwerkes," in *Holzwege*, 41.

18. The report—no doubt of little reliability, but nonetheless symptomatically interesting—originates in Bettina von Arnim's 1840 *Die Günderode*. It dates back, however, to letters from 1806. *Die Günderode*, ed. Elisabeth Bronfen (Munich: Matthes & Seitz 1982), 224–28.

19. Maurice Blanchot, *L'entretien infini* (Paris: Gallimard, 1969), 42; *The Infinite Conversation*, trans. Susan Hanson (Minneapolis: University of Minnesota Press, 1993), 30 (trans. modified).

FREY *Hölderlin's Marginalization of Language*

1. The word *Rand*, a significant one for this essay, is rendered as either "margin" or "edge," depending on context.—Trans.

2. For "Aber die Sprache . . . ," the sheet is available in *FA* 4: 131 and 14: 331 in facsimile form. For "Im Walde," the best reproduction is found in *SF* 34. The sheet was reproduced and analyzed by Peter Szondi in "Der andere Pfeil: Zur Entstehungsgeschichte des hymnischen Spätstils," in *Schriften*, ed. Jean Bollack et al. (Frankfurt am Main: Suhrkamp, 1978), 1: 289–324. Beißner gives a description of the sheet in *SA* 2; 2: 663–64.

3. On this see "Textrevision bei Hölderlin," in Hans-Jost Frey, *Der unendliche Text* (Frankfurt am Main: Suhrkamp, 1990), 77–123.

4. With the exception of references to the texts quoted, this essay contains no notes. This can be explained most simply by saying that it was written without

using any secondary literature and is based only on the manuscript materials that have become available thanks to D. E. Sattler's Frankfurter Ausgabe. But although this explanation is correct, it remains unsatisfactory to the extent that the study of manuscripts from the perspective of their importance for an understanding of the text is no original undertaking, but rather has become considerably more intense—for example in France with the *critique génétique*—since improvements in the technology of reproduction have made more manuscripts available in complete form. Thus, an attempt such as the one made here has a kind of surrounding atmosphere within which more or less similar reflections can take shape. By neglecting to refer to such works in notes, one contravenes an academic convention. One can submit oneself to this convention or reject it. But one can also, more fundamentally, ask what it means. I will pursue this question briefly in what follows, because it is closely related to the thematics of the text's margin and the marginal text.

The purpose of referring to other literature in notes is to situate oneself. This is a standing, that is, too little questioned demand of scholarly work, in which one is supposed to be able to prove one's familiarity with the current research and to determine what is one's own in one's contribution. Thus the note in the scholarly text is directed in two different directions from the beginning. With the note, the author integrates himself in a community of colleagues who are preoccupied with similar questions, but this integration takes place by setting the author off; its purpose is to differentiate his performance from other, more or less similar, ones, and to secure it as original.

As an example let me mention an essay by Dieter Burdorf that also discusses the role of the graphic appearance of the text in its interpretation: "Der Text als Landschaft: Eine topographische Lektüre der Seiten 73 bis 76 des Homburger Folioheftes," in *Neue Wege zu Hölderlin*, ed. Uwe Beyer (Würzburg: Schriften der Hölderlin-Gesellschaft 18, 1994). The title of the essay is followed by this note: "In this essay I take up again several aspects of my dissertation while formulating the question differently. See my *Hölderlins späte Gedichtfragmente 'Unendlicher Deutung voll'* (Stuttgart; Weimar, 1993). The manuscript was completed (with the exception of individual additions) in November 1991." The note makes the text dependent on another from the beginning; however, since the other text is one's own, the note fulfills the task of proving the originality of the present text, whose central thought is backdated by three years, so that it is clear that nothing is owed to works on the topic that might have appeared in the meantime. The importance given here to situating the thought in time is clear also from the nineteenth footnote of the same essay (132): "I obtained this result in my dissertation (see footnote 1, p. 350–54), independently of Groddeck (1990/91 [see footnote 9], p. 312), who published his suggestion in 1991." By mentioning the work of another, one signals one's belonging to the circle of Hölderlin scholars, but by stressing the question of precedence one also, at the same time, makes known one's claim to originality and independence. Nothing can be objected to either gesture. However, the fact that both occur together indicates the difficult relation of one's own text to that of others, the relation that notes in scholarly texts tend to secure.

This relation corresponds exactly to the one between the note and the main text to which it belongs and from which it distinguishes itself. The status of the note is undecidable. Already its graphic appearance makes it impossible to situate. As a marginal text, it does not respect the border of the block of text: it crosses it and pushes into view what is outside of it. But by appearing beyond the block of text, it also reaffirms it at the same time and safeguards the differentiation of inside and outside, own and foreign, which however is also immediately put back in question, since the note demands to be inserted at a specific point in the text, where it disturbs its orderly flow and lets foreign material pour in through the gap that it opens. Of course one can avoid such forks in the text by reading on without paying attention to notes, but one pays for leaving them out with the feeling that one is missing something. The note may be superfluous, but one misses the superfluous when one skips it. What this means is that the note both belongs and does not belong to the text. It shares this undecidable, intermediate position with other marginal texts such as the title, the preface, the afterword, and the epigraph.

The decision not to supply a text with notes does not free the text from its ambivalent position with respect to its surroundings. At first blush such a text seems to respond to a stronger need to isolate itself and affirm its independence. This presumption, however, has a reverse side. Together with the reference to otherness, something else is lost, namely the gesture of power at work in the note, whose aim is to bring otherness under control. Although the note is an opening toward otherness, its goal is to define oneself against it. The text without notes does not explicitly establish a relation to its surroundings, but it also does not attempt to bring this relation under control and defuse it. It is less protected and exposes itself more. It speaks closer to the margin.

The marginal gloss is the predecessor of the scholarly note. In "The Monument and the Margin," *The South Atlantic Quarterly* 91, no. 4 (Autumn 1993), 835–64, Laura Kendrick has shown that medieval authors often comment on themselves in marginal notes, and this fencing in of the text monumentalizes it and is what really turns the author into an author:

> Such medieval self-authorizing annotation may be compared with modern scholarly self-annotation in the form of footnotes (the more powerfully self-canonizing format) or endnotes (at the "margins" of the book as a whole, rather than of the page). By taking it upon ourselves to supplement or add corroborations or modifications of the central text, but only in an explicitly extratextual space, in the fine print below the visible or invisible footnote line, we visibly defer to the originality and authority of our own central text. (859)

This note is not protected against any of the observations made or presented in it.

SCHESTAG *The Highest*

1. *Pindari carmina cum fragmentis*, ed. C. M. Bowra (Oxford: Clarendon Press, 1947), 242. About the context of this fragment in Pindar see, among others, Max Treu, "ΝΟΜΟΣ ΒΑΣΙΛΕΥΣ: Alte und neue Probleme," *Rheinisches Museum für Philologie*, n.s., 106 (1963): 193–214; and Martin Ostwald, "Pindar, ΝΟΜΟΣ, and Heracles," *Harvard Studies in Classical Philology* 69 (1965): 109–38 (reprinted in *Pindaros und Bakchylides*, ed. William Musgrave Calder III and Jacob Stern [Darmstadt: Wissenschaftliche Buchgesellschaft, 1970], 194–231).

2. Friedrich Hölderlin, *SA* 6; 1: 246; *Essays and Letters on Theory*, trans. Thomas Pfau (Albany: State University of New York Press, 1988), 150 (trans. modified). Further references will be given in the main body of the text, as *Essays* with reference to the Pfau edition indicated after a semicolon.

3. On this see Jacob Grimm, "Über die wechselseitigen Beziehungen und die Verbindung der drei in der Versammlung vertretenen Wissenschaften" (1846): "Let me begin with the simple question: what is a people? and answer just as simply: a people is the embodiment of persons who speak the same language." *Kleinere Schriften* (Berlin: Dümmler, 1884), 7: 556–63.

4. Martin Heidegger, " 'Wie wenn am Feiertage . . . ,' " *Erläuterungen zu Hölderlins Dichtung*, Gesamtausgabe (Frankfurt am Main: Klostermann, 1981), 4: 62.

5. Paul de Man, "Les exégèses de Hölderlin par Martin Heidegger," *Critique* 100–101 (1955): 814; "Heidegger's Exegeses of Hölderlin," trans. Wlad Godzich, in Paul de Man, *Blindness and Insight: Essays in the Rhetoric of Contemporary Criticism*, 2d ed. (Minneapolis: University of Minnesota Press, 1983), 261.

6. De Man, "Les exégèses," 815; *Blindness and Insight*, 261.

7. De Man, "Les exégèses," 809; *Blindness and Insight*, 254–55. On the discussion of Heidegger's explanation of the Pindar Fragment "The Highest," from the tension between being and naming—close to taking—see Hans-Jost Frey, *Studien über das Reden der Dichter* (Munich: Fink, 1986), 135–44; *Studies on Poetic Discourse: Mallarmé, Baudelaire, Rimbaud, Hölderlin*, trans. William Whobrey (Stanford, Calif: Stanford University Press, 1995), 178–91.

8. Heidegger, *Erläuterungen*, 4: 104.

9. Ibid., 149.

10. Martin Heidegger, *Wegmarken* (Frankfurt am Main: Klostermann, 1978), 357; "Letter on Humanism," trans. Frank Capuzzi, in Martin Heidegger, *Basic Writings*, ed. David Farrell Krell (San Francisco: Harper San Francisco, 1977), 238 (trans. modified).

11. Martin Heidegger, *Erläuterungen*, 4: 167. In the essay "Moira (Parmenides Fragment 8, lines 34–41)," written around 1951–52 as a not presented part of the lecture "Was heißt Denken?", *Moira*, a word that means "part" and "lot," close to the interpretation of *nomos* as "attribution," is glossed as apportionment [*Zustellung*]:

> Parmenides speaks of ἐόν, of the presencing (of what is present, and of duality, and in no sense of "Being." He names the *Moîra*, the apportionment [Zuteilung], which allots [verteilt] by bestowing and so unfolds the twofold. The apportionment dispenses [beschickt], (provides and presents) through the duality. Apportionment is the dispensation of presencing, as the presencing of what is present, which is gathered in itself and therefore unfolds of itself. *Moîra* is the destiny [Geschick] of "Being," in the sense of ἐόν. ("Moira [Parmenides Fragment 8, lines 34–41]," trans. Frank Capuzzi, in Martin Heidegger, *Early Greek Thinking: The Dawn of Western Philosophy* [San Francisco: Harper San Francisco, 1975], 97)

It must matter for this interpretation to fade out the remembrance of the unregulated divisibility of the apportionment [*Zuteilung*] or act of providence [*Schickung*], but also, first of all, the consciousness of a cutting from which the (precarious) possibility of *Moira*, of the part *as* part and lot, comes in the first place; repeated otherwise, it must matter for it to fade out the remembrance of the splitting up [*Zerteilung*] (*nomos*), in order not to deinterpret [*entdeuten*] it as *nemein*, but, avoided for a second time, to subordinate it to the asserted precedence of *legein*. The next paragraph of the essay, which speaks about "what it is that Moîra metes out" and thus about the *nomos*, shows this. It declines the *Vernehmen* ("to hear") and pulls in the *name*.

> What is the significance of the fact that destiny releases the presencing of what is present into the duality? . . . The unfolding of the twofold reigns as φάσις, saying as bringing-forward-into-view. . . . What is taken up [vernommen] in thinking, however, is the presencing of what is present. The thoughtful saying that corresponds to the duality is the λέγειν, the letting-lie before of presencing. . . . But what becomes of the φάσις ("saying") reigning in revealing destiny if this destiny should abandon what is unfolded in the twofold to the everyday perception [Vernehmen] of mortals? Mortals accept [nehmen auf] (δέχεσθαι, δόξα) whatever is immediately . . . offered to them. . . . But what becomes of their speech if it is not capable of being a λέγειν, a letting-lie before? The ordinary speech of mortals, insofar as they do not consider presencing, that is, insofar as they do not think, ends up as a speaking of names [Namen] in which vocalization and the immediately perceptible form of the word, as spoken or written, are stressed. The unequivocal restriction of speech (of letting-lie before) to word-signs shatters the gathering taking-heed-of [in-die-Acht-Nehmen]. ("Moira," in *Vorträge und Aufsätze* (Pfullingen: Neske, 1954), 243–46; *Early Greek Thinking*, 98–99)

The closeness to each other of *Name* ("name") and *Nehmen* ("to take") is touched, but banned. They restrict taking—*nemein*—another, more precise word for *thinking*, but also for *reading*, to a *gathering* taking-care or taking in attentively (before the—more splitting, more cutting—taking.) They attempt, according to the subordination of the *nomos* under the Moira, to have put taking (*nemein*) into the hands of letting exist (*legein*), to protect the outline of the doubleness. For the *more taking* discourse does not so much correspond [*entspricht*] to doubleness, as it, quietly splintering the outline of the *double*ness, de-speaks [*entspricht*] it. It attracts to mortals (or to immortals) the impossibility of interpreting mortals unambiguously, because they "hear" [*vernehmen*], yet this taking is not supported by any rule which could not be taken, for its part—otherwise. In this discourse, the condition of possibility, but impossibility, neither of the *talking* about being, nor of *being*, appears.

In this context let me refer to the peculiar closeness to one another of *Sein* and *Nehmen*, which Jacob Grimm mentions in a letter to Karl Lachmann on July 9, 1833:

> The impersonal expressions that are formed with *nemen* and *hân* and take the accusative have something conspicuous and heavy about them. *Mich nimt wunder*, *mich nimt hœle*, *mich nimt tiure*, *untiure*; and *mich hât wunder* as well. . . . For if here *nimt* should mean "catches," *capit*, *occupat*, "holds," "takes in," like the French *il me prend envie*, then that is appropriate to *wunder*, but we do not know if it is appropriate to the obscure *hœle*, and not at all to the adjective *tiure*. . . . If you are familiar with other nouns in such connections, please refer me to the passages. The oddest thing however, is the use in Old High German and Gothic of the verb *esse*. Otfried constructs in several passages *mih ist wuntar* exactly in the sense of the Middle High German *mich nimt wunder*. . . . The thing is always in the genitive in these expressions, *mich ist es wunder*, and still today: *mich nimmts wunder*. But how can one conceive the *accusative* of the person with the verb *ist*? I know of nothing quite the same in other languages. . . . Do you know any more Old High German nouns that are connected in such a way with *ist*? and do you know any Middle High German examples for: *mich ist wunder*? (*Briefwechsel der Brüder Jacob und Wilhelm Grimm mit Karl Lachmann*, ed. Albert Leitzmann [Jena: Frommann, 1927], 2: 619)

On this see also Jacob Grimm, *Deutsche Grammatik* (Göttingen: Dieterichsche Buchhandlung, 1837), 4: 246–49, and the oscillation there between *mich nimt Wunder* and *mich gibt Wunder*.

12. For a less literal translation, see *Essays*, 113–14.—Trans.

13. In the range of meanings of the word *Geschichte* ("history"), the *Dentsche Wörterbuch*, close to the *Geschick* in the sense of destiny, refers, among other things, to "that which is one's lot, which is due to one." *Deutsches Wörterbuch* (Leipzig: Hirzel, 1897), 5: 3859.

14. Which a letter to his brother—Homburg, June 4, 1799—talks about: ". . . that original drive, the drive toward idealizing, furthering, reworking, de-

veloping, completing nature" (*SA* 6; 1: 326–30), and which a letter to Schelling, in July of the same year, states more precisely: "My thinking and my studies were mostly restricted to that which I practiced at first, poetry. . . . This led me to reflect about education and the drive toward education [Bildungstrieb] in general, about its cause and determination, to the extent that it is idealistic and actively forming [bildend]" (*SA* 6; 1: 345–49; *Essays*, 145).

15. In his text from 1917, "Über das Programm der kommenden Philosophie," Walter Benjamin writes: "Cognizing man, the cognizing empirical consciousness, is a type of insane consciousness." *Gesammelte Schriften*, ed. Rolf Tiedemann and Hermann Schweppenhäuser (Frankfurt am Main: Suhrkamp, 1974–89), 2; 1: 162; "Program of the Coming Philosophy," trans. Mark Ritter, *Philosophical Forum* 15, no. 1–2 (Fall–Winter 1983–84): 45.

16. On the complicity between the tragic and the speculative-dialectical scheme see also Philippe Lacoue-Labarthe, "La césure du spéculatif," in *L'imitation des modernes*, Typographie II (Paris: Galilée, 1986), 39–69; "The Caesura of the Speculative," in *Typography: Mimesis, Philosophy, Politics*, ed. Christopher Fynsk (Cambridge: Harvard University Press, 1989), 208–35.

17. On this see also the lines in the Homburger Folioheft: "Der Rosse / Leib war / der Geist" ("The horse / was body / spirit") (*HF* 100), which do not only take up differently the word of a prophet, a passage in Isaiah 31: 1–3, turning the words less *against* each other, but rather split the outline of both words, the countering [*gegenwendige*] word of the prophet and the unturned, straighter word of God:

> Woe to them that go down to Egypt for help; and rely on horses, and trust in chariots, because *they* are many; and in horsemen, because they are very strong; but they look not unto the Holy One of Israel, neither seek the LORD!
>
> Yet he also he is wise, and will bring evil, and will does not call back his words; but will arise against the house of the evildoers, and against the help of them that work iniquity.
>
> Now the Egyptians *are* men, and not God; and their horses flesh, and not spirit. (Authorized King James Version)

On the irritating, irresolvable interpenetration of the sensuous and the spiritual body, not only in the political theology of the Middle Ages, see also Ernst Kantorowicz, *The King's Two Bodies: A Study in Medieval Political Theology* (Princeton, N.J.: Princeton University Press, 1957).

18. On the reversal, or the discussion, of the figure of inversion, see also Werner Hamacher, "Die Sekunde der Inversion: Bewegungen einer Figur durch Celans Gedichte," in *Paul Celan*, ed. Werner Hamacher and Winfried Menninghaus (Frankfurt am Main: Suhrkamp, 1988) 81–126; "The Second of Inversion: Movements of a Figure Through Celan's Poetry," trans. Peter Fenves, in Aris Fioretos, ed., *Word Traces: Readings of Paul Celan* (Baltimore: Johns Hopkins University

Press, 1994), 219–63; and the "Catastrophe" section in Philippe Lacoue-Labarthe, *La poésie comme expérience* (Paris: Bourgois, 1986), 61–103; "Catastrophe," trans. Andrea Tarnowski, in *Word Traces*, 130–56.

19. Three lines in the seam of the Hamburger Folioheft start out drawing the outline of the (syntactical) chain by describing this outline, yet relate both tendencies—toward outline, toward cohesion—because they are detrimental to both, to one *another*: "Leicht fanget aber sich / In der Kette, die / Es abgerissen, das Kalblein" ("easily however gets caught / in the chain / it has torn off / the calf") (*HF* 116).

20. This traditional, tradition-founding conception of *transport*, in another word, of *metaphor*, is what the "Remarks on *Oedipus*" call tragic, underlining: "The tragic *transport* is, namely, actually empty, and the most unbound" (*FA* 16: 250).

21. Sophocles, *Dramen*, Greek and German, ed. and trans. Wilhelm Willige, reworked by Karl Bayer (Munich and Zürich: Artemis, 1985), 592–93. The (more immediately factual) translation there formulates, avoiding the *dysmoros*: "*Oedipus*. Do not be apprehensive about anything I say! / *Chorus*. O sorrow! Oh! Oh! / *Oedipus*. Wretched! / *Chorus*. Oh! Oh!"

22. Whose understanding as the *last*, and as a *line* defined by a beginning and end of the line, is held out, but held up, by the expression *nicht ende* ("not end")—an expression that unsettles the decision about what *Mein Eigentum* may be, as well as how it is to be comprehended or abandoned—even whether it *is*, and *there*, at all.

23. For a less literal translation see Hölderlin, *Poems and Fragments*, trans. Michael Hamburger (Ann Arbor: University of Michigan Press, 1967), 101.—Trans.

24. On this see also Gerhard Baudy, "Hierarchie oder: Die Verteilung des Fleisches," in *Neue Ansätze in der Religionswillenschaft*, ed. Burkhard Gladigow and Hans G. Kippenberg (Munich: Kösel, 1983), 131–74, esp. 153–67. See also the references to the expression *moiras nemein* in Homer (*Odyssey* 14.436 and 15.140) in Bořivoj Borecký, *Survivals of Some Tribal Ideas in Classical Greek* (Prague: Charles University Press, 1965), 76–88.

25. Johann Arnold Kanne, *Erste Urkunden der Geschichte oder allgemeine Mythologie*, preface by Jean Paul (Bayreuth: Bey Johann Andreas Lübecks Erben, 1808), 1: 242–43. The following paragraph ties the *moira* more intimately to language:

> The prophetic Moira spun and sang (*θεσποδειν*), and was a muse long before the art of poetry: namely, she sang as Memnonic bird of time and Canicula *προκυη*, from which the destinies of souls and men came with the rising star. Already in Hebrew, the word פרט, from which comes *partiri*, means to sing and play. In Latin a poem was called something spun, *camen*, from *carere lanam*, the Muse was called *Carmenta* and *Carmena*, the old Roman *Casmena* was a *Carmena*." (243–44)

The insight about the *Nemesis* as "distributor" attributed to Herder refers to Herder's text *Nemesis: Ein lehrendes Sinnbild* (1785), at whose entrance a note explicitly does not bring back the "double meaning" of Nemesis to *nemein*, but, under the word *genommen* (how the Nemesis should be taken) opens up the "distributing" [*Vertheilen*] toward the signification of divisibility inside the word *nehmen* ("taking")—νεμειν:

> This double meaning does not depend on the fact that *Nemesis* can be derived from νεμειν or νεμεσαειν or is related to the two concepts, the *just distribution* and the *resentment over an injustice*; it depends on the use of the word itself, which, taken in good or bad spirit, must also give a different meaning. Its main concept, however, is νεμειν, "just distribution"; νεμεσαειν, "to rage about an unfair distribution," derives from it. (*Sämmtliche Werke*, ed. Berhard Suphan [Berlin: Weidmann, 1888], 15: 396)

26. The words recorded by Wilhelm Waiblinger (*Friedrich Hölderlins Leben, Dichtung und Wahnsinn* [Leipzig: Brockham, 1831])—"Once he said: 'I have now become orthodox. Your Holiness! No, no! At present I am studying Mr. Kant's third volume, and am very involved with the newer philosophy,'" (which gave cause for the supposition of Hölderlins "palinodic thought")—present, *almost in reverse*, precisely that "Greek-tragic," speculative-idealistic thought that is less a mode of proceeding in Hölderlin's writings than, more literally, and still unsettling the belief in literality, it is *discussed*. A passage in Benjamin's "Schicksal und Charakter" voices, at first glance, that supposition that does not see in palinodic thought, or thought through opposition, the object of the discussion as Hölderlin's mode of proceeding, the word, when it says:

> In the Greek classical development of the idea of fate, the happiness granted to a man is understood not at all as confirmation of an innocent conduct of life but as a temptation to the most grievous offense, hubris. There is, therefore, no relation of fate to innocence. And—this question strikes even deeper—has fate any reference to good fortune, to happiness? Is happiness, as misfortune doubtless is, an intrinsic category of fate? Happiness is, rather, what releases the fortunate man from the embroilment of the Fates and from the net of his own fate. Not for nothing does Hölderlin call the blissful gods "fateless" [schicksallos]. (*Gesammelte Schriften* 2; 1: 174; "Fate and Character," trans. Rodney Livingstone, eds. Marcus Bullock and Michael W. Jennings, in Walter Benjamin, *Selected Writings*, gen. ed. Michael W. Jennings, vol. 1 [1913–26] [Cambridge: Harvard University Press, 1996], 203)

The fateless ones, the gods, are, however, only opposed to those bound to destiny, the earthly ones: they remain bound to them as their negative, or vice versa. The reference to happiness [*Glück*], however, does hit "more deeply": *Glück*, which one etymological supposition connects to *Lücke*, an interruption, reaches that semantic layer out of which the Greek words—*moira, moros,* and others—which correspond inexactly to *Schicksal* ("destiny"), *Fügung* ("chance"), and *Los* ("lot"), originate as part and lot [*Anteil*]: the verbs, among others, *meiromai* and *nemo,* in which the divisibility of the part, the impossibility of a reconciling interpretation of the part *as* part, and thus also the invalidity of the opposition of men and gods, those bound to destiny and those without destiny, which builds on the indivisibility of the lot, or destiny, appears. Moreover the memory of the unregulated divisibility of part and lot cuts also through the chain of meanings of the noun *moira,* which has as its correspondence, among others, also *Glück.*

27. Grimm, *Deutsche Mythologie,* 2: 540. About the separating trait in *Abnehmen* see also the remarks in the reworked edition of the *Deutsches Wörterbuch* (Leipzig: Hirzel, 1983), 1: 603–6. Among other things, the quote from Albrecht Dürer, *VNderweysung der messung,* which talks about the cutting of the letter *A*: "Cut [nim ab] the letter on top [the tip of the *A*] with a slender line."

28. Friedrich Hölderlin, *Friedensfeier,* ed. Wolfgang Binder and Alfred Kelletat (Tübingen: Mohr, 1959). An English translation can be found in Hölderlin, *Poems and Fragments,* trans. Michael Hamburger (Ann Arbor: University of Michigan Press, 1967), 439.

29. The translation is slightly modified from Hamburger's in *Poems and Fragments,* 415. That the point of the turn toward *participation* is to "hit" the sketch of destiny as the kernel of the Greek-tragic, more mediate understanding of language, is indicated by Hölderlin in a parenthesis in a letter to Leo von Seckendorff, Nürtingen, March 12, 1804: ". . . or, more questionably, how they conduct themselves in it [*destiny*]. . . . The fable, poetic view of history, and architectonic of the heaven, is what I am mostly interested in at the moment, especially the national, to the extent that it is different from the Greeks. . . . The different destinies of heroes, knights and princes, how they serve destiny, or more questionably, how they conduct themselves in it, I have comprehended in general" (*SA* 6; 1: 437–38).

Theilnehmen ("participating"), one of the words that returns with most frequency in Hölderlin's letters, is here first of all brought in the vicinity of the *Nahme* ("taking, name"). Thus also in the previously mentioned letter to Schelling: ". . . it will be easy to put yourself into my limited perspective and to sanction with your name and your participation an enterprise which is designed to bring people closer to one another *without insincerity and syncretism*" (*SA* 6; 1: 347; *Essays,* 146). In another, earlier letter, in the "Sendbrieff D. M. Luthers: Von Dolmetzschen vnd Fürbitt der Heiligen" (Nürnberg, 1530), written near the end of Luther's stay in Coburg in 1530 we read: "And they should not take [nemen] my name [namen]" (D. Martin Luther, *Biblia/Das ist die gantze Heilige Schrifft/Deudsch auffs new zugericht/Wittenberg 1545,* ed. Hans Volz [Munich: Rogner and Bernhard], 248).

The memory of the interpenetration of *-nehmen* and *Nahmen* is alive in Paul Celan as well. In an inconspicuous place in his writings, in the poem "Todtnauberg," dedicated to the meeting with Martin Heidegger in June 1967, we find: "die in das Buch / —wessen Namen nahms auf / vor dem meinen?—, / die in dies Buch / geschriebene Zeile von / einer Hoffnung, heute, / auf eines Denkenden / kommendes / Wort / im Herzen . . . (the line / —whose name did the book / register before mine?— / in that book about / a hope, today, / of a thinking man's / coming / word / in the heart . . .). *Gesammelte Werke*, ed. Beda Allemann (Frankfurt am Main: Suhrkamp, 1983), 2: 255; *Poems of Paul Celan*, trans. Michael Hamburger (New York: Persea, 1972), 293 (trans. modified).

30. *Deutsches Wörterbuch* 5: 3839.

Hölderlin in English: A Bibliography

This bibliography—purposely select—lists book-length translations of Hölderlin's literary writings and the more significant translations of his theoretical texts and letters. (Several entries record second—revised and/or enlarged—editions.) Also listed are most critical studies, devoted in part or in whole to Hölderlin, available in English. Unpublished dissertations, prefaces, introductions, or afterwords to book-length translations, and dictionary or handbook entries have not been listed. Generally, reprints are listed only when texts have been notably revised. The essays included in this volume do not appear.

TRANSLATIONS

Poetry and Prose

Short Poems. Trans. Pierre Loving. Girard, Kans.: Haldeman-Julius, 1925.

Hölderlin's Madness. Trans. David Gascoyne. London: Dent, 1938.

Poems. Trans. Michael Hamburger. London: Nicholson & Watson, 1943.

Some Poems. Trans. Frederick Prokosch. New York: New Directions, 1943.

Selected Poems. Trans. James Blaire Leishman. London: Hogarth, 1944.

The Archipelago. Trans. James Blaire Leishman. London: Critic, n.d.

Poems, 2d enl. ed. Trans. Michael Hamburger. London: Harvill, 1952.

Selected Poems, 2d rev. and enl. ed. Trans. James Blaire Leishman. London: Hogarth, 1954.

Selected Verse. Plain prose trans. Michael Hamburger. Harmondsworth, England: Penguin, 1961.

Alcaic Poems. Trans. Elizabeth Wilhelm Henderson. New York: Ungar, 1962.

The Death of Empedocles, 3d version. Trans. Michael Hamburger. *Quarterly Review of Literature* 13 (1964): 98–121.

Hyperion; or The Hermit in Greece. Trans. Willard Ropes Trask. New York: Signet, 1965.

Poems and Fragments. Trans. Michael Hamburger. London: Routledge & Kegan Paul, 1966.

In Friedrich Hölderlin, Edouard Mörike. *Selected Poems.* Trans. Christopher Middleton. Chicago: University of Chicago Press, 1972.

Poems and Fragments, 2d enl. ed. Trans. Michael Hamburger. Cambridge: Cambridge University Press, 1980.

Bordeaux Memories: A Poem Followed by Five Letters. Trans. Kenneth White. Bordeaux, France: Blake, 1984.

Hymns and Fragments. Trans. Richard Sieburth. Princeton, N.J.: Princeton University Press, 1984.

"Pindar Fragments." Trans. Jeremy Adler. *Comparative Criticism* 6 (1984): 41–46.

"Hyperion" and Selected Poems. Ed. Eric L. Santner. New York: Continuum, 1990.

Selected Poems. Trans. David Constantine. Newcastle, England: Bloodaxe, 1990.

"A Map of Mnemosyne." Trans. Richard Sieburth. *Talisman* 6 (1991): 51–64.

Poems and Fragments, 3rd ed. Trans. Michael Hamburger. London: Anvil Press, 1994.

Selected Poems. 2d exp. ed. Trans. David Constantine. Newcastle, England: Bloodaxe, 1996.

Theoretical Texts and Letters

The Poet's Vocation: Selections from Letters of Hölderlin, Rimbaud, and Hart Crane. Ed. and Trans. William Burford and Christopher Middleton. Austin: University of Texas Press, 1968.

"On the Process of the Poetic Mind." Trans. Ralph R. Read. In *German Romantic Criticism*, ed. Leslie Wilson, 219–37. New York: Continuum, 1982.

"Notes on the *Antigone*." Trans. Jeremy Adler. *Comparative Criticism* 5 (1983): 237–44.

"Notes on the *Oedipus*." Trans. Jeremy Adler. *Comparative Criticism* 5 (1983): 231–37.

"The Ground of the *Empedocles*." Trans. Jeremy Adler. *Comparative Criticism* 7 (1985): 157–67.

"On the Process of Becoming in Passing Away." Trans. Jeremy Adler. *Comparative Criticism* 7 (1985): 168–73.

"Judgment and Being." Trans. Pierre Adler. *Graduate Faculty Philosophy Journal* (New York) 11, no. 1 (1986): 17–18.

Essays and Letters on Theory. Ed. and Trans. Thomas Pfau. New York: State University of New York Press, 1988.

SECONDARY LITERATURE

Abrams, M. H. "Some Other Educational Traveler: Hölderlin's 'Hyperion,' Goethe's 'Faust,' the Romances of Novalis." In *Natural Supernaturalism: Tradition and Revolution in Romantic Literature*, 237–52. New York: Norton, 1971.

Adler, Jeremy. "Introduction [to 'Notes on the *Oedipus*' and 'Notes on the *Antigone*']." *Comparative Criticism* 5 (1983): 205–30.

———. "Philosophical Archaeology: Hölderlin's 'Pindar Fragments.' A Translation with an Interpretation." *Comparative Criticism* 6 (1984): 23–40.

———. "Introduction [to 'The Ground of the *Empedocles*' and 'On the Process of Becoming in Passing Away']." *Comparative Criticism* 7 (1985): 147–56.

Adorno, Theodor W. "Parataxis: On Hölderlin's Late Poetry." In *Notes to Literature*, trans. Shierry Weber Nicholsen, vol. 2, 109–49. New York: Columbia University Press, 1992.

Ahearn, Edward J. "The Search for Community: The City in Hölderlin, Wordsworth, and Baudelaire." *Texas Studies in Literature and Language* 13 (1971): 71–89.

Allen, Robert van Roden. "Pathway to Hölderlin." *Clio* 17, no. 4 (1988): 311–21.

Arend, Jutta. "Crazy Horse in Mecklenburg: *Indianervisionen* in Joochen Laabs' *Der Schattenfänger*." *German Quarterly* 65, nos. 3–4 (1992): 407–13.

Aylesworth, Gary. "Heidegger and Hölderlin." *Philosophy Today* 32 (1988): 143–55.

Bach, Emmon. "The Syntax of Hölderlin's Poems: I." *Texas Studies in Literature and Language* 3 (1960): 383–97.

———. "'In lieblicher Bläue': Hölderlin or Waiblinger?" *Germanic Review* 36 (1961): 27–34.

———. "The Syntax of Hölderlin's Poems: II." *Texas Studies in Literature and Language* 3 (1961): 444–57.

———. "*Einst* and *Jetzt* in Hölderlin's Works." *Deutsche Beiträge zur geistigen Überlieferung* 5 (1965): 143–56.

Baeumer, Max Lorenz. "Nietzsche and the Tradition of the Dionysian." In *Studies in Nietzsche and the Classical Tradition*, ed. James C. O'Flaherty, Timothy F. Sellner, and Robert M. Helm, 165–89. Chapel Hill: University of North Carolina Press, 1976.

Bahti, Timothy. "Lightning Bolts, Arrows, Weather Vanes, and the Crux of Soullessness: Directions and Ends of the Poetic Image in Three Hölderlin Poems." In *Ends of the Lyric: Direction and Consequence in Western Poetry*, 97–145. Baltimore: Johns Hopkins University Press, 1996.

Baker, John Jay. "The Problem of Poetic Naming in Hölderlin's Elegy 'Brod und Wein.'" *Modern Language Notes* 101, no. 3 (1986): 465–92.

———. "Poetic Calling, Poetic Failure: Self-Reflexivity in Texts of Hölderlin, Wordsworth, and Shelley." *Modern Language Notes* 105, no. 5 (1990): 914–38.

Barrows, Herbert. "Hölderlin and the English-Speaking Reader." In *Friedrich Hölderlin: An Early Modern*, ed. Emery E. George, 27–53. Ann Arbor: University of Michigan Press, 1972.

Bayley, John. "Lowell and Hölderlin: A Note and Suggestion." *Agenda* 18, no. 3 (1980): 30–33.

———. "The Poetry of Existence." *Agenda* 32, no. 2 (1994): 125–28.

Beare, Robert Livingston. "'Patmos. Dem Landgrafen von Homburg.'" *Germanic Review* 28 (1953): 5–22.

Beckmann, G. H. A. "Some Observations on Michael Hamburger's Art of Translating Poetry." *The Journals of Pierre Menard* 1 (1969): 33–43.

Behler, Ernst. "Jena and Early Romanticism." In *German Romantic Literary Theory*, 14–24. Cambridge: Cambridge University Press, 1993.

Benn, Maurice Bernard. "Hölderlin." *Proceedings of the Australian Goethe Society* 6 (1954–55): 55–65.

———. "Hölderlin's Optimism: A Reply to Dr. L. J. Ryan." *German Life and Letters* 15 (1961–62): 189–93.

———. *Hölderlin and Pindar*. The Hague: Mouton, 1962.

———. "The Dramatic Structure of Hölderlin's *Empedokles*." *Modern Language Review* 62 (1967): 92–97.

———. "Hölderlin and the Greek Tradition." *Arion* 6 (1967): 495–516.

———. "Hölderlin's 'Achill': The Genesis of a Poem." *German Life & Letters* 23 (1969–70): 295–303.

Berman, Antoine. "Hölderlin: The National and the Foreign." In *The Experience of the Foreign: Culture and Translation in Romantic Germany*, trans. S. Heyvaert, 157–74. Albany: State University of New York Press, 1992.

Bernd, Clifford A. "The Formative Qualities of Hölderlin's 'Wink für Darstellung und Sprache.'" *Modern Language Review* 60 (1965): 400–404.

Bernhardt, Walter. "Setting a Poem: The Composer's Choice For or Against Interpretation." *Yearbook of Comparative and General Literature* 37 (1988): 32–46.

Bersier, Gabrielle. "A Lone Hellenist: Johann Friedrich Hölderlin." In *A Concise History of German Literature to 1900*, 188–89. Columbia, S.C.: Camden House, 1992.

Blackall, Eric Albert. "Toward a Poetic Novel: Jean Paul and Hölderlin." In *The Novels of the German Romantics*, 65–106. Ithaca, N.Y.: Cornell University Press, 1983.

Blanchot, Maurice. "Hölderlin's Itinerary." In *The Space of Literature*, trans. Ann Smock, 269–76. Lincoln: University of Nebraska Press, 1982.

———. "The 'Sacred Speech' of Hölderlin." In *The Work of Fire*, trans. Charlotte Mandell, 111–31. Stanford, Calif.: Stanford University Press, 1995.

Blunden, Allan. "The Question of Time: Notes on Hölderlin's 'Sonnenuntergang' and Lessing's 'Laokoon.'" *Modern Language Review* 69 (1974): 821–28.

Bonney, William. "Hölderlin's *Empedokles*: Discontinuous Metaphor as Character." *Neophilologus* 74 (1990): 240–48.

Böschenstein, Bernhard. "Peter Szondi: 'Studies on Hölderlin.' Exemplarity of a Path." Trans. Kristine Anderson and Marilyn Gaddis-Rose. *Boundary 2* 11, no. 3 (1982–83): 93–110.

Bowie, Andrew. "Reflections on the Subject: Fichte, Hölderlin, and Novalis." In *Aesthetics and Subjectivity: From Kant to Nietzsche*, 81–103. Manchester, England: Manchester University Press, 1990.

Bowra, Cecile Maurice. "Hölderlin's Hymns." In *Inspiration and Poetry*, 130–52. New York: St. Martin's Press, 1955.

Braun, Wilhelm Alfred. "Hölderlin." In *Types of Weltschmerz in German Poetry*, 9–34. New York: Columbia University Press, 1905.

Brown, Calvin S. "Josef Winheber's Hölderlin-Variations: A Comment and Translation." *Southern Humanities Review* 2 (1968): 468–79.
Brown, Marshall. "The Eccentric Path." In *The Shape of German Romanticism*, 161–70. Ithaca, N.Y.: Cornell University Press, 1979.
Bruton, Kevin J. "Luis Cernuda's Debt to Hölderlin: Symbolical Reference and Internal Rhythm." *Revue de littérature comparée* 58, no. 1 (1984): 37–49.
Burwick, Fred. L. "Hölderlin and Arnold: Empedocles on Etna." *Comparative Literature* 17 (1965): 24–42.
Butler, Eliza Marian. "The Martyr: Hölderlin." In *The Tyranny of Greece over Germany: A Study of the Influence Exercised by Greek Art and Poetry over the Great German Writers of the Eighteenth, Nineteenth, and Twentieth Centuries*, 201–40. Cambridge: Cambridge University Press, 1935.
Campos, Haroldo de. "Hölderlin's Red Word." Trans. Albert G. Bork. *Twentieth Century Studies* 11, no 6 (1973–74): 4–10.
Chancellor, Gary. "Hölderlin, Brecht, Anouilh: Three Versions of *Antigone*." *Orbis Litterarum* 34 (1979): 87–97.
Christiansen, Rupert. "Despondency and Madness." In *Romantic Affinities: Portraits from an Age, 1780–1830*, 53–94. London: Bodley Head, 1988.
Chytry, Josef. "The Early Hölderlin, Hegel, and Schelling: Dialectics, Revolution, and the 'Theocracy of the Beautiful.'" In *The Aesthetic State: A Quest in Modern German Thought*, 125–47. Berkeley: University of California Press, 1989.
———. "Hölderlin: Dialectics of Tragedy." In *The Aesthetic State*, 148–77.
Closs, August. "Hölderlin." In *The Genius of the German Lyric: An Historic Survey of Its Formal and Metaphysical Values*, 264–72. London: Allen & Unwin, 1938.
———. "Hölderlin's Tragic Vision of Hellas and Germany." *The Gate* 2, nos. 3–4 (1949): 27–31.
———. "Hellas, Christianity, Germany." In *Medusa's Mirror: Studies in German Literature*, 120–33. London: Cresset, 1957.
Coates, Paul. "Form and Formlessness in the Poetry of Hölderlin: The Implications of His Theory of 'Wechsel der Töne.'" In *Words After Speech: A Comparative Study of Romanticism and Symbolism*, 43–65. London: St. Martin's Press, 1986.
———. "Hölderlin, Rilke, Celan: Towards an Understanding of the Late Work." In *Words After Speech*, 145–51.
Constantine, David. "Journeying and Homecoming in the Life and Poetry of Hölderlin." *Oxford German Studies* 5 (1970): 28–47.
———. "Hölderlin's Pindar: The Language of Translation." *Modern Language Review* 73 (1978): 825–34.
———. "The Meaning of a Hölderlin Poem." *Oxford German Studies* 9 (1978): 45–67.
———. *The Significance of Locality in the Poetry of Friedrich Hölderlin*. London: MHRA, 1979.

———. *Early Greek Travellers and the Hellenic Ideal.* Cambridge: Cambridge University Press, 1984.
———. "Translation and Exegesis in Hölderlin." *Modern Language Review* 81, no. 2 (1986): 388–97.
———. *Hölderlin.* Oxford: Clarendon, 1988.
———. "Hölderlin's Translations." *PN Review* 15, no. 5 (1989): 21–22.
Cook, Albert Spalding. "Hölderlin's Brink." In *Thresholds: Studies in the Romantic Experience*, 83–96. Madison: University of Wisconsin Press, 1985.
Corngold, Stanley. "Hölderlin and the Interpretation of the Self." *Comparative Criticism* 5 (1983): 187–200.
———. "Hölderlin and the Question of Self." In *The Fate of the Self: German Writers and French Theory*, 21–53. New York: Columbia University Press, 1986.
———. "Hölderlin's Poetry and the Persistence of the Self." In *Literature as Philosophy/Philosophy as Literature*, ed. Donald G. Marshall, 205–31. Iowa City: University of Iowa Press, 1987.
———. "Nietzsche's Moods." *Studies in Romanticism* 29, no. 1 (1990): 67–90.
Dallmayr, Fred R. "Heidegger, Hölderlin, and Politics." *Heidegger Studies* 2 (1986): 81–95.
DelCaro, Adrian. "The Columbus Poems of Hölderlin and Nietzsche." *Colloquia Germanica* 21, nos. 2–3 (1988): 144–58.
———. *Hölderlin: The Poetics of Being.* Detroit: Wayne State University Press, 1991.
———. "A Hölderlinian Background to Hesse's *Morgenlandfahrt.*" *German Life & Letters* 46, no. 3 (1993): 254–65.
———. "Paul Celan's Uncanny Speech." *Philosophy and Literature* 18, no. 2 (1994): 211–24.
Derrida, Jacques. *Memoires: For Paul de Man.* Trans. Cecile Lindsay, Jonathan Culler, and Eduardo Cadava. New York: Columbia University Press, 1986.
———. *Of Spirit.* Trans. Geoffrey Bennington and Rachel Bowlby, 73–82. Chicago: University of Chicago Press, 1989.
———. "Schibboleth: For Paul Celan." Trans. Joshua Wilner. In *Word Traces: Readings of Paul Celan*, ed. Aris Fioretos, 3–72. Baltimore: Johns Hopkins University Press, 1994.
Detweiler, Robert. "Martin Heidegger: 'Hölderlin and the Essence of Poetry.'" In *Story, Sign, and Self: Phenomenology and Structuralism as Literary Critical Methods*, 72–76. Philadelphia: Fortress, 1978.
Dibble, Gisela V. "Antigone: From Sophocles to Hölderlin and Brecht." In *Legacy of Thespis*, ed. Karelisa V. Hartigan, 1–11. Lanham, Md.: University Press of America, 1984.
Düsing, Wolfgang. "Schiller Versus Hölderlin: The Condemnation of the Classics in Peter Weiss' Hölderlin-Drama." Trans. Regina Knecht. In *Friedrich von Schiller and the Drama of Human Existence*, papers from the Hofstra University Conference on Friedrich von Schiller, 164–70. New York: Greenwood, 1988.

Else, Gerald F. "Sophocles the Elusive." In *Friedrich Hölderlin: An Early Modern*, ed. Emery E. George, 119–34. Ann Arbor: University of Michigan Press, 1972.

Emad, Parvis and Kenneth Maly. "Poetic Saying as Beckoning: The Opening of Hölderlin's 'Germanien.'" *Research in Phenomenology* 19 (1989): 121–38.

Enright, D. J. "'That the Existing Be Well Interpreted': Hölderlin in English." In *Man Is an Onion: Reviews and Essays*, 117–22. London: Chatto and Windus, 1972.

Fehervary, Helen. *Hölderlin and the Left: The Search for a Dialectic of Art and Life*. Heidelberg: Winter, 1977.

Feldman, Burton, and Robert D. Richardson. "Friedrich Hölderlin." In *The Rise of Modern Mythology: 1680–1860*, 328–32. Bloomington: University of Indiana Press, 1972.

Fiedler, Theodore. "Hölderlin and Trakl's Poetry of 1914." In *Friedrich Hölderlin: An Early Modern*, ed. Emery E. George, 87–105. Ann Arbor: University of Michigan Press, 1972.

Findlay, L. M. "From Helikon to Aetna: The Precinct of Poetry in Hesiod, Empedocles, Hölderlin, and Arnold." In *The Existential Coordinates of the Human Condition, Poetic, Epic, Tragic: The Literary Genre*, 119–40. Dordrecht, Holland: Reidel, 1984.

Fineman, Clarissa. *Hölderlin's Philosophy with Particular Reference to Plato*. Detroit: Harlo, 1981.

Fioretos, Aris. "Nothing: History and Materiality in Celan." In *Word Traces: Readings of Paul Celan*, ed. Aris Fioretos, 295–341. Baltimore: Johns Hopkins University Press, 1994.

Fitzell, Henry John. "The Hermit and Nature: Hölderlin." In *The Hermit in German Literature: From Lessing to Eichendorff*, 95–101. Chapel Hill: University of North Carolina Press, 1961.

Fitzgerald, William. *Agonistic Poetry: The Pindaric Mode in Pindar, Horace, Hölderlin, and the English Ode*. Berkeley: University of California Press, 1987.

Fleming, Ray. *Keats, Leopardi, and Hölderlin: The Poet as Priest of the Absolute*. New York: Garland, 1987.

Fóti, Véronique M. "Mnemosyne's Death and the Failure of Mourning." In *Heidegger and the Poets: Poiésis, Sophia, Technē*, 60–77. Atlantic Highlands, N.J.: Humanities Press, 1992.

———. "Textuality and the Question of Origin: Poiesis, Sophia, Techne." In *Heidegger and the Poets*, 44–59.

Fredén, Gustaf. "Mother of All—Mother of God." In *Orpheus and the Goddess of Nature*, 154–76. Göteborg, Sweden: Elanders, 1958.

Frey, Hans-Jost. "Hölderlin." Trans. William Whobrey. In *Studies in Poetic Discourse: Mallarmé, Baudelaire, Rimbaud, Hölderlin*, 135–91. Stanford, Calif.: Stanford University Press, 1996.

Friedländer, Benjamin. "'Aporia' after Friedrich Hölderlin." *Talisman* 6 (1991): 93–95.

Frye, Lawrence O. "Seasonal and Psychic Time in the Structuring of Hölderlin's *Hyperion.*" In *Friedrich Hölderlin: An Early Modern,* ed. Emery E. George, 148–79. Ann Arbor: University of Michigan Press, 1972.
Furness, Raymond. "Hölderlin and the Problem of Solipsism and Abandonment." *German Life and Letters* 22 (1968–69): 121–26.
———. "The Death of Memory: An Analysis of Hölderlin's Hymn 'Mnemosyne.'" *Publications of the English Goethe Society* 40 (1969–70): 30–68.
Fürst, Norbert. "Vigny, Leopardi, Hölderlin." *Modern Language Forum* 24 (1939): 98–105.
———. "Three German Novels of Education: 1. Hölderlin's *Hyperion*; 2. Stifter's *Nachsommer*; 3. R. M. Rilke's *Malte Laurids Brigge.*" *Monatsschrifte für deutschen Unterricht, deutsche Sprache und Literatur* 38 (1946): 339–47; 413–25; and 463–78.
Fynsk, Christopher. "Hölderlin's Testimony: An Eye Too Many Perhaps." In *Heidegger: Thought and Historicity,* 174–229. Ithaca, N.Y.: Cornell University Press, 1986.
Gabriel, Norbert. "Paul de Man on Hölderlin." Trans. John Jay Baker in collaboration with author. In *(Dis)continuities: Essays on Paul de Man,* ed. Luc Herman, Kris Humbeeck, and Geert Lernout, 111–32. Amsterdam: Rodopi, 1989.
Galinsky, Gotthard Karl. *The Herakles Theme: The Adaptions of the Hero in Literature from Homer to the Twentieth Century.* Oxford: Blackwell, 1972.
Gallie, W. B. "The Religious Poetry of Hölderlin." *The Dublin Review* 438 (1946): 59–75.
Garber, Frederick. "The Adequacy of Imagination." In *The Autonomy of the Self from Richardson to Huysmans,* 122–49. Princeton, N.J.: Princeton University Press, 1982.
———. "Self, Society, Value." In *The Autonomy of the Self from Richardson to Huysmans,* 150–73. Princeton, N.J.: Princeton University Press, 1982.
Gaskill, Peter Howard. "Some New Hölderlin Decipherments from the 'Homburger Folioheft.'" *Publications of the Modern Language Association* 80 (1965): 123–40.
———. "The 'Fürst des Fests' in Hölderlin's 'Friedesfeier.'" *Modern Language Review* 65 (1970): 94–115.
———. "Hölderlin and Revolution." *Forum for Modern Language Studies* 12 (1976): 118–36.
———. "Edwin Muir as a Critic of Hölderlin." *Forum for Modern Language Studies* 14 (1978): 345–64.
———. "Meaning in History: 'Chiliasm' in Hölderlin's 'Patmos.'" *Colloquia Germanica* 11 (1978): 19–52.
———. "Hölderlin and the Poetry of Edwin Muir." *Forum for Modern Language Studies* 16 (1980): 12–32.

———. "Hölderlin's 'Friedensfeier' and the 'Living Building.'" *New German Studies* 8 (1980): 169–88.

———. "'Ich seh,' ich sehe, wie das enden muß . . .': Observations on a Misunderstood Passage in Hölderlin's *Hyperion*." *Modern Language Review* 76 (1981): 612–18.

———. "Some Recent Trends in Hölderlin Criticism." *German Life and Letters* 36 (1982–83): 166–81.

———. *Hölderlin's* Hyperion. Durham, N.C.: University of Durham Press, 1984.

———. "Open Circles: Hoffmann's *Kater Murr* and Hölderlin's *Hyperion*." *Colloquia Germanica* 1, 19 (1986): 21–46.

———. "'Ossian' Macpherson: Towards a Rehabilitation." *Comparative Criticism* 8 (1986): 113–46.

———. "'Diss ist die Zeit der Könige nicht mehr': Hölderlin and Republicanism." *Strathclyde Modern Language Studies* 11 (1991): 39–56.

———. "Hölderlin and Ossian." *London German Studies* 4 (1992): 147–65.

George, Emery Edward. "A Family of Disputed Readings in Hölderlin's Hymn 'Der Rhein.'" *Modern Language Review* 61 (1966): 619–34.

———. "On the Authorship of a Variant in Hölderlin's Hymn 'Der Rhein.'" *Texas Studies in Literature and Language* 9 (1967): 391–414.

———. "Hölderlin's Hymn 'Patmos': Comments." In *Friedrich Hölderlin: An Early Modern*, 258–76. Ann Arbor: University of Michigan Press, 1972.

———. *Hölderlin's "Ars Poetica": A Part-Rigorous Analysis of Information Structure in the Late Hymns*. The Hague: Mouton, 1973.

———. "An Unknown Hymn by Hölderlin?" *Journal of English and Germanic Philology* 86, no. 2 (1987): 165–74.

———. "Hölderlin and His Biographers." *Journal of English and Germanic Philology* 89, no. 1 (1990): 51–85.

———. *Hölderlin and the Golden Chain of Homer: Including an Unknown Source*. Lewiston, N.Y.: Mellen, 1992.

———, ed. *Friedrich Hölderlin: An Early Modern*. Ann Arbor: University of Michigan Press, 1972.

Gethmann-Siefert, Annemarie. "Heidegger and Hölderlin: The Overusage of 'Poets in Impoverished Time.'" Trans. Richard Taft. *Research in Phenomenology* 19 (1989): 59–88.

Gladstone, Alice. "Hölderlin's 'Patmos': Voyage as Homecoming." *Quarterly Review of Literature* 10 (1959): 64–76.

Golb, Joel. "Reading Celan: The Allegory of 'Hohles Lebensgehöft' and 'Engführung.'" In *Word Traces: Readings of Paul Celan*, ed. Aris Fioretos, 185–218. Baltimore: Johns Hopkins University Press, 1994.

Goodson, Alfred Clement. "Hölderlin and the Bounds of Romantic Metaphor." In *Deutsche Romantik and English Romanticism*, ed. Theodore G. Gish, 42–51. Munich: Fink, 1984.

Goozé, Marjanne E. "Hölderlin's Sophocles: An (Un)authorized Translation." *Studies in the Modern and Classical Languages and Literatures* 1 (1988): 135–44.

Green, André. "Hölderlin: Drunken Knowledge and the God-Man Coupling." In *The Tragic Effect*, trans. Alan Sheridan, 225–30. Cambridge: Cambridge University Press, 1979.

Grigson, Geoffrey. "A Poet's Tower." *The Observer*, 3 February 1957.

Grugan, Arthur A. "Heidegger: Preparing to Read Hölderlin's 'Germanien.'" *Research in Phenomenology* 19 (1989): 139–67.

———. "Questions from Heidegger's Hölderlin Interpretations." Philosophy Today 36, no. 2 (1992): 114–21.

———. "The Instress of Being and the Chronach of Dasein in Hölderlin's 'Germanien.'" *Heidegger Studies* 9 (1993): 72–95.

Grundlehner, Philip Eicks. "Friedrich Hölderlin: 'Heidelberg.'" In *The Lyrical Bridge: Essays from Hölderlin to Benn*, 27–40. Rutherford, N.J.: Fairleigh Dickinson University Press, 1979.

Haar, Michel. "Heidegger and the God of Hölderlin." *Research in Phenomenology* 19 (1989): 89–100.

Halliburton, David. "The Essence of Poetry: Hölderlin." In *Poetic Thinking: An Approach to Heidegger*, 77–112. Chicago: University of Chicago Press, 1981.

Hamburger, Michael. "The Sublime Art: Milton and Hölderlin." In *Reason and Energy: Studies in German Literature*, 43–65. London: Weidenfeld & Nicolson, 1970.

———. "Hölderlin." In *Art as Second Nature: Occasional Pieces 1950–74*, 57–63. Manchester, England: Carcanet, 1975.

———. "Purity and Punishment." *Times Literary Supplement*, 10 September 1976.

———. "Brief Afterthoughts on Versions of a Poem by Hölderlin." In *Translating Poetry: The Double Labyrinth*, 51–56. Basingstoke, England: Macmillan, 1989.

Hamlin, Cyrus. "'Hälfte des Lebens': An Essay in Criticism." *Helicon* 1 (1966): 36–62.

———. "Hölderlin in Perspective: 1770–1970." *Seminar* 7 (1971): 123–43.

———. "Hölderlin's 'Heidelberg' as Poetic Myth." *Seminar* 7 (1971): 175–90.

———. "Hölderlin's Elegy 'Homecoming': Comments." In *Friedrich Hölderlin: An Early Modern*, ed. Emery E. George, 232–45. Ann Arbor: University of Michigan Press, 1972.

———. "The Poetics of Self-Consciousness in European Romanticism: Hölderlin's *Hyperion* and Wordsworth's *Prelude*." *Genre* 6 (1973): 142–77.

———. "The Temporality of Selfhood: Metaphor and Romantic Poetry." *New Literary History* 6 (1974–75): 169–93.

———. "Tragedy and Modernity: Hölderlin, Hegel, and the Theory of Modern Tragedy." In *Tragique et tragédie dans la tradition occidentale*, ed. Pierre Gravel and Timothy J. Reiss, 151–65. Montreal: Presses de l'Université de Montréal, 1979.

———. " 'Stimmen des Geschiks': The Hermeneutics of Unreadability." In *Jenseits des Idealismus: Hölderlins letzte Homburger Jahre (1804–1806)*, ed. Christoph Jamme and Otto Pöggeler, 252–76. Bonn: Bouvier, 1988.

———. "Literary History after New Criticism: Paul de Man's Essays on Romanticism and Modernity." In *(Dis)continuities: Essays on Paul de Man*, ed. Luc Herman, Kris Humbeeck, and Geert Lernout, 133–48. Amsterdam: Rodopi, 1989.

Hanak, Miroslav John. "Transition from Classicism to Romanticism: Hölderlin's Mythic Poetry and Dialogue with a Mute Transcendence." In *A Guide to Romantic Poetry in Germany*, 71–91. New York: Lang, 1987.

Hargreaves, Raymond. " 'On a Fine Day': Hölderlin in English." *Agenda* 32, no. 2 (1994): 129–34.

Harries, Karsten. "Heidegger and Hölderlin: The Limits of Language." *The Personalist* 44 (1963): 5–23.

———. "The Root of All Evil: Lessons of an Epigram." *International Journal of Philosophical Studies* 1, no. 1 (1993): 1–20.

Harris, Henry Silton. "Appendix: The 'Earliest System-Programme of German Idealism.' " In *Hegel's Development: Toward the Sunlight, 1770–1801*, 249–57. Oxford: Clarendon, 1972.

Harris, Sylvia C. "The Figure of Melitta in Grillparzer's Sappho." *Journal of English and Germanic Philology* 60 (1961): 102–10.

Harrison, Robert B. *Hölderlin and Greek Literature*. Oxford: Oxford University Press, 1975.

———. "Hölderlin's Titans and the Book of Revelation: An Eschatological Interpretation of History." *Publications of the English Goethe Society* 46 (1975–76): 31–64.

Hartley, Catherine. "In Pursit of Past Lives: Four Biographies." *Journal of Australasian Universities Language and Literature Association* 67 (1987): 115–23.

Hatfield, Henry. "The Greek Gods and Christ: Hölderlin." In *Aesthetic Paganism in German Literature: From Winckelmann to the Death of Goethe*, 142–65. Cambridge: Harvard University Press, 1964.

Haverkamp, Anselm. "Error in Mourning—A Crux in Hölderlin: 'dem gleich fehlet die Trauer' ('Mnemosyne')." Trans. Vernon Chadwick. *Yale French Studies* 69 (1985): 238–53.

———. *Leaves of Mourning: Hölderlin's Late Work, With an Essay on Keats and Melancholy*. Trans. Vernon Chadwick. Albany: State University of New York Press, 1996.

Hayden-Roy, Priscilla. *A Foretaste of Heaven: Friedrich Hölderlin in the Context of Württemberg Pietism*. Amsterdam: Rodopi, 1994.

———. "New and Old Histories: The Case of Hölderlin and Württemberg Pietism." *Clio* 21, no. 4 (1994): 369–79.

Heidegger, Martin. *Hölderlin's Hymn "The Ister"*. Trans. William McNeill and Julia Davis. Bloomington: Indiana University Press, 1996.

Heller, Erich. "Thinking about Poetry, Hölderlin, and Heidegger." In *Herkommen und Erneuung: Essays für Oskar Seidlin*, ed. Gerald Gillespie and Edgar Lohner, 168–84. Tübingen: Niemeyer, 1976.

Heller, Erich and Anthony Thorlby. "Idealism and Religious Vision in the Poetry of Hölderlin." *Quarterly Review of Literature* 10, nos. 1–2 (1959): 23–40.

Henrich, Dieter. *The Course of Remembrance and Other Essays on Hölderlin*. Trans. Abraham Anderson and Taylor Carman. Ed. Eckart Förster. Stanford, Calif.: Stanford University Press, 1997.

Herminghouse, Patricia. "Trends in Literary Reception: Coming to Terms with Classicism. Goethe in GDR Literature of the 1970's." *German Quarterly* 56 (1983): 273–84.

Herrmann, Friedrich-Wilhelm von. " 'The Flower of the Mouth' ": Hölderlin's Hint for Heidegger's Thinking of the Essence of Language." Trans. Parvis Emad and Kenneth Maly. In *Martin Heidegger: Critical Assessments*, ed. Christopher Macann, 277–92. London: Routledge, 1992.

Hilton, Ian. "Vernon Watkins and Hölderlin." *Poetry Wales* 12, no. 4 (1977): 101–17.

Hobson, Irmgard. "Hölderlin, Our Contemporary: *Empedokles* and German Intellectuals of the 1970's." *Journal of European Studies* 8 (1978): 258–73.

Hoffmeister, Donna L. "Hölderlin-Biography, 1924–1982: Transformations of a Literary Life." *Seminar* (Toronto) 21, no. 3 (1983): 207–31.

Hofmannsthal, Hugo von. "Vienna Letter." *The Dial* 75 (1923): 271–77.

Honig, Edwin. "A Conversation with Christopher Middleton." *Modern Language Notes* 91 (1976): 1588–1602.

Hopper, Stanley Romaine. "On the Naming of Gods in Hölderlin and Rilke." In *Christianity and the Existentialists*, 148–90. New York: Scribner, 1956.

Hutinet, Eva. T. "Pascal, Diderot, Hölderlin: A New Analysis." *Neohelicon* 18, no. 1 (1991): 219–23.

Jakobson, Roman, and Grete Lübbe-Grothues. "The Language of Schizophrenia: Hölderlin's Speech and Poetry." *Poetics Today* 24 (1980): 137–44.

———. "Two Types of Discourse in Hölderlin's Madness." Trans. Susan Abrams. In *Cognitive Constraints on Communication: Representations and Processes*, ed. Jaako Hintikka and Lucia Vaina, 115–36. Dordrecht, Holland: Reidel, 1984.

Jamme, Christoph. "Hegel und Hölderlin." *Clio* 15, no. 1 (1985): 359–77.

Jennings, Michael W. "Benjamin as a Reader of Hölderlin: The Origins of Benjamin's Theory of Literary Criticism." *German Quarterly* 56, no. 4 (1983): 544–62.

———. "The Mortification of the Text: The Nature of the Work of Art and Its Criticism." In *Dialectical Images: Walter Benjamin's Theory of Literary Criticism*, 164–211. Ithaca, N.Y.: Cornell University Press, 1987.

———. "The Permanent Catastrophe: Benjamin's Philosophy of History." In *Dialectical Images: Walter Benjamin's Theory of Literary Criticism*, 42–81.

Jens, Walter and Hans Küng. "Friedrich Hölderlin, *Hymns*." In *Literature and*

Religion: Pascal, Gryphius, Lessing, Hölderlin, Novalis, Kierkegaard, Dostoyevsky, Kafka, trans. Peter Heinegg, 107–43. New York: Paragon House, 1991.

Juhl, Peter D. "The Doctrine of 'Verstehen' and the Objectivity of Literary Interpretations." In *Interpretation: An Essay in the Philosophy of Literary Criticism*, 239–300. Princeton, N.J.: Princeton University Press, 1980.

Kahn, Ludwig W. "Romanticism: Community. Hölderlin." In *Social Ideals in German Literature 1770–1830*, 68–75. New York: AMS Press, 1938.

Kavanagh, R. J. "Friedrich Hölderlin and Jacobinism." *1789: Reading, Writing, Revolution*, proceedings of the Essex Conference on the Sociology of Literature, July 1981, 198–208. Colchester, England: University of Essex, 1982.

Kerkhoff, Manfred. "Timelessness and Translation." *Texte* 4 (1985): 41–52.

Kermode, Frank. "Dwelling Poetically in Connecticut." In *Wallace Stevens: A Celebration*, ed. Frank Doggett and Robert Buttel, 256–73. Princeton, N.J.: Princeton University Press, 1980.

Kirkland, John D., Jr. "Reading-Writing the Labyrinth of Reading-Writing." *Boundary 2* 5 (1977): 723–42.

Klabes, Günter. "Political Reality and Poetic Mission: Hölderlin's and Shelley's Heterocosm." In *English and German Romanticism: Cross-Currents and Controversies*, ed. James Pipkin, 301–21. Heidelberg: Winter, 1985.

Kling, Wilfred Ludwig. "*Lesearbeit*: Friedrich Hölderlin's 'Der Winkel von Hahrdt' and the *Nachtgesänge* in the Frankfurt and Stuttgart Editions of His Works." *Modern Language Notes* 94 (1979): 587–600.

Kohns, Hans. "The Eve of German Nationalism (1789–1812)." *Journal of the History of Ideas* 12 (1951): 256–84.

Kramer, Lawrence. "The Return of the Gods: Keats to Rilke." *Studies in Romanticism* 17 (1978): 483–500.

Krell, David Farrell. "The Oldest Program Towards a System in German Idealism." *The Owl of Minerva* 17, no. 1 (1985): 5–19.

———. "Hölderlin's 'Hymn to Serenity': A Newly Discovered Poem." *Graduate Faculty Philosophy Journal* 11, no. 1 (1986): 3–15.

———. "Nietzsche, Hölderlin, Empedocles." *Graduate Faculty Philosophy Journal* 15, no. 2 (1991): 31–48.

———. "The Sensuality of Tragedy, the Tragedy of Sensuality." In *Lunar Voices: Of Tragedy, Poetry, Fiction, and Thought*, 3–23. Chicago: University of Chicago Press, 1995.

———. "Stuff • Thread • Point • Fire: Hölderlin's Dissolution." In *Lunar Voices*, 24–51.

Kudszus, Winfried. "Acknowledgments: 'An Georg Trakl' by Robert Walser." In *The Dark Flutes of Fall: Critical Essays on Georg Trakl*, ed. Eric Williams, 264–83. Columbia. S.C.: Camden House, 1991.

Kurz, Gerhard. *The Great Drama: Germany and the French Revolution.* Trans. Robert Elsie. Bonn: Inter Nationes, 1989.

Kuzniar, Alice A. "Hölderlin," In *Delayed Endings: Nonclosure in Novalis and Hölderlin*, 133–90. Athens: University of Georgia Press, 1987.

Lacoue-Labarthe, Philippe. "The Caesura of the Speculative." Trans. Robert Eisenhauer. In *Typographies: Mimesis, Philosophy, Politics*, ed. Christopher Fynsk, intro. Jacques Derrida, 208–35. Cambridge: Harvard University Press, 1989.

———. "Hölderlin and the Greeks." Trans. Judi Olson. In *Typographies: Mimesis, Philosophy, Politics*, 237–47.

———. "Catastrophe." Trans. Andrea Tarnowski. In *Word Traces: Readings of Paul Celan*, ed. Aris Fioretos, 130–56. Baltimore: Johns Hopkins University Press, 1994.

Ledes, Richard C. "From a Screenplay Entitled *Anti* (Going, Going) Gone (Based on *Antigone* by Sophocles)." *Talisman* 6 (1991): 47–50.

Leeder, Karen. "Towards a Profane Hölderlin: Representations and Revisions of Hölderlin in Some GDR Poetry." In *Neue Ansichten: The Reception of Romanticism in the Literature of the GDR*, ed. Howard Gaskill, Karin McPherson, and Andrew Barker, 212–31. Amsterdam: Rodopi, 1990.

Lernout, Geert Maria Jan. *The Poet as Thinker: Hölderlin in France*. Columbia, S.C.: Camden House, 1994.

Lindsay, J. M. "The Education of Hölderlin and Mörike." *Modern Languages* 45 (1964): 61–68.

Llewelyn, John. "Ontological Responsibility and the Poetics of Nature." *Research in Phenomenology* 19 (1989): 3–26.

Low, Peter. "Jaccottet and German Literature: The Influence of Rilke, Musil, and Hölderlin." *Swiss-French Studies* 4 (1983): 52–65.

MacCall, Tom. "The Case of the Missing Body." *Le Pauvre Holterling* 8 (1988): 53–72.

———. "Plastic Time and Poetic Middles: Benjamin's Hölderlin." *Studies in Romanticism* 31, no. 4 (1992): 481–99.

Malsch, Wilfried. "The Concept of Enlightenment in Hölderlin's Poetry." In *Friedrich Hölderlin: An Early Modern*, ed. Emery E. George, 135–47. Ann Arbor: University of Michigan Press, 1972.

Man, Paul de. "Heidegger's Exegeses of Hölderlin." In *Blindness and Insight: Essays in the Rhetoric of Contemporary Criticism*, 2d enl. ed., intro. Wlad Godzich, 246–66. Minneapolis: University of Minnesota Press, 1983.

———. "The Image of Rousseau in the Poetry of Hölderlin." In *The Rhetoric of Romanticism*, 19–45. New York: Columbia University Press, 1984.

———. "Intentional Structure of the Romantic Image." In *The Rhetoric of Romanticism*, 1–17.

———. "Wordsworth and Hölderlin." In *The Rhetoric of Romanticism*, 47–65.

———. "Keats and Hölderlin." Trans. Dan Latimer. In *Critical Writings, 1953–1978*, ed. Lindsay Waters, 41–60. Minneapolis: University of Minnesota Press, 1989.

———. "The Riddle of Hölderlin." In *Critical Writings, 1953–1978*, 198–213.

———. "The Temptation of Permanence." In *Critical Writings, 1953–1978*, 30–40.

———. "Heaven and Earth in Wordsworth and Hölderlin." In *Romanticism and Contemporary Criticism: The Gauss Seminar and Other Papers*, ed. E. S. Burt, Kevin Newmark, and Andrzej Warminski, 137–46. Baltimore: Johns Hopkins University Press, 1993.

———. "Hölderlin and the Romantic Tradition." In *Romanticism and Contemporary Criticism*, 123–36.

———. "Patterns of Temporality in Hölderlin's 'Wie wenn am Feiertage . . .'" In *Romanticism and Contemporary Criticism*, 50–73.

Markiewicz, Barbara. "Reason and Mythology, or, 'Das älteste Systemmprogram des deutchen Idealismus.'" Trans. Lech Petrowicz. *Dialectics and Humanism* 2, (1987): 147–61.

Mason, Eudo Colecesta. 'Hölderlin and Goethe." *Publications of the English Goethe Society* 22 (1953): 64–83.

———. "Hölderlin's 'Goethe-Erlebnis.'" In *Hölderlin and Goethe*, ed. Peter Howard Gaskill, 77–139 Bern: Lang, 1975.

Masson, David I. "Thematic Analysis of Sounds in Poetry." *Leeds Philosophical and Literary Society* 9 (1959–60): 133–47.

Mauderli, Max O. "Goethe's Evaluation of Hölderlin." *Germanic Review* 25 (1950): 95–108.

McClain, William H. "Symbolic Extensions of the Hyperion-Myth." In *Traditions and Transitions: Studies in Honor of Harld Jantz*, ed. Lieselotte E. Kurth, William H. McClain, and Holger Homann, 177–93. Munich: Delp, 1972.

McCormick, Peter. "Heidegger on Hölderlin." *Philosophical Studies* 22 (1976): 7–15.

McLean, William Scott. "Friedrich Hölderlin and Gary Snyder: *Die Wege zur Wildnis.*" *Exempla* 3, no. 1 (1977): 31–37.

———. "Private Song and the Public Sphere: Some Remarks on the Development of Hölderlin's Later Poetry." In *Goethezeit: Studien zur Erkenntnis und Rezeption Goethes und seiner Zeitgenossen*, ed. Gerhart Hoffmeister, 265–80. Bern: Francke, 1981.

Michael-Titus, C. "The Poetic Mirage of the Absolute." Trans. Leon Levitchi. *Romanian Review* 44, no. 1 (1990): 64–77.

Middleton, Christopher. "Hölderlin. Pindar's Twelfth Pythian. Comment." *Arion* 7 (1968): 243–47.

———. "Syntax and Signification in Hölderlin's 'Andenken.'" In *The Pursuit of the Kingfisher*, 98–119. Manchester, England: Carcanet, 1983.

Miles, David H. "The Past as Future: *Pfad* and *Bahn* as Images of Temporal Conflict in Hölderlin." *Germanic Review* 46 (1971): 95–118.

Miller, Ronald Duncan. *A Study of Hölderlin*. Harrogate, England: Duchy, 1958.

Mitchell, P. M., and Muriel M. Brown. "Recurrent Descriptive Words in Hölderlin." *Germanic Review* 21 (1946): 257–67.

Montgomery, Marshall. "Hölderlin and 'Diotima.'" *Modern Language Review* 7 (1912): 193–207.

———. *Friedrich Hölderlin and the German Neo-Hellenic Movement.* Oxford: Oxford University Press, 1923.

———. "Herder's Influence on the Metrical Version of Hölderlin's *Hyperion.*" In *Studies in the Age of Goethe*, 79–101. Oxford: Oxford University Press, 1931.

———. "Hölderlin's Ideals as Reflected in *Hyperion* and *Empedokles.*" In *Studies in the Age of Goethe*, 102–20.

Moore, Richard. "Fanatical Poets and Reasonable Poets." *Stanford Literature Review* 4, no. 2 (1987): 175–92.

Morford, Mark. "The Riddle of Heracles: Hölderlin and the Trachinae." *Semasia* 2 (1975): 239–47.

Morse, David. "Romanticism and the Infinite." In *Perspectives on Romanticism: A Transformational Analysis*, 213–54. London: Macmillan, 1981.

Most, Glenn Warren. "Hölderlin and the Poetry of History." *Germanic Review* 61, no. 4 (1984): 154–67.

Muir, Edwin. "A Note on Hölderlin." *The Freeman* 7 (1923): 488–90.

———. "Friedrich Hölderlin." *Essays on Literature and Society*, 83–89. London: Hogarth, 1949.

———. "Hölderlin's 'Patmos.'" *Essays on Literature and Society*, 90–102.

Murray, Michael. "Heidegger's Hermeneutic Reading of Hölderlin: The Signs of Time." *The Eighteenth Century* 21 (1980): 41–66.

Nägele, Rainer. "Text, History, and the Critical Subject: Notes on Peter Szondi's Theory and Praxis of Hermeneutics." *Boundary 2* 11, no. 3 (1982–83): 29–51.

———. "The Discourse of the Other: Hölderlin's Voice and the Voice of the People." In *Reading After Freud: Essays on Goethe, Hölderlin, Habermas, Nietzsche, Brecht, Celan, and Freud*, 47–65. New York: Columbia University Press, 1987.

———. "Benjamin's Ground." In *Benjamin's Ground*, ed. Rainer Nägele, 19–37. Detroit: Wayne State University Press, 1988.

Nalbantian, Suzanne. "Wordsworth, Hölderlin, and Their Contemporaries: The Imperial Soul." In *The Symbol of the Soul from Hölderlin to Yeats: A Study in Metonymy*, 13–37. New York: Columbia University Press, 1977.

Nancy, Jean-Luc. "Hyperion's Joy." Trans. Christine Laennec and Michael Syrotinski. In *The Birth to Presence*, 58–81. Stanford, Calif.: Stanford University Press, 1993.

Nasta, Michael M. "On Translating Greek Tragedy: Problems of Poetic Equivalence." Trans. Liana Nasta. *Schriften der Sektion für Altertumswissenschaft* 55, no. 3 (1969): 15–87.

Nauen, Franz Gabriel. *Revolution, Idealism, and Human Freedom: Schelling, Hölderlin, Hegel, and the Crisis of Early German Idealism.* The Hague: Mouton, 1971.

Nicholson, Graeme. "Heidegger on Thinking." *Journal of the History of Philosophy* 13 (1975): 491–503.

Niebylski, Dianna C. "Totality of Fragmentation: The Poet's Dilemma." In *The Poem on the Edge of the Word: The Limits of Language and the Uses of Silence in the Poetry of Mallarmé, Rilke, and Vallejo*, 13–33. New York: Peter Lang, 1993.

O'Brien, William Arctander. "Getting Blasted: Hölderlin's 'Wie wenn am Feiertage . . .'" *Modern Language Notes* 94 (1979): 569–86.

Ogden, Mark. "*Amor dei Intellectualis*: Hölderlin, Spinoza, and St. John." *Deutsche Vierteljahrsschrift für Literaturwissenschaft und Geistesgeschichte* 63, no. 3 (1989): 420–60.

———. "Hölderlin's Wager on Transcendence." *Comparative Criticism* 12 (1990): 273–81.

———. *The Problem of Christ in the Work of Friedrich Hölderlin*. London: Modern Humanities Research Association, 1991.

Olson, Alan M. "Renunciation and Metaphysics: An Examination of Dialectic in Hölderlin and Hegel During Their Frankfurt Period." *Man and World* 15 (1982): 123–48.

Orr, Leonard. "Thinking and Poetry: Heidegger and Literary Criticism." *Studia neophilologica* 60, no. 2 (1988): 207–14.

Payne, Robert. "The Titans." In *The Wanton Nymph: A Study of Pride*, 225–64. London: Heinemann, 1951.

Peacock, Ronald. "Hölderlin's Poetry." *Publications of the English Goethe Society* 12 (1937): 48–64.

———. *Hölderlin*. London: Methuen, 1938.

Peckham, Morse. "Friedrich Hölderlin (1770–1843)." In *The Birth of Romanticism: 1790–1815*, 157–63. Greenwood, Fla.: Penkevill, 1986.

Pelmas, Sarah. "Literary Reading: Hölderlin through Heidegger through de Man." *Surfaces* 3, no. 5 (1993): 1–24.

Perotti, James L. "The Poet and the Thinker." In *Heidegger on the Divine: The Thinker, the Poet, and the God*, 94–114. Athens: University of Georgia Press, 1974.

Peter, Agnes. "An Essay in Romantic Typology: Hölderlin and Shelley: An Inquiry Based on Heidegger's Lectures on Hölderlin." *Neohelicon* 21, no. 2 (1994): 71–86.

Peucker, Brigitte. "Afterword" In *Arcadia to Elysium: Preromantic Modes in 18th Century Germany*, 123–27. Bonn: Bouvier, 1980.

Pike, Burton. "The Idealistic Element in Hölderlin's *Empedokles*." *Germanic Review* 32 (1957): 178–85.

Pitte, Frederick van de. "The Role of Hölderlin in the Philosophy of Heidegger." *The Personalist* 43 (1962): 168–79.

Prawer, Siegbert Salomon. "Approaching the German Lyric." In *German Lyric Poetry: A Critical Analysis of Selected Poems from Klopstock to Rilke*, 9–25. London: Routledge & Keegan Paul, 1952.

Raffel, Burton. "Friedrich Hölderlin: 'Der gute Glaube.'" In *The Forked Tongue: A Study of the Translation Process*, 169–71. The Hague: Mouton, 1971.

Raine, Kathleen. "Germany's Golden Age." *Agenda* 32, no. 2 (1994): 135–43.

Raleigh, Peter J. "Hölderlin: Peter Weiss's Artist in Revolt." *Colloquia Germanica* 7 (1973): 193–213.

Rannit, Aleksis. "Hölderlin and the Twentieth Century: Is Elevated Style in Poetry Possible Today?" In *Friedrich Hölderlin: An Early Modern*, ed. Emery E. George, 106–18. Ann Arbor: University of Michigan Press, 1972.

Rauch, Irmengard. "Dimensions of Sound Change in Relation to an Early Hölderlin Poem." *Linguistics* 34 (1967): 46–54.

Reed, Jeremy. "Realities Ancient and Modern: Mythology." In *The Classical Centre: Goethe and Weimar, 1775–1832*, 194–222. Totowa, N.J.: Barnes & Noble, 1980.

———. "The White Wall." In *Madness: The Price of Poetry*, 137–54. London: Owen, 1989.

Regina, Umberto. "Phenomenology and the Salvation of Truth: Heidegger's Shift in the *Beiträge zur Philosophie*." In *Manifestations of Reason: Life, Historicity, Culture—Part II: Phenomenology in the Adriatic Countries*, 273–305. Ed. Anna-Teresa Tymieniecka. Dordrecht, Holland: Kluwer, 1993.

Richardson, William J. "As When Upon a Day of Rest . . ." In *Heidegger: Through Phenomenology to Thought*, 423–33. The Hague: Mouton, 1963.

———. "'Homecoming,' 'Recollection.'" In *Heidegger*, 440–72.

———. "The Origin of a Work of Art: Hölderlin and the Essence of Poetry." In *Heidegger*, 403–17.

———. "'. . . Poetically Doth Man Dwell . . .'" In *Heidegger*, 588–94.

Roche, Mark William. "Narration and 'Ruhe' in Hölderlin's *Hyperion*: A Reinterpretation of the Novel's Conclusion." In *Dynamic Stillness: Philosophical Conceptions of "Ruhe" in Schiller, Hölderlin, Büchner, and Heine*, 63–119. Tübingen: Niemeyer, 1987.

Ronnell, Avital. "The Televised Metaphysics." In *The Telephone-Book: Technology, Schizophrenia, Electric Speech*, 191–200. Lincoln: University of Nebraska Press, 1989.

Rosenfeld, Alvin. "Hölderlin, Heidegger, and Stevens: 'What Are Poets For?'" *The American Poetry Review* 3, no. 1 (1974): 33–36.

Rosolowski, Tacey A. "Specular Reciprocity and the Construction of the Feminine in Friedrich Hölderlin's *Hyperion*." *Modern Language Studies* 25, no. 3 (1995): 43–75.

Ruin, Hans. Introduction to *Enigmatic Origins: Tracing the Theme of Historicity through Heidegger's Works*, 1–34. Stockholm: Almqvist & Wiksell, 1994.

———. "The Other Beginning." In *Enigmatic Origins*, 247–73.

Ryan, Lawrence. "Hölderlin's Hellenism." *Journal of Australasian Universities Language and Literature Association* 14 (1960): 51–60.

———. "Hölderlin's Tragic Drama." *German Life and Letters* 14 (1960–61): 141–50.

———. "Hölderlin's *Hyperion*: A Romantic Novel?" In *Friedrich Hölderlin: An Early Modern*, ed. Emery E. George, 180–91. Ann Arbor: University of Michigan Press, 1972.

———. "Friedrich Hölderlin (20 March 1770–7 June 1843)." In *German Writers in the Age of Goethe, 1789–1832*, ed. James Hardin and Christoph E. Schweitzer, 171–82. Detroit: Gale Research, 1989.

Ryan, Thomas E. *Hölderlin's Silence*. New York: Lang, 1988.

Ryder, Frank G. "Hölderlin's Spondee—True or False?" In *Essays on European Literature: In Honor of Liselotte Dieckmann*, ed. Peter Uwe Hohendahl, Herbert Lindenberger, and Egon Schwartz, 133–46. St. Louis: Washington University Press, 1972.

———. "The Ebbing of Style: Hölderlin's Late Poetry in the Perspective of Meter." *Texte und Studien zur deutschen und vergleichenden Literaturwissenschaft* (1973): 161–76.

Salm, Peter. "Poetic Fulmination from Klopstock to Hölderlin." *Germanic Review* 57 (1982): 78–81.

Salzberger, L. S. *Hölderlin*. New Haven, Conn.: Yale University Press, 1952.

Santner, Eric L. *Friedrich Hölderlin: Narrative Vigilance and the Poetic Imagination*. New Brunswick, N.J.: Rutgers University Press, 1986.

Schabert, Ina. "The Poetics of Fictional Biography: 1. Reaching into Otherness." In *In Quest of the Other Person: Fiction as Biography*, 9–30. Tübingen: Francke, 1990.

Schier, Rudolf Dirk. "Trees and Transcendence: Hölderlin's 'Die Eichenbäume' and Rilke's 'Herbst.'" *German Life & Letters* 20 (1966–67): 331–41.

Schultz, Karla L. "Hölderlin and Beckett: From Song to Silence." In *Mimesis on the Move: Theodor W. Adorno's Concept of Imitation*, 97–141. Bern: Lang, 1990.

Schürmann, Reiner. "Situating René Char: Hölderlin, Heidegger, Char, and the 'There Is.'" *Boundary 2*, no. 4 (1975–76): 513–34.

Schutter, Dirk. "Words Like Stones." In *(Dis)continuities: Essays on Paul de Man*, ed. Luc Herman, Kris Humbeeck, and Geert Lernout, 99–110. Amsterdam: Rodopi, 1989.

Schuwer, André. "Nature and the Holy: on Heidegger's Interpretation of Hölderlin's Hymn 'Wie wenn am Feiertage . . .'" *Research in Phenomenology* 7 (1977): 225–37.

Seidler, Ingo. "*Stifter einer weiteren Ahnenreihe?*: Hölderlin's Influence on German Poets of the Twentieth Century." In *Friedrich Hölderlin: An Early Modern*, ed. Emery E. George, 64–86. Ann Arbor: University of Michigan Press, 1972.

Sewell, Elizabeth. "Poetry and Madness, Connected or Not?—And the Case of Hölderlin." *Literature and Medicine* 4 (1985): 41–69.

Seyhan, Azade. "Friedrich Hölderlin's *Hyperion*." In *Representation and Its Discon-*

tents: The Critical Legacy of German Romanticism, 107–16. Berkeley: University of California Press, 1992.

Shaffer, Elinor Stoneman. "Hölderlin's 'Patmos' Ode and 'Kubla Kahn': Mythological Doubling." In *"Kubla Kahn" and "The Fall of Jerusalem": The Mythological School in Biblical Criticism and Secular Literature, 1770–1880*, 145–90. Cambridge: Cambridge University Press, 1975.

Shapiro, Gary. "The Owl of Minerva and the Colors of the Night." *Philosophy and Literature* 1 (1977): 276–94.

Shelton, Roy C. *The Young Hölderlin*. Bern: Lang, 1973.

Sieburth, Richard. "The Guest: Second Thoughts on Translating Hölderlin." In *The Art of Translation: Voices from the Field*, ed. Rosanna Warren, 237–43. Boston: Northeastern University Press, 1989.

Silver, Philip W. "De Man Faces 'Destruction': An Ontology of Poetry." *Dispositio* 9, nos. 24–26 (1984): 107–21.

Silz, Walter. "Hölderlin." In *Early German Romanticism: Its Founders and Heinrich von Keist*, 106–18. Cambridge: Harvard University Press, 1929.

———. "Hölderlin's Ode 'Heidelberg.'" *Germanic Review* 37 (1962): 153–60.

———. *Hölderlin's Hyperion: A Critical Reading*. Philadelphia: University of Pennsylvania Press, 1969.

———. "Hölderlin and Wordsworth: Bicentenary Reflections." *Germanic Review* 45 (1970): 259–70.

———. "Hölderlin: Two Domestic Poems." *Germanic Review* 48 (1973): 5–20.

Simon, John Ivan. "German Prose Poetry: The Poem that Dare Not Speak Its Name." In *The Prose Poem as a Genre in Nineteenth-Century European Literature*, 539–621. New York: Garland, 1987.

Simon, Martin F. A. *Friedrich Hölderlin, the Theory and Practice of Religious Poetry: Studies in the Elegies*. Stuttgart: Heinz, 1988.

Simon, Ulrich. "Utopianism around A.D. 1800." In *The Interpretation of Belief: Coleridge, Schleiermacher, and Romanticism*, ed. David Jasper, 1–15. London: Macmillan, 1986.

Spender, Stephen. "Hölderlin, Goethe, and Germany." *Horizon* 8 (1943): 273–80.

Stahl, Ernst Ludwig. *Hölderlin's Symbolism: An Essay*. Oxford: Blackwell, 1945.

———. "Hölderlin's Poetic Mission." *German Life & Letters* 2 (1948–49): 45–61.

———. "Hölderlin's Idea of Poetry." In *The Era of Goethe: Essays Presented to James Boyd*, 148–62. Oxford: Blackwell, 1959.

———. "Hölderlin's 'Friedensfeier' and the Structure of Mythic Poetry." *Oxford German Studies* 2 (1967): 55–74.

Stansfield, Agnes. "*Das Dritte Reich*: A Contribution to the Study of the Idea of the 'Third Kingdom' in German Literature from Herder to Hegel." *Modern Language Review* 29 (1934): 156–72.

———. *Hölderlin*. Manchester, England: Manchester University Press, 1943.

Stavenhagen, Lee. "The Transparent Riddle: The Technique of Hölderlin's Poetic Maturity." *College Language Association* 11 (1967–68): 228–35.

Steiner, George. *After Babel: Aspects of Language and Translation*. London: Oxford University Press, 1975.

———. *Antigones*. Oxford: Clarendon, 1984.

Stierlin, Helm. "Lyrical Creativity and Schizophrenic Psychosis as Reflected in Friedrich Hölderlin's Fate." In *Friedrich Hölderlin: An Early Modern*, ed. Emery E. George, 192–215. Ann Arbor: University of Michigan Press, 1972.

Stott, Douglas Wayne. *Idealism and Consciousness in the Work of Friedrich Hölderlin*. Ann Arbor: University of Michigan Press, 1978.

Szondi, Peter. "Hölderlin's Overcoming of Classicism." Trans. Timothy Bahti. *Comparative Criticism* 5 (1983): 251–70.

———. "The Notion of the Tragic in Schelling, Hölderlin, and Hegel." In *On Textual Understanding and Other Essays*, ed. Michael Hays. Trans. Harvey Mendelsohn, 43–55. Minneapolis: University of Minnesota Press, 1986.

———. "On Textual Understanding." In *On Textual Understanding and Other Essays*, 3–22.

———. "The Other Arrow: On the Genesis of the Late Hymnic Style." In *On Textual Understanding and Other Essays*, 23–42.

Taylor, Ronald. "Hölderlin. 'Der Abscheid.' Second Version; 'Die Wanderung.'" In *The Intellectual Tradition of Modern Germany: A Collection of Writings from the Eighteenth to the Twentieth Century*, vol. 1, 134–44. London: Bell, 1973.

Tonsor, Stephen. "Hölderlin and the Modern Sensibility." In *Friedrich Hölderlin: An Early Modern*, ed. Emery E. George, 54–63. Ann Arbor: University of Michigan Press, 1972.

Torno, Timothy. *Finding Time: Reading for Temporality in Hölderlin and Heidegger*. New York: Peter Lang, 1995.

Trabant, Jürgen. "Memories and Perspectives of Literary Semiotics." *Kodikas/Code* (Tübingen) 13, nos. 1–2 (1990): 41–60.

Turner, Barnard. "Hölderlin's 'Deutscher Gesang' over 'Patmos': A Romantic Pindaric Ode in the Light of Modern Criticism." *Neophilologus* 79, no. 1 (1995): 119–33.

Unger, Richard. *Hölderlin's Major Poetry: The Dialectics of Unity*. Bloomington: University of Indiana Press, 1975.

———. *Friedrich Hölderlin*. Boston: Twayne, 1984.

Ursachi, Mihai. "The Poetry of Being: From Hölderlin to Celan." In *Kulturlandschaft Bukowina: Studien zur deutschsprachigen Literatur des Buchenlandes nach 1918*, ed. Andrei Corbea and Michael Astner, 126–34. Iasi, Romania: Editura Universitéa, 1990.

Vycinas, Vincent. "Festivity." In *Earth and Gods: An Introduction to the Philosophy of Martin Heidegger*, 288–96. The Hague: Mouton, 1961.

Waidson, Herbert Morgan. "Vernon Watkins and German Literature." In *Affinities: Essays in German and English Literature, Dedicated to the Memory of Oswald Wolff (1897–1968)*, 110–25. London: Wolff, 1971.

Waite, Geoffrey. "The Politics of 'the Question of Style': Nietzsche/Hölderlin."

In *Identity of the Literary Text*, ed. Mario J. Valdes and Owen Miller, 246–73. Toronto: University of Toronto Press, 1985.

Walther, Bo. "One God among the Gods: Traces of Hölderlin in Adorno and de Man." *Orbis Litterarum* 51, no. 1 (1996): 1–10.

Warminski, Andrzej. "Endpapers: Hölderlin's Textual History." In *Readings in Interpretation: Hölderlin, Hegel, Heidegger*, 3–22. Minneapolis: University of Minnesota Press, 1987.

———. "Heidegger Reading Hölderlin." In *Readings in Interpretation*, 45–71.

———. "Hölderlin in France." In *Readings in Interpretation*, 23–44.

———. " 'Patmos': The Senses of Interpretation." In *Readings in Interpretation*, 72–92.

Weber, Joseph. "The Poetics of Memory." *Symposium* 4, 33 (1979): 293–98.

Weilnbock, Harald. "Who's Afraid of Dorbeck? The Romantic Seduction of the Reader and the Implosion of Bourgeois Perception: A Psychoanalytic Reading of Willem F. Hermans' *De donkere Kamer van Damokles* and Its Reference to Hölderlin's *Hyperion*." *New German Review* 5–6 (1989–90): 110–25.

Weisinger, Kenneth. "Hölderlin's 'Wie wenn am Feiertage . . .' and Its Pindaric Model." *Monatshefte für deutscehn Unterricht, deutsche Sprache und Literatur* 68 (1976): 257–69.

———. "Hölderlin's 'Die Wanderung': Encomium and Prophecy." *German Studies Review* 7, no. 3 (1984): 399–422.

Weiss, Richard. "(Life-Rhythms.) Vittoria Colonna—Susette Gontard—Lou Salome." In *The Secret of Individuality: Reflected in 100 Historical Lifes*, 439–441. Sydney: Angus and Robertson, 1957.

Weisstein, Ulrich. "Imitation, Stylization, and Adaption: The Language of Brecht's *Antigone* and Its Relation to Hölderlin's Version of Sophocles." *German Quarterly* 46 (1973): 581–604.

Wellbery, David E. "Benjamin's Theory of the Lyric." In *Benjamin's Ground*, ed. Rainer Nägele, 39–59. Detroit: Wayne State University Press, 1988.

Wellek, René. "Max Kommerell as Critic of Literature." In *Teilnahme und Spiegelung: Festschrift für Horst Rüdiger*, ed. Beda Allemann and Erwin Koppen, 485–98. Berlin: de Gruyter, 1975.

Wells, Frederic Lyman. "Hölderlin: Greatest of 'Schizophrenics'?" *Journal of Abnormal Psychology* 41 (1946): 199–207.

Wilkinson, Elizabeth M. "Group-Work in the Interpretation of a Poem by Hölderlin." *German Life and Letters* 4 (1951): 248–60.

Willson, Amos Leslie. "Friedrich Hölderlin." In *A Mythical Image: The Ideal of India in German Romanticism*, 170–86. Durham, N.C.: Duke University Press, 1964.

Wood, Hugh. "Britten's Hölderlin Songs." *Musical Times* 104 (1963): 781–83.

Zagano, Phyllis. "Poetry as the Naming of the Gods." *Philosophy and Literature* 13, no. 2 (1989): 340–49.

Zimmermann, Michael E. "Hölderlin and the Saving Power of Art." In *Heidegger's Confrontation with Modernity: Technology, Politics, and Art*, 113–33. Bloomington: Indiana University Press, 1990.

Ziolkowski, Theodore. "The Generic Norm: The Classical German Elegy." In *The Classical German Elegy, 1795–1950*, 111–34. Princeton, N.J.: Princeton University Press, 1980.

———. "The Romantic Obsession with Madness." In *German Romanticism and Its Institutions*, 202–6. Princeton, N.J.: Princeton University Press, 1990.

Index of Names

In this index an "f" after a number indicates a separate reference on the next page, and an "ff" indicates separate references on the next two pages. A continuous discussion over two or more pages is indicated by a span of page numbers, e.g., "57–59." *Passim* is used for a cluster of references in close but not consecutive sequence. For obvious reasons, the only important name not indexed is that of Friedrich Hölderlin.

Cultural Memory in the Present

Aris Fioretos, ed., *The Solid Letter: Readings of Friedrich Hölderlin*

Peter Schwenger, *Fantasm and Fiction: On Textual Envisioning*

Didier Maleuvre, *Museum Memories: History, Technology, Art*

J. Hillis Miller / Manuel Asensi, *Black Holes / J. Hillis Miller; or, Boustrophedonic Reading*

Mieke Bal, ed., *The Practice of Cultural Analysis: Exposing Interdisciplinary Interpretation*

Miryam Sas, *Fault Lines: Cultural Memory and Japanese Surrealism*

Niklas Luhmann, *Love as Passion: The Codification of Intimacy*

Jacques Derrida, *Monolingualism of the Other; or, The Prosthesis of Origin*

Andrew Baruch Wachtel, *Making a Nation, Breaking a Nation: Literature and Cultural Politics in Yugoslavia*

Jacques Derrida and Gianni Vattimo, eds., *Religion*

Library of Congress Cataloging-in-Publication Data
The solid letter : readings of Friedrich Hölderlin / edited by Aris Fioretos.
p. cm. — (Cultural memory in the present)
"Some of the contributions in this book have previously appeared elsewhere, in part or in whole"—Introd.
Includes bibliographical references and index.
ISBN 0-8047-2942-5 (alk. paper). —
ISBN 0-8047-2943-3 (pkb. : alk. paper)
1. Hölderlin, Friedrich, 1770–1843—Criticism and interpretation.
I. Fioretos, Aris. II. Series.
PT2359.H2S65 1999
831'.6—dc21 99-21525

Original printing 1999

Designed by Eleanor Mennick
Typeset by Tseng Information Systems, Inc. in 11/13.5 Adobe Garamond text, with the same font family used as display